KU-024-673

This book is dedicated to my children

JASON AND JENNIFER

and my hope that, no matter where they are, they will
so love the natural world around them that they
will give of their time and energy to
not only enjoy it, but
also preserve it.

Contents

III ANALYZING THE RESULTS   345

Preface to the second edition

This revision does not abandon the format of the first edition, and the objective of presenting a logical, meaningful and practical approach to the study of animal behavior remains. However, its contents have been expanded, enriched and I hope improved. The book is still a compilation of procedures, techniques and tools, and with the continued use of personal and borrowed philosophy it offers ways of thinking as well as doing. The chapters progress through the steps in ethology, proceeding from research questions through data collection and analysis to interpretation of results. This provides beginning students with a stepwise introduction while allowing experienced researchers to delve selectively into topics of special or timely interest.

Although the book has been updated, I have attempted to retain contact with the roots of ethology. For example, the data analysis sections have been updated and expanded, but I have emphasized that careful description of behavior is the mandatory precursor to rigorous experimentation and analysis. I retained some of the classic examples and figures that remain clearly illustrative of concepts and techniques. While I am confident that beginning researchers will encounter more recent examples, I fear that many of the time-honored examples which layed the foundations of ethology, and remain valid today, might too soon be forgotten.

Some areas have changed little since the first edition and others have shown ballistic morphogenesis, especially the advancements in electronic technology and computer software that can be applied to data collection and analysis. Although some of the electronic equipment described will be technologically outdated by the time of publication, it will not be outdated in terms of usefulness to most ethologists. With both technology and techniques, I have described 'modern sophisticated' approaches without abandoning the discussion of traditional methods; much can still be accomplished by ethologists with the motivation to develop the observational and data recording skills necessary to better utilize their binoculars, notebooks and pencils.

Two topics have been greatly amplified in this edition. First, I expanded the statistics chapter in response to requests by students and other colleagues. They expressed a continuing need for a more complete, yet simple and straightforward, presentation of experimental design and nonparametric statistics that is illustrated with ethological examples. To that end, I have unabashedly described the selection and use of experimental designs and nonparametric statistical tests in a 'cookbook' format; for advanced researchers, there is an abundance of textbooks with more complex and theoretical treatments.

Secondly, I took the model of animal behavior introduced in the first edition and expanded the discussion of its development and application into an entire chapter. With continued thought and application, the prototype evolved into the conceptual Model for a Behavioral Act. Over several years, the model has helped students in my introductory animal behavior class better understand 'how behavior works', and assisted graduate students and colleagues in conceptualizing, designing and conducting ethological research. I acknowledge that I have merely reassembled 'bits-and-pieces' from various sources in constructing the model. I cannot count myself among those very few individuals who are blessed with a handful of unique ideas in their lifetime – ideas that change the course of inquiry and understanding or move them ahead in leapfrog fashion. Most of us are destined to rediscover, reformulate and build on our ideas one brick at a time; but as Goethe said, we can still be part of the game:

> All truly wise thoughts have been thought already thousands of times;
> but to make them truly ours, we must think them over again honestly,
> until they take root in our personal experience.

I started gathering material for this revision several years ago, but the enormous growth in literature describing the use of various methods and new equipment seemed destined to outpace my ability to select, sort, compile, distill and discuss that information. I probably would have stumbled to a halt without the encouragement of numerous novice and experienced ethologists, many of whom I know only through correspondence. I thank all of you.

In the final stages, I received valuable criticism and comments on various topics and chapters from the following: Marc Bekoff, Ivan Chase, Vic DeGhett, Jim Ha, Jack Hailman, Dennis Henry, Suzanne Johnson, Dale Lott, Barb Maynard, Ray Sterner, and Bill Wright. A special thanks goes to my 1994 Ethological Methods class for their valuable advice on clarification and presentation. As always, the responsibility for all errors of omission and commission is mine. I am also indebted to my editor Tracey Sanderson who has guided me throughout in ways too numerous to relate. Special thanks to Christy Kimpo for the indexing.

Throughout the book I have occasionally injected snippets of levity which are not only illustrative (in the manner of a Gary Larson cartoon), but also provided oases of frivolity for me as I was writing. When you encounter these passages, underscore the message, not the mirth, for in no case is the amusement meant to devalue the topic. With that in mind, a final acknowledgement goes to my special friends, George and Jack from Tennessee and Jim from Kentucky, who never failed to comfort me after an arduous day in the field and who added an inner glow to the warmth of the many campfires we shared whenever there was a 'nip' in the air.

Philip N. Lehner

1 Introduction

One does not meet oneself until one catches the reflection from an eye
other than human.   *[Eiseley, 1964:24]*

In choosing to study the behavior of animals you are setting off on a voyage across
waters that are often rough and, in some areas, poorly charted. This book is basically a compilation of practical information that is intended to help smooth the
waters and assist you in charting your course. I have taken the liberty of infusing it
with personal and borrowed philosophy. These philosophical interludes are meant
to cement together concrete blocks of facts and invite you to stop and ponder what
you have learned and what it means.

The most disheartening (and sometimes disarming) question that an uninitiated
ethologist can face is: 'So what?' You can be confronted with that question at technical meetings or cocktail parties, and answers vary from informative discourses to
obscene threats. I hope that some of the discussions found in this book help you to
wrestle with that question.

> The study of behavior encompasses all of the movements and sensations
> by which animals and men mediate their relationships with their
> external environments – physical, biotic and social. No scientific field is
> more complex, and none is more central to human problems and
> aspirations.   *[Alexander, 1975:77]*

One of the greatest challenges you face is to keep your own studies of animal
behavior in perspective. Why did you choose that particular species or concept to
study (sections 1.4, 2.1, 2.2)? What is the focus of your study? Three dimensions of
these questions are discussed in section 1.3, below. What is already known about the
subject (section 4.3)? Are you replicating someone else's research, testing someone
else's hypothesis, or are you answering your own questions? Will your results be
limited to the individuals you are observing or will you consider them a representative sample of a larger population?

Earlier I said that the study of animal behavior is a fascinating voyage; remember to continually assess where you are, where you have been, and where you are
going.

1.1 WHAT IS ETHOLOGY?

Everyone seems to have their own definition of ethology, although a precise and widely accepted definition eludes ethologists. A verbal exchange among Konrad Lorenz, Niko Tinbergen, Theodore Schneirla and others on the definition of ethology makes interesting history and reading (Schaffner 1955:77–78). Eisner and Wilson (1975:1) define ethology as the:

> . . . study of whole patterns of animal behavior under natural conditions, in ways that emphasize[d] the functions and the evolutionary history of the patterns.

However, for our purposes, let us focus on Lorenz's (1960a) definition: 'Ethology can be briefly defined as the application of orthodox biological methods to the problems of behavior'. As a basis for this definition, Lorenz (op. cit.) described the geneology of ethology as follows: '. . . biology . . . is its mother . . . whereas . . . for a father, a very plain zoologist Charles Darwin'. Ethology's heritage should be kept firmly in mind.

Ethology is a nearly limitless discipline which operates basically along the three dimensions illustrated in Figure 1.1. These are continuous dimensions along which studies can be focused.

1.2 WHY STUDY ANIMAL BEHAVIOR?

Reasons for studying animal behavior might emphasize either obtaining or applying knowledge. Drickamer and Vessey (1992) listed some of the reasons we study animal behavior:

1 Curiosity about the living world.
2 Learn about relationships between animals and their environments.
3 Establish general principles common to all behavior.
4 Better understand our own behavior.
5 Desire to preserve and maintain the environment.
6 Conserve and protect endangered species.
7 Control economically costly animal pests.

When asked why she became an ethologist, Jane Goodall replied 'curiosity' (Anonymous, 1988). Curiosity probably drives most ethologists, but in order to get a broader perspective on the question I recommend perusing Dewsbury's (1985) compilation of the autobiographies of 19 well known ethologists.

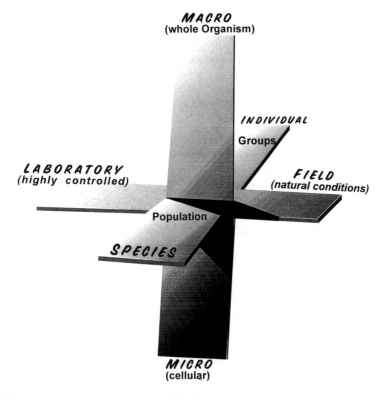

Fig. 1.1   Three dimensions of ethology (drawing by Brenda Knapp).

1.3   WHAT TO STUDY?

For some researchers this question is seemingly unimportant since they have fixed most of their attention on a chosen species. Choice of species, however, is only a partial answer to the question: 'What to study?' Others want to emulate the media's portrayals of Dian Fossey, Jane Goodall, Konrad Lorenz, and Niko Tinbergen. They do not really care what they study as long as it is adventurous. However, what is presented in an hour-long television program about an ethologist's research are the highlights of thousands of hours of data collection. Powers (1994), while discussing the positive impact of television on the environmental movement, offered the following, somewhat overstated assessment:

> ... TV programs tend to distort nature's rhythms and interactions, to say nothing of its scope and diversity. Animals in the wild are mostly inert – sunning or grooming themselves. Television abhors inertia. Thus, even the good shows drastically edit and condense the movement patterns of animals to the point where, as author Bill McKibben puts it,

'nature films are like the highlight clips they show on the evening sportscasts, all rim-bending slam dunks and home runs and knee-crumpling knockout punches.' As for diversity, there are about 1.4 million known species on earth. Only a minuscule fraction of these pass television's audience-appeal standards.

The serious student of ethology must be prepared to observe inactive animals for many uncomfortable hours under adverse conditions, when even the most avid ethologist would admit to dreariness and physical misery. Ethology has often been glamorized, but it is not always glamorous (See Lott's, 1975, 'Protestations of a field person'). Even determining exactly what it is you want to study can sometimes be tedious, but this initial step is exceedingly important in the overall process. You can first define your research interests within the *levels of behavior, areas of study* and *categories of questions*, discussed below.

1.3.1 Levels of behavior

Behavior occurs at various organizational levels along the individual-species and macro-micro dimensions (Figure 1.1), a perspective which the ethologist must obtain and maintain (Menzel 1969). Figure 1.2 gives an example of how these levels can be viewed for Canada geese (*Branta canadensis*). Another example of levels of behavior, using the turkey (*Meleagris gallopavo*), is provided in Lehner (1987). The concept of levels of organization is important to ethologists, and they should be able to 'zoom in' and 'zoom out' (Menzel 1969) to and from the aspect of behavior they are studying (Figure 6.2). We can approach a real understanding of behavior only if we maintain an accurate overview. Crews (1977) illustrated this point in his study of the reproductive behavior of the American chameleon (*Anolis carolinensis*) (Figure 1.3).

Ethologists must decide at what level they will be conducting research and how their study will integrate with what is already known at the other levels. This book will concentrate on methods applicable to the study of whole organisms at the individual or group level under field conditions.

1.3.2 Areas of study

Many ethologists (e.g. Shettleworth 1983) subscribe to Tinbergen's (1963) categorization of four areas of study in ethology: *function, causation, ontogeny*, and *evolution*.

   • *Function* – This can include the study of proximate and/or ultimate function, or what Hinde (1975) has called the 'weak meaning' and 'strong

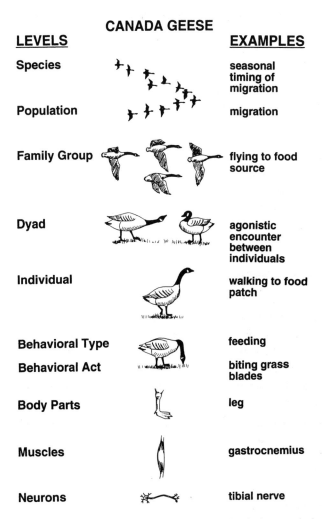

CANADA GEESE

LEVELS	EXAMPLES
Species	seasonal timing of migration
Population	migration
Family Group	flying to food source
Dyad	agonistic encounter between individuals
Individual	walking to food patch
Behavioral Type	feeding
Behavioral Act	biting grass blades
Body Parts	leg
Muscles	gastrocnemius
Neurons	tibial nerve

Fig. 1.2 Levels of behavior in Canada geese. At the upper levels there can be interspecific and intraspecific interactions with other groups and individuals (drawing by Lori Miyasato).

sense' of function, respectively. The proximate function refers to the immediate consequence of the behavior on that animal, other animals, or the environment. Through many observations correlations can be established which lead to conclusions of cause and effect (i.e. proximate function). The ultimate function refers to adaptive significance of the behavior in terms of improving the individual's fitness, and how natural selection operates to maintain the behavior.

Levels of
Analysis

species

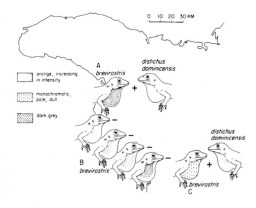

organismal

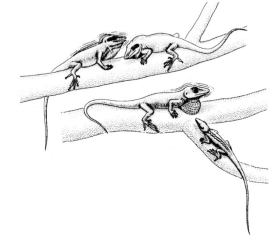

physiological

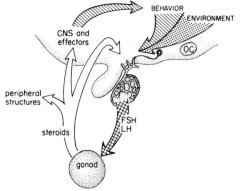

- *Causation* – What are the mechanisms that underly the behavior? What are the contexts in which it occurs, and what are the external (exogenous) and internal (endogenous) stimuli that elicit the behavior? (See Figure 2.2)
- *Ontogeny* – How does the behavior develop in the individual? What maturational and learning processes are important in the development of the behavior?
- *Evolution* – How did the behavior develop in the species? This includes a phylogenetic comparative approach in which the behavior is studied in closely related species to reveal differences which may reflect evolutionary change (Lorenz, 1951; Kessel, 1955). This often leads to the development of an 'ethocline' which presents the differences in behavior among several species along an evolutionary continuum (Evans, 1957). Harvey and Pagel (1991) provide an excellent overview of the new rigor which has been brought to comparative studies of behavior in an evolutionary context by the development and refinement of molecular techniques. They describe several comparative methods to test different models of the evolution of behavioral traits.

Tinbergen's four areas of study can be grouped into two sets which address questions of 'proximal' or 'ultimate' causation (Wilson, 1975), or as Alcock (1989) stated, questions of 'how' or 'why', respectively. The Table 1.1 below illustrates levels of study at which research on the two sets of questions is often directed, as well as how the two sets can be divided into the study of causes and origins.

In addition, Marler and Hamilton (1966) suggested the following five broad areas of investigation of animal behavior that overlap those of Tinbergen: motivation, ecology, social communication, phylogeny and ontogeny.

Fig. 1.3   Levels of research on *Anolis carolinensis* reproductive behaviour. At the species level, pure populations of *A. brevirostris* (polulation B) exhibit clinical variation in dewlap color, whereas populations to the north and south, which are sympatic with *A. distichus*, have uniform dewlap colors. At the organismal level, aggresive posturing between male *A. carolinensis* (on the top branch) involves erected nuchal and dorsal crest, lateral compression of body black spot behind eye, engorged throat, and lateral orientation, whereas during the courtship display (bottom branch), the male has a relaxed body posture and extended dewlap, and the crest and eye spot are absent. The physiological level shows some of the principal intrinsic and extrinsic factors mediating reproductive events in vertebrates and their feedback relationships. (Adapted from Crews, 1977).

Table 1.1. *The relationships of Tinbergen's four areas of study to questions of* '*proximal' and 'ultimate' causation*

	'Proximal causation' (Wilson, 1975)	'Ultimate causation' (Wilson, 1975)
Type of question:	'How questions' (Alcock, 1989)	'Why questions' (Alcock, 1989)
Research directed at:	Level of individual	Level of individual or population
Origins of the behavior:	Ontogeny	Phylogeny (evolution)
Causes of the behavior:	Control (causation)	Function

1.3.3 Categories of questions

Behavior is *what* an animal does. However, ethologists do not restrict themselves to examining only what an animal does. Nielsen (1958) stated that ethology also includes the study of *when, how and why*, and I would add the study of *where* and *who*.

- *What* – A description of the behavior of the animal.
- *Who* – Behavior may differ within and between species, sexes, age groups, dominance ranks and individuals. It is often important to know who performs a behavior, who is present, to whom behavior is directed, etc. The refinement of DNA fingerprinting techniques has allowed researchers to determine the genetic relatedness of individuals, such as in the paternity of nestling house martins (Riley *et al.*, 1995), red-winged blackbirds (Westneat 1995) and dunnocks (*Prunella modularis*) (Burke *et al.*, 1989), and of black vultures roosting in aggregations (Parker *et al.*, 1995). It is extremely important to know the genetic relatedness (kinship) of individuals involved in altruistic behavior (e.g. givers and receivers of care and alarm calls). An excellent overview of DNA fingerprinting molecular techniques and their application in determining kinship is provided by Avise (1994).
- *When* – The temporal component of the behavior. This can include the occurrence of the behavior with respect to the animal's lifetime, the season, time of day, or position in a sequence. The duration of a behavior and its contribution to an animal's time budget are also considered under the study of *when*.
- *Where* – This is the spatial aspect of a behavior. Studies include where a

behavior occurs geographically or relative to other animals or environmental parameters. Keep in mind that spatial characteristics are three dimensional.

- *How* – This includes the motor patterns used to accomplish a goal-oriented behavior (e.g. flying from one tree to another) as well as the underlying physiological mechanisms (an aspect not covered in this book). Studies of the relevant stimuli associated with behavior are included in this category. The evolutionary and phylogenetic determinants of behavior are also studied to answer *how* questions. For example: How did flight evolve in birds?

- *Why* – Two basic concepts underlie the study of the *why* of behavior. These are motivation and ecological adaptation. These separate, but related, concepts are generally treated in different disciplines: motivation in psychology and ecological adaptation in behavioral ecology. However, the latter is sometimes incorporated into ethological studies, and the former has figured prominently in conceptual behavior models (e.g. displacement behavior).

    The focus of research designed to answer a *why* question depends on the emphasis the researcher puts on different parts of the question. McFarland (1985) illlustrated this with the question: Why do birds sit on eggs?

    > Why do birds sit on *eggs*?
    > Why do birds *sit* on eggs?
    > Why do *birds* sit on eggs?
    > *Why* do birds sit on eggs?

    The different emphases all relate to the adaptiveness of the behavior, but the focus of the research is different (see McFarland, 1985:1–3 for further discussion).

The *what* question is the starting point for all research; that is, we must determine what the animal does before we can address any of the other questions. The *who* question is often addressed next, if we can determine sex, age, etc. The spatial and temporal aspects (*where* and *when* questions) are usually relatively easy to measure and are addressed next. The *how* is more difficult to answer, and *why* questions are the most difficult.

The discussion above should give you a general idea of the various aspects of animal behavior that you might study. The following sections provide an initial outline of the steps an ethologist takes in studying animal behavior.

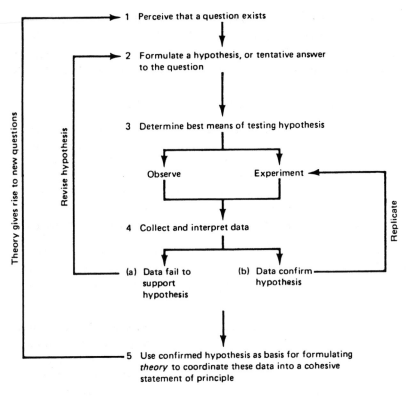

Fig. 1.4   Flow diagram of scientific inquiry (from Jessop 1970). Serendipity (fortuitous discovery) can operate at all stages of the process.

1.4 SCIENTIFIC METHOD

Ethology is a science (McFarland, 1976), and as such, serious studies should adhere to established guidelines. The 'orthodox biological methods' mentioned in Lorenz's definition of ethology include the scientific method. The scientific method (Figure 1.4) is a logical, stepwise approach to research in all sciences including ethology (Tinbergen, 1963; however, see Bauer, 1992 and Hailman, 1975, 1977 for other viewpoints). This book is organized around the scientific method adapted for use in ethology (see section 1.5).

Now that I have focused in on the mechanistic approach that will serve as the foundation for this book, let me state that this does not exclude the pleasure, excitement and artistry inherent in ethological studies. I also want to champion the researchers who get to 'know' and empathize with their animals, and who recognize and appreciate the harmony that has evolved between the animals being studied and the environment. Several researchers have described their state of mind and thought processes while studying animals in their natural habitat. I would especially recommend reading Tinbergen (1958), Schaller (1973) and the compilation of autobiographies by Dewsbury (1985). Dethier (1962) also gives an illustrative and entertaining description of the development of his laboratory research on blowflies. Ethologists should be more than collectors and analyzers of data; they should seek to 'understand' their animal subjects at a level higher than quantitative analysis can provide.

> When we watch animals at different levels on the evolutionary scale, as when Seitz watches fishes, when Dr. Tinbergen watches gulls, when Dr. Lorenz watches ducks, when Dr. Schneirla watches ants, or when I watch doves, the observer can get a feeling of what is going to happen next, which is compounded in different degrees of the intellectual experience of relationships that are involved on the one hand, and, on the other hand, of building yourself into the situation.   *[Lehrman, 1955]*

Darling (1937) believed that in order to gain insight into behavior, the observer must become 'intimate' with the animals under observation. Lorenz (1960b) suggested that an even higher level of empathy with the animals under study is necessary for a true understanding of their behavior.

> It takes a very long period of observing to become really familiar with an animal and to attain a deeper understanding of its behaviour; and without the love for the animal itself, no observer, however patient, could ever look at it long enough to make valuable observations on its behaviour.   *[Lorenz, 1960a:xii]*

The ethologists quoted above were not encouraging rampant anthropomorphism (see Chapter 4) at the expense of objectivity (Carthy, 1966). Rather, they were stating that genuine interest in the animal *per se* will foster additional insight into its behavior beyond mechanistic (machine or machine-like) data collection and analysis. Ultimately, valid research is conducted only by ethologists who have found a proper balance between empathy and objectivity.

1.5 ETHOLOGICAL APPROACH

> Ethology . . . is characterized by an observable phenomenon (behavior, or movement), and by a type of approach, a method of study (the biological method). . . . The biological method is characterized by the general scientific method.   *[Tinbergen, 1963:411]*

The ethological approach (Figure 1.5) is the result of fitting the scientific method to ethology. It consists of a careful stepwise procedure by which data are gathered and analyzed using descriptive statistics (descriptive research) or test statistics in

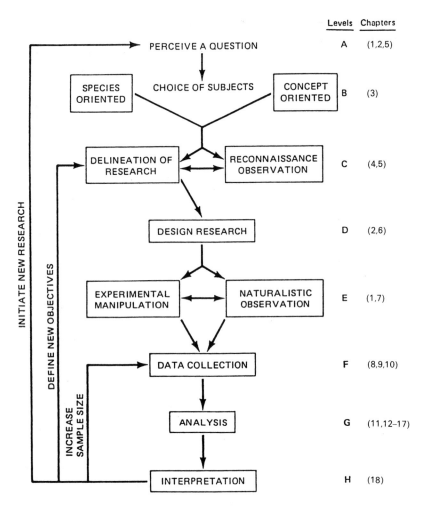

Fig. 1.5   The ethological approach. The numbers in parentheses indicate the chapters in this book which discuss the respective steps in the ethological approach.

order to test hypotheses (experimental research). The ethological approach expands on the four basic operations described by Delgado and Delgado (1962): 1. the isolation of objects or variables to be measured; 2. the establishment and definition of units of measurement; 3. the comparison of actual events and units; and 4. the interpretation of the meaning of the observed events.

Even during descriptive research and the earliest stages of experimental research (i.e. reconnaissance observation), the ethological approach generally includes ill-defined hypothesis testing. When we first begin observing an animal, the fact that we are surprised by an unexpected behavior reveals that we had an ill-defined hypothesis about the animal's behavioral repertoire. Most people's approach to life is through constant hypothesis testing. For example, the very first time we go to work we travel a specific route which we hypothesize will get us there. After a series of successful trips our hypothesis that our office is located in Room 300 in the building at the corner of Oak and Center Streets gains considerable support. However, should our boss move our office to another building while we are on vacation we will have to reject our hypothesis and either revise it or formulate a new one.

More particularly, ethologists often perceive questions which arise from the rejection of our long-standing theories about the way animals should behave.

> . . . biologists are drawn to study events that seem to contradict what we
> have been taught to expect on the basis of our knowledge of non-living
> things. It is this discrepancy between what an animal 'ought to do' and
> what it is actually seen to do that makes us wonder. Like a stone
> released in mid-air, a bird ought to fall; yet it flies away.
>
> *[Tinbergen, 1972:20]*

That is, we constantly seek the order which we believe exists in the universe (chaos theorists notwithstanding), and when events occur which seem to upset our preconceived ideas of the ways things should be, we either ignore the situation or as scientists seek explanations through study.

1.6 DESCRIPTIVE VERSUS EXPERIMENTAL RESEARCH

These two types of research will be discussed in more detail in Chapter 6; however, at this point I want to stress the importance of both approaches (Figure 1.4). In the first edition of this book they were designated as naturalistic observation and experimental manipulation following a traditional distinction (e.g. Schneirla, 1950; Siever, 1968) which reflected a field versus laboratory dichotomy. Both descriptive and experimental studies are conducted all along the continuum that connects the laboratory and field (Figure 1.1; Chapter 7).

Description and experimentation become connected out of curiosity and necessity. Descriptive research often generates hypotheses which lead to experimental research (Bakeman and Gottman, 1986), and experimental research is preceded by complete descriptions and definitions of what an animal does (descriptive research). The following excerpts from Tinbergen illustrate how his study of the begging behavior of herring gull chicks evolved from descriptive to experimental research.

> When the chicks are a few hours old they begin to crawl about under the
> parent, causing it to shift and adjust every so often. Sometimes the

parent stands up and looks down into the nest, and then we may see the first begging behaviour of the young. They do not lose time in contemplating or studying the parent, whose head they see for the first time, but begin to peck at its bill-tip right away, with repeated, quick, and relatively well-aimed darts of their tiny bills. Their remarkable 'know-how,' not dependent on experience of any kind never fails to impress one as an instance of the adaptedness of an inborn response. It seems so trivial and common at first sight, but the longer one watches it the more remarkable it appears to be. . . . The reaction is innate, and it is obviously released by very special stimuli which the parent bird alone can provide, and which enable the chick to distinguish the parent's bill-tip from anything else it may encounter in its world.

. . . In the literature we had found some observations which seemed to show that here again was a reaction dependent on only very few 'sign stimuli.' . . . Heinroth . . . wrote (1928) that his Herring Gull chicks had the habit of pecking at all red objects, especially when they were kept low, so that they could peck downwards. . . . The special sensitivity to red was further demonstrated by the fact that Goethe could elicit responses by red objects of various kinds, and of an appearance that was rather different from a Herring Gull's bill: such as cherries, and the red soles of bathing shoes!

It seemed to us worthwhile to go into this problem a little deeper. That the chicks were responding to the red patch was obvious; however, as the bill without red did also elicit some response, there must be more in a parent bird's bill that stimulated the chick. Also, the downward tendency had to be explained. As regards opportunity for experimental work, the reactions to cherries and bathing shoes showed that it should be easy to design dummies capable of eliciting responses. Further, the very fact that reactions to crude dummies were not rare, showed that the chick's sensory world must be very different from ours, for we would never expect a bathing shoe to regurgitate food.

Therefore, when in the summer of 1946 no war conditions prevented us any longer from working in the field, I took my zoology students out for a fortnight's work in one of the Herring Gull colonies on the Dutch Frisian Isles. We carried with us an odd collection of Herring Gull dummies and thus started a study which was to occupy and fascinate us during four consecutive seasons.   *[Tinbergen, 1960b:178, 184–186]*

Historically, it was believed that ethologists focused their observational skill only on 'does' questions ('. . . intensive observation and description of organisms in species-typical contexts.', Miller, 1985). Conversely, psychologists were believed to spend most of their efforts on experiments involving 'can' questions ('. . . ascertain the capabilities of organisms . . .', Miller, 1985). Further, the ethologists' realm was thought to be naturalistic studies in the field, and the psychologists' purview was believed to be experimentation in the laboratory.

The physiological approach to animal behavior has become more prominent over the last few decades, especially with the merging of neurobiology and ethology into neuroethology (Camhi, 1984; Guthrie, 1987; Hoyle, 1984). Neuroethology techniques are not discussed in this book, but an introductory overview can be found in Camhi (1984) and Ewert (1980). Examples of neuroethology can be found in the synopses of the neural bases of bird song (Kolata, 1985; Konishi, 1989; Nottebohm, 1989), barn owls capturing mice (Konishi *et al.*, 1988), bats capturing mice (Suga, 1990), and toads capturing flies (Ewert, 1987). Neuroethology is an example of one field which provides ample opportunity for cross-fertilization between levels of study within ethology (Figure 1.2).

> It is not only that behavioural analysis can lead to analysis of neural events, but also that neurobiology should become fruitful in suggesting directions for behavioural studies. *[Bateson, 1987:301–302]*

All areas, sub-disciplines and levels of study within ethology are important to a full understanding of animal behavior.

The astute researcher oscillates between refined descriptive research and experimentation, using skilled observation in both, better to define and test his hypotheses. Lorenz expressed his suspicion that experimental psychologists were being guided solely by remotely recorded data without observing the behavior of animals placed in operant-conditioning chambers.

> I can never help a shrewd suspicion that the worshipper of quantification and despiser of perception may occasionally be misled into thinking that two goats plus four oxen are equal to six horses. Counting pecks of pigeons in Skinner Boxes without observing what the birds inside really do, might occasionally add up to just this. *[Lorenz, 1960a:72]*

Lorenz's comment somewhat overstated the case, but is probably still valid in many instances today. The 'balance between the two' (description and experimentation) is the key to good ethological research. However, my purpose here is not to campaign for more direct observation in the experimental psychologists' methods (Hutt and Hutt, 1970, have effectively done this), but rather to insist that ethologists recognize the importance of both descriptive and experimental research, both of which rely on observation and quantification.

In 1963, while discussing the history of ethology, Tinbergen expressed concern that the balance was shifting too far towards experimentation.

We must hope that the descriptive phase is not going to come to a premature ending. Already there are signs that we are moving into an analytical phase in which the ratio between experimental analysis and description is rapidly increasing.   [*Tinbergen, 1963:412*]

As we approach the twenty-first century, a further caution is relevant. Theories and models should not greatly outdistance descriptive and experimental research. One theory and one predictive model each beg for a multitude of experimental studies which test their validity.

Ethology and its allied disciplines are now racing forward in high gear. Experimentation and quantification are no longer the tools limited to psychologists and observation the only tool of ethologists. The disciplines have long since overlapped, merged, borrowed from other disciplines, and fractured into subdisciplines. There is no longer a territorial battle claiming that methodologies belong solely to one discipline or the other (Hutt and Hutt, 1970; Willems and Raush, 1969). Ethologists and psychologists no longer work in their private vacuums, but rather they are united with strong subdisciplines in the quest for knowledge about behavior. See Burghardt (1973) for an extensive account of the development of ethological and psychological concepts and methods into a more holistic approach.

I Getting started

2 A conceptual model of animal behavior

The model for a behavioral act I develop and discuss below is an extension of the model I presented in the first edition of this book. I have found the model to be simple enough to be a useful research tool, yet complex enough to accomodate most behaviors and environmental contexts. It has assisted students of animal behavior in studying, understanding and explaining what an animal is doing now, as well as in predicting what it will do in the future.

The model will not meet with universal approval amongst those who forage through this book. However, the model is not presented as a 'general law of behavior' nor is it designed to quell or exacerbate controversy. The purpose of the model is to:

1 Provide a framework on which to design and conduct ethological research, and
2 Assist researchers in putting areas of study and types of question into an overall perspective while at the same time focusing on their specific research objective.

Those readers who do not agree with the assumptions I make in building and applying the model are encouraged to make their own modifications. To that end, I discuss each portion of the model in sufficient detail to explain how the model works and to allow each researcher to modify and expand those portions which are most important to their study (see section 2.4.4). Also, one should examine other models with varying degrees of similarity that have been offered by several ethologists (e.g. Hinde, 1982; Lorenz, 1981).

2.1 CONTINUOUS STREAM OF BEHAVIOR

Animals are always behaving. They perform a continuous stream of behavior from the moment when movement can first be detected in the embryo until their death.

It is impossible (at least impractical and inefficient) to study all of an animal's behavior throughout its lifetime. However, since an animal's behavior is not random, the relative frequency and duration of different behaviors can be approximated through sampling. Normally, our sample is a continuous, but restricted, segment (or segments) of the stream of behavior, or a series of samples at points in time (see section 8.3). The behavior being performed by an animal at any point in time can be considered a behavioral act (see Chapter 6). The model developed below is for a point in time (behavioral act), but a series of models can be strung together to analyze a segment of the continuous stream of behavior.

2.2 PREDISPOSITION TO BEHAVE: DIFFERING CONTRIBUTIONS

An animal's predisposition to behave (i.e. probability of responding to specific stimuli with specific behaviors) is the result of a combination of factors expressed below.

Behavioral predisposition $=(G_B+E_B+I_B)+(A_B+P_B)$

Where: G_B=contribution of the *genotype*, primarily.

E_B=contribution of the *environment* (including experience), primarily.

I_B=contribution from the *interaction* of the genotype and Environment.

A_B=behavioral capacity provided by the animal's *anatomy*.

P_B=behavioral propensity and capacity provided by the animal's *physiological* mechanisms.

Using a computer analogy, $(G_B+E_B+I_B)$ can be thought of as creating, modifying, and containing the *software* that consists of 'closed' and 'open' programs (Mayr, 1974). Closed programs are primarily a result of the *genotype* and do not allow appreciable modification; open programs allow for modification as a result of *environmental* input that provides *experience*. The (A_B+P_B) is the *hardware* which stores the programs (e.g. brain) and provides the machinery for their action (e.g. sensory, nervous and muscular systems).

2.2.1 Genotype, environment (experience) and interaction

The $(G_B+E_B+I_B)$ portion of the *predisposition to behave model* is the same as the formula used by population geneticists (e.g. Futuyama, 1986) to express the total phenotypic variation in a population (V_t) as the sum of the variation due to genotype (V_G), environment (V_E), and the interaction (V_I) of genotype and environment:

$$V_T=V_G+V_E+V_I$$

Almost all ethologists (e.g. Bateson, 1983) and psychologists (e.g. Tarpy, 1982) have concluded that an animal's behavior is the result of both its genotype and the environment, as well as their interaction (see below). There is no direct correspondence between a gene and an individual behavior pattern (Bateson, 1987), and the interaction of the genotype and environment is likely to be so complex that it is very difficult to identify the genetic and environmental components of an individual behavior (Bateson, 1983), or the extent of their contribution. For example, Lyons *et al.* (1988:1323), in their study of the development of temperament in dairy goats, reached only the general conclusion that 'An individual's genotype and its early postnatal environment both contributed to processes underlying the development of stable individual differences in temperament [timidity] . . .'.

Nevertheless, it is still heuristic to make some generalizations about the *relative* contribution of the genotype and the environment as functions of both phylogeny and ontogeny.

For those species in which the environment (experience) plays a relatively large role in the life history of a species (e.g. birds and mammals), the relative contribution of genotype and environment is believed to change ontogenetically (Figure 2.1)

Early in life, in addition to the effect of it's prenatal experience, an animal's predisposition to respond to 'releasers' and 'unconditioned stimuli' (see section 2.3.5a) is primarily genetically determined. Also, as the animal matures its predisposition to learn more readily in response to certain stimuli and reinforcers may be genetically controlled; this has been called 'preparedness' (Seligman, 1970).

2.2.1a Genotype

An axiom at the roots of ethology is that there is a genetic basis for variation in behavior between individuals (e.g. Partridge, 1983). The entire field of behavioral genetics is based on this axiom. Behaviors that result from genetically fixed, closed programs are called *innate* (e.g. egg laying in a snail, Scheller and Axel, 1984); however, the environment provides for the proper development and expression of those innate behaviors.

Anyone familiar with breeds of dogs selectively bred for various tasks, such as retrieving game and herding livestock, recognizes that an animal's genotype contributes to its behavior. It is also evident that to realize its full potential, that predisposition to retrieve or herd must be shaped in the proper environment. The *genotype* also provides the blueprint for the development of the *anatomy* and *physiology*.

> Genes contribute to the observed differences between individuals,
> in their behavior as in other things, but their contribution is not
> sacrosanct. It can be lessened or enhanced in just the same way as
> the contribution from the environment can.   *[Dawkins, 1986:53]*

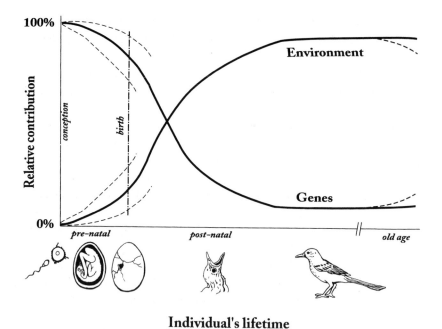

Individual's lifetime

Fig. 2.1   The relative contribution of the genotype and environment during an animal's lifetime. Variation between species and individuals indicated by dashed lines (drawing by Brenda Knapp).

2.2.1b   Environment and experience

As described above, the *environment* (biotic and abiotic) provides for the proper development and expression of innate behaviors and predispositions. This is also true for behaviors that are primarily learned. That is, the environment also provides experience and the proper context for expression. For example, squirrels have to learn through experience how to efficiently crack open and eat a hazelnut (Eibl-Eibesfeldt, 1963); they can develop and later express this behavior only in the presence of hazelnuts, or similar nuts. Experience results from the various associations of stimuli, behavior and consequences which are the bases for learning and modifying future behavior (Davey, 1989; also see section 2.3.5). The environment also affects the anatomy and physiology through factors such as injury, disease, climate, and nutrition.

2.2.1c   Interaction of genotype and environment (experience)

Although we attribute 'innate' behaviors primarily to the genotype and 'learned' behaviors primarily to the environment (experience), we also know that the genotype and environment interact. Gould and Marler (1987) referred to 'innately guided learning'. This interaction is reflected in: 1. Hailman's (1969) title, 'How an instinct is learned', for the report on his study of food begging in laughing gull chicks; and 2. Ewert's (1987:337) conclusion that prey-catching in toads is 'mediated by innate releasing mechanisms (IRMs) with recognition properties partly modifiable by experience'.

Also, the genotype, along with the animal's anatomy and physiology place 'biological constraints on learning' (e.g. Hinde and Stevenson-Hinde, 1973; Shettleworth, 1972). That is, '. . . learning, rather than occurring indiscriminately, is subject to species-specific and task-specific limitations and predispositions' (Roper, 1983:185). The classic example is Garcia and Koelling's (1966) demonstration that rats readily associate taste (internal cue) with illness and exteroceptive cues (light/sound) with painful electric shock, but they don't readily associate taste with shock or light/sound with illness.

In summary, behavior that is initially programmed in the genome can be developed and expressed only in the proper environment and can be modified by experience (learning) only within constraints provided by the genotype, environment, anatomy and physiology.

2.2.2 Anatomy and physiology

Anatomy and Physiology both enable and constrain behaviors. Wings (along with other adaptations) allow a bat to fly, just as the lack of wings contribute to a mouse's inability to fly. From the earlier analogy, an animal's anatomy and physiology is the *hardware* necessary to perform behavior and store feedback from the environment (memory), allowing for modification of future behavior. Anatomy (including morphology) and physiology develop according to a genetic blueprint, but that development is affected by the environment. For example, Zeki (1993:218–219) summarized the results of Hubel and Wiesel's extensive research in the 1960s and 1970s on the effect of sensory deprivation on the development of the visual system of cats and monkeys by stating: 'at the cellular level, there is a critical period, during which adequate visual stimulation is mandatory if the animal is to be able to see at all, even if to all appearances the visual pathways and cortex appeared to be intact'.

Researchers have identified the brain centers and neural pathways necessary for the development, memory and performance of bird song (e.g. Nottebohm, 1989). The brain centers for bird song development arise in part from a genetic blueprint and in part from endogenous and exogenous stimulation. The hormones (endogenous stimuli) necessary for stimulating development are themselves triggered by exogenous stimuli (e.g. photoperiod and temperature). In some species those brain centers are modified by exogenous stimulation in the form of songs from other males of the same species in order to develop the 'innate template' into an 'acquired template'.

Therefore, bird song occurs in response to stimulation from the appropriate neural pathways (physiology), and its form results from a genetic blueprint (genotype) and learned modifications (experience) stored in the brain centers (anatomy). For example, the genotype provides the 'innate template' in the young swamp sparrow's brain that allows it to filter out all but swamp sparrow syllables from the songs it hears in the environment (Marler and Peters, 1977). The adult male swamp sparrow songs it hears provide the syllables to create the 'acquired template' which will be the basis for comparison when the young swamp sparrow begins singing its subsong, listening to itself, and improving it vocal output to create its crystallized primary song.

2.3 A MODEL FOR A BEHAVIORAL ACT

2.3.1 The animal

The basis for the *animal* in the model is the formula for it's predisposition to behave described above. That is, all the important components that predispose an animal to perform specific behaviors under a given set of environmental conditions are provided by G_B, E_B, I_B, A, and P. Also incorporated into the *animal* portion of the model are *endogenous stimuli* (discussed below). Other parts of the model (Figure 2.2) are *exogenous stimuli, behavior, consequences, contingencies, feedback and feedforward*, all of which are discussed below.

2.3.2 Stimuli

Stimuli are changes in the environment (internal or external) to which an animal normally responds. However, the environmental context in which the stimulus occurs will determine whether it is effective in eliciting a response. Therefore, an *effective stimulus* is a change in the environment which elicits a response at that point in the animal's stream of behavior when the environmental context (internal and external) is appropriate. Stimuli that are effective in one context may be filtered (centrally or peripherally) in another context. For example, cat food kibbles will normally stimulate feeding only when the cat is hungry (internal environment) and not involved in a higher priority activity (e.g. chasing a mouse – external environment).

Stimuli have both *activational* and *organizational* functions. That is, they not only *initiate* and *orient* behavior directly, but they also *facilitate* and *maintain* behavior and *predispose* an animal to particular behaviors.

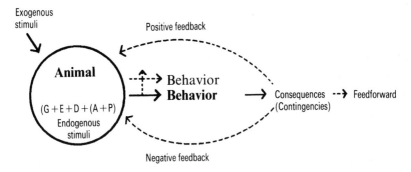

Fig. 2.2   The model for a behavioral act.

2.3.2a *Endogenous stimuli*

Endogenous stimuli arise from the animal's internal environment. An assumption of the model is that all behavior is the result of exogenous and/or endogenous stimulation. That is, we operationally define behavior as that which an animal does as the result of stimulation. We observe behavior being emitted, but we assume that all behavior is elicited. Like Kuo, the Model assumes that 'there is no such thing as spontaneous behavior' (*behavior that is not stimulated*) (quoted in Marler and Hamilton, 1966:605). However, there are behaviors that occur for which we cannot determine the source of stimulation, and these are often called 'spontaneous' behaviors (e.g. Hinde, 1966).

Although behavior does not occur spontaneously, it may be elicited by endogenous stimuli that arise spontaneously.

> . . . every cell in the nervous system is not just sitting there waiting to be
> told what to do. It's doing it the whole darn time. If there's input to the
> nervous system, fine. It will react to it. But the nervous system is
> primarily a device for generating action spontaneously
>
> *[Graham Hoyle quoted in Allport, 1986]*

For example, Bekoff (1978a) found neural pattern generators in embryonic chicks that stimulate coordinated limb movements and develop in the absence of any patterned sensory input.

There are also behaviors that occur in the absence of the exogenous stimuli that normally elicit them (e.g. birds 'nest building' without materials, cats 'prey catching' without prey, dogs 'burying' and 'covering' food in a concrete floor). Lorenz explained the occurrence of these 'vacuum' behaviors as an excessive buildup of 'action specific energy' (which forced open the valve in his Psychohydraulic Model; Lorenz, 1950). However, another possibility is that vacuum behavior is elicited and oriented by 'imaginary exogenous stimuli' created by the animal (also suggested by

Lorenz, 1981); for example, we refer to 'hallucinogenic' play in cats, implying that the 'prey' is in the cat's 'mind's eye'.

Physiological needs are signaled through endogenous stimuli that elicit behaviors designed to meet those needs. For example, low blood sugar leads to the 'state of hunger' which stimulates the animal to seek, find and ingest food.

There also appears to be a psychological need to perform a normal range of behaviors. Various terms have been applied to this psychological need, including 'ethological need' (Hughes and Duncan, 1988), 'drive' and 'instinct'. For example, Garcia et al. (1973:3) state that 'drive is the psychological concomitant of physiological need', and Dawkins (1986:64) states that 'instinct . . . refers to the inner drive or motivational force that leads an animal or person to behave in a certain way'. The model acknowledges that behaviors can be elicited by endogenous stimuli regardless of how you choose to perceive and label the basis for those stimuli.

2.3.2b Exogenous stimuli

Exogenous Stimuli come from the external environment (biotic and abiotic), take many forms (e.g. light, sound, odor) and arrive via the animal's various sensory receptors (e.g. eyes, ears, nose). The animal's *anatomy* and *physiology* are responsible for receiving, processing and responding to exogenous stimuli, as well as generating behaviors which produce exogenous stimuli. Several terms have been applied to exogenous stimuli depending on the role they play in different behavioral paradigms. *Sign stimuli, releasers, unconditioned stimuli, neutral stimuli, conditioned stimuli*, and *discriminative stimuli* are all discussed in the contexts of the ethological model of innate behavior and the learning paradigms below.

2.3.3 Behavior

Behavior results when an effective stimulus is received or generated by the animal. As shown in the model, when one behavior is elicited, an ongoing behavior may be inhibited. This is required if the the the two behaviors are mutually exclusive; that is, they cannot occur at the same time (e.g. sitting and walking). A behavior may also stop because it is no longer stimulated. For example, sleeping is *inhibited* when an animal is ingesting food; ingestion will cease when the animal is satiated and eating is *no longer stimulated* by the sight and smell of food.

2.3.4 Ethological model of innate behavior

The ethological model of innate behavior (below; e.g. Lorenz, 1981) is easily incorporated into the model (Figure 2.3) since it involves exogenous stimuli, anatomy

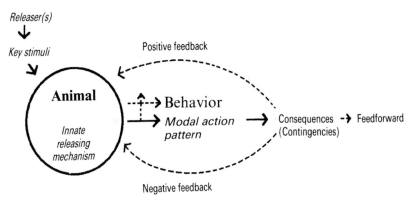

Fig. 2.3   Where the ethologists' model of innate behavior fits into the model for a behavioral act.

and physiology, and behavior. This hypothetical construct is diagrammed as follows:

Releaser(s)→key stimuli→innate releasing mechanism (IRM)
↓
modal action pattern (MAP)

A morphological structure and/or movement (*releaser*) emit one or more *key stimuli*. The key stimuli operate on the *innate releasing mechanism* (at present, an unidentified part of the animal's anatomy and physiology) which 'releases' a *modal action pattern*, a relatively stereotyped behavior pattern. Since these modal action patterns are very similar between individuals of the same species, ethologists believed that they are primarily under genetic control and originally referred to them as *fixed action patterns*. However, it is now recognized that experience also plays an important role. For example, Gerard Baerends, who began his ethological research six decades ago, recently concluded from his many years of studying herring gulls that 'The information encoded in the IRM [*innate releasing mechanism*] and the acquired information [*learning*] were found to work in combination.' (italics mine; 1985:37).

Historically, ethologists and psychologists have been branded as having diametrically opposed viewpoints on the relative contribution of the genotype and learning to behavior. However, ethologists and psychologists are increasingly borrowing from each other's research and theories, and overlap between ethology and learning theory can more readily be found in the literature. For example, adjunctive behaviors of learning theorists 'are very similar dynamically and functionally to what have been known in the ethological literature as displacement activities' (Davey, 1989:89).

2.3.5 Associative learning paradigms

Learning can be defined as the 'adaptive' modification of behavior as the result of experience (e.g. Lorenz, 1965). Natural selection has shaped various types of learning in animals (Staddon, 1983), but each type of learning incurs selective costs (in terms of fitness), as well as selective benefits (Johnston, 1981). Also, the benefits of a specific type of learning may be limited to the environmental contexts in which it evolved (McNamara and Houston, 1980). Possible routes of evolution of the primary associative learning paradigms, *classical and instrumental conditioning*, are discussed by Weisman and Dodd (1980) and Skinner (1988), respectively. Both of these learning paradigms are incorporated directly into the model.

2.3.5a *Classical conditioning*

The *classical conditioning* paradigm (below) describes how neutral stimuli become conditioned through association, thus gaining the ability to elicit specific behaviors.

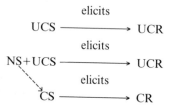

An unconditioned stimulus (UCS) elicits an unconditioned response (UCR). When a neutral stimulus (NS) is paired with the UCS, over time it becomes a conditioned stimulus (CS) capable of, by itself, eliciting a conditioned response (CR), a reasonable facsimile of the UCR.

Note the similarity between classical conditioning and the ethologists' model of innate behavior (discussed above). The UCS is essentially the same as the ethologist's *releaser*, and the UCR is essentially the ethologists' MAP(FAP).

2.3.5b *Instrumental (operant) conditioning*

The *instrumental conditioning* paradigm (below) states that a *behavior* is emitted and is followed by *consequences* (S^{R+}, S^{R-}) that either maintain, increase or decrease the probability of that behavior occuring again (see Consequences section below).

Discriminative stimuli
(S^{D+} or S^{D-})

Behavior ⟶ Consequences
(emitted)

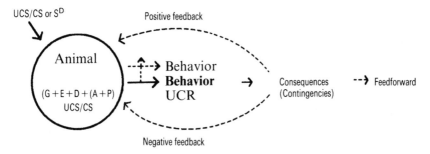

Fig. 2.4   Where classical and instrumental conditioning fit into the model for a behavioral
act.

The paradigm begins with a behavior being *emitted* and does not consider
stimuli that might *elicit* that behavior. However, the paradigm does include *discriminative stimuli*. Figure 2.4 illustrates how both classical and instrumental conditioning fit into the model.

Discriminative stimuli

Positive discriminative stimuli (S^{D+}) predict that a specific behavior will be followed
by positive consequences, and Negative discriminative stimuli (S^{D-}) predict that a
specific behavior will be followed by aversive consequences (see discussion below
about specific types of consequences). Since exogenous stimuli in the model include
both conditioned and discriminative stimuli as types of exogenous stimuli, it is
interesting to note how Davey summarizes their similarity in function.

> . . . Pavlovian CSs or instrumental discriminative stimuli (S^Ds) elicit a
> motivational state appropriate to the reinforcer and . . . this
> motivational state in some way mediates the emission of the
> instrumental response.   *[Davey, 1989:210]*

The ability to use the environmental context (including S^Ds) to predict the consequences of behavior is a major benefit of learning. Tarpy presents a psychologist's
perspective without using the term discriminative stimuli.

> Learning is fundamentally a process whereby the animal comes to
> expect a future event based upon the patterns of stimuli in the
> environment or upon its own behavior.   *[Tarpy, 1982:18–19]*

Likewise, Lorenz presents an ethologist's perspective on predicting the outcome of
behavior, also without using the term discriminative stimuli.

Table 2.1. *During or immediately following the behavior the reinforcing stimulus is:*

Reinforcing stimuli	Presented	Omitted
Positive (S^{R+})	Positive reinforcement	Extinction
Aversive (S^{R-})	Punishment	Negative reinforcement

> . . . a bird that 'wants' to carry out the beautiful motor pattern of nest building . . . learns to recognize the situation in which performing the nest-building movements gives the maximum satisfaction.
>
> *[Lorenz, 1981:291]*

Consequences

Proximate consequences   The four *proximate consequences*, in the matrix shown in Table 2.1, are the immediate result of a behavior. They are determined by the *type of reinforcing stimulus* and whether the reinforcing stimulus is presented or omitted.

Reinforcing stimuli and consequences are defined by their effect on the animal's behavior. *Positive reinforcing stimuli* (S^{R+}) are often called rewards. They are reinforcing stimuli which the animal perceives as 'good'. That is, in the appropriate contexts, the animal will perform the behaviors necessary to receive those S^{R+}s. The S^{R+}s are received by the animal under different *contingencies* of behavior called *schedules of reinforcement* (discussed below).

An S^{R+} may be a single stimulus (e.g. food item) or the opportunity to perform a chain of behaviors (e.g. search, stalk, capture, kill, consume). Lorenz (1965) considered the consummatory act in a chain of behaviors as a reinforcer for antecedent behaviors (appetitive behaviors). Also, behaviors may be organized in a 'preference hierarchy' (Premack, 1965) so that an S^{R+} could be the opportunity to perform a preferred behavior.

Although we often think of S^{R+}s as meeting obvious basic physiological needs (e.g. food and water), ethological needs (discussed earlier) may stimulate an animal to perform behaviors in order to obtain subjective rewards.

> . . . the opportunity to manipulate, to explore, or to merely observe is labeled a reward, reflecting the assumption that if learning has occurred there must have been some reward, even though it cannot be empirically specified.   *[Garcia et al., 1973:33]*

Animals may perform behaviors that are primarily innate (e.g. courtship and mating) because they receive inherent, subjective rewards. For example, in Sigmund's (1993:208) view, 'play is its own reward'. Collias (1962) noted that domestic chicks, during the process of socialization, received 'self-reinforcement' from social experience during the early sensitive period. Positive reinforcing stimuli that meet these ethological needs are said to have an underlying 'hedonic value' (Tarpy, 1982); that is, they possess a *subjective* quality of a positive affective state (Toates, 1988). Whether this is pleasure as we know it, or what Lorenz calls 'feeling good' and 'satisfaction' (Nisbett, 1977:138, 289), does not affect the operant conditioning paradigm or the model. Even reinforcers that meet basic physiological needs might be considered hedonistic.

> I have seen that rats drink more of stuff that tastes good. And I do not see anything bad in thinking of this hedonistically; they like it, they do it.   *[Bolles, 1988:450]*

· *Aversive reinforcing stimuli* (S^{R-}) are stimuli which the animal perceives as 'bad' (e.g. pain). That is, under those conditions the animal will perform behaviors necessary to avoid those S^{R-}s.
· *Positive reinforcement* results when the consequence of performing a behavior is receipt of S^{R+}. The probability of the behavior occurring again is *maintained* or *increased*. The S^{R+}s which result in positive reinforcement are received under different *contingencies*, discussed below.
· *Extinction (omission)* can only occur after an animal has been positively reinforced for a behavior. If that same behavior now does not result in the animal receiving the S^{R+}, omission is the consequence and the probability of the behavior occurring again *decreases* until it is extinguished. Extinction is not simply a dissipation of the response, but is an active learning process (Davey, 1989).
· *Punishment* results when the consequence of performing a behavior is receipt of S^{R-}. The animal initially *escapes* the S^{R-}, and the probability of the behavior occurring again *decreases*.
· *Negative reinforcement* can only occur after an animal has been punished for a behavior. If, under the same conditions in which the original behavior resulted in punishment, a different behavior results in *avoiding* the S^{R-}, then the probability of that behavior is *maintained* or *increased*.

These four proximate consequences probably rarely operate alone, but rather in some combination. For example, Balph (1968) found that individual Uinta ground squirrels (*Spermophilus armatus*) differed in the probability their being retrapped one or more times. Based on his observations, he concluded that each individual squirrel differentially weighed the positive reinforcement of the bait versus the punishment of being trapped. Likewise, hitting your barking dog with a newspaper (S^{R-}) is not punishment unless the barking stops. If the dog maintains or increases its barking, then the *overall interaction* is perceived by the dog as positive reinforcement. That is, the dog's interaction with you (S^{R+}) outweighs the aversive aspects of being hit with the newspaper (S^{R-}).

Reinforcers and reinforcement as information   Consequences in instrumental conditioning provide not only immediate reinforcement, but also information (e.g. Cherfas, 1980). Garcia *et al.* (1973) argued that since natural selection 'has favored' animals that seek and incorporate information, we should think in terms of animals acquiring information rather than receiving reinforcement.

I prefer to retain the terms positive and negative reinforcement while using the term 'information' in two contexts: 1. as a reinforcing stimulus; and 2. as feedback about the relationship between stimuli (classical conditioning), and stimuli, behavior and consequences (instrumental conditioning).

First, I agree with Garcia *et al.* (1973) that *information* can, by itself, serve *as a reinforcing stimulus*. This concurs with Lorenz (1981) who stated that animals are motivated by curiosity to gather information. For example, an animal might seek more information about a novel object it sees in its environment by listening, smelling, tasting and touching it. What it hears, smells, tastes and feels are reinforcing stimuli for the respective behaviors. The reinforcing stimuli all provide information, but they can also be perceived as positive (e.g. smells 'good'=positive reinforcement), aversive (e.g. tastes 'bad'=punishment) or neutral (e.g. feels rough=no consequence (see below); provides only information about its texture).

Secondly, the animal obtains *information via feedback* about the relationship between the stimuli eliciting the behavior, the behavior, and the consequences (including information about the object). In all cases the animal is acquiring information (via feedback) about information (reinforcing stimuli). In the case where touching the object results only in learning that it feels rough (neither positive nor aversive), that information alone is a reinforcing stimulus, and the animal has obtained information about gathering information. That is, they have learned that when a novel object appears in their environment they can learn more about it by touching it. In this context, the use of the term information agrees with Plotkin (1988), who considers learning to be the acquisition of information, and most cognitive theorists who think of animals as acquiring information about their environment (Bolles, 1988). Likewise, classical conditioning is seen by many as the acquisition of information about stimulus relationships (Davey, 1989).

The use of the term information in these contexts in no way negates the utility of applying the term reinforcement to consequences in instrumental conditioning.

Consequences of no value   Some behaviors appear to result in consequences that have no reinforcing value (i.e. they are seemingly perceived as neither positive nor aversive). One example is the *habituation* of unconditioned (primarily innate) behaviors such as the predisposition to show fear/avoidance of novel stimuli. The response wanes and disappears (habituates) if it is neither immediately inhibited by punishment nor maintained by positive reinforcement.

Satiation of a drive such as hunger can also be perceived in a similar manner. Initially, the consequence of ingestion is positively reinforcing stimulation (S^{R+}); however, the reinforcing value decreases until satiation is reached; at this point its reinforcing value can be considered zero (neither positive nor aversive) and the behavior would normally cease. Should the animal continue to ingest, the food would begin to become aversive (S^{R-}) and punishment would be the consequence of further ingestion; hence the behavior would cease.

Ultimate Consequences   Proximate consequences can be translated into *ultimate consequences* within the context of natural selection and evolution. We can use the model to draw a parallel between instrumental conditioning and natural selection (Skinner, 1981; Rosenberg, 1984) by envisioning positive reinforcement in the sense of increasing individual fitness and punishment in the sense of decreasing fitness.

> . . . species-specific patterns are shaped by natural selection as operant
> behavior is shaped by reinforcement   *[Garcia* et al.*, 1973:5]*

That is, natural selection will positively reinforce adaptive behaviors by allowing the genes that contribute to those behaviors to reproduce themselves in future generations (i.e. selfish gene hypothesis; Dawkins, 1976b). Conversely, genes that contribute to disadvantageous behavior will be punished by decreasing their ability to replicate. The most extreme punishment is death, which immediately reduces that individual's future fitness to zero (unless a molecular biologist decides to clone its DNA).

This concept of *'Selection by Consequences'* (Skinner, 1981) has been developed by several theorists, including Campbell (1956), Ghiselin (1973), Pringle (1951), Skinner (1953, 1981) and Simon (1966). Also, Lorenz (1965) discussed the relationship between reinforcement and natural selection, and Pulliam and Dunford (1980) designed a model to demonstrate how it would operate in a hypothetical predator. Pulliam and Dunford (1980) developed their mathematical model based on consequences, feedback, experience and evolution. Their hypothetical cybernetic lizard is able to classify neural input from eating red ants (toxic=punishment) and black ants (pleasant taste=positive reinforcement), store the information and recall it when it next encounters an ant.

$$P_{n+1}=aP_n+(1-a)^{Vn}$$

Where: P_{n+1}=prob. of eating ant on trial $n+1$
   a=learning parameter (0–1):

0=guided most by recent experience (good if ants are in discrete patches)

1=guided most by earlier experience

P_n=probability of eating ant on trial n=residual innate tendency+accumulated reinforcement experiences

V_n=Reinforcement experienced on trial n (0=ant toxic; 1=ant palatable)

Pulliam and Dunford assume that a lizard that has 'a genetic program producing this neural architecture . . . is favored by natural selection, because it leads to the development of lizards that avoid poisoning themselves' (Pulliam and Dunford, 1980:13)

As Maynard Smith (1982) and Sigmund (1993) have pointed out, it is difficult to imagine how an animal can translate proximate consequences into fitness. In fact, natural selection makes that translation; hence, Maynard Smith operationally defined the process, as follows:

> An animal which performs the 'correct' action – correct in fitness-maximizing terms – when simultaneously experiencing hunger, thirst and sexual motivation, will, by definition, leave [the] most offspring.
>
> *[Maynard Smith, 1982]*

Staddon expressed the same concept in slightly different terms.

> . . . things that are 'reinforcing' in the definitional sense of 'strengthening' behavior, must also be things that in the history of the species have promoted inclusive fitness; that is, if it feels good, it probably is good (or was good, at least, for one's ancestors)   *[Staddon, 1980:xviii]*

There are potential pitfalls to avoid when using the model to conceptualize selection by consequences, including:

1 Not all characteristics of a species (especially, all behavior acts of an individual) are adaptive and have been selected for ('Adaptionist Fallacy'; Lewonton, 1979; Johnston, 1985). In other words, not all behaviors that are positively reinforced are necessarily adaptive (Skinner, 1981).

2 It is the genotype which leads (at least in part) to the behavior which is selected, not the behavior which leads to the consequences (e.g. Dawkins, 1984).

3 It then follows that 'Consequences may be beneficial, yet not provide material for the action of natural selection . . .' (Hinde, 1982: Fig.14).

Selection by consequences (and contingencies – see below) is the basic premise behind optimal foraging theory. That is, animals which carry genes promoting proper assessment of energy expenditure versus positive reinforcement (energy gain

Table 2.2. *Contingencies effective in producing the four basic consequences of behavior*

Consequences	Effective contingencies
Extinction	S^{R+} *does not follow* any occurrence of the behavior
Punishment	S^{R-} *follows* every occurrence of the behavior
Negative reinforcement	S^{R-} *does not follow* any occurrence of the behavior
Positive reinforcement	S^{R+} *follows* after various time intervals or occurrences of the behavior; these include: *Fixed ratio* *Fixed interval* *Variable ratio* *Variable interval*

in food) and punishment (predator risk) will leave more offspring. Of course, there have been critics of the concept of *selection by consequences* (e.g. Catania and Harnad, 1984). Finally, I will pass along a caveat sent to me by Dale Lott after he had reviewed this chapter: 'There is selection for a population of genes, and there is selection of a repertoire of operant behavior patterns, but the mechanisms are only superficially similar and pressing the analogy this far will confuse students'.

Contingencies (schedules of reinforcement)

Contingencies are the ways in which behavior and reinforcing stimuli are paired (Staddon, 1980). Contingencies are also called *schedules of reinforcement* (Ferster and Skinner, 1957). Only certain contingencies are effective in modifying behavior by producing the *consequences* discussed above (see Table 2.2).

These four basic contingencies for positive reinforcement produce different rates of behavior. *Ratio* contingencies are based on the number of occurrences of a behavior. *Interval* contingencies are based on the amount of time that has passed since the last reinforced occurrence of a behavior. The number of occurrences (ratio) or the time elapsed (interval) can be either *fixed* or *variable*.

These four contingencies are combined (in the laboratory and, perhaps, in nature) in various ways to create, for example, multiple, mixed, chained and tandem schedules of reinforcement (Fantino and Logan, 1979). When different contingencies apply at the same time to different behaviors, or the same behavior in different contexts, they are referred to as concurrent schedules of reinforcement. Concurrent schedules are the basis of studies of optimal foraging in patches of prey with different schedules of reinforcement.

2.3.5c *Learning paradigms in the model*

I agree that 'the world of animals is not a gigantic "Skinner Box" in which they gradually learn, by trial and error, what to do and what not to do' (Dawkins, 1986:60). Trial and error learning (*operant conditioning*) and classical conditioning are only part of the animal's capacity to deal with its world. Other learning related phenomena, such as insight, latent learning, discrimination learning, habituation and sensitization, also occur and can be encompassed by the model. Also, the Model is not subsumed within *general process theory* (e.g. Skinner), rather its flexibility concurs with Roper's (1983) statement that,

> . . . observations of natural learning tend to encourage the view that learning consists, not of a unitary general capacity, but of a collection of specialized abilities which have evolved independently in particular species to do specific jobs   [*Roper, 1983:205*]

The model does not assume that the types of learning involved can always be clearly distinguished. For example, Davey (1989) argued that in a typical classical conditioning study of salivation (e.g. Pavlov, 1927), it cannot be determined whether the dog salivates to the bell because the bell is associated with food or because the food pellets reinforce salivation. Likewise, the relative role of classical and instrumental conditioning in filial imprinting remains unclear (e.g. Davey, 1989; Rajecki, 1973).

Also, the model does not assume that classical and instrumental conditioning are mutually exclusive and work independently. In fact, the model encourages the researcher to consider both in behavioral analysis, as the following illustrates:

> The sight of a stimulus associated with food elicits, by a process of classical conditioning, an appropriate set of consummatory responses, but aspects of these responses may then be modified as the animal learns the correlation between variations in its behavior and variations in its success. Both processes are adaptive, for the opportunity for successful instrumental conditioning would probably not arise without appropriate classical conditioning in the first place . . .
>
> [*Mackintosh 1983:168*]

Researchers should also realize that we cannot directly observe learning but can only infer it from a change in behavior, called 'performance' (Davey, 1989). We also recognize that learning results in a change in the animal's anatomy and physiology; that is, the nervous system is used for the receipt, storage (memory), and retrieval of information.

2.3.6 Feedback

Feedback makes the model a closed-loop system (Pringle, 1951; Manning and Dawkins, 1992; McFarland, 1971; Toates, 1980) which is necessary for any comprehensive behavior model.

> . . . all the phyla of animals which have evolved a centralized nervous system have hit on the 'invention' of feeding back to the mechanism initiating a behavior the consequences of its performance.
>
> *[Lorenz, 1981:70]*

The model provides *feedback* from the *consequences* of behavior to the animal in the form of *information* which allows it to update its *experience* (association between stimuli, behavior and consequences) and modify *expectancies* through some 'internal representation' (Hinde, 1982; Ethology, Davey, 1989) which we call memory.

> . . . S* [a reinforcing stimulus] acts backward, like a feedback, I suppose, to produce expectancies. . . . the quality of what the animal expects importantly influences its behavior.   *[Bolles, 1988:450]*

Positive and negative feedback form two basic types of expectancy which direct animals to perform behaviors which result in receiving positive reinforcement and avoiding punishment, respectively. Expectancies come into play when the animal is motivated to perform a particular behavior.

> . . . both associative and motivational factors may contribute to making some activities readily performed when they are reinforced or readily suppressed when they are punished.   *[Shettleworth, 1983:23]*

2.3.7 Feedforward

The model recognizes that the next behavioral act in the continuous stream of behavior can be elicited by the consequence of the behavior that precedes it (i.e. 'reinforcer-elicited behavior', Davey, 1989:207). For example, aversive reinforcing stimuli generally elicit behaviors which allow the animal to escape the punishment, such as spitting out distasteful food.

Feedforward also occurs in a chain of behaviors (e.g. search, capture, consumption of prey) as described by Hineline (1988) in his commentary on Gardner and Gardner (1988):

> As described [by Gardner and Gardner 1988], 'feedforward' concerns behavior–environment relations in which one situation evokes an

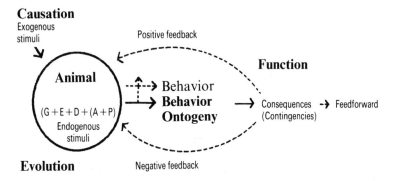

Fig. 2.5   Where Tinbergen's areas of study fit into the Model for a behavioral act.

organized pattern of action resulting in a new situation that evokes a different pattern of action resulting in yet another situation, and so on. This type of behavior–environment relation has long been described as applying to behavioral chains based on positive reinforcement.

[Hineline, 1988:457]

My use of the term feedforward is somewhat different, but not exclusive of its use by Toates (1980) and McFarland (1971:102) as a 'phenomenon which enables an animal to anticipate the long-term consequences of behaviour and to take appropriate action to forestall such consequences'. That is, an animal that is not thirsty might still drink in anticipation of a future water deficit (McFarland, 1971).

2.4 APPLICATION OF THE MODEL

Below are examples of how the model can be used as a basis for: 1. conceptualizing the areas of study (Chapter 1); 2. organizing the types of question (Chapter 1); 3. focusing research; 4. developing an expanded or more focused model; and 5. diagnosing and treating behavior problems. Chapter 5, on delineation of research, describes how the model can assist in defining questions, objectives and hypotheses.

2.4.1 Conceptualizing the areas of study

Tinbergen's four areas of study (Chapter 1) can be located in portions of the model (Figure 2.5, Table 2.3). This assists the researcher in seeing how his particular research interest fits into the 'big picture'; that is, how it integrates with other areas of study.

Examples of research that focuses on portions of the model relevant to each area of study are provided below.

Table 2.3. *Where Tinbergen's four areas of study are incorporated in the model for a behavioral act*

Areas of study	Relevant portions of model
Function	Proximate Consequences; Feedback
Causation	Exogenous and endogenous stimuli; feedforward
Ontogeny	Behavior changes during maturation; these result from changes in environment, experience, anatomy, physiology, stimuli, and consequences.
Evolution	Genotype; often studied using comparative studies of the behavioral phenotype (i.e. Behavior)

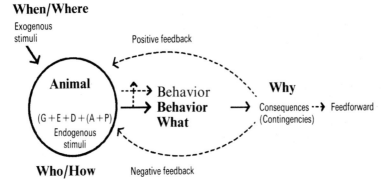

Fig. 2.6   Where the six types of questions about behavior fit into the model for a behavioral act.

2.4.2 Organizing the types of question

Figure 2.6 illustrates where the types of questions discussed in Chapter 1 are focused in the model. The first question to be answered in any research project is always 'What does the animal do?'. The model demonstrates that all the other questions are directed at factors that affect *what* an animal does (see Table 2.4)

2.4.3 Focusing research

Having determined the area(s) to be studied and type(s) of Questions to be addressed, research then becomes focused on one, or more, parts of the model. Regardless of what portion we focus on, research results (by necessity) always relate back to the animal's behavior, as illustrated in the examples below.

Table 2.4. *Where the six types of questions are incorporated in the model for a behavioral act*

Types of question	Relevant portions of model and examples
What?	Behavior (e.g. courtship and territorial display)
Who?	Anatomy, physiology, environment, genotype (e.g. large, healthy and motivated to breed, dominant, male sage grouse)
Where?	Exogenous stimuli (in central territory, on breeding ground in North Park, Colorado with other males and females)
When?	Exogenous stimuli, endogenous stimuli (spring breeding season; Courtship behavior in presence of females; Territorial displays in presence of males; reproductive 'drive')
How?	Anatomy and physiology (endocrine and nervous system stimulate and integrate use of wings, tail, legs and gular sacs in 'strutting' and 'booming' displays)
Why?	Proximate consequences, ultimate consequences (Hedonistic value of displaying and mating to the individual male; a male whose displays allows it to hold a central territory and attract and mate with females has a higher fitness)

- *Genotype*: Ralph and Menaker (1988) found a mutation in hamsters in which *wild-type* (normal genotype) animals have a circadian locomotor rhythm of about 24 hours; *heterozygous* animals have rhythms of about 22 h, and animals *homozygous* for the mutation have rhythms close to 20 h.
- *Environment*: Holekamp and Smale (1993) found that the *presence of mother* spotted hyaenas (*Crocuta crocuta*) during their offspring's aggressive interactions strongly influenced the outcomes for juveniles less than 6 months of age.
- *Anatomy*: Choudhury and Black (1993) studied mate selection in captive barnacle geese (*Branta leucopsis*). They found that *heavier* females and those with *darker face patterns* sampled significantly more potential mates.

- *Physiology*: Landsman (1993) studied sex differences in *electric organ discharge* in a weakly electric fish (*Gnathonemus petersii*).
- *Exogenous stimuli*: Nicoletto (1993) studied the effect of *male ornamentation and display rate* on the sexual response of female guppies.
- *Endogenous stimuli*: Dethier (1966) describes how distention in the foregut of a blowfly sends a *message via the recurrent nerve* to the brain, which inhibits extension of the proboscis, thus inhibiting further feeding.
- *Behavior*: Fraser and Nelson (1984) described the frequencies and sequencing of 16 *courtship behaviors* in a Madagascan cockroach (*Gromphadorhina portentosa*).
- *Consequences*: Shettleworth (1978a,b) found that scrabbling behavior in hamsters increased in rate when *positively reinforced* with seeds and decreased in rate when *punished* with mild electric shock.
- *Contingencies*: Krebs *et al.* (1978) determined the foraging strategies of great tits (*Parus major*) in two food patches with different densities of mealworms (i.e. variable ratio schedules of reinforcement).
- *Feedback*: Brown and Gass (1993) demonstrated that rufous hummingbirds (*Selasphorus rufus*) learn spatial associations between cues and rewarding feeders.
- *Feedforward*: Lawhon and Hafner (1981) fed seeds (millet) and seed mimics (glass beads) to kangaroo rats (*Dipodomys*) and pocket mice (*Perognathus*). They found that glass beads were always rejected (rather than pouched) after unsuccessful attempts to husk them.

2.4.4 Developing an expanded or more focused model

The model can be expanded to include additional factors or focused on portions of more importance to the researcher.

2.4.4a *Expanding the model*

Hinde (1982), in his discussion of behavioral development, proposed a model of the relationship between ontogenetic and causal factors and consequences (Figure 2.7). Hinde's model incorporates a route of action for ontogenetic factors that I have not specified in my model. However, I have identified ontogenetic factors in section 2.4.1. Hinde has also included (but not labeled) feedback (from functions to ontogenetic factors) and feedforward (from goals to eliciting factors).

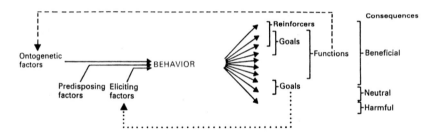

Fig. 2.7   Relationships between ontogenetic and causal factors and consequences. The distinctions between ontogenetic and predisposing and between predisposing and eliciting factors are often somewhat arbitrary. Among the consequences, the categories of reinforcers, goals and functions only partially overlap. Although a goal is normally achieved as a consequence of a behaviour, an internal representation (anticipation) may contribute to causation (dotted line). Consequences may be beneficial, yet not provide material for the action of natural selection (functions, strong sense). Exceptionally, harmful consequences can be goals. The dashed line indicates evolutionary consequences on the next generation (from Hinde, 1982).

2.4.4b   Focusing the model

1. The model can incorporate the 'behavior systems' approach (e.g. Timberlake, 1983; Timberlake and Lucas, 1989) to modeling behavior by relating; i. innate behaviors (modal action patterns); ii. 'motivational states' (behavioral systems); and iii. exogenous stimuli (Davey, 1989). For example, Davey's (1989:158) organization of feeding behavior in the rat (Figure 2.8) is incorporated by recognizing that the behavior system and subsytems are stored, modified, processed and activated (via endogenous stimuli) within the animal's anatomy and physiology. The model can then, for instance, assist in visualizing the proximate and ultimate consequences of each of the behaviors.

Davey's model is similar in structure to Baerends' (1976) classic model of interruptive behavior during incubation in herring gulls (see Davey, 1989:320).

2. Alcock (1993) presented a model for the control of circadian rhythms that incorporates exogenous stimuli, anatomy, physiology, and rhythmic behavior patterns (Figure 2.9). It focuses more explicitly on what I have labeled in a general way as anatomy and physiology.

3. It is possible to focus the model on internal motivational states and decisions by incorporating more explicit control theory and including comparators, controllers and servomechanisms as part of the anatomy and physiology (e.g. Toates, 1980; McFarland, 1971). An example of a model which uses a comparator is Huntingford's (1984) modification of Archer's (1976) model of fear and aggression

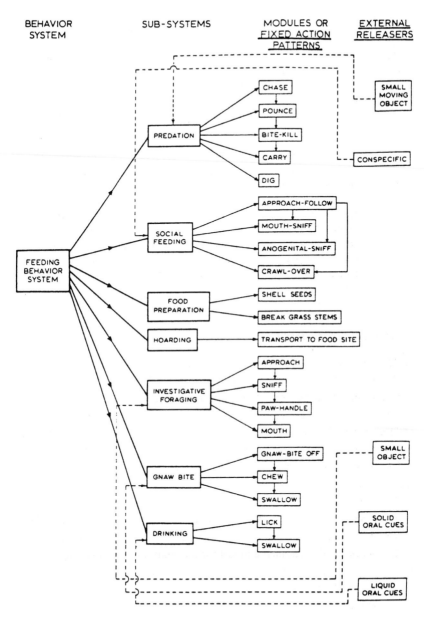

Fig. 2.8   The functional organization of a putative feeding behavior system for the rat (from Davey 1989).

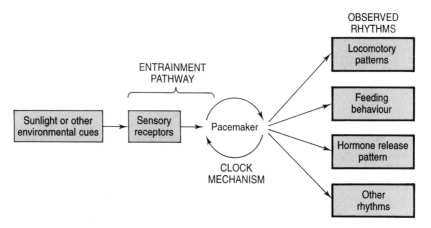

Fig. 2.9   A master clock may, in some species, act as a pacemaker to regulate the many circadian rhythms of an individual (from Alcock, 1993, after Johnson and Hasting, 1986).

in vertebrates (Figure 2.10). Huntingford (1984) should be consulted for a thorough discussion and examples of various types of models of motivation.

2.4.5 Diagnosing and treating behavior problems in animals

The model can be used by applied animal behaviorists as the basis for determining the etiology of a behavior problem, what is maintaining it, and what manipulations can be made to mitigate or eliminate it. For example, Figure 2.11 illustrates how the model can be used to organize some of the factors to be considered in diagnosing and treating an aggression problem in dogs.

2.5 SUMMARY

The model presented above is the result of my attempting to provide a structure on which ethological research could be designed and conducted. I have found it useful in assisting graduate students in designing research by getting them to recognize the various factors, past and present, which contribute to behavior.

For the behavior (*what*) of interest, you can 'plug into' the model what you have learned from the literature about the variables that affect that behavior. For the types of question (e.g. where? when?) you are attempting to answer, you can begin to identify the variables (e.g. genotype, endogenous stimuli) that are likely to have a major effect on your results. You can put the partially completed model on the blackboard and brainstorm with colleagues to identify additional variables (e.g. exogenous stimuli) which should be considered and potential methods of manipulating, measuring or eliminating those variables.

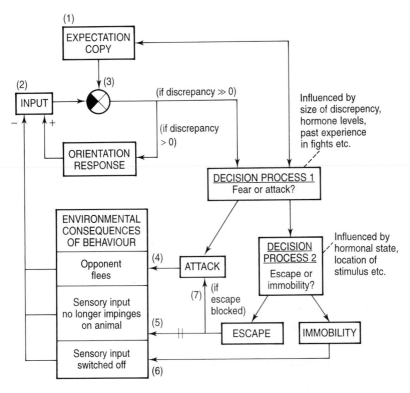

Fig. 2.10 Motivational models using control theory; a simplified version of a control theory model of aggression and fear in vertebrates (from Huntingford, 1984).

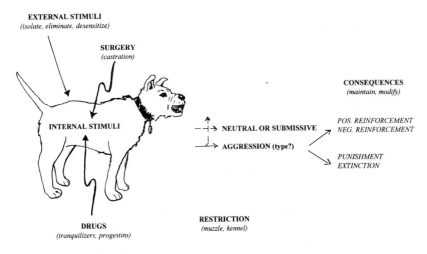

Fig. 2.11 Some of the factors an applied ethologist might consider in diagnosing and treating an aggression problem in a dog (drawing by Brenda Knapp).

LIBRARY, UNIVERSITY COLLEGE CHESTER

In time, as most of the terms in the model become replaced with real variables, you begin to have more confidence in your grasp of the complexity of the research you are proposing. You can pose more clearly defined questions, formulate more exacting hypotheses, and begin the task of determining how you will answer those questions and test those hypotheses. That is what the rest of this book is about.

3 Choice of subjects

Ethologists come to study particular species for various reasons, but all generally travel via two routes: 1. their special interest in a particular *species*; or 2. using a species that is suitable for investigating a particular *concept*. These two routes intertwine so that a researcher may become interested in pursuing a concept after having studied various aspects of a species' behavior. Likewise, a researcher may initially study a species in pursuit of answers to a conceptual problem and become fascinated with other aspects of the species' behavior.

3.1 SPECIES-ORIENTED RESEARCH

Many ethologists who settle on a particular species to study are naturalists who discover a fascinating species while spending time in the field. The following excerpt from *Curious Naturalists* describes how Tinbergen began his many years of research on the digger wasp:

> On a sunny day in the summer of 1929 I was walking rather aimlessly over the sands, brooding and a little worried. I had just done my finals, had got a half-time job, and was hoping to start on research for a doctor's thesis. I wanted very much to work on some problem of animal behaviour and had for that reason rejected some suggestions of my well-meaning supervisor. . . .
> While walking about, my eye was caught by a bright orange-yellow wasp the size of the ordinary jam-loving Vespa. It was busying itself in a strange way on the bare sand. With brisk, jerky movements it was walking slowly backwards, kicking the sand behind it as it proceeded. The sand flew away with every jerk. I was sure that this was a digger wasp. . . .
> I watched these wasps at work all through that afternoon, and soon became absorbed in finding out exactly what was happening in this busy insect town. . . .
> As I was watching the wasps, I began to realize that here was a wonderful opportunity for doing exactly the kind of field work I would like to do. Here were many hundreds of digger wasps exactly which species I did not know yet, but that would not be difficult to find out. . . .
> My worries were over; I knew what I wanted to do. This day, as it

turned out was a milestone in my life. For several years to come I was to
spend my summers with these wasps . . .   *[Tinbergen, 1958:5–8]*

Dethier, who spent many years pursuing the behavioral biology of the blowfly,
describes how he settled upon the blowfly as a research animal in the following
excerpt from his delightful book, *To Know a Fly*:

> When choosing an experimental animal, therefore, why settle for
> anything so prosaic as the laboratory rat, so giddy as the guinea pig,
> so phlegmatic as the frog, so reptilian as the chicken, so cousinly as the
> chimpanzee? Why not choose an excitingly different creature like the
> aardvark or the dugong? Why not choose the fly?
> With so many kinds of flies in nature's burgeoning storehouse of
> life, how does one choose a proper species for study? The answer is
> simple. Let the species choose you. This was how our laboratory came
> to work with the black blowfly fifteen years ago.   *[Dethier, 1962:7–8]*

The rest of Dethier's account of his early years of research on the blowfly makes
heuristic and enjoyable reading.

Von Frisch, a nobel laureate along with Lorenz and Tinbergen, began his many
productive years on the study of bees through his attempt to 'set the record straight'
regarding the color vision of bees. Von Frisch describes the launching of his career
of bee study in the preface to his fascinating book, *The Dancing Bees*:

> . . . some forty-five years ago . . . a distinguished scientist, studying the
> colour sense of animals in his laboratory, arrived at the definite and
> apparently well-established conclusion that bees were colour-blind. It
> was this occasion which first caused me to embark on a close study of
> their way of life; for once one got to know, through work in the field,
> something about the reaction of bees to the brilliant colour of flowers,
> it was easier to believe that a scientist had come to a false conclusion
> than that nature had made an absurd mistake. Since then I have been
> constantly drawn back to the world of the bees and ever captivated
> anew. I have to thank them for hours of the purest joy of discovery,
> parsimoniously granted, I admit, between days and weeks of despair
> and fruitless effort.   *[Von Frisch, 1953:iii]*

Some researchers have begun their studies in animal behavior on a single species
and then become interested in a limited type of behavior which they have generalized to other species. For example, Allee recounts how he began studying freshwater
isopods and then became interested in social behavior in general.

> Almost forty years ago as a graduate student in zoology I was engaged
> in studying the behavior of some common small fresh-water animals
> called isopods . . . day after day I put lots of five or ten isopods into

shallow water in a round pan. . . . When a current was stirred in
the water the isopods from the streams usually headed against it; but
those from ponds were more likely either to head down current or to
be indifferent in their reaction. . . .

Rather cockily I reported after a time to my instructor that I had
gained control of the reaction of these animals to a water current. By
the judicious use of oxygen in the water, I could send the indifferent
pond isopods hauling themselves upstream, or I could induce the stream
isopods to going with the current. I had not reckoned with another
factor that presently caught up with me.

After a winter in the laboratory it seemed wise as well as pleasant to
take my pan out to a comfortable streamside one sunny April day and
there check the behavior of freshly collected isopods in water dipped
from the brook in which they had been living. To my surprise, the
stream isopods, whose fellows all winter had gone against the current,
now went steadily downstream or cut across it at any angle to reach
another near-by isopod. When I used five or ten individuals at a time,
as I had done in the laboratory, they piled together in small close
clusters that rolled over and over in the gentle current. Only by testing
them singly could I get away from this group behavior and obtain a
response to the current; and even this reaction was disconcertingly
erratic.

It took another year of hard work to get this contradictory behavior
even approximately untangled; to find under what conditions the
attraction of the group is automatically more impelling than keeping
footing in the stream; and that was only the beginning of the road that
I have followed from that April day to this time, continuing to be
increasingly absorbed in the problems of group behavior and other
mass reactions, not only of isopods, but of all kinds of animals, man
included.

As the years have gone on, aided by students and other collaborators
and by the work of independent investigators, I have tried to explore
experimentally the implications of group actions of animals.

[Allee, 1938:5–7]

Allee pursued his interest in the social behavior of animals and became an early
and influential ethologist. The unusual events that led to E.O. Wilson's selection of
insects, especially ants, as his focus of study are recounted in his revealing autobi-
ography (Wilson, 1994). Insight into how several other noted ethologists selected
their research species can be found in Dewsbury's (1985) compilation of autobiographies.

Ethologists often become deeply interested in particular species and pursue different lines of questions as they arise while studying other aspects of the species'

behavior. The cliche that 'research generally provides more questions than it answers' keeps some ethologists studying one, or a few, species for their entire career. This is often efficient in that they can build on past experience with the species and are able to make effective use of their time.

3.2 CONCEPT-ORIENTED RESEARCH

Although Lorenz came to study birds, particularly waterfowl, as an outgrowth of his apprenticeship with Heinroth, he selected species with an eye to unraveling analogous and homologous relationships in patterns of behavior. Nisbett describes Lorenz's choice of subjects this way:

> However his interest in an animal may have arisen in the first place – and this may in part have been by the interplay of change and curiosity – his chosen subjects did in fact form a coherent and rational array. The different species fell into several groups. First, there were those which were in their own right the central objects of his study; initially the jackdaws, then the herons, and now the geese. Second, were the closely related species: ravens for comparison with jackdaws, or mallard ducks to watch out of the corner of an eye while looking at geese. These showed not only what the ducks had in common with their geese cousins but also what they had developed differently: he could ask himself 'why?' Heron society was markedly different from that of jackdaws or geese; again 'why?' Then there were the species unrelated to jackdaws or geese, but which had similar elements of behaviour. This allowed him to look for patterns of behaviour to which evolution came independently in different species.   *[Nisbett, 1977:44]*

In some cases, researchers discover concepts that can be tested on their favorite species. In other cases, previous research on a species reveals aspects of its behavior which make it suitable for testing a concept; this is illustrated by Alcock's (1973) use of red-winged blackbirds to test L. Tinbergen's (1960) 'search image hypothesis'. Earlier work with the species had suggested that they might selectively search in patches where they had found food previously, but more importantly for Alcock they were omnivorous and available. Therefore, he used this species in an experimental food maze to answer the questions 'do birds learn where they are likely to find food and come to use locational cues to direct their searching and/or do birds learn what food they are likely to find and come to search preferentially for that item on the basis of visual cues associated with the food?' The bird in this case was the red-winged blackbird. Other researchers have tested L. Tinbergen's hypothesis on carrion crows (*Corvus corone*; Croze, 1970), domestic chicks (*Gallus gallus*; Dawkins, 1971), and great tits (*Parus major*; Royama, 1970).

3.2.1 August Krogh principle

Researchers who pursue answers to conceptual questions or concentrate their efforts on a particular type of behavior attempt to study species that best represent the concept or type of behavior under study. In 1929 August Krogh stated that 'For a large number of problems there will be some animal of choice, or a few such animals, on which it can be most conveniently studied'. Krebs (1975) has labeled this statement 'The August Krogh Principle'.

The indigo bunting was selected by Carey and Nolan (1975) to test the 'Verner–Willson–Orians hypothesis' that polygyny would evolve in avian species where critical resources are distributed in widespread patches, if the advantages of one male mating with several females offset the disadvantages of reduced parental attention and possibly increased attraction of predators and depletion of food resources. Preliminary study of an indigo bunting population in Indiana had led Carey and Nolan to predict that the population would be polygynous and therefore provide a good opportunity to test the hypothesis.

Experimental psychologists traditionally use rats, pigeons, or selected primates as their subjects in studying learning theories that they believe will then be applicable to (or at least worthy of testing on) humans. Two reasons are generally given for pursuing concepts of learning on such few species. First, many psychologists suggest that learning processes are basically the same in all species, differing in amount rather than kind. These species are being used as biological models in essentially the same way that the drug industry uses rats and mice to test drugs designed for human consumption. Second, there is a wealth of background data already available on these species. It is questionable whether the August Krogh principle was invoked when the rat was selected for the early psychology studies. It was probably convenience that initiated and maintained the momentum that established the rat as the psychologist's choice subject (Beach, 1950).

Several years ago while cooling off in a local pub (after a hot afternoon observing yellow-headed blackbirds in a mosquito-infested marsh), Dr. Gordon Orians explained to J. R. Watson and myself that he had become very interested in the slugs that foraged in his backyard garden in Seattle. His interest was more in testing ecological theory than in saving his vegetables. Later Cates and Orians reported on their test of the hypothesis that 'early successional plant species make a lesser commitment of resources to defend against herbivores, and should then provide better food resources for generalized herbivores than later successional and climax plants' (Cates and Orians 1975:410). What would be a good research animal on which to test these predictions? Cates and Orians explain:

> Generalized herbivores are required for testing these predictions: we
> have used in our experiments two species of slugs, both known to graze

Table 3.1. Credit–debit sheet of some characteristics to be considered when selecting a subject species

Characteristic	Questions	Credit	Debit
Suitability	Is the species suitable for the concept being studied (August Krogh principle)? Can you recognize individuals by natural marks or can they be easily marked? Does it engage in interesting behavior which you can observe repeatedly? Can you make the necessary manipulations on this species?		
Availability	Is the species found locally or will you have to travel to study it in the field? If it is found in a foreign country, what are the political ramifications?		
	If you want to observe the species in the wild, is it accessible in its habitat? Can observations be easily made without altering its behaviour? Is it nocturnal or diurnal?		
	If you want to bring the animal to you, can it be easily obtained? Is it on the rare and endangered species list? Can it be easily captured? Can subjects be replaced if they die? Can individuals be returned to the point of capture when the research is done?		
Adaptability	How will the animal adapt to life in captivity? Can you simulate its natural environment and provide for its special needs? Are its habits compatible with yours?		
Background information	What is already known about the species? Is there a reasonable backlog of data on which to build? Has someone already done the research you are planning? How does it help you answer the questions above and anticipate other problems?		
Summary	Total the credits, assess the financial commitment, and accept or reject the species as your subject. Will you be able to follow the guidelines for animal welfare and ethics in research? (see Appendices C and D)		

on a wide variety of plants, and both abundant in and around Seattle, where they are active most of the year.   *[Cates and Orians, 1975:411]*

Following the August Krogh principle may not be a simple task, since we often don't know enough about the behavior of different species to make a wise choice of the best species for concept-oriented research (Gans, 1978). However, as Cates and Orians pointed out (above), if we can identify a suitable species, it also helps if the species is available.

3.2.2 Suitability versus availability

It would be ideal to study our species of choice, either because of our interest in the species *per se* or because it is the most appropriate species for studying a concept (August Krogh principle). However, this is often not possible or desirable due to a multitude of factors, of which suitability and availability are only two. For example, Leuthold (1977:13) states that 'Ease of observation is probably the main factor that has influenced the choice of species [among African ungulates] for field studies'.

In choosing a subject species, you will find it valuable to construct a credit–debit sheet of the desirable and undesirable characteristics of the particular species for your study. Table 3.1 contains a partial list of characteristics to be considered.

Referring to the selection of subjects at the class, rather than the species level, Konishi *et al.* (1989) state that birds have provided important models for developing and testing new ideas in biology. These include the evolutionary significance of variations in social organization, encoding of information in animal communication, and the sensory basis for migration and navigation. They argue that the reasons for the widespread use of birds is that they 'are widely distributed, highly diversified, and exhibit behavior and social organizations equal in complexity to mammals, yet they are generally more conspicuous and approachable in natural environments' (Konishi *et al.*, 1989:465).

Regardless of how you come to choose your subject animals, do it carefully. Your investment in studying that species may be years or your entire career. Your colleagues will associate you with the species whether you like it or not. Select wisely and spend your time well.

4 Reconnaissance observation

4.1 HOW TO OBSERVE

An early step in the study of animal behavior involves intensive reconnaissance observation. This may occur before you have decided what aspect of behavior to study and probably before you have formulated any hypothesis (Lorenz, 1935). This early stage in which you become familiar with the animal's behavior 'is the most arduous and demanding aspect of behavioral study' (Marler, 1975:2). It is extremely important, for no successful research can be launched without this background knowledge. Methods have been suggested for evaluating sampling plans for longitudinal studies of behavior based on 'pilot samples' taken during reconnaissance observations (e.g. Kraemer *et al.*, 1977).

Besides helping you to design your research, these initial observations also provide an important source of additional questions and hypotheses.

> Having myself always spent long periods of exploratory watching of natural events, of pondering about what exactly it was in the observed behaviour that I wanted to understand before developing an experimental attack, I find this tendency of prematurely plunging into quantification and experimentation, which I observe in many younger workers, really disturbing, unless, as happens to come, they do, from time to time, return, more purposefully than before, to plain, though more sophisticated watching.
>
> *[Tinbergen, 1951:vi]*

Tinbergen (above) talks about 'exploratory watching', 'more sophisticated watching', and 'observed' behavior. The terms watching and observing are commonly interchanged (see quote from Huxley on p. 113), but I believe ethologists should draw a distinct dichotomy.

4.1.1 Watching versus observing

Baby-sitters *watch* children; developmental psychologists *observe* them. *Watching* is a casual endeavor; *observing* is a rigorous process. Ethologists enjoy both watching and observing animals. They receive pure enjoyment from *watching* animals, and while watching they may develop interest in a particular species, behaviors and questions; but obtaining answers to those questions requires careful *observations*. I recently watched a goshawk glide from it's perch in an aspen tree and snatch a red squirrel off a log using the talons of its left foot. It was a chance occurrence which was exciting to watch and raised some questions, such as: Do goshawks usually employ only one set of talons when capturing red squirrels? If so, are individual goshawks primarily left-footed or right-footed? Answers to initial questions, such as these, usually lead to additional questions.

Most visitors to zoos occasionally stop and watch animals. When they leave they can tell you about many of the animals they saw and some that they watched; few can relate any in-depth observations of particular behaviors. Many bird-watchers see a bird, identify it, and rush on hoping to add more species to their lists. Observers take the time to study the behavior of the birds, describing in their note-books the intricate details of individual and social behaviors and perhaps recording the birds' vocalizations for later reference and enjoyment. Intense observation is generally more rewarding than superficial watching. It is also necessary in the study of animal behavior. Observation may be as much a state of mind and awareness as it is a clearly defined technique. This is reflected in the quotes in Chapter 3.

Observers must be more than a visual recorder; they must also be aware of input to their other senses and must think. One must be disciplined enough to know when to be a machine-like recorder of data and when to contemplate what is happening or has happened. The experienced and astute observer often develops and 'tests' hypotheses mentally while keeping animals under observation. Obviously, attention to rapidly occurring behavior is not consistent with theorizing, hence priorities must be established.

Researcher priorities while observing various behaviors and species in various habitats are generally the same, as follows:

1 Record data accurately and completely.
2 Check equipment to make sure it is working properly and repair it if necessary.
3 Think about what is happening, put it in perspective, and formulate hypotheses. Think about the hypotheses, discard the easily disproven ones, and write down those that call for additional thought and testing. This constant consideration of hypotheses is a procedure to be used only when making initial reconnaissance observations for the formulation of questions, objectives and hypotheses. However, some sampling methods (Chapter 8) and/or inactivity of the animal(s) under observation can provide periods during data collection during which you can allow yourself to think about additional questions and hypotheses; but never allow

your mind to wander to where it affects Priority 1 (above) and results in *observer error* (Chapter 8).

4 Satisfy basic bodily needs. Plan your sampling periods so that your basic bodily needs do not suddenly appear to be priority number one; however, should the urgency of urination and/or defecation prevail, carefully consider the valuable protocols provided by Meyer (1989). Meyer has provided carefully researched recommendations suitable for killer whale observers in sea kayaks, peregrine falcon observers suspended on a cliff face, and desert tortoise observers trudging the sands.

Once you have the above priorities firmly established in your research protocol, remember that periods spent in the field are likely to be some of the most treasured times of your existence. They should be both productive and enjoyable, but they won't be constant fun. Lott (1975) reflected the feelings of many ethologists when he wrote 'Protestations of a field person':

> 'Welcome back! Have a good vacation?' 'I wasn't on vacation. I was in the field.' Well that's not what I mean. Being in the field isn't a vacation; it's hard work, a hard life, and besides . . . But hold your tongue. People who spend months at a time noting the behavior of animals in odd corners of the world are usually greeted that way. We're happy with our work, of course, but for several reasons it doesn't qualify as a vacation.
>
> Sometimes just living there is a problem. If you can't find or afford a convenient house, camping out becomes living in a tent by the second week. And the kind of stick-to-your-ribs food that stores well in a burlap bag or metal box soon starts to stick in your throat. Some colleagues and I went to visit Patti Moehlman's burro study in Death Valley a couple of years ago, and brought along some steak. Patti had been without refrigeration for weeks. Her response was succinct and eloquent: 'GOLL-EE REDMEAT.' . . . So welcome was the steak that it hardly mattered that the water from a desert cloudburst streamed into our plates as we ate crouching under a picnic table.
>
> But food and housing are far from the worst of it. You can get used to eating almost anything and sleeping almost anywhere. The worst of it is that you get to be a little bit batty.
>
> To be more specific, you get to be sort of manic-depressive. You experience mood swings that increase as a direct function of the number of seven-day weeks you've spent on the project. . . . The most salient symptom is that your evaluation of the study gets to be wildly unrealistic.
>
> High noon may find your pulse racing as your mind forms the kind of modest, contained, but penetrating remarks that persuade the National Academy of Sciences plenary session that the Nobel Committee did indeed know what it was doing when it cited your

analysis of the distribution of deer droppings, the grumpiness of goose gatherings, or the ballistics of bison bellowing as the intellectual link that completed the conceptual chain from molecule to mastodon. Your lips move a little as you take on a set of bored, cynical journalists who came to the press conference to play it for laughs and a chance to get in a dig at the granting agency that spent nearly $1,750 in support of your research. A basic stock of fine ironic wit, a dash of captivating candor, an irresistibly lucid illumination of The Link in layman's terms and they are first sobered, then entranced. When you release them from your spell, they will sprint to their typewriters and set their two forefingers to banging out near poetry in praise of basic research and (blush) you.

That evening you may be so sunk in shame that you want to change not only your study but your name. How could you have committed yourself to a study so barren and one for which you are so ill prepared? What will you say to that granting agency when they ask what became of more than $1,750 intended to support significant basic research? If you take the entire blame you'll never get another chance. Besides it wasn't all your fault; but how do you make them understand that fate has thwarted you at every turn, that your field glasses fogged up during nearly three goose gatherings, that the microphone salesman was lying, lying when he told you how far away it would pick up bison bellows? Yes, who will bear witness that your failure was not really your fault now that God has turned his face from you?

And so you go on, ever more sublimely happy, ever nearer suicide. During your more lucid moments you realize, of course, that you're getting to be a little bit batty, and you come to crave some stabilizing influence to dampen your oscillations. Contact with an old friend becomes so welcome that you hold your tongue even if he says something stupid like, 'Welcome back! Have a good vacation?'

[See also Hailman's (1973) discussion of fieldism]

The following quotation carries a hidden message: '. . . as the *work* [italics mine] by Cain and Sheppard has shown . . . ' (Tinbergen, 1958). Whether intended or not, the word work is indicative of what ethological studies can, at times, become. Even though the overall experience is enjoyable and the results rewarding, it can become tiresome, and you are often confronted with having to convince yourself that the end justifies the means. Cheney and Seyfarth describe the 'syndrome' this way:

> Anyone who has ever studied animals in the natural habitat recognizes that even the most stimulating project includes moments of unrelieved
> tedium.   *[Cheney and Seyfarth, 1990:313]*

For example, Estes (1967:45) admitted to near boredom when he wrote, 'A week later, at 9:20 P.M., while I was out alone, the rallying call of hyenas again distracted

me from *the rather tedious job of recording gnu activity patterns* [italics mine].'
Schaller's description of part of his study on the Serengeti lion is also illustrative:

> My existence revolved around lions, I was wholly saturated with them,
> talked and wrote about them, and thought about them. . . . A few times,
> though, I saw too much of lions. Once Bill . . . and I decided to track a
> lion continuously by radio for several weeks. . . . The first few days were
> rather pleasant. . . . As the days passed this delight vanished, and we
> went about our task with grim determination. . . . We stayed with the
> male for twenty-one consecutive days and suffice it to say that for
> once I had a surfeit of lions.   *[Schaller, 1973:90–91]*

The descriptions of behavior of wild animals that you read in the literature are
often the result of weeks, months and years of careful stalking, hiding and
painstaking observations (e.g. Packer, 1994). Often hours are spent in a blind under
less than ideal conditions, with inclement weather making you physically uncomfortable and your view of the animals poor, and the inactivity of the animals
becomes frustrating. Your binoculars get beaten about and rained and snowed
upon, and the pages of your field notes become limp and stuck together. Field
research can be trying at times, but you can make the best of it by being physically
and mentally prepared. Expect Murphy's Law, 'If anything can go wrong it will', to
take effect from time to time. Allow for some slack in your schedule to absorb days
when you cannot collect data because of poor weather or equipment breakdowns.
Often, you can release much frustration merely by recording the disasters in your
field notes, realizing that in years to come you will remember even those days fondly
as you reflect on the data-rich days with pride.

Successful data collection through observations necessitates your: 1. having
developed the skills necessary for effective and efficient observation (see exercise
below); 2. having the proper equipment (e.g. binoculars and spotting scopes);
3. understanding the various ways to describe behavior; 4. having a well-designed
system for recording your field notes; and 5. knowing when your data are sufficient.
The observations we are discussing here are either *ad libitum* samples (Chapter 8) or
initial reconnaissance observations on which a future well-designed study will be
based. Regardless, the skills used are the same; only the relative emphasis on the
data collected is likely to be different.

4.1.2 An exercise in observing

Learn to see what you are looking at   *(L'Amour, 1982:59)*

For several years I had struggled to find an effective way of teaching students
how to observe animal behavior. Then some ten years ago serendipity took a hand,

and I discovered some delightful insights and an effective method in Frederick Franck's (1973, 1979) books: *The Zen of Seeing* and *The Awakened Eye*. First, I have to admit that, like Hofstadter (1979:246) 'I'm not sure I know what Zen is'. But, for our purposes, that doesn't matter; Franck's technique of teaching drawing through seeing is what is important.

> While I am SEEING/DRAWING, I take hold of the thing, until it fills my total capacity for experience. Once I have thus taken possession of a hill, a body, a face, I let go, let it go free again, as if I were releasing a butterfly. Yet it remains mine forever. After much SEEING/DRAWING my eye goes on drawing whether my hand draws or not.
>
> *[Franck, 1973:125]*

Although Franck developed his technique for teaching drawing through seeing, I found it equally effective in teaching seeing/observing through drawing. Based on Franck's description of his seeing/drawing procedure, I developed the following exercise:

- *Exercise in developing observational skills through drawing*
- Objective: To develop observational skills, not to create an artistic rendition of the animal. However, it is important to attempt to recreate with the pencil what you see/observe with your eyes.

> Please realize you are not 'making a picture,' you are *not* being creative. We are just conducting an experiment in SEEING, in undivided attention! The experiment is successful if you *succeed* in feeling you have become that leaf or that daisy, regardless of what appears on the paper.
>
> *[Franck, 1979:xvii]*

- Materials: 1. An easily observed animal, perhaps in captivity. It helps if the animal is not overly active. You might find it better to begin with a stuffed animal and go from Step 1 to Step 4; then go through the entire procedure with a live animal.
- 2. Paper, clipboard and pencil.

- Procedure:
- Step: 1. Find a comfortable place to sit and conduct yourself as an individual alone with the animal. Pay no attention to anything else around you, besides the animal.
- 2. Spend a few minutes freely observing the animal to get 'a feel' for its general behavior patterns.
- 3. At some point in time 'freeze' a mental image of the animal and *close your eyes*. Keep your eyes closed and hold the mental image until it fades. Repeat this until you can draw the outline of the animal with your eyes

closed; that is, trace the outline of the mental image. It's not important whether your drawing looks like the animal. Forcing yourself to draw forces you to form a more stable mental image. Repeat this process until the mental image is fairly stable and more detailed.

- 4. Again 'freeze' the animal in time and space, but *keep your eyes open.* Stare at the space where you 'froze' the image and hold it regardless of what the animal does and where it goes. Repeat this process until you can hold the image strongly and long enough to draw it while focusing on the image in space.

- Treat your mental image of the animal as the object that Franck is referring to below:

> Allow the image on your retina to set off the reflex arc that goes directly from your eye, through your body, to the fingers that hold the pencil. . . . There is no thinking, judging, labeling in this reflex arc: it goes from the eye to the hand and skips the thinking, judging, discriminating brain: just allow the reflex to work, to take over! . . . Don't be surprised: in the beginning your concentration span will be short. When your attention flags, stop for a moment.   [*Franck, 1979:35,38*]

- 5. After you have become reasonably good at Step 4, begin to observe the animal in motion, first as a series of disjunct 'freeze frames' and then as continuous motion. Locomotion (e.g. different gaits of horses) provides an excellent opportunity to develop this skill. At this stage you should be able to close your eyes and 'replay' a segment with mental images.
- 6. Write a description of a segment of the animal's behavior. Repeat the process until you see and retain more.

It is most important that you strive to draw in Steps 3 and 4. Just as recopying class notes helps you retain the information, drawing the animal compels you to focus your energy into observing and retaining the mental image. This is reflected in Walther's (1984:xi) account of why he draws animals.

> I draw animals not to compete with Leonardo da Vinci but to make sure that I have seen them correctly. As one of my anatomy teachers phrased it: you have only seen something when you have made a sketch of it.

Drawing the animal focuses your attention on its morphology. Movements and posturing result in changes in the relative positions of parts of the animal's morphology. Observing the details of the postures, and then the movements, gradually becomes easier. At first, you will probably observe primarily in black and white, but later you will observe and remember color. Colors and their changes can be an important means of communication in animals. If recording colors is an important part of your research, you should use standards, such as Smithe's (1972) 86 standard colors which are based on Ridgway (1912) and Palmer (1962). After learning to observe through seeing postures, movements and colors, you can learn to 'observe' sounds through hearing in a similar manner; sounds should then be incorporated into your descriptions.

Each step in the exercise outlined above takes time, and Steps 3, 4 and 5 will need to be repeated several times. That time and effort will serve as a good measure of your disposition and aptitude for ethological research. If you aren't comfortable with the concentrated effort required in this exercise, you shouldn't try to conduct research that requires extensive observations of an animal's behavior.

For additional exercises in observing you can consult Roth (1982); he offers seven exercises involving increasingly more difficult activities in a chapter entitled 'The art of seeing'.

4.1.3 Field notes

Field notes are often the best, and sometimes the only, record you have of your activities and observations in the field. The method of taking field notes described below is primarily applicable to informal field trips and reconnaissance observations. However, for some studies, data can be collected in a field note format which incorporates previously designed and prepared data forms (Chapter 9).

Good field notes are the end result of developing a skill into an art, and the basis for learning those skills is a knowledge of fundamentals.

> Note taking . . . is an art requiring sensitivity and skill. Perfection is
> unattainable, a goal to be sought and pursued through selective training
> and persistent practice just as beauty or reality are sought by the
> sculptor, the musician or the poet.   *[Emlen, 1958:178]*

The system of taking notes that you decide upon will determine their value to you and other researchers in the future. Most ethologists with whom I have spoken use some variation of the system developed by Dr Joseph Grinnell of the Museum of Vertebrate Zoology, University of California, Berkeley. The usual format is to divide the notebook into three sections: 1. journal; 2. species accounts; and 3. catalog. The species-accounts section is used to keep records of your observations by species, in contrast to the journal, where your observations are recorded by date and time. The catalog section is used to record specimens collected in the field. The discussion to follow is based upon Grinnell's system, but will be concerned only with the journal section.

Good-quality white, lined paper should be used. It should be a bond, having a rag content of 50–100%. The size should be approximately 6¼ inches by 8½ inches

(16 centimeters by 21½ centimeters) preferably looseleaf and kept in a six- or three-ring binder. An important reason for a ring binder notebook is that many fieldworkers prefer to use two notebooks. One is taken into the field for the day's records, and the other contains past records for the year and is kept safely in camp or at home.

Some field ethologists prefer to use bound notebooks which are designed for outdoor use, such as durable, 4⅝″ × 7¼″ notebooks with 80 lined pages and polyethylene covers, or polyethylene covered ring binders (4⅝″ × 7″) and relatively 'waterproof' loose leaf paper.

Black waterproof ink is usually recommended; Forestry Suppliers sell a ballpoint pen which will write underwater, upside down and in temperatures to $-50\,°F$ ($-45.5\,°C$). A hard pencil is the second best choice. You should always carry a pencil, however, for both a backup and for writing in heavy mist or rain; ink is not waterproof until it is dry. If you use a Rapidograph pen, try to use a 'nonclogging' drawing ink. The following three inks are recommended: 1. Pelikan Drawing Ink; 2. Higgins Eternal Black Ink; and 3. Koh-I-Noor Rapidograph Ink.

A sample field note (journal) page is shown in Box 4.1.

Box 4.1   The field note journal

A sample field note (journal) page is shown below:

	Page no.
	Name
	Locality
	Date

Beginning mileage
Mileage at stop points
Other observers
Weather
Habitat type
Time into field

Time: Observations and remarks

Time: Observations and remarks

Time out of field
Ending mileage

Page number: Number consecutively for the year.

Box 4.1 (*cont.*)

Locality: Specific locality, direction and estimated distance from known point (e.g. E shore of Cobb Lake, 6 km NE of Ft. Collins, Colorado); section, township and range are also useful information.

Date: Date of observations, travel, meetings, etc.

Mileage (beginning, stop points, and ending): This provides a record to confirm route of travel and may later provide useful additional information when giving directions to others (or conferring with the IRS).

Other observers: Provides for later verification and elaboration (Remsen, 1977).

Weather: Precipitation, percent cloud cover, wind speed and direction, temperature, etc. *Insert changes in your notes as they occur.* Temperature can be measured by carrying a small, metal-shielded thermometer with you into the field. For most studies it will be sufficient, and worthwhile (e.g. Mrosovsky and Shettleworth, 1975), to estimate measures of the other weather variables. In 1805, Commander Francis Beaufort of the British Navy devised a scale of nine categories for classifying wind force at sea. More recently a scale was developed (based on Beaufort's scale) for use on land. The scale's 13 categories are convenient for *estimating* wind speed while in the field:

Beaufort's scale for wind speed

Scale	Wind velocity	Environmental indicators
0	*Calm*	Movement of the air is less than 1 mile (1.6 km) per hour. Smoke rises vertically; bodies of water are mirror-smooth.
1	*Light air*	1–3 miles (1.6–4.8 km) per hour. The drift of smoke indicates the direction of the breeze.
2	*Light breeze*	4–7 miles (6.4–11.3 km) per hour. Leaves begin to rustle.
3	*Gentle breeze*	8–12 miles (12.9–19.3 km) per hour. Leaves and twigs in motion; crests on waves begin to break.
4	*Moderate breeze*	13–18 (20.9–29.0 km) per hour. Small branches move; dust rises; many whitecaps on large bodies of water.

Box 4.1   (*cont.*)

5	*Fresh breeze*	19–24 miles (30.6–38.6 km) per hour. Small trees in leaf begin to sway.
6	*Strong breeze*	25–31 miles (40.2–49.9 km) per hour. Large branches begin moving.
7	*Moderate gale*	32–38 miles (51.5–61.1 km) per hour. Whole trees in motion.
8	*Fresh gale*	39–46 miles (62.8–74.0 km) per hour. Twigs on trees break off.
9	*Strong gale*	47–54 miles (75.6–86.9 km) per hour. Foam blows in dense streaks across water at sea.
10	*Whole gale*	55–63 miles (88.5–101.4 km) per hour. Trees uprooted; huge waves build up with overhanging crests.
11	*Storm*	64–75 miles (103.0–120.7 km) per hour.
12	*Hurricane*	Wind velocities above 75 miles (120.7 km) per hour.

If you need accurate ($\pm 3\%$) measures of wind velocity, small hand-held, digital anemometers which read in three scales (fpm, m/s, mph, and knots) are available for example from Cole-Palmer Instrument Co., 7425 North Oak Park Ave., Chicago, Illinois 60648.

> *Habitat type*: General topography and vegetative cover; note prominent physiographic features.
>
> *Time into field*: Time at which you begin your field-related activities (e.g. 1345–1510). Midnight=0000; Noon=1200; 6:00 pm=1800.

Observations and remarks:

1 Record the species, number, age, and sexes if possible.
2 Describe behavior as accurately, clearly, concisely, and completely as possible (see discussion below).
3 Include remarks such as unusual occurrences (Short, 1970), thoughts, and ideas about the behavior. Lindauer (1985:7) reports that Karl von Frisch 'had a weakness for *little things*, for unexpected results, for so called *singularities*'.

Record observations at once – do not trust to memory.

> *Time out of field*: It is desirable to record time at which you leave the study site, as well as when you reach home or camp.
>
> *General comments*: At the end of each day's field observations it is often desirable to head a new page *General Comments* and list:

- Route of travel
- Hours of observation   ·   Weather
- Species observed
- General impressions about the day's observations

Some fieldworkers recopy their field notes after a day's observations (e.g., Schaller, 1973; Remsen, 1977; P. Johnsgard, pers. commun.). Others believe that your first impressions are the most accurate and that in recopying you are prone to edit them, thereby making them less accurate. Past journals should be kept by year. Field notes are extremely valuable, and loss or destruction in the field, home or office must be avoided.

Regardless of how you choose to take and store your field notes, remember that they are a record of your fieldwork – *be complete*. They are a source of data and hypotheses and can be the basis for future research – *be accurate*. They may be used by other researchers for similar purposes – *be clear and concise*. Last, but not least, your field notes will be a diary and a source of memories; insert thoughts you will enjoy having again, 20 or 40 years later.

4.1.4 Equipment

This section will deal with the three basic tools used in field ethology. More sophisticated, high-tech data collection and analysis equipment will be discussed in Chapters 9 and 13–17, but for reconnaissance observations and some descriptive studies a blind, binoculars and a field notebook will often suffice. Huxley's (1968:15,76) account of his classic study in 1914 could easily be used to recount a descriptive study of today.

> A good glass, [*binoculars*] a notebook, some patience, and a spare
> fortnight in the spring – with these I not only managed to discover many
> unknown facts about the crested grebe, but also had one of the
> pleasantest of holidays. . . . Some of the watching was done concealed in
> the boat-houses, and some from a screened punt, but the major part
> from the bank. This in many ways the most useful . . . every action can
> be easily followed, the birds are not scared, the field of view is
> uninterrupted, and it is far easier to follow the actions of the same pair
> of birds for a long period of time.

4.1.4a *Blinds*

The purpose of a blind (or hide) is to allow observation of animals with as little disturbance as possible (e.g. Sorenson, 1994). That is, you hope you are observing behavior which is unaffected by your presence, even if your presence is perceived by the animals. In situations where animals are accustomed to human presence, or where you can habituate them to yourself, no blind is necessary. At the other extreme are species with whom great caution must be employed. You can think of too much activity in the field as having the following effect:

> A walker displaces the territory as a swimmer does water, but a quiet sitter is a dropped stone and his ripples subside and water laps back in: submergence.
>
> [Heat-Moon, 1991:367]

Blinds are of two general types – natural and artificial – and may be in, on or beside water; or they may be below, at or above ground level. Knowledge of the animal's reactions to novel objects at various places in their environment is necessary for selecting the proper blind. For example, sitting in a tree will sufficiently hide you from many species that are not prone to be look vigilantly far above ground level (e.g. deer).

Artificial blinds are generally designed to blend in with the habitat as closely as possible. An unobtrusive blind is less likely to disturb the animals and attract curious humans. However, rather than selecting a natural blind (or constructing an artificial blind) with as little disturbance as possible, you might be able to use an obvious structure (e.g. vehicle, tent) and allow the animals time to habituate to it. For example, since coyotes in the National Elk Refuge were accustomed to seeing vehicles along the dirt roads, Ryden (1975) rented a bright yellow van to use as a blind for her observations. The same ploy was used by Kucera (1978) in his study of mule deer in Big Bend National Park, by Walther (1978) in his observations of oryx in Serengeti National Park, by Renouf (1989) in his study of harbor seals, and by Laurenson and Caro (1994) in their study of cheetahs. Even with domestic animals, similar procedures are often necessary to ensure that the animals under observation are not disturbed by the researcher's presence. For example, Baldock *et al.* (1988) described their use of the same technique in their study of domestic sheep:

> All observations of behaviour were made from a parked vehicle overlooking the field containing the ewes. Other vehicles were regularly parked nearby. We obtained access discreetly and quietly, making every effort not to disturb the sheep. As far as we could tell, the behaviour of the sheep was not affected by the observer's presence.
>
> [Baldock, et al. 1988:36]

Nylon or canvas tents often make good artificial blinds. Small, lightweight tents can easily be moved between observation sites, and larger tents can be staked into the ground to provide a more permanent blind. If you decide to build a blind, there are several sources of instructions for building various types, as well as ideas for creating your own design. For example, Woodin (1983) describes construction of a portable umbrella blind, and Rodenhouse and Best (1983) describe construction of a portable tower-blind. Figure 4.1 illustrates the design of two types of blind.

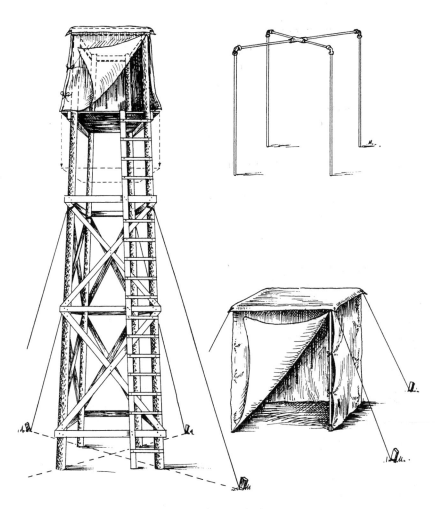

Fig. 4.1 Examples of observation blinds (from Pettingill 1970).

The following characteristics should be considered when selecting a natural blind or constructing an artificial blind:

1 Behavior of animal
   a. Reaction to strange objects; approach or avoid?
   b. Spatial distribution of behavior patterns; are you at the right spot to observe the behavior?
2 Observational capability
   a. Number and size of openings for observation and filming.
   b. Capacity for number of observers anticipated.
3 Permanence

a. Can it be permanent for several weeks or years? Will the animals be there during your observation periods?

b. Should the blind be temporary and portable (e.g. Winkler, 1994)? How often and how rapid will the moves be? A camouflage suit can be considered the most easily and rapidly moved artificial blind.

4 Climatic conditions

a. Severity; must it be built to withstand severe wind and precipitation?

b. Prevailing winds may be important for locating the blind downwind to reduce your olfactory and auditory stimuli from reaching the animals.

c. Comfort and portability are often conflicting objectives. Remember that you can function effectively as an observer/recorder only if you are reasonably comfortable.

Further descriptions and discussions of blinds and their use can be found in Hanenkrat (1977), Roth (1982) and outdoor photography literature, such as Baufle and Varin (1972), Ettlinger (1974), and Marchington and Clay (1974).

4.1.4b *Binoculars and spotting scopes*

Binoculars

One of the most important pieces of equipment to the ethologist is a good pair of binoculars. In fact, as Tinbergen (1953:132) has said, they '... are almost indispensable'

Humans are a visually oriented species; therefore, ethologists tend to rely very heavily on what they see. A very large percentage (probably over 95%) of what we record as observations are what we see. Equipment that will provide us with better vision or in other ways make the animals more visible, such as a strategically located blind, will pay off immensely. Always consider the species to be studied and the distances over which they will be observed when selecting binoculars or spotting scopes (see below).

Bergman (1981) has listed several criteria for evaluating the optical quality of binoculars, including the following:

1 *Brightness* of the image (see below for a measure of relative brightness).

2 *Resolution* of the image which can be affected by the following defects:
a. *Edge-of-field defects*. The margins of the field of view should approximate the sharpness of the center.
b. *Pincushion distortion*. Parallel lines crossing the field of view may appear to curve toward the center.
c. *Curvature of the image*. A straight surface, such as a wall may appear concave or convex.

d. *Spherical aberration*. Results in the inability to focus the binoculars
sharply.

3 *Range of resolution*. Good optics should have a wide range of distance
over which they are in focus.

4 *Alignment*. The images seen through each barrel of the binocular should
align and merge perfectly into a single image.

5 *Eye relief*. When the binoculars are held comfortably to the eyes, you
should see a full field of view. Eyeglass wearers should check the effect of
the retractable cups found on many binoculars.

There are two numbers engraved on all binoculars. These are commonly used to
designate the 'type' of binoculars, such as '7 by 35':

$$7 \qquad \times \qquad 35$$
magnification   diameter of objective lens (mm)

Magnification tells you the number of times greater than normal that an object
being viewed will appear. Although increased magnification would appear to be
desirable, it often carries with it some problems. Generally, the following occurs
with increased magnification:

1 The field of view becomes smaller (see below).
2 Clarity is lost, since the more powerful the lens the more the imperfections in the lens are also magnified.
3 Light transmission is decreased.
4 Increased blurring results from movement of the binoculars.

The first three problems can, of course, be overcome by the manufacturer, but
this will result in higher-priced binoculars. For most field studies, 6 to 8× magnification will probably be sufficient. When greater magnification is necessary, spotting
scopes are commonly used (discussed later). However, Huxley (1968) used 12×
binoculars in his classic study in 1914 of the courtship behaviors of great crested
grebes (*Podiceps cristatus*), even though the optics at that time were not nearly as
good as they are today. Flegg (1972) can be consulted for recommendations on the
type of binoculars (mag.×objective lens diameter) to be used for observing birds in
different light levels and habitat types, and over different distances.

Objective lens diameter affects the amount of light that enters the binocular.
What is important, however, is the amount of light which finally passes through the
ocular lens (light transmission or relative brightness). Everything else being equal,
the larger the objective lens the greater the relative brightness; therefore, binoculars
with large objective lens are better under low light conditions (e.g. dawn, dusk). To
compare light transmission for uncoated binoculars use the following formula:

$$\text{Relative Brightness} = \frac{\text{diameter of objective lens (mm)}^2}{\text{magnification}}$$

For example, 7×50 binoculars have about twice the relative brightness of 7×35 binoculars.

$$\frac{50^2}{7} = 51 \qquad \frac{35^2}{7} = 25$$

One important disadvantage of large objective lens binoculars is that they are heavy; this weight is overcome, to some degree, in roof-prism binoculars (especially in ultra-light binoculars; discussed later).

You can also check for relative brightness by holding the binoculars eight to ten inches in front of your eyes and pointed at the sky. The circle of light which you see in the ocular lens is the exit pupil. Relative brightness increases as the square of the exit pupil:

$$\text{Relative Brightness} = (\text{exit pupil})^2$$
$$= \frac{(\text{diameter of the objective lens})^2}{\text{magnification}}$$

A large exit pupil can be wasted on a human eye in which the pupil is constricted on a bright day. The human pupil may be constricted to less than 3 mm on bright days, whereas the binocular's exit pupil may often exceed this (Table 4.1). Also, the pupil diameter of a human's dark adapted eye decreases with age. Therefore, the binoculars usually become the primary limiting factor only under conditions of low environmental light intensity.

Another measure of a binoculars' light transmission which takes into account relative brightness, as well as the percentage of light transmitted, is relative light efficiency (Table 4.1). This is calculated according to the formula:

$$\text{Relative light efficiency} = 2 \times \text{relative brightness} \times$$
$$\text{percentage of light transmitted}$$

One factor that has a great effect on the percentage of light transmitted is coating. Light passing through uncoated binoculars loses about 5% of its brightness at each surface it passes through. This can be considerable when ten surfaces are normally involved. Coating the surfaces (usually with magnesium flouride) can reduce this figure by 0.5% at each surface. It generally increases the relative light transmission by 50%. Therefore, it is wise to insist that all the optics are coated in any binocular you purchase or intend to use.

Field of view is another characteristic of binoculars which is very important to the ethologist. This is limiting to different degrees depending on the type of research to be conducted. However, everything else being equal, you should select binoculars with the largest field of view. Usually the field of view is expressed in feet at 1000

Table 4.1. *Some specifications for representative binoculars*

Model	Diameter exit pupil (mm)	Relative brightness	Relative light efficiency (RLE) Percentage of coating			Field (at 1000 Yds)	Field (angular degrees)	Relative field (ft)
			20%	80%	100%			
Binoculars–standard field								
6×15	2.5	6.2	6.8	8.7	—	370	7	2220
6×30	5	25	27	35	37.5	450	8.5	2700
7×18	2.6	6.8	7.5	9.5	—	325	6.2	2275
7×35	5	25	26.5	35	37.5	380	7.3	2660
7×50	7.1	50	55	70	75	380	7.3	2660
8×30	3.75	14	15.5	20	—	330	6.3	2640
Binoculars–semi-wide field and wide field								
6×24	4	16	—	22	—	636	12	3816
6×25	4.2	17.5	—	—	26.2	577	11	3462
7×35	5	25	27.5	35	37.5	525	10	3675
8×30	3.75	14	15.5	20	21	390	7.4	3120
8×30	3.75	14	15.5	20	21	450	8.5	3600
8×40	5	25	27.5	35	37.5	375	7.2	3000
9×35	3.9	15	—	21	23	390	7.4	3510
10×50	5	25	27.5	35	37.5	370	7	3700

Source: From Reichert and Reichert (1961).

yards (900 meters); that is, the width of the scene you can see at 1000 yards (900 meters). The field of view is controlled primarily by the field lens. The normal field of view can be increased (semi-wide or wide field) by the manufacturer by using a different ocular system, especially a larger field lens. This, of course, also increases the price of binoculars. The following formula converts field of view in degrees (Table 4.1) to an approximation of field of view in feet at 1000 yards (Robinson, 1989).

Field of view = Field of view (degrees) × 53 (ft at 1000 yds)

Generally, the greater the magnification the smaller the field of view. Reichert and Reichert (1961) provide a formula to calculate the relative field for different types of binoculars.

Relative field = magnification × field at 1000 yards

The larger the relative field, the closer the binocular approaches the ideal (i.e. high magnification and large field of view).

Binocular focusing systems are of two types. With individual eyepiece focus the observer focuses each eyepiece separately while keeping the other eye closed. If you focus them while viewing a distant object, your binoculars will be in focus at all distances beyond about 30 feet (9.1 meters). You must refocus for closer objects. Binoculars constructed with individual eyepiece focus are generally better sealed against moisture and dirt than are center focusing binoculars.

When adjusting center focus binoculars the observer uses the center focus wheel to focus the left eyepiece while the right eye is closed, and then the right eyepiece is focused while the left eye is closed. Now the observer can focus the binoculars for varying distances by using the center focus alone. This is an advantage over individual eyepiece focus binoculars.

It should be clear from this overview of binoculars that the choice of the proper binocular is always going to be a trade-off of advantageous and disadvantageous characteristics. These are summarized in Table 4.2.

An advance in prism binoculars has been the development of roof-prisms. Leitz was the first to introduce roof-prism binoculars, and they still produce high-quality optical instruments. The optical characteristics of roof-prism binoculars, although essentially the same as those for porro-prism binoculars, are in some instances better. Representative figures are given in Table 4.3.

The greatest advantages of roof-prism binoculars are their ease of handling and light weight, which result from their slim shape and small size compared to porro-prism binoculars. Ultra-compact roof-prism binoculars have become popular with many field ethologists now that their optical characteristics are comparable to larger, heavier models (Table 4.4). The weight of ultra-compact roof-prism binoculars can be only 15–25% that of comparable standard sized binoculars. Roof-prism binoculars are more expensive than porro-prism binocu-

Table 4.2. *Summary of advantages and disadvantages that accrue from features in binoculars*

Features	Advantages	Disadvantages
1. Greater magnification	Viewing animals more closely	Smaller field of view Lower light transmission Poorer image clarity Increased movement of binoculars Greater weight
2. Larger objective lens	More light transmission	Heavier
3. Larger field of view	Increased size of scene	Increased price
4. Coating	Increased light transmission	Increased price
5. Individual eyepiece focus	Better sealing against moisture and dirt	Individual focus of eyepieces at close distance
6. Center focus	Use of center focus alone for varying distances	Less well sealed against moisture and dirt
7. Rubber 'armored' covering	Better withstands 'hard' use Less reflection of light	Inceased price

Table 4.3. *Some representative optical specifications for roof-prism binoculars*

Model	Diameter exit pupil (mm)	Relative brightness	Field (at 1000 yds) (ft(m))	Field (Angular degrees	Relative field (ft(mm))
6×18	3	9	420 (128)	8	2520 (768)
7×21	3	9	372 (113)	7.6	2604 (794)
8×24	3	9	366 (111)	7	2928 (892)

lars, but only the individual consumer can properly weigh differences in features versus the differences in cost.

Binoculars that electronically zoom (e.g. Copitar zoom from 7× to 15×) are available from a few distributors; however, these binoculars probably have poorer optical characteristics than most ethologists will find acceptable for long-term observations.

Two good discussions of the general characteristics of binoculars and their use for fieldwork can be found in Bergman (1981) and Robinson (1989).

Table 4.4. Some representative specifications for ultra-light roof-prism binoculars (ft(m))

Description	Weight (oz.(g))	Field (deg.)	Field (at 1000 Yds)	Eye relief (mm)	Near focus (ft(m))	Exit pupil (mm)	Relative brightness index
8x25 Celestron WA Monocular	4 (113.4)	8.7	456 (139)	5	15 (4.5)	3.1	9.8
8x30 Pentax Mono/Micro	6 (170.1)	6.2	325 (99)	10	6 (1.8)	3.7	14.0
7x20 Pentax Mono/Micro	2.5 (70.8)	7.5	393 (120)	11	11 (3.3)	2.8	8.2
6x20B Zeiss Monocular	1.7 (48.2)	6.9	362 (110)	15	6.6 (2.0)	3.3	11.1
8x21 Orion Super Compact	6 (170.1)	7.0	368 (112)	10	13 (3.9)	2.6	6.9
10x25 Orion Super Compact	9 (255.1)	5.5	289 (88)	9	15 (4.5)	2.5	6.3
8x25 Celestron WA Mini	10 (280.3)	8.7	457 (139)	5	15 (4.5)	3.1	9.8
7x20 Pentax Compact	7.4 (209.8)	7.5	394 (120)	12	8 (2.4)	2.9	8.2
9x20 Pentax Compact	7.4 (209.8)	6.2	326 (99)	12	8 (2.4)	2.2	4.9
8x20 Nikon Sportstar	7.5 (212.6)	6.3	331 (101)	10	9 (2.7)	2.5	6.3
8x21 Celestron Waterproof	14 (396.9)	6.0	315 (96)	15	18 (5.5)	2.6	6.9
8x20B Swarovski Habicht	7.6 (215.5)	6.6	346 (105)	13	13 (3.9)	2.5	6.3
10x25B Zeiss Mini	7.5 (212.6)	5.4	284 (86)	14	18 (5.5)	2.5	6.3

Note:
From Orion catalog.

Spotting scopes

The only *spotting scopes* that are useful to the ethologist are prism scopes which are constructed on the same principle as prism binoculars. Therefore, their operation is basically the same as one half of a prism binocular. The same characteristics of brightness and field of view are applicable to scopes and are calculated in the same way.

Since spotting scopes are used for greater magnification than binoculars, an important characteristic is resolving power (i.e. clarity). It is difficult to manufacture optics with high magnification which also produce a clear image. Reichert and Reichert (1961) described a method for measuring resolving power of a spotting scope which is useful for comparisons between scopes:

$$\text{Resolving power} = \frac{\text{maximum distance at which you can distinctly see specific details through the scope}}{\text{maximum distance at which you can distinctly see them with the unaided eye } (20 \times 20)}$$

For example, if you can distinctly see a small bird at a maximum distance of 20 feet (6.1 meters) with the naked eye, but with a 20–power scope you can see it clearly at a maximum of 360 feet (108 meters), then:

$$\text{Resolving power} = \frac{360}{20} = 18$$

For a very high-quality spotting scope, the resolving power should equal the magnification. Although the optical characteristics vary between manufacturers, the figures shown in Table 4.5 are common.

A feature which is often very useful is a 'zoom' lens. This can be purchased as a permanent feature of the scope or through a variable-power eyepiece. The effective viewing distances will depend on the size of the animal and the specific behaviors being observed. For example, Goss-Custard and Sutherland (1984) used a $15 \times –60 \times$ scope to observe feeding methods of oystercatchers (*Haematopus ostralegus*, a shorebird) at distances up to 100 meters, and Creswell (1993) used a $15 \times –60 \times$ scope to observe escape responses in redshanks (*Trigona totanus*, a smaller shorebird) at distances up to 1000 meters.

Spotting scopes should be tested and compared before a choice is made for use or purchase. As with binoculars, there is the inevitable compromise between magnification, brightness, and clarity. Table 4.6 provides some characteristic specifications.

Scopes are greatly affected by two factors: 1. heat waves from the ground; and 2. movement. The first can be overcome best by observing from the highest point available. If possible, use the top of a hill or rock outcropping. If you are not concerned about concealment, just getting up on the roof of a vehicle will often reduce the effect of heat waves considerably.

Table 4.5. *Examples of common combinations of*
magnification and resolving power in spotting scopes

Magnification	Resolving power
20×	18
30×	20
40×	22

Movement is reduced best by the use of a sturdy tripod (Figures 6.2, 9.22). These come in many designs and price ranges. There are four questions to keep in mind when you examine a tripod for use or purchase:

1 How rugged is it?
2 Does the height adjust to a comfortable eye level?
3 Do the legs extend smoothly and rapidly for quick extension and occasional adjustment.
4 How portable is it?

Gunstock mounts can be a useful compromise between a good tripod and no tripod at all. They can be adapted from an old gunstock, carved and custom fitted (Figure 4.2), or purchased. Also, window mounts are extremely useful if most of your observations will be made from a vehicle.

Two good discussions of spotting scopes, their characteristics, and uses for field-work, as well as evaluations of various models, can be found in Bergman (1986) and Robinson (1989).

Photography through binoculars and spotting scopes is mentioned here only because it is possible, not because it is recommended. Although manufacturers do provide adapters for mounting a camera on binoculars or spotting scopes, I recommend that it be avoided if possible. It is best to treat your observational and photographic equipment as two different and important types of equipment. The problems of assembly and disassembly, coupled with the generally poorer quality of photographs, does not warrant the extensive use of binoculars and scopes for photography. If you choose to investigate this type of photography further, I suggest you consult Reichert and Reichert (1961), who extol its virtues.

Sources of binoculars, spotting scopes and tripods
Robinson (1989) lists several binoculars, spotting scopes, and tripod manufacturers. There are several excellent sources of binoculars and spotting scopes including local camera and sporting goods stores, as well as by mail.

Table 4.6. Some specifications for representative spotting scopes

Model	Diameter exit pupil (mm)	Relative brightness	Relative light efficiency (RLE) Percentage of coating			Field (at 1000 yds) (ft(m))	Field (angular degrees)	Relative field (ft(m))
			20%	80%	100%			
20×40	2	4	4.5	5.6	6	107 (32.6)	2.1	2140 (652.2)
16×50	3.1	9.6	—	—	14.5	200 (60.9)	3.8	3200 (975.3)
20×50	2.5	6.2	7	8.8	9.4	118 (35.9)	2.3	2360 (719.3)
15×60	4	16	17.5	22.5	24	150 (45.7)	3	2250 (685.8)
20×60	3	9	10	12.5	13.5	112 (34.1)	2.1	2240 (682.7)
20×60	3	9	—	—	13.5	170 (51.8)	3.3	3400 (1036.3)
30×60	2	4	4.5	5.6	6	80 (24.3)	1.5	2400 (731.5)
40×60	1.5	2.2	2.4	3.1	3.3	61 (18.5)	1.2	2440 (743.7)

Source: From Reichert and Reichert (1961).

Fig. 4.2   Andy Sandoval observing bighorn sheep through a spotting scope mounted on a gunstock.

Viewing instruments for nocturnal observations

Night vision binoculars were developed by the US Department of Defense and the Soviet Union in the 1960s for military use under conditions of extremely low light levels. The technology has improved from Generation 0, which required an infrared light source, to Generation II, with improved photo-intensifier capacity and decreased size. Generation II is available for civilian use, but Generation III, which uses a super-sensitive photocathode, is only available to military and government agencies.

These binoculars photomultiply the available light by approximately 30 000×, but they magnify the image only approximately 2.5×. The Russian technology binoculars are available commercially from a few manufacturers and distributors, such as Damark (7101 Winnetka Ave. N., PO Box 29900, Minneapolis, MN 55429–0900). The primary American technology binoculars are ITT's Night Mariner and Bausch and Lomb's Night Ranger 250; both these binoculars were developed by ITT. The American technology provides better field of view (40° versus 10°), waterproof construction, buoyancy, lightweight and bright source protection (Anonymous, 1995). Additional information can be obtained from:

>Bausch & Lomb
>Sports Optics Division
>9200 Cody
>Overland Park, KS 66214, USA

ITT Night Vision
7635 Plantation Rd.
Roanoke, VA 24091, USA

Night vision monoculars and binoculars are sold commercially by numerous out-doors sporting equipment dealers.

Night vision scopes were developed along with binoculars. They photomultiply the environmental illumination by 20 000–50 000× and have proved useful for nocturnal observations (e.g. Clapperton, 1989; Waser, 1975a). Bausch and Lomb's Night Ranger 150 (developed by ITT) can be fitted with a 3-in-1 magnification lens which extends the normal 2.2× magnification to 6×. Additional specifications and ordering information can be obtained from the sources listed above. Also, Noctron Electronics and Javelin Electronics (6357 Arizona Circle, Los Angeles, California 90045) manufacture starlight scopes which are relatively small and practical for ethological observations (e.g. Randall, 1994). They are available in different models which vary in physical dimensions, light amplification, and magnification and have options for use with still and motion-picture cameras, as well as video recorders. An excellent review of instruments for nocturnal observations is provided by Hill and Clayton (1985).

Infrared viewing devices can also be used for nocturnal observations. They flood an area with infrared light, and an infrared sensitive scope is used to view the area and observe the animals' behavior (Figure 4.3)

Night vision equipment can often be obtained on loan from the US Army and Navy. Inquiries should be directed to:

Army Technology Transfer Program
Night Vision Directorate
ATTN AMSEL RD NV TSD PET (Miller)
10221 Burbeck Ste 430
Ft. Belvoir, VA 22060–5806, USA

Naval Air Warfare Center Weapons Division
Measurements and Support Systems Branch (Code P2391)
Point Mugu, CA 93042–5001, USA

Miscellaneous observational devices

Several types of special miscellaneous observational devices have been developed by individual researchers to meet their particular needs. For example, Parker (1972) and Smith and Spencer (1976) developed mirror and pole devices which allowed them to look into high birds' nests. Moriarity and McComb (1982) developed a fiber optics system for observing in tree cavities. Descriptions of various devices are scattered throughout the technical and popular literature. However, time can often be better spent applying your own ingenuity to developing a device to meet your

Fig. 4.3   Linda Pezzolesi and the infrared night scope she used to observe burrowing owl behavior (photo by R. Scott Lutz).

needs than in searching for something suitable in the literature. Also, you can probably save yourself a great deal of money.

4.2 HOW TO DESCRIBE BEHAVIOR

At the heart of the modern approach to the analysis of behavior in
animals is the problem of description.   *[Marler, 1975:2]*

When developing a catalog of behaviors, you will be describing and applying names to the behaviors you observe. For descriptive studies, your catalog (discussed below) can contain names for behaviors which carry implicit descriptions, as well as *descriptions* which, by themselves, serve as terms for those specific behaviors. As you shape your catalog of behaviors into an ethogram (discussed below) you will probably apply *terms* to the behaviors you have named and described, primarily for ease of data collection. For further ease of data collection, you may replace the terms with code letters and numbers (Chapter 8). For experimental studies, especially, you will further sharpen the descriptions into operational definitions (Chapter 6).

The discussion below applies primarily to the early phases of a study in which you are making reconnaissance observations, taking *ad libitum* field notes, or beginning to compile a catalog of behaviors. Therefore, I have interspersed the words 'term', 'name' and 'description' as synonyms.

4.2.1 Empirical versus functional descriptions

As you first observe the behavior of an animal you will likely be confused by the complexity of what the animal does; but in time some order will appear in the types of behavior engaged in, the contexts in which they appear, and the movements and postures that are involved (Marler, 1975). Familiarity with an animal's behavior and insight into its function are continuing processes that generally lead to revision of both hypotheses and terminology.

Nevertheless, in time, it will be necessary to describe what you have observed in terms which are clear yet unassuming. The problem of description is resolved through experience in observing the animal's behavior and your ability to select terminology that will assist, not hinder, future analysis.

There are two basic types of behavioral description (see Tables 4.7 and 4.8):

- *Empirical description*: description of the behavior in terms of body parts, movements and postures (e.g. baring the teeth).
- *Functional description*: incorporation of reference to the behavior's function, proximally or ultimately (e.g. bared- teeth threat).

Table 4.7. *Examples of empirical and functional descriptions (terms) for observation of a flying mourning dove (see text)*

Type of description	Behavior description
	a. Behavior X
Empirical descriptions	b. Rapid alternate contraction and relaxation of the pectoralis muscle
	c. Wing flapping
Functional descriptions	d. Flying
	e. Escape flight

These types are nearly synonymous with the two types used by Hinde (1970): 1. description by spatio-temporal patterns of muscular contraction, including patterns of limb and body movement; and 2. 'description by consequence', respectively. Wallace (1973) calls Hinde's first type 'description by operation'.

The type of description selected will depend in part on your knowledge of the animal's behavior and type of study you wish to pursue. Descriptions can be thought to lie along a continuum of information conveyed. At some stage, conveying additional information generally entails drawing conclusions from data about function. After careful study the researcher may be able to use a term which more clearly describes the context of a behavior. For example, W. J. Smith (1968) studied the use of the 'kit-ter' call by the eastern kingbird (*Tyrannus tyrannus*) and concluded from observational data that it provides information relative to the caller's indecision about flying versus staying put, flying towards versus flying away, or flying versus landing. Hence, he labeled the call the 'locomotory hesitance vocalization.'

As another example, let us say we are walking through a wheat stubble field, and 50 m ahead of us a mourning dove flies up out of the stubble and lands in a tree 50 m to our right. We can describe the behavior of the dove in flight using, at least, five different levels of description (Table 4.7).

Describing (i.e. naming) the behavior as 'behavior X' provides us with no information unless we have access to a definition of the ethogram code being used. Rapid alternate contraction and relaxation of the pectoralis muscle tells us something about the mechanics of the behavior but does not provide the ethologist with much useful information. Wing flapping creates an image in the mind of the ethologist; but we do not know if the dove was standing and flapping its wings (perhaps an intention movement) or actually flying. By describing the behavior as flying we get a clearer picture of the behavior and still are not assuming anything about underlying motivation. By describing it as escape flight we are assuming that the dove was responding to a stimulus from which it was motivated to escape. We probably do not really know if that was the true function of the flight or if, for example, it had finished feeding and was merely flying to the tree where it could rest with relatively greater safety.

This example illustrates that the same behavior may be used in several contexts. Mounting may occur in sexual or dominant–subordinate contexts in dogs, just as urination may be marking or merely elimination (Bekoff, 1979b). Functional descriptions should be avoided, except when the function is intuitively obvious (see below) or supported by data, since they can be confusing and misleading (Marler, 1975) and lead to changes in terminology as the study progresses (Tinbergen, 1959).

The type of behavior, as well as the type of data being collected, often force the use of both empirical and functional descriptions. Hinde (1970) suggests that since threat and courtship behavior in birds involves both relatively stereotyped motor patterns and an orientation with respect to the environment, both description by operation (empirical) and consequence (functional) are necessary.

Eisenberg (1967) provided a list of behaviors for rodents (Table 4.8) that included both empirical and functional terms for convenience of presentation. It is useful to examine the list and identify those terms that are borderline, as well as those that are clearly empirical or functional.

As Table 4.8 illustrates, the distinction between empirical and functional descriptions is not always clear-cut, so that the problem is generally resolved in terms of the observer's intent. For example, does 'sniffing' imply searching for olfactory stimuli or merely wiggling the nose and vibrissae. This type of confusion over the observer's intent is clarified through the definition of behavior units (discussed in section 6.3.3).

Some descriptive terms are clearly functional, but they are readily accepted since the motivation and goal of the behavior appears obvious. For example, the terms 'nest building' and 'egg retrieval' are accepted in ethological parlance, but they still must be clearly described and/or defined for each species.

Your descriptions should inform others of your observations in an objective way without bias to your own experiences or personal beliefs. *Anthropomorphism*, the attribution of human characteristics to nonhuman animals, is often considered one of the gravest sins that an ethologist can commit (Carthy, 1966); recently, the use of anthropomorphism by ethologists has become a more controversial topic. Anthropomorphism is a form of 'functional description' (described below), but how can we categorize its various forms of usage? Is it a fatal flaw when used in research? Can it, in fact, be useful to ethologists? Anthropomorphism, as I've defined it above, is only one of three forms described by Topoff (1987) and is in the category 'interpretive anthropomorphism' according to Fisher's (1990) scheme (Fisher should be consulted for a philosopher's perspective on anthropomorphism).

Table 4.8. *List of rodent general maintenance behaviors utilizing both empirical and functional terms*

Sleeping and resting
  Curled
  Stretched
    On ventrum
    On back
  Sitting
Locomotion
  On plane surface
    Diagonal
    Quadrupedal saltation
    Bipedal walk
    Bipedal saltation
    Jumping
  Climbing
    Diagonal coordination
    Fore and hind limb alteration
  Swimming
Care of the body surface and comfort movements
  Washing
    Mouthing the fur
      Licking
      Nibble
    Wiping with the forepaws
    Nibbling the toenails
  Scratching
  Sneezing
  Cough
  Sandbathing
    Ventrum rub
    Side rub
    Rolling over the back
    Writhing
  Stretch
  Yawn
  Shake

Care of the body surface and comfort movements
  Defecation
  Urination
  Marking
    Perineal drag
    Ventral rub
    Side rub
Ingestion
  Manipulatin with forepaws
  Drinking (lapping)
  Gnawing (with incisors)
  Chewing (with molars)
  Swallowing
  Holding with the forepaws
Gathering foodstuffs and caching
  Sifting
  Dragging, carrying
  Picking up
    Forepaws
    Mouth
  Hauling in
  Chopping with incisors
  Digging
  Placing
  Pushing with forepaws
  Pushing with nose
  Covering
    Push
    Pat
Digging
  Forepaw movements
  Kick back
  Turn and push (forepaws and breast)
  Turn and push (nose)
  Molding

Table 4.8. (*cont.*)

Nest building	*Isolated animal exploring*
Gathering	Elongate, investigatory
Stripping	Upright
Biting	Testing the air
Jerking	Rigid upright
Holding	Freeze (on all fours)
Pushing and patting	Escape leap
Combing	Sniffing the substrate
Molding	Whiskering
Depositing	

Source: From Eisenberg (1967).

Regardless of how strongly one might attempt to avoid anthropomorphism, it is very difficult, if not impossible, to do so (Crocker, 1981). It can be argued that we cannot have knowledge of anything which we have not ourselves experienced either directly or indirectly; therefore, researchers sometimes unconsciously slip into its use (Kennedy, 1992). As Rioch (1967) has remarked, we are both limited and directed by our vocabulary (symbolic behavior) in describing observed behavior.

> I will readily admit that observation has one great drawback; it is hard to convey to others. Experimental conditions can be reproduced, pure observation unfortunately cannot. Therefore it does not have the same objective character. The observer who studies and records behavior patterns of higher animals is up against a great difficulty. He is himself a subject, so like the object he is observing that he cannot be truly objective. The most 'objective' observer cannot escape drawing analogies with his own psychological processes. Language itself forces us to use terms borrowed from our own experience. *[Lorenz, 1935:92]*

Lorenz (1974) has also suggested that in some instances the use of terms like 'falling in love,' 'friendship' or 'jealousy' is not anthropomorphic, but rather refers to functionally determined concepts. In this regard, anthropomorphism might be useful as a metaphor for describing what an animal does (Kennedy, 1992; Ristau, 1986) and what its 'emotional' and motivational states appear to be (e.g. fear), without implying that some level of conscious thought is involved. For example, the phrase 'the ship *plows* the sea' provides us with a visual image analogous to a farmer's plow pushing aside the soil. Wiley (1983:167) notes that 'In thinking about opportunities for manipulation in animal communication, analogies drawn from human interactions tend to dominate'; commonly used terms include 'deceit', 'selfishness', and 'spite'. Wiley (1983:167) concludes that the use of 'these familiar words make visualization of technical discussions easier', but we should provide technical definitions of the terms in order to 'guard against misleading inferences'

Where and how does the beginner draw the line? The safest approach is to avoid using terms that could be misinterpreted and use only empirical descriptions. You should especially consider avoiding terms which may be inflammatory and offensive, as well as misleading. For example, Estep and Bruce (1981) argue against the use of the term 'rape' by ethologists. An integral part of their argument goes beyond Wiley's (1983) call for technical definitions to the issue of redefining terms, stated as follows:

> Beach (1978, 1979) has warned both of the danger of taking words from common usage and applying specialized meaning to them without definition, and of resorting to Humpty-Dumptyism (taking a word from common usage and redefining it to mean only what you want it to mean). Both of these problems exist in the current application of the term rape to non-human behavior.   *(Estep and Bruce, 1981:1272)*

What should we call behavior in nonhuman species that we know as rape in humans? Estep and Bruce (1981) suggest the term 'resisted mating' as a purely descriptive term, or we could use the term 'forced copulation' (e.g. Sorenson, 1994).

In summary, anthropomorphism can be perceived as a gradient:

1 No anthropomorphism.
2 Human terms technically defined.
3 Human terms used as a metaphor.
4 Human terms freely used with all the underlying implications.

You should give considerable thought to choosing at what point along the gradient you will report your observations.

Descriptions are often quantified to delineate more accurately and completely delineate what the animal does when it performs the behavior. Some examples of these quantitative descriptions are illustrated in Chapter 10.

4.2.2 Catalog, repertoire and ethogram

4.2.2a Catalog and repertoire

We begin every ethological study by compiling a catalog of behaviors for the species. The *catalog* is a list of all the behaviors that we have observed, listened to, or have knowledge of. Catalogs can be restricted to only the specific type of behaviors (e.g.

courtship), sex or age group we are interested in studying. The catalog is only a portion of an animal's *repertoire* – all the behaviors that the animal is capable of performing. We call the catalog an *ethogram* when we believe that it closely approximates the complete repertoire. The size of the repertoire will, of course, vary from species to species as well as between individuals, depending on sex, age and experience.

One decision that you must make during reconnaissance observations is when to stop. When do you have sufficient information to ask incisive questions, formulate precise hypotheses, and design a sound research project? At what point do you have a reasonably complete ethogram for the animal(s)?

If we were to observe an individual animal continuously for an extended period of time and record the behaviors that it showed, we could then plot the cumulative number of observed behaviors by the time (Figure 4.4a).

An asymptote is reached after many hours of observation (arrow, Figure 4.4a), beyond which few additional behaviors are seen for each unit of time spent observing. This asymptote may take tens, hundreds or thousands of hours to reach, depending on the species studied. Nolan (1978), for example, spent 5524 hours observing the behavior of prairie warblers (*Dendroica discolor*), yet he saw only nine copulations. If only one type of behavior (e.g. agonistic) is under study, then the time to the asymptote will generally be shortened. Fagen and Goldman (1977:268) concluded that 'familiarity with an animal's behavior will tend to require years of experience if the animal is a mammal or bird with a complex repertory. But if the animal's behavior is simple and relatively stereotyped such familiarity may be gained in a few months'. The objective of descriptive studies of a species (and to a certain degree of developing an ethogram) is to determine the true frequency of rare or unusual behaviors; short-term studies record too many unusual behaviors. The result is that we often overestimate the importance of some unusual behaviors since we lack the perspective provided by long-term studies (Weatherhead, 1986).

Hailman and Sustare (1973) described an interesting laboratory exercise in 'the analytical power of biological observations'. The objective was to deduce the 'behavioral organization' of a talking, stuffed toy elephant – Horton. The first step consisted of listening to Horton's total vocal repertoire by pulling the string and listing the vocalizations emitted. These data were transferred to a cumulative graph (Figure 4.5) and examined for an asymptote to determine if the entire repertoire had been recorded after 100 successive vocalizations.

Another way to look at the behavioral repertoire of an animal is through the time devoted to particular behaviors (i.e. time budget) or by frequency of occurrence (Hutt and Hutt, 1970). Since the frequency of occurrence varies for the behaviors in an animal's repertoire, a plot of cumulative percentages of total time spent in the various behaviors against their rank order by frequency of occurrence will show

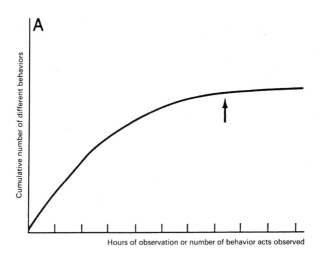

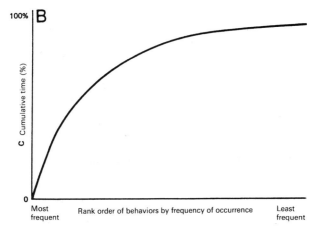

Fig. 4.4   A hypothetical example of cumulative number of different behaviors plotted against hours of observation. The arrow denotes the approximate asymptote. b. Conceptual representation of an animal's repertoire plotted as a cumulative percentage of the time spent in the various behaviors (adapted from Hutt and Hutt, 1970).

a curve which reaches an asymptote at the less frequently occurring behaviors (Figure 4.4b).

Fagen and Goldman (1977) researched methods of analyzing behavioral catalogs and concluded that most distributions (types of behavioral act/number of acts observed) could be described by a logarithmic regression slope of approximately 0.3 (e.g. Figure 4.6).

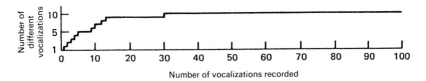

Fig. 4.5   The cumulative number of different vocalizations as a function of the total
number of vocalizations recorded (from Hailman and Sustare, 1973).

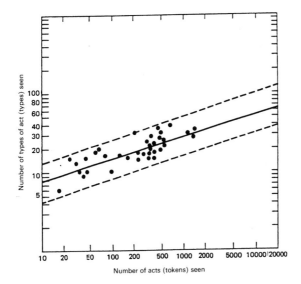

Fig. 4.6   Plot of cat behavior data with fitted regression line $Y=4.01X^{0.29}$ (solid line) and
95% confidence bounds for regression line (dashed lines) (from Fagen and
Goldman, 1977).

Since the regression line has no finite asymptote, it does not allow the observer to
predict repertoire size. However, this procedure did encourage Fagen and Goldman
(1977:263) to recommend the following rule: 'A ten fold increase in the total number
of acts will, on the average, double the number of behaviour types in the catalog'.

We can estimate our *sample coverage* ($\hat{\theta}$) by calculating the probability that the
next behavioral act will be a new type (Fagen and Goldman, 1977) . If $\hat{\theta}$ approaches
1, the probability of observing a new behavioral act is low.

$$\hat{\theta}=1-\frac{N_1}{I}$$

N_1 is the number of behavior types seen only once, and I equals the total number
of acts seen. When N_1 is small relative to I then $\hat{\theta}$ will approach 1; the closer that $\hat{\theta}$

approaches 1, the more complete the sample coverage. For example, S. Altmann (1965) observed 5507 acts in rhesus monkeys and saw 32 behavioral types only once.

$$\hat{\theta} = 1 - \frac{32}{5507} = 0.9942$$

This indicates that Altmann's sample coverage was essentially complete. Fagen and Goldman (1977) caution that this method emphasizes the significance of rare behavioral acts. They provide a more complex procedure for estimating the repertory fraction which properly weighs the frequency of occurrence of different behavioral types.

Of primary importance to any ethological study and particularly to a consideration of catalogs and repertoires is a determination of a behavioral unit. The size of the catalog will vary according to the way in which behavioral units have been defined. The more inclusive (i.e. lumping several different behavioral acts, such as threat and submission, into a single behavioral unit – agonistic behavior), yet mutually exclusive (e.g., agonistic versus ingestive; see below), the smaller the catalog will be. The selection and definition of behavior units will vary according to the objectives and logistics of the individual study (Chapter 6).

4.2.2b Ethogram

An ethogram is a set of comprehensive descriptions of the characteristic behavior patterns of a species (Brown, 1975). It is the result of refining your catalog of behaviors after many hours of observation (in some cases audio recording) and description, and it should be the starting point for any ethological research, especially species-oriented research. Schleidt *et al.* (1984) provide a brief history of the use of ethograms by ethologists.

When concept-oriented research is conducted, researchers may compile an ethogram of only those behaviors within, or closely related to, the category in which they are interested. For example, Fraser and Nelson (1984) compiled an ethogram of the courtship behaviors of male and female Madagascan hissing cockroaches (*Gromphadorhina portentosa*) (Table 4.9).

Limiting our knowledge, as well as our research, to only one type of behavior in a species does pose potential hazards. For example, Tinbergen (1953) argued that the more we restrict our view of the animal's total behavior patterns, the greater the probability of misinterpreting results.

> The need for a broad, observational approach cannot be stressed too much. The natural tendency of many people, particularly of young beginners, is to concentrate on an isolated problem and to try to penetrate into it. This laudable inclination must be kept in check or else it leads to an accumulation of partial, disconnected results, to a

Table 4.9. *Courtship behaviors performed by male and female Madagascan hissing cockroaches*

Behavioral unit	Description
Approach	One animal (either male or female) moves forward and makes contact with the other animal
Antennate	Two behaviors can be distinguished: mutual antennation by both male and female where contact is antennae–antennae; and solo male or female antennation of the other animal's dorsal surface. The latter form of antennation can take the form of 'tapping' or 'horizontal rubbing' movements. These movements are very different from the rapid vertical 'fencing' movements of aggressive behavior (Nelson and Fraser 1980)
Hiss	This audible sound results from the forceful expulsion of air through a specialized abdominal spiracle. Male courtship hissing can be separated by its acoustic characteristics and behavioral context into type-1 and type-2 hissing. Type-1 hisses are isolated, soft hisses, whereas type-2 hisses are shorter and occur in trains (for a more detailed description of the acoustics of sound production, see Nelson and Fraser, 1980)
Mount and palpate	One animal puts one or more legs on the other and taps the other animal's body surface, usually the dorsum, with the labial or the mandibular palps
Posture	Male stands high off the substrate with the abdomen curved upward and extended. This is sometimes accompanied by extrusion of the phallomeres and/or type-1 hissing. During posturing a distinct odour is noticeable to the (human) observer
Cross-over	Female crawls over the posterior tip of the male's abdomen, dragging her abdomen over his
Copulation attempt	Male attempts to copulate by rapidly thrusting the tip of the abdomen towards the female's abdomen. Usually the male starts thrusting towards the lateral ventral surface of the female abdomen, moving to an opposed position for copulation
Stand	Female stands with legs braced laterally, body close to the ground, the abdomen flexed downward at the tip. This normally accompanies copulation attempts by the male
Move away	One or both animals walk or run away from each other

Note:
From Fraser and Nelson (1984)
Source: Copyrighted by Bailliere Tindall.

collection of sociological oddities. A broad, descriptive reconnaissance of the whole system of phenomena is necessary in order to see each individual problem in its perspective; it is the only safeguard for a balanced approach in which analytical and synthetical thinking can cooperate. This, of course, is true not only of sociology, it is true of each science, but in ethology and sociology it is perhaps forgotten more often than in other sciences.   *[Tinbergen, 1953:130]*

Descriptions of the behaviors in the ethogram should be clear and concise, yet complete. A useful adjunct to a written description is a photograph or line drawing. Figure 4.7 shows how Enquist *et al.* (1985) used line drawings to supplement the written descriptions in their ethogram of behaviors performed by fulmars (*Fulmarus glacialis*) when competing for fish.

Schleidt *et al.* (1984) pointed out the large variation in description, format and completeness of published ethograms. They designed a 'standard ethogram' which they hoped would serve as a prototype for future ethograms of birds and, perhaps, other taxa. The ethogram, which consists of the 60 most commonly observed visual behavior patterns, was tested and refined in a study of the bluebreasted quail (*Coturnix chinensis*).

The discussion above dealt with ethograms for descriptive studies or experimental studies of normal behaviors in a natural environment. An ethogram of behaviors to be measured is also compiled when conducting a manipulation experiment (e.g. Godwin, 1994); in this case, the behaviors are operationally defined, instead of being described (Chapter 6), and they are normally mutually exclusive.

4.2.2c *Mutually exclusive behaviors*

Mutually exclusive behaviors are those that cannot occur simultaneously, either because the animal cannot perform them simultaneously or we have defined them so as to eliminate two behaviors being recorded simultaneously. The behaviors described in your ethogram *must* be mutually exclusive if you are determining time budgets (i.e. the amount and percentage of time devoted to each behavior). Time budget studies also require that your ethogram be *exhaustive* (i.e. the animal must always be engaged in one of the behaviors in your catalog). Also, the behaviors measured in an experimental study are almost always defined so that they are mutually exclusive, so that the animal is recorded as responding with only one of several possible behaviors (see Chapter 6).

The behaviors in your ethogram *should* be mutually exclusive if you are unable to record accurately more than one behavior at once, either because of your lack of experience or because you are using a data logger (or software) that will not allow for the recording of simultaneous behaviors (Chapter 9).

Bill-pointing. The bird points with the bill against the opponent. This behaviour varies in intensity from a turn of the head to an unsuccessful attempt to peck the opponent. The mouth is often open and a sharp sound is utterred. The owner also directs this behavior to flying birds. Bill-pointing is always combined with Wing-raising.

Breast-to-breast. The two birds meet, orientating the body somewhat upright with their breasts touching. Breast-to-breast is nearly always preceded with rushing behavior performed by one, or both, birds.

Fig. 4.7   Line drawings and descriptions of two behavior patterns performed by fulmars when competing for food (from Enquist *et al.*, 1985). Copyright by Bailliere Tindall.

Using only mutually exclusive behaviors may provide a high enough level of resolution to answer some research questions, but it is unrealistic to believe that it will accurately reflect the animal's true behavior. Probably all animals are capable of, and do perform, simultaneous behaviors.

4.3 INFORMATION RESOURCES

The usually accepted dogma states that before embarking on your research you should learn as much as you can about your subject animal, especially about its behavior; however, this might not always be true (see section 4.3.1). You should collect information, in addition to your initial reconnaissance observations, from all the sources at your disposal, including available literature, data from other researchers' efforts (personal interviews, logs, diaries, field notes, etc.), and films and videotapes.

4.3.1 Literature

How much time should you devote to reading and reviewing the literature? The answer depends on several factors. First, for descriptive studies, a review of the literature might provide you with biases and a restricted, myopic view of the behavior, and prejudice your observations. Secondly, beginning researchers can spend too much time acquiring a breadth of knowledge in their field. P.B. Medawar, a Nobel Prize winning experimental pathologist, offers the following in his *Advice to a Young Scientist*:

> . . . considerations apply to a novice's inclination to spend weeks or months 'mastering the literature.' Too much book learning may crab and confine the imagination, and endless poring over the research of others is sometimes psychologically a research substitute. . . .The beginner must read, but intently and choosily and not too much. Few sights are sadder than that of a young research worker always to be seen hunched over journals in the library; by far the best way to become proficient in research is to get on with it . . .   *[Medawar, 1979:16,17]*

Researchers' inability to read and absorb the enormous body of published information available, while still conducting their own research, is a widespread phenomenon. Thomson, in his informative discourse on the state of scientific literature said:

> The ultimate irony is that a major proportion of the 'literature,' so expensively and nerve-wrackingly produced, is rarely read or cited. . . . if only because of the simple fact that so much is published, even in the most narrowly defined fields, that one cannot both read it all and still have time to add one's own peerless contributions.   *[Thomson, 1984:185]*

However, for experimental studies it is generally advisable to be current on the body of knowledge available about the concept or species you are studying. This will help you to generate new hypotheses and avoid unknowingly duplicating someone else's research. Even this approach may consume much of the time you would prefer spending on your own research. For example, S.E. Luria (1984), a Nobel Prize winning microbiologist, acknowledges that he is not 'an eager seeker of information'.

> I like to operate with only a fraction of the enormous accumulation of knowledge. When I open a new issue of a scientific journal I do not scan the table of contents looking for exciting novelty; on the contrary, I hope that there will be nothing in it that I must read.   *(Luria, 1984:115)*

When you reach the level of experience and stature of a Nobel laureate, your research will be the 'cutting edge'; perhaps then your mind will generate sufficient ideas and direction to keep your research on a productive course. I assume that the reader of this book has not reached that level.

There are four classes of literature which can be read and reviewed (Fenner and Armstrong, 1981):

1 Pre-primary literature includes 'in-group' memos and technical reports. These are produced primarily in state and federal agencies and industry. Beginning ethologists, especially those in academia, will have limited access to the pre-primary literature.

2 Primary literature includes all the professional journals in animal behavior or related disciplines. Journal articles will be the best written source of detailed information about research on species or concepts of interest.

3 Secondary literature includes indexes and abstracts of the primary literature that are compiled by various services, such as *Animal Behaviour Abstracts*.

4 Tertiary literature includes syntheses of the literature such as textbooks, encyclopedias and handbooks. Tertiary literature is often where a beginning ethologist's interest in particular species and concepts begins.

Access to primary literature (journal articles) can be sought in three ways. First, refer to a recent article on the topic. You can find one by looking in a recent secondary source (see below), in *Current Contents Life Sciences*, by searching one of the journal online catalogs such as *Uncover*, *Article1st* and *Contents1st*, or by asking a researcher in that area. Look at the articles referenced by the author. These articles will often be of interest to you. By reading these articles and using their reference lists, you can go back many years, sometimes to the original paper on the subject.

Second, you can use the secondary literature (indexing and abstracting tools) which provide references to journal articles by subject. Since most of these tools are now produced by computer, they can be searched in three ways: through the printed issues, online through commercial vendors such as Dialog and STN, and on CD-ROM in libraries or on your own computer using Internet, Bitnet or other accesses to the information 'superhighway'. Some are now available as part of online catalogs of libraries. Online and CD-ROM versions generally do not cover the entire range of years that have been published. For example, the online version of *Biological Abstracts* (Biosis) covers 1969 to the present; the CD-ROM version covers 1985 to the present; the print version covers 1926 to the present. Dates of electronic versions, especially in the CD-ROM format, are likely to change in the future as more older printed 'issues' are transcribed to a computer-readable format.

Several steps are involved in conducting a literature search, either using printed sources or electronic formats:

1 Write a short sentence or paragraph stating your need.
2 Decide if you want a few relevant articles or 'everything'.
3 Divide your search topic into concepts and label them.
4 Under each concept label, write appropriate terms (key words).
5 Decide on appropriate abstracting and indexing services or databases to search.
6 Review instructions on access points in each service or database (title words, assigned subject headings, subject codes, etc.).
7 For electronic formats, select appropriate Boolean operators (and, or, not). For example, 'dogs and wolves, not prairie dogs'.
8 Do the search.
9 Look at your results and go through steps 1–8 again with new words or terms suggested in articles identified.

There are usually several ways of conducting any literature search. Your success will often depend on how well you know the area of research and the researchers that are publishing.

Third, you can scan the yearly indexes in the professional journals where you suspect articles on your specific topic may have been published. These, if they exist, can usually be found at the back of the last issue in each volume. Also, search the subject indexes in the tertiary literature (textbooks) on animal behavior.

Fourth, if you know a classic or original paper on a subject, you can find out who is currently citing it by using the citation index portions of *Science Citation Index* (SCI) and *Social Science Citation Index* (SSCI). SCI and SSCI work on the same principle as the first method above, by telling you who is citing what. SCI and SSCI also have author (name) indexes and key-word-in-title subject indexes.

Other sources of information about current research are programs, published abstracts, and published proceedings of scientific meetings. Addresses of presenters are usually provided, and you can write the author directly for more information or a copy of the manuscript.

Monographs and books are also important sources of information on animal behavior. It is not unusual for long-term studies to be published in this format. Jane Goodall, for example, first published her observations on the chimpanzees in the Gombe Stream Reserve as one volume in the *Animal Behaviour* Monograph series, but her most well-known work is the book *In the Shadow of Man*, published in 1971.

Books can be found by searching subjects in online catalogs or the traditional card catalogs of libraries. In card catalogs, efficient access is best obtained by first referring to the Library of Congress Subject Headings (15th edn. 1992) under the

phrase 'Animals, habits and behavior of' following the name of the species of interest. Online catalogs allow you to access single words in titles, plus subject headings
and authors.

The United States government is the world's largest publisher. Many important
publications are produced by various agencies within the Departments of
Agriculture and Interior. These can be identified in online catalogs and in the publication from the Government Printing Office entitled *Monthly Catalog of United
States Government Publications*. Federal and state documents may be available from
local or state offices of governmental agencies. They should also be available in state
government depository libraries.

The world of information is changing almost daily. Indexes and abstract journals were once available only through the printed format which required tedious,
laborious and time consuming 'manual' searching. Today, the same indexes and
abstracts can be searched (with the results printed) in a few minutes. Tomorrow,
they will be readily and widely available through the electronic networks or online
library catalogs using your office and home computer. A reference librarian at a university or public library will be able to tell you the current status of availability of
these services.

Once you have a growing list of citations of references you should consider using
a microcomputer (see Appendix B) and a reference manager software package for
easy storage and retrieval of references. Most software packages will allow you to
retrieve references by various key words in the title, journal and author. Some will
print your selected list of citations in the format required by the professional journal
where you will submit your publication. Four of the more popular reference managers are *ProCite* (Personal Bibliographic Software, Inc. P.O. Box 4250, Ann Arbor,
MI 48106, USA; or Woodside, Hinksey Hill, Oxford 0X1 5AU, England), *Reference
Manager* (Research Information Systems, Inc. 2355 Camino Vida Roble, Carlsbad,
CA 92009–1572, USA, *EndNote* (Niles and Associates, Inc. 2000 Hearst St.,
Berkeley, CA 94709, USA), and *Papyrus* (Research Software Design, 2718 S. W.
Kelly St., Portland, OR 97201, USA).

4.3.2 Other researchers

Contact other researchers who are working (or have worked) on concepts or species
in which you are interested. Discuss your proposed research relative to your own
ideas. Do not parasitize and ask questions that imply 'Tell me everything you know
about . . .'. Most researchers are glad to discuss ongoing projects, but are unwilling
to provide mini-lectures for people who have not already attempted to learn as much
as they can through other sources.

4.3.2a *Professional meetings*

Professional meetings, such as the Annual Meeting of the Animal Behavior Society, are excellent opportunities not only to expand your knowledge about the various behaviors of a wide variety of species, but also to meet and talk with other researchers. Question and answer sessions follow the technical paper presentations, and hallways and lobbies always 'buzz' with more informal discussions. Always, realize and respect the fact that most professionals are cautious about divulging the crucial details of their research until it is nearing completion, has been presented at a meeting, or is in published form. Young ethologists are strongly encouraged to attend, present papers, and join in the discussions at every professional meeting possible.

4.3.2b *Telecommunications*

It is now possible to quickly contact many researchers through electronic mail (E-mail). Some researchers are willing to carry on extended discussions about research questions, designs, procedures, results, and numerous other topics over E-mail. Keep in mind the caveat given above that most professionals are cautious about discussing crucial aspects of their own research during its early stages.

There are also several electronic bulletin boards and newsletters devoted to animal behavior. For example, you can subscribe to ABSnet by contacting Jim Ha via E-mail: jcha@milton.u.washington.edu. See Appendix B for further discussion of telecommunications.

4.3.3 Films and videotapes

Films and videotapes are additional sources of preliminary information about an animal's behavior. Viewing films before embarking on a project is a strategy perfected by successful football coaches. In most cases, the more familiar you are with the animal, the greater the probability of success in your project.

Other researchers are often willing to discuss not only their research and your proposed project with you, but will often loan motion-picture footage and videotapes to responsible investigators.

Another source of films and videotapes is film libraries, where unedited footage or polished productions can be rented for a nominal fee. Four of these sources are listed below:

> Encyclopaedia Cinematographica Archive
> The Penn State University
> 211 Mitchell Building
> University Park, PA 16802, USA

Audio Visual Services
The Penn State University
17 Willard Building
University Park, PA 16802, USA

Rockefeller University Film Service
Box 72
1230 York Avenue
New York, NY 10021, USA

UCLA Media Center
Instructional Media Library
405 Hilgard Avenue
Royce Hall No. 8
Los Angeles, CA 90024, USA

Companies that provide film and videotape footage to production companies for incorporation into programs are another potential source for ethologists. Three of these companies are:

Dreamlight Images 932
N. LaBrea Ave., Suite C
Hollywood, CA 90038, USA

Energy Production
2690 Beachwood Dr.
Los Angeles, CA 90068, USA

Fabulous Footage, Inc.
19 Mercer St.
Toronto, Ontario M5V 1H2
Canada

A reference for approximately 140 000 commercial videotape programs and nearly 2000 sources is Klisz (1995). The Video Source Book. 2 vols. Gale Research Inc., 835 Penobscot Bldg., Detroit, MI. 3988pp. Additional sources of films and videotapes can be obtained by contacting the chair of the Animal Behavior Society's Film Committee and educational media departments at colleges and universities.

5 Delineation of research

5.1 CONCEPTUALIZING THE PROBLEM

A separate chapter has been devoted to the concepts discussed below because of their relatively great importance. Proper delineation of research is the cornerstone of any successful study. A common fault among many beginning researchers is their inability to state clearly the questions they are attempting to answer and the objectives they are striving to meet (see also Barnard *et al.*, 1993). Before the actual research can begin you must decide what you are trying to accomplish. As mentioned in Chapter 1, most of our activities in ethology will be concerned with hypotheses-testing in a broad or specific manner (discussed later). Broadly defined, hypotheses arise from a process of observation–question–hypothesis. Questions and hypotheses are a natural product of our thought processes and cannot be turned off and on. Therefore, while we are reading, observing animals, or listening to other researchers, we are constantly generating questions and formulating hypotheses. Medawar (1960:73), in a discussion of the scientific method, stated his belief that hypotheses arise as follows:

> So far as I can tell from my own experience and from discussion with my colleagues, hypotheses are thought up and not thought out. One simply 'has an idea' and has it whole and suddenly, without a period of gestation in the conscious mind. The creation of an hypothesis is akin to, and just as obscure in origin as, any other creative act of mind.

The thought processes from which questions and hypotheses are born should be allowed to run free during our initial reconnaissance observations, but they must be held in check during actual data collection when bias may creep in (Chapter 8).

5.1.1 What are the questions?

Ethologists are never really at a loss for questions. We are constantly generating questions as we observe animals (Lorenz, 1991) and pursue our own research. Reflecting on his many years of research on herring gulls, Tinbergen (1960b:xiv) stated:

> Soon after starting such a study of a social bird's community life, one begins to realize how little one knows. Even an hour's careful

observation of the goings-on in a gullery faces one with a great number of problems – more problems, as a matter of fact, than one could hope to solve in a lifetime.

The real problem arises when we attempt to isolate and define a question, or a group of related questions. A clear statement of the question (problem) in scientific research is perhaps one of the most difficult steps in a continuing process (see Bronowski, 1973).

It is imperative that your research question clearly addresses the knowledge you are attempting to obtain (see also Barnard *et al.*, 1993). For example, Miller (1985) states that in studies of learning there are two types of question which reflect different concepts: 1. *Can* questions address learning capabilities of animals, such as 'Can species *X* learn by higher-order classical conditioning?'; 2. *Does* questions are more ecologically oriented, such as 'Does species *X* learn behavior *Y* by higher-order classical conditioning?' Examples of clear and concise ethological research questions are quoted in section 5.2 from Crump's (1988) study of aggression in harlequin frogs (*Atelopus varius*).

A duplicative type of research can result because you have an interest in a question that has already been addressed by another researcher. Perhaps, you are not convinced, based on your own observations, that you would get the same results if you *replicated* their study. There are two types of replication studies (Martin and Bateson, 1986): 1. *Literal replication* is an attempt to duplicate the study exactly using the same methods and species under the same conditions; this would provide a test of the internal validity of the other researcher's results; 2. *Constructive replication* involves the use of other measures, similar conditions, or other species to determine whether the same (or similar) results provide external validity for the other researcher's results (internal and external validity are discussed in Chapter 6). Although replicating someone else's research may seem mundane to an established researcher, it can be a valuable experience for a beginning scientist, as described by Medawar (1979:17):

> It is psychologically most important to *get* results, even if they are not original. Getting results, even by repeating another's work, brings with it great accession to self-confidence; the young scientist feels himself one of the club at last, can chip in at seminars and at scientific meetings with 'My own experience was . . .' or 'I got exactly the same results'. . .

Whatever your question, you must be committed to it and genuinely interested in answering that question; otherwise, you might be prone to making observer errors, errors of recording and/or computational errors (Chapter 8). That is, if you are content to merely find *an* answer to the research question, rather than *the* answer, you may allow yourself to conduct invalid research.

5.1.2 Stating objectives

The objective of research, as stated above, is to answer one or more questions. The research objective should be stated as a goal, not a method.

For example, an ethologist may be studying social behavior in domestic chickens. It is noticed that the frequency of agonistic encounters appears to vary with flock size. But does it vary systematically? A suitable research question is: What is the relationship between flock size and frequency of agonistic encounters? The research objective would be stated as:

- To *determine* whether there is a relationship between flock size and frequency of agonistic acts in domestic chickens;
- *not as*:
- To *measure* the frequency of agonistic acts in flocks of different sizes in domestic chickens.

The overall objective of experimental studies, such as that just mentioned, is to answer a question through tests of an hypothesis.

5.1.3 Research hypotheses

As discussed in Chapter 1, hypotheses may be of a broad undefined nature or specific and defined in such a way that they can be experimentally tested. Specific hypotheses are of two types: *research hypotheses* and *statistical hypotheses*. Statistical hypotheses are the outgrowth of research hypotheses and are the basis for statistical tests; they will be discussed in Chapters 11 to 17.

Research hypotheses are conjectural statements about behavior and other related variables. They are what you perceive to be the 'true situation'. From the example in the previous section, your research hypothesis might be that agonistic encounters increase as flock size increases. This is the relationship that you believe exists between the behavior (agonistic encounters) and an independent variable (flock size). In science we progress from *possibilities* (questions), to *probabilities* (research hypotheses, based on limited observations), to *acceptable probabilities* (based on experiments, statistical hypotheses, and statistical tests).

Generally, the transition from question to research hypothesis is based on thought and careful consideration of observations and, in some cases, previous experiments. For example, in 1966, Hamilton introduced the following research hypothesis:

> The current explanation of V-formations in bird flocks is that they establish favorable air currents, reducing flight energy requirements. An

alternative hypothesis is proposed here, that this flock structure is a form of orientation communication, enabling the individuals of these flocks to take maximal advantage of the collective orientation experience of the group.   *[Hamilton, 1966:64]*

Lissaman and Shollenberger (1970), provided evidence in support of the hypothesis of reduced flight energy, but Hamilton's alternative hypothesis remains untested (as far as I am aware).

Unlike the thinking that went into Hamilton's hypothesis (above), the transition from a question to a research hypothesis can sometimes be rather sudden, with the researcher not fully aware that they are formulating in their mind a probable answer (research hypothesis) to a question they have been pondering. In fact, Medawar (1960) has suggested that this sudden leap of insight is the normal mechanism for generating hypotheses.

5.2 USING THE MODEL FOR A BEHAVIORAL ACT

A further look at Crump's (1988) study of aggression in harlequin frogs (*Atelopus varius*) (mentioned above) provides an example of how the Model for a behavioral act (Chapter 2) can be used to help delineate research. During preliminary studies in Costa Rica, Crump: 1. found that individual frogs were abundant along a stream throughout the year; 2. observed site fidelity and aggression in both males and females; 3. observed that after intrasexual aggression the loser left the area resulting in more even spacing among individuals; and 4. observed unusual intersexual aggression. Therefore, this population of harlequin frogs provided Crump with the opportunity to address several questions about the function and causation of their aggression that could be compared to other anuran species and other animal groups; that is, this population met both the *suitability* and *availability* criteria (Chapter 3).

The numbers of the statement and questions below correspond to the numbers in Figure 5.1 showing where they fit into the model. The only type of question not addressed in this research was a 'How' question.

1 Crump first had to define operationally 10 behavioral patterns involved in the frogs' aggressive encounters.

2 'Are there seasonal differences in behavioral interactions among frogs, and if so, do they relate to timing of breeding?'

3 'What are the apparent functions of the various forms of aggression?'

4 'Which individual wins aggressive encounters: resident or intruder?'

5 'Is success rate related to body size?'

6 'Does the level of aggression vary with changes in population density?'

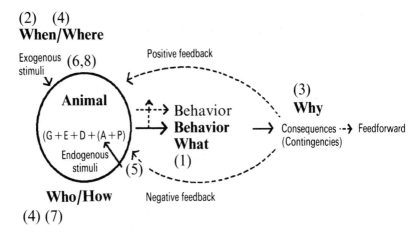

Fig. 5.1   The application of the model for a behavioral act to Crump's (1988) study of aggression in harlequin frogs. The numbers in parentheses refer to questions addressed by Crump (see text for explanation).

7   'Are subadult (less than 1 year) males and adult males aggressive towards each other? If so does success rate vary with age?'

8   'Is aggression related to the availability of prey in the habitat?'

Crump's research objectives were to answer these research questions. Her research hypotheses (i.e. what she believed would be the answers to the questions) were based on her reconnaissance observations and review of the literature (Chapter 4). Her research questions and hypotheses then led to statistical hypotheses (Chapter 11), the research design (Chapter 6), and a consideration of which factors to measure and manipulate (see Chapters 6 and 7). These factors included sampling arthropod populations, observing natural aggression, and placing 'intruder' frogs near resident frogs (called 'originals' because she couldn't document their previous activity). Crump then proceeded with data collection (Chapters 8 and 9), data analysis (Chapters 11 to 17) and her interpretation of the results (Chapter 18).

6 Design of research

6.1 DESCRIPTION VERSUS EXPERIMENTATION

The dichotomy of description versus experimentation was introduced in Chapter 1 and will be expanded on in this chapter. Also, reconnaissance observations were previously discussed (Chapter 4) as a means of gathering background information about an animal (group or species) before designing a research project. Reconnaisance observations help not only in formulating questions and defining objectives, but also in determining what aspects of behavior can be measured, what manipulations are feasible, and the degree of variability that is to be expected. These initial observations occur at Level C in the Ethological Approach (Fig. 1.5); they may occur before, or simultaneously with, delineation of research (Chapter 5). Once the research has been delineated, it must be designed (Level D), and an approach (description or experimentation) selected (Level E).

Descriptive studies usually involve observations under natural conditions since we want to describe normal behavior; therefore, in the first edition of this book I referred to descriptive studies as naturalistic observation (see Chapter 1), and I contrasted those with experimental manipulation; this is the same dichotomy used by Martin and Bateson (1986). In contrast, I am now convinced that the clearest and most useful primary dichotomy should be between describing behavior (*descriptive research*) and testing one or more hypotheses (*experimental research*). Experiments are further divided into *Mensurative* and *Manipulative*, which are the observational and experimental approaches, respectively, of Martin and Bateson (1986). My definition of an experiment, a test of a hypothesis, is the same as that given by Medawar (1960). The division of experiments into two types, mensurative and manipulative, follows Hurlbert (1984).

Both descriptive and experimental research should include quantification, but quantification and statistical analysis must not become the overlords of good-quality observation. Selected examples quantifying descriptions of behavior are presented in Chapter 10. Sometimes initial observations are difficult to quantify, but their value is not necessarily greatly diminished.

> Nowadays, it is regarded as modern to set experimentation above observation (no matter how assumption-free) and to see quantification as a more important source of understanding than description. We tend

to forget that description is the foundation of all science. I do not mean
to question the value of experiments, but observation must come first, in
order to generate questions for experimentation to answer. The
emphasis on blind, quantitative experimentation without prior
observation is based on the erroneous assumption that scientists already
know the questions to ask about the natural world. A theoretical
interest and patience are not the sole qualifications for arriving at the
principles that govern the patterns of social behavior in higher animals.
That aim can be achieved only by those of us whose attention is riveted
on the behavior of our animal subjects because of the great pleasure we,
amateurs and dilettantes, take in our work!   *[Lorenz, 1991:7]*

Descriptive research involves observing and describing the behavior of animals
as it occurs naturally with as little human intrusion as possible. Naturalistic observation does not have to be conducted in the wild. Often an environment can be
created in the laboratory (e.g. for insects or fish) or captivity which closely approximates the natural habitat of the animal. Gibbons *et al.* (1992) discuss the design and
use of naturalistic environments in captivity for animal behavior research; they
include a timely and relevant discussion about concerns for animal welfare and regulations (also see Appendices C and D).

Description was the approach used by Tinbergen in his initial scientific study of
herring gulls (*Larus argentatus*), building on informal observations made throughout the earlier years of his life.

Throughout the years of my boyhood watching the life in the large
gullery was complete happiness . . . watching the snow-white birds
soaring high up in the blue sky . . . It was this sentiment that sent me
back to the gulls in later years, when I returned with a matured scientific
interest, intent on exploring the secrets of their community life.

                                              [Tinbergen, 1960b:xiii]

Tinbergen's initial objective could be stated as: To describe the individual behavior,
social behavior and social organization of herring gulls living in a colony. Later, his
long-term studies evolved into experimental and comparative studies conducted by
himself and his students (Tinbergen, 1973).

Descriptive research involving naturalistic observation was the basic approach
used by Kruuk (1972:5) in his study of the spotted hyena (*Crocuta crocuta*) in the
Serengeti National Park. The approach is reflected in the behavioral questions he
asked:

1 How is the species organized socially, and how does this compare with
  other carnivores?
2 What are its foraging habits, and how do these compare with the feeding
  habits of its carnivorous relatives?

3 How are the questions in 1 and 2 related?

Description using naturalistic observation was the approach used by Nice (1937, 1943) in her study of song sparrows, Evans (1957) in his studies of wasps, Geist (1971) in his study of mountain sheep, Schaller (1972, 1973) in his studies of the mountain gorilla and African lion, and Darling (1937) in his study of red deer.

Although experimentation (discussed in Chapter 7) and description represent a dichotomy, they also complement each other, and most long-term research programs oscillate between the two approaches (Tinbergen 1951). Menzel (1969) discussed the role of the two approaches in primate studies (additional discussion is found in Mason 1968); note that he used the term 'naturalistic' instead of descriptive.

> I am, in fact, convinced that naturalistic and experimental studies can be compatible with each other; that their respective methods can be applied to any situation (instead of being linked to a given situation such as laboratory or field); that both types of information are necessary to a meaningful general science of primate behavior; that any sharp division between naturalistic and experimental methodology is not only undesirable but impossible; and that, finally, disputes as to which method or situation is intrinsically best are nonsensical.
>
> *[Menzel 1969:78]*

The further splitting of both descriptive and experimental research into field and laboratory studies is discussed in the next chapter; whichever approach is selected, a prescribed plan for collection of data is necessary. If experimental research is selected in order to test a hypothesis, then a careful consideration of variables and experimental designs is necessary.

6.2 VARIABLES

A *variable* is a property that takes on different values (e.g. duration of feeding bouts). Two types of variable are involved in experimental studies (also see discussion in section 6.5):

- *Independent variable*: the property that changes (or is manipulated; e.g. light level) and is believed (research hypothesis) to affect the dependent variable (e.g. activity).

  An independent variable is the presumed cause of the dependent variable, the presumed effect.   *[Kerlinger 1964:39]*

- *Dependent variable*: the property that is believed to be affected by a change in the independent variable. It is sometimes called the measured variable or assigned variable.

Behavior will most often be treated as the dependent variable. However, it can be an independent variable, as well (e.g. the effect of one animal's behavior on another's).

The independent variable is generally designated as X and the dependent variable as Y. Some relationship is presumed to exist between the two such that changes in Y can be predicted from changes in X. Correlations are often much more complex than these linear correlations, but discussion of those correlations is beyond the scope of this book.

Variables can be further divided into *quantitative variables* and *qualitative variables*. Quantitative variables vary in *amounts*; qualitative variables vary in *kinds*. The type of dependent variable measured will, in part, determine the scale of measurement you can use. For example, qualitative variables require a nominal scale, whereas ordinal, interval, and ratio scales of measurement can be used with quantitative variables (Chapter 8). The scale of measurement is important in selecting appropriate statistical tests (Chapters 12–17).

Many experimental designs used by ethologists measure only one dependent variable at a time; therefore, selection of the most appropriate dependent variable (e.g. 'sitting' versus 'inactive') is very important (see discussion of behavior units below). For example, Schleidt (1982) argues that among the many variables that can be measured in a given behavior pattern, those that are the most stereotyped are the most useful in characterizing that behavior pattern. Selection of dependent variable(s) to measure should be based on five factors:

- *Validity* – the variable should accurately reflect an effect of the independent variable and not some other extraneous variable(s) (see discussion later in this chapter).
- *Sensitivity* – the variable should have a high probability of showing an effect due to a change in the independent variable(s).
- *Reliability* – the variable chosen should provide the most consistent results; those results should be repeatable.
- *Distribution* – it is valuable if the variable will produce normally distributed measurements; parametric statistics can normally only be applied when this condition is met.
- *Practicality* – if several dependent variables emerge as candidates after consideration of the four factors above, choose one or more that are easily measured; how much cost can you afford in terms of equipment, time and wages?

All experiments are also affected by *nuisance variables*. These are undesired sources of variation which can bias the results. Psychologists recognized their inability to control all the variables and coined the term *intervening variable* (Tolman, 1958) to account for internal, not directly observable psychological processes that in turn affect behavior; Kerlinger (1967:434) calls intervening variables 'in-the-head' variables. All potential sources of undesired variation are nuisance variables (e.g. previous experience, perceptual ability). Nuisance variables can be controlled in four ways:

1 Eliminate the nuisance variable, if possible.
2 Hold the nuisance variable constant for all subjects.
3 Include the nuisance variable as one of the variables in the experimental design.
4 Treat the effects of the nuisance variable statistically through the use of covariance analysis (not considered in this book).

Besides nuisance variables there is also *nondemonic intrusion* which occurs in all experimental studies. Hurlbert (1984:192) defines nondemonic intrusion as 'the impingment of chance events on an experiment', such as weather changes. If these intrusions do not affect all samples equally, they will increase the experimental error.

Examples of independent, dependent, and nuisance variables to be considered in ethological research will be discussed in the next chapter.

6.3 BEHAVIOR UNITS

Measurements in ethology are made on carefully selected, described, and defined (in experiments) behavior units. Whether you are conducting a descriptive study or an experiment, one or more categories of behavior (e.g. reproductive behavior) will be selected for measurement. For example, you could *describe* reproductive behavior in species X, or *test the hypothesis* that increasing temperature decreases the frequency reproductive behaviors in species X. In either case, you must clearly describe, and in experiments define, the units of reproductive behavior that you will measure (e.g. courtship, mating, copulation, birth).

The choice of appropriate behavior units to be measured is at once one of the most important and difficult decisions to be made (Barlow, 1977). As we shall see later in this chapter, this can be crucial to the validity of your research. The choice of an appropriate behavior unit is generally based on experience, tradition, logistics and intuition.

> Even where there is agreement about what general kind of description is
> appropriate, there may be disagreement between 'lumpers' and 'splitters'
> about what should be counted as a *unit of behavior* for quantitative
> purposes. Where some tally fights, songs and journeys, others tally
> blows, notes and footsteps.   *[Beer, 1977:158]*

6.3.1 Classification of behavior units

Behavioral units can be classified in several ways (e.g. Bekoff, 1979a; Hinde, 1966, 1970). In the report of his field study of social communication in rhesus monkeys, S.A. Altmann (1965) stated that 'categorizing the units of social behavior involves two major problems: when to split and when to lump'. He goes on to point out that there are natural units, 'Thus, the splitting and lumping that one does is, ideally, a reflection of the splitting and lumping that the animals do'. The aim in categorizing behavior units is to be as objective as possible; however, as Fentress (1990) points out, ethologists' personal perceptions, constructs and methods of categorization all affect their separation and recombination of behavior units from a stream of behavior.

A reasonable approach to classifying behavior units is to work from the general to the specific, using a scheme which closely follows the level of organization along the species-individual dimension discussed in Chapter 1 (Figure 1.1). Throughout this book I will be using the following hierarchy of categories:

- General category: This is the broadest level of classification; Scott calls this category a 'Type'(below; e.g. *agonistic*)
- Behavior type: A type of behavior within a general category (e.g. *aggression*)
- Social Interaction: This follows part of Delgado and Delgado's scheme (below; e.g. *chasing*)
- Behavior pattern: This refers to the linking of several behavioral acts together into a reasonably predictable and stereotyped pattern; it coincides with part of Delgado and Delgado's scheme (below; e.g. *threat–chase*)
- Behavioral act: This is either an element of a behavior pattern (e.g. *threat*) or a single act that normally occurs alone
- Component part: This is one portion of a behavioral act (e.g. *bareing teeth* during threat)

One general classification scheme was developed by J.P. Scott (1950; Table 6.1). He suggested that, 'The list provides a convenient guide for the description of behavior in a new species, but in any particular case certain types [general categories] may be absent' (Scott, 1963:23).

The *general categories* and *types of behavior* selected for study will be dictated, in large part, by the questions being asked and the approach selected. In descriptive studies we usually gather data on many general categories of behavior, while we often measure behavior units within only one type of behavior in experimental research.

Table 6.1. *Scott's (1950) classification of behavior*

General types of adaptive behavior	Definition
Ingestive	Eating and drinking
Investigative*	Exploring social, biological, and physical environment
Shelter-seeking*	Seeking out and coming to rest in the most favorable part of the environment
Eliminative	Behavior associated with urination and defecation
Sexual	Courtship and mating behavior
Epimeletic*	Giving care and attention
Et-epimeletic*	Soliciting care and attention
Allelomimetic*	Doing the same thing, with some degree of mutual stimulation
Agonistic	Any behavior associated with conflict, including fighting, escaping, and freezing

* Note that these are functional terms which must be supported by data before they are applied in any particular study.

Behavior types can be further classified according to *complexity* and *social interaction* by following a scheme prepared by Delgado and Delgado (1962); this scheme was an outgrowth of their studies of modifications in the social behavior of monkeys. Their classification of behavior units 'evolved from a system of definition' and was used within the framework of an explanatory model of behavior which they also developed. Briefly, their classification scheme is as follows:

A Simple behavior units
  1 Individual
    (i) *Static or postural units* can be identified by static relations (e.g. sleeping alone)
    (ii) *Dynamic or gestural units* can be defined and identified only by a sequence of spatial relations (e.g. climbing a wall)
      (a) *localized* – involving only part of the system (e.g. moving legs)
      (b) *generalized* – involving a change of position of the whole system in relation to its environment (e.g. walking on ground)
  2 Social
    (i) *Static* (e.g. monkeys sleeping, embracing each other)
    (ii) *Dynamic* (e.g. a monkey chasing another)

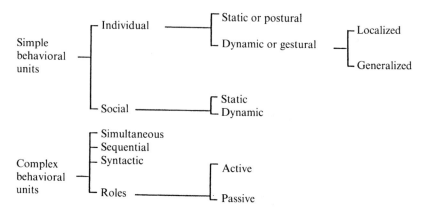

Fig. 6.1   Delgado and Delgado's (1962) scheme of classification of behavioral units.

B   Complex behavior units
1   Simultaneous
2   Sequential
3   Syntactic – the significance of the behavioral unit may vary with context
4   Roles (e.g. groomer–groomed, threatener–threatened)
   (i)   *Active*
   (ii)   *Passive*

Delgado and Delgado's (1962) scheme of classification of behavior units is illustrated diagrammatically in Figure 6.1.

At the next level in the hierarchy, specific *behavior patterns* can be isolated. For example, a duck can move (locomotion) from one place to another by using one of four behavior patterns: walking, swimming, diving, or flying. These patterns could be broken down further according to social, spatial and temporal variables such as flocking and height and speed of flight.

The next level specifies *behavioral acts* within a given behavior pattern. For example, flying can be broken down into the acts of taking off, flight and landing.

Acts can be further classified into *component parts*. Taking off can be divided into movements of various parts of the body (e.g. head, wings, legs), anatomical structures (e.g. muscles and bones), and neurological activity. Study of the internal, physiological, component parts of behavior are beyond the scope of this book; however, several excellent references are available (e.g. Camhi, 1984; Ewert, 1980; Guthrie, 1980, 1987; Kandel, 1977).

Ethologists should not concentrate only on behavior units at the particular level selected for intensive study; they can gain additional understanding through the process of 'focusing-in' and 'focusing-out' from social interactions to component

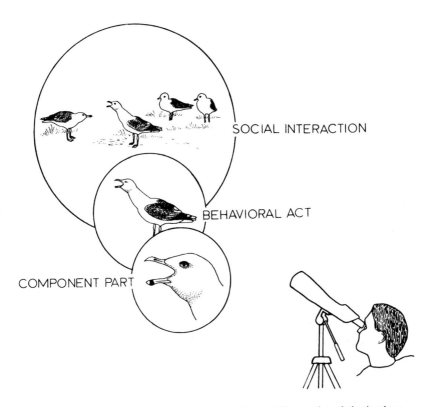

Fig. 6.2 An observer focusing in and out between the social interactions, behavioral act and component parts of a herring gulls' oblique long call (drawing by Dan Thompson; based on Tinbergen, 1959, 1960b).

parts of behavior (Figure 6.2). This is important not only during reconnaissance observations when you are deciding on the particular behavior units you want to study, but also when you are trying to gain additonal insight into the causation or function of specific behavior patterns you have been observing. Based on his descriptive studies of great crested grebes (*Podiceps cristatus*), conducted almost a century ago, Huxley (1968:77; reprint of book published in 1914) made suggestions which hold true today.

> ... *watching at close quarters* must be adopted in order to work out the exact meaning of each separate bit of behaviour. Only when the general course of events has been roughly traced and some hypothesis, however vague, framed concerning it in the watcher's mind, can the fine shades of behaviour have any meaning for him. It is impossible to notice or record everything, and only when some general idea has been gained can the value of any fact be properly appreciated. It is on this account that I

would say, *always begin by distant watching*; otherwise you will not be able to see the wood for the trees.

In Figure 6.2, the observer is focusing in on the alarmed herring gull described by Tinbergen:

> Some of these movements and postures are not difficult even for the human observer to appreciate, though the detection of most of them requires careful study. There are a multitude of very slight movements, most, if not all, of them characteristic of a special state of the bird. The student of behavior is to a high degree dependent on his ability to see and interpret such movements. In the beginning, he will notice them unconsciously. For instance, he will know very well on a particular occasion that a certain gull is alarmed, without realizing exactly that he knows it. Upon more conscious analysis of his own perception (an important element in behaviour study), he will notice that the alarmed gull has a long neck. Still later, he will see another sign, the flattening of the whole plumage, which makes the bird look thinner. Upon still closer study, he will see that the eye of an alarmed bird has avery special expression, due to the fact that it opens its eyes extremely wide.
>
> *[Tinbergen, 1960b:7]*

6.3.2 Spatial and temporal aspects

In Chapters 1 and 2 we discussed the what, when, where, who, how and why questions of behavior. In this section, we are dealing specifically with the *when* and *where* questions. These two dimensions are inherent in all behaviors (Chapter 2; Figure 2.6); however, they may either be an important aspect of our research question, or they may be ignored relative to other aspects (e.g. who, how). Examples of the study of spatial and temporal patterns will be provided in Chapter 10.

Spatial and temporal parameters of behavior are usually *relative dimensions*. This will become clearer as we consider the questions below, which focus from the general to specific spatial and temporal dimensions.

Spatial questions:
1 In what geographical location is the animal found?
2 In what habitat is it located?
3 At what position (vertically and horizontally)?
4 What is its location relative to other members of its group or population?
5 How do its movements vary relative to where it (and other members of its group or population) are located?
6 How do the movements of one part of the body correlate with movements of other body parts (e.g. spatial placement of feet during locomotion)?

Temporal questions:

1 At what ages do different behaviors appear in the animal?
2 When is the individual observed (year, month, day)?
3 How does its occupation of a particular location correlate with season (e.g. migration)?
4 How does its behavior vary on a daily cycle?
5 How is its daily activity broken up into different general categories of behavior or behavior types (i.e. time budget)?
6 What is the duration of occurrences of specific behavior acts (or bouts of behavior)?
7 What is the relative timing of behavioral acts in a behavior pattern (e.g. synchronization of leg movements during locomotion)?
8 What is the relative timing of different component parts during a behavioral act (e.g. movements of parts of the body during threat)?

It can be seen that the spatial and temporal questions merged together in Questions 6 and 7. These two dimensions are interrelated and separated by the ethologist either because of the emphasis of the research question(s), or for convenience and efficiency. Spatial aspects are an integral part of Drummond's Domains of Regularity (described below).

6.3.3 Descriptions and definitions of behavior units

Once behavior units are chosen for study they must be clearly *described* (Chapter 4), and especially for experimental studies they must be clearly *defined*. Clear and complete descriptions are not only necessary for *your* research, but they also provide an accurate picture of the behavior for other researchers; this sometimes results in the published descriptions of behaviors being more complete than the actual criteria used by the researcher to determine when a behavior unit is occurring.

Behavior units used in experimental studies are generally *operationally defined* (e.g. Giles and Huntingford, 1984; see below) in order to increase reliability (both intra- and inter-observer). For example, we can say that when behavioral acts X and Y occur together, we will record an occurrence of behavior pattern Z; that is, behavior pattern Z is operationally defined as the simultaneous occurrence of behavioral acts X and Y. We may have chosen to operationally define behavior pattern Z in this way for several reasons: 1. we consider them the 'most important' behavioral acts in behavior pattern Z; 2. X and Y are the most easily observed behavior acts in behavior pattern Z; 3. X and Y are the least arbitrarily defined acts; and 4. the simultaneous occurrence of X and Y has been used as the criterion by other researchers.

Table 6.2. *Some of the behaviors recorded during the pike tests and heron tests of anti-predator behavior in sticklebacks* (Gasterosteus aculeatus)

Behavior	Definition
Position in tank	
* At surface	Remaining at the water surface for more than 0.5 s
* At bottom	Laying upon the substrate, usually still, fore more than 0.5 s
* In weed	Remaining within artificial weed clump provided in experimental tank, for more than 0.5 s
* Open water	Remaining at least 1 cm away from the water surface, substrate or weed clump for more than 0.5 s
Locomotion type	
* Sneaky swimming	Smooth swimming usually along the bottom using caudal and pectoral fins with dorsal and ventral spines lowered
Jerky swimming	Fast agitated swimming using pectoral fins typified by abrupt stopping and starting
* Normal swimming	Slow bouts of pectoral swimming with frequent pauses and stoppages
* Still	Remaining stationary in any part of the experimental tank for more than 0.5 s
Barrage balloon	Slow vertical ascent to water surface with little or no fin movements, facilitated by swim bladder expansion
* Jump away	A rapid body flexure and stroke of the caudal fin causing a fast 'leap' through the water usually to a place of cover

Note:
* Behaviors included in principal component analyses (see Chapter 16).
Source: From Giles and Huntingford (1984). Copyrighted by Bailliere Tindall.

> It is obvious that we cannot measure what we cannot define. It is equally true that the way we define and record behavioral elements will be affected by the types of measurement we wish to subsequently apply to them.   *[Hutt and Hutt, 1970:33]*

As an example, Table 6.2 lists some of the operational definitions used by Giles and Huntingford (1984) in their study of the anti-predator behavior of sticklebacks in response to pike and herons.

When defining behavior units for experimental studies, one should consider Drummond's (1981) *Domains of Regularity*. Drummond (1981:5,6) states that 'If we examine the very broad range of phenomena regarded as behavior patterns, it is apparent that the identity of each one resides in certain regularities, in those properties which are common to all instances, and that the regularities lie within a limited number of domains', as follows:

1 *Location* of the animal in relation to its environment (see below)
2 *Orientation* of the animal to the environment (see section 10.2.1, on displays)
3 *Physical topography* of the animal (see section 10.2.1, on displays)
4 *Intrinsic properties* of the animal (e.g. changes in color, temperature, and electrical and chemical properties)
5 *Physical effects* induced in the environment by the animal (see section 7.1)

6.3.3a *States and events*

Determining the exact duration of a behavior is often very difficult, not because the instrumentation is not available, but rather we often do not have the necessary skill and observational experience; that is, it is often difficult to determine when a behavior begins and ends. Nevertheless, after observing animals for only a short period of time it becomes obvious that most behaviors can be divided into two categories based on their relative duration (Altmann, 1974):

· *State* – the behavior an individual, or group, is engaged in; an ongoing behavior (e.g. a robin flying); a behavior you can time with a stopwatch; a *duration meaningful* behavior (Sackett, 1978).
· *Event* – a change of states (e.g. a robin taking off); it approaches an instantaneous occurrence that happens so fast that you just count its occurrence; a *momentary* behavior (Sackett, 1978).

For example, in Table 6.2 'sneaky swimming' would be a state, but 'jump away' could be considered an event. Likewise, 'normal swimming' (state) can be interrupted with frequent 'pauses' (events).

Throughout an animal's life it is constantly cycling through states and events. In studying animal behavior we are merely sampling selected states and events, either as they occur 'naturally' or are induced by the experimenter.

Events and states can be measured in various ways. The most frequent measurements are listed in Table 6.3.

6.3.3b *Bouts*

The term *bout* is generally applied to: 1. a repetitive occurrence of the same behavioral act (e.g. a bout of pecking); or 2. a relatively stereotyped sequence of behaviors

Table 6.3. *Measures for states and events*

Type of measure	Definition	Usual application
Total frequency	Number of occurrences per sample unit	Events, states
Partial frequency	Unknown percentage of total occurrences per sample unit	Events, states
Rate	Number of occurrences per unit time	Events
Duration	Amount of time per behavior unit	States

that occur in a behavior pattern (e.g. a courtship-display bout). Analysis of the latter type of bout will be considered in Chapter 15.

There are two criteria (qualitative and quantitative) for defining bouts of behavior:

1 *Change in behavior.* 'If more than one behaviour is being observed, then a bout of one behaviour is said to end when a different type of behaviour begins' (Machlis 1977:9). For example:

Bout: a consecutive series of songs which may vary in minor ways but nevertheless conforms to a particular song type.   *[Mulligan, 1963:276]*

2 *Intervals between occurrences.* 'A criterion interval X_t is chosen to separate one bout from another. All intervals equal to or greater than X_t are classified as between bout intervals and all those less than X_t as within bout intervals' (Machlis 1977:9). For example:

If mouth activites of the same kind followed each other in intervals of less than 16 seconds, they are considered to be a 'bout'.

[Heiligenberg, 1965:164]

Figure 6.3 illustrates how the use of different criteria affects the analysis of bouts. Behavior A appears to occur in discrete bouts that seem obvious from the record. Both bout criteria (above) could be applied equally well to defining these obvious bouts. Behavior B, however, will fall differently into bouts depending on the criterion chosen. If we choose criterion 1 then there are two bouts. If we use criterion 2 with a bout criterion interval (BCI) of 30 seconds, then there are three bouts; if we use a BCI of 10 seconds then there are five bouts.

Separation of a stream of behavior into bouts should reflect what we consider appropriate for the species we are studying. For example, Rosenblum (1978) discussed two dimensions of human behavior which he found useful in delineating

Fig. 6.3 Hypothetical record from an event recorder for behavioral events A and B. Time
marked every 10 seconds. See text for explanation.

bouts: 1. a change in the level (intensity) of motor output; and 2. change in orientation. A bout criterion interval should only be designated after the observer has
become acquainted with the species' behavior.

After becoming familiar with the social behaviors of laboratory rats, Grant and
Mackintosh (1963) selected three seconds as the maximum amount of time that
could elapse between behavioral acts of the same rat in order to consider them part
of the same sequence (bout). Dane and Van der Kloot (1964) studied inter-individual courtship display sequences in groups of male goldeneye ducks (*Bucephala clangula*). From their observations, they set a maximum time of five seconds which
could elapse between displays of two ducks in order to consider them a
stimulus–response sequence (bout).

A more objective method of defining the BCI is to examine the frequency histogram of intervals between behaviors, or the log survivorship curve (Figure 6.4).
The log survivorship curve describes the probability of an behavioral act occurring
relative to the time elapsed since the last act. When behaviors occur in bouts, the
slope of the curve is steep initially and then becomes gradual as the intervals
lengthen. The curve is generally considered to break into two portions: 1. a steep
section of short within-bout intervals and 2. a gradual section of longer between-
bout intervals. The break point between the steep and gradual sections of the curve
can usually be approximated by visually inspecting the curve (Slater 1974) (Figure
6.5).

The gradual portion of the curve can usually be fitted with a straight line,
making it an exponential function. Machlis (1977:14) devised a procedure 'to determine the maximum number of long intervals which can be incorporated into the
"tail" of the curve but still have this tail fit reasonably well to an exponential function'. This provides a very objective procedure for defining the bout criterion interval. When your primary interest is in the behavior which occurs at the start of bouts,
Slater and Lester (1982) recommend selecting a time interval slightly longer than
that at which the slope changes most rapidly. When your primary interest is in the

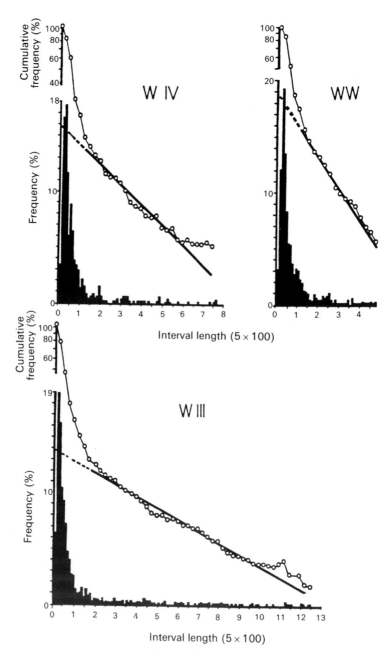

Fig. 6.4 Bouts of pecks by Cobb's Highland male chicks at ladies' hat pins. Frequency histograms of intervals between pecks, and log survivorship curves W III = experiment 3 using a single white pin (*n*=39); W IV = experiment 4 using a single white pin (*n*=20); WW = experiment using two white pins (*n*=19). (from Machlis, 1977, Figure 7).

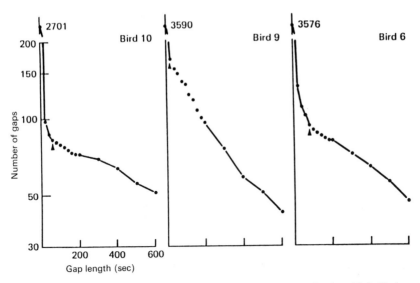

Fig. 6.5 Log survivor functions for intervals between pecks at food for three birds. Each point gives the number of intervals longer than gap length shown on the abscissa. Arrows indicate the point chosen for bout definition for each bird (from Slater, 1974).

time allocated to different behaviors (i.e. time budget), Slater and Lester (1982) recommend selecting the time interval at which the slope changes most rapidly in order to reduce the number of within and between bout intervals being assigned to the wrong category.

When long-term records of behavior are examined, bouts may be found clustered into 'super-bouts' (Machlis, 1977). The log survivorship curve may then be a composite of three types of bout intervals: 1. the initial steep portion of within-bout intervals; 2. the intermediate slope reflecting between-bout intervals; and 3. the very gradual slope representing between-cluster intervals (Figure 6.6). The 'super-bouts' may reflect diurnal rhythms, such as morning and evening feeding periods. We then have a hierarchy of behavior units as in Figure 6.7.

Sibly *et al.* (1990) recommend using log frequency rather than log survivorship curves for splitting behavioral acts into bouts. The data points in log frequency curves are independent of each other so that non-linear curve-fitting procedures can be applied and analysis of variance tests can be used to determine whether the behavioral acts are properly split into bouts. Since this is a more complex procedure, Sibly *et al.* (1990) should be consulted for the applicable methods, as well as how to apply least-squares estimates to the formula provided by Slater and Lester (1982).

Broom (1979) can be consulted for methods to analyze series of a behavior in which periodicity is suspected (i.e. rhythms). These methods include auto-correla-

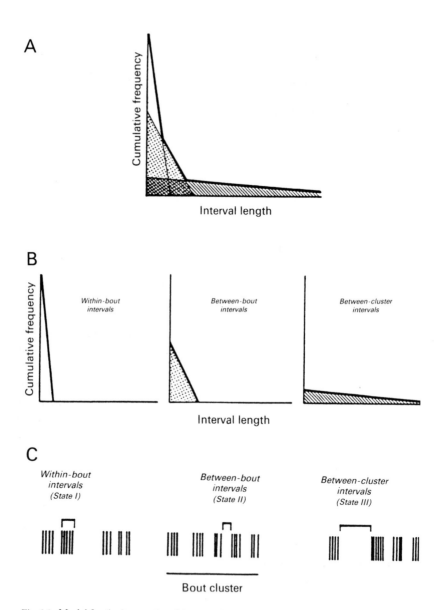

Fig. 6.6 Model for the log survivorship curve based on the assumption that the intervals between pecks represent three different states within the chick (C) and that these intervals are Poisson generated. Such intervals will be exponentially distributed and will have log survivorship functions as shown in (B). The composite of these distributions is shown in (A) (from Machlis, 1977).

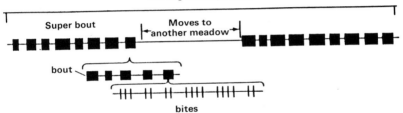

Fig. 6.7 Hypothetical record of a portion of an elk's spring feeding pattern broken into a feeding period, super bouts, bouts and bites. Super bout=time in meadow spent primarily in feeding; bout=time with head bent to ground biting vegetation or seeking other clumps, separated by head-raised posture.

tion (e.g. Roberts, 1994) and Fourier-type spectral analysis. The analysis of bouts of a group of different behaviors is discussed in detail in Haccou and Meelis (1992).

6.3.4 Number of behavioral units to measure

Selection of the number of behavioral units to measure will be affected by several factors. The *minimum* number will be dictated by a determination of what units are important in answering the research question; that is, what is necessary for a valid test of the hypothesis. The *maximum* number will be determined by the experimental design, sampling method, data collection procedure (i.e. equipment), the ability and experience of the observer, and logistics. Hutt and Hutt (1970) concluded that an observer is unable to cope reliably with more than 15 separate behaviors. However, many examples can be found in the literature where more than 15 behaviors have been recorded; nevertheless, reliability (Chapter 8) may decrease as the number of behavioral units measured increases.

Logistics may restrict the number of behaviors that a single observer can measure. Behaviors of interest may be distributed throughout a 24 h period such that a single observer would have to make observations continuously; an efficient, if not impossible proposition if extended over several days. In these situations, several observers are responsible for recording data during different periods of the 24 h period. Likewise, the behaviors of interest might be spread out spatially such that no single observer can record all the data. For example, Krebs (1974) used three observers in his study of colonial nesting and social feeding in great blue herons in order to gather complete and accurate data on feeding flights during the peak of the nesting period.

> One observer was stationed in the observation hide in the colony and was in radio contact with a second observer on the shore line close to

the heronry . . . These two observers were able, between them, to record which part of the colony each heron departed from, and in which direction it flew off after passing over the shore. . . .The third observer was situated on Iona Island . . . and his job was to separate birds flying 'south' into those which landed at Iona Island (north or south) and those flying to Lulu Island or further.   *[Krebs, 1974:105]*

6.3.5 Validity

Selection of the type of research (descriptive or experimental), variables, behavior units, number of units to measure, and accuracy of the measurements will all have an effect on the validity of your research.

Validity refers to how well the design of your research and the methods you employ answer your research question. For example, if you are not convinced that the behavior units you have selected to measure are appropriate to answer your research question, then your research is not valid, your objectives cannot be met, and you should go no further. All of your methods, including data collection, analysis and interpretation (Chapters 8 to 18), will affect the validity of your study. At each step in the preparation of your research you must reflect on the validity of what you intend to do and what you are doing.

There are two types of validity:

- *Internal validity*: how well your research methodology answers your research question for your chosen sample.
- *External validity*: how applicable your results are to other situations (environmental contexts), populations or species.

If you are conducting *species-oriented* research you will be concerned primarily with internal validity. However, if your research is *concept oriented* you will also have to demonstrate external validity.

Kazdin (1982) and Kratochwill (1978) provide a list of threats to internal and external validity directed primarily towards research using single-case, time-series designs; however, most are also applicable to animal behavior research using other experimental designs (Table 6.4).

6.4 RESEARCHER-STATISTICIAN

Few researchers are both good ethologists and good statisticians. Consequently, many maintain a basic understanding of statistics while developing their greatest skills in ethology. Although this is the age of specialization, computers and statistics software packages, the ethologist should have a basic knowledge of experimental

Table 6.4. *Threats to internal and external validity*

Threats to internal validity		Threats to external validity	
1. History	Any event (other than the intervention)	1. Generality across subjects	The extent to which the results can be extended to subjects whose characteristics may differ from those included in the investigation
2. Maturation	Any change over time that may result from processes within the subject		
3. Testing	Any change that may be attributed to the effects of repeated assessment.	2. Generality across settings	The extent to which the results extend to other situations
4. Instrumentation	Any change that takes place in the measuring instrument or assessment procedure over time	3. Generality across response measures	The extent to which the results extend to behaviors not included in the experiment
5. Statistical regression	Any change from one assessment occasion to another that might be due to a reversion of scores toward the mean	4. Generality across times	The extent to which the results extend beyond the times that the invervention is in effect
6. Selection biases	Any differences between groups that are due to the assignment of subjects to groups	5. Generality across behavior change agents	The extent to which the intervention effects can be extended to other researchers
7. Attrition	Any change in overall scores	6. Pretest sensitization	The possibility that assessing the subjects before treatment in some way sensitizes them to the intervention that follows
8. Diffusion of treatment	Occurs when the intervention is inadvertently provided to part or all of the control group or at the times when teatment should not be in effect.	7. Multiple-treatment interference	The results may only apply to other subjects who experience both of the treatments in the same way or in the same order

Source: Adapted and abridged from Kazdin (1982). Coyrighted by 1982 Oxford University Press. Inc. Reprinted by permission.

designs (below) and statistical procedures (Chapters 11–17) before setting out to collect data. However, if you are uncomfortable with your choices of experimental design and statistical tests, you should review them with a statistician.

> ... to write of the 'experimenter' and the 'statistician' as though they are separate persons is often convenient; the one is concerned with undertaking a piece of research comprehensively and accurately yet with reasonable economy of time and materials, the other is to provide technical advice and assistance on quantitative aspects both in planning and in interpretation ... the statistician can produce good designs only if he understands something of the particular field of research, and the experimenter will receive better help if he knows the general principles of design and statistical analysis. Indeed, the two roles can be combined when an experimenter with a little mathematical knowledge is prepared to learn enough theory of design to be able to design his own experiments.   *[Finney, 1960:3]*

As Finney has stated above, just as the researcher relies on the statistician, the statistician relies on the researcher to provide the proper data.

The discussions of experimental designs and statistical analysis in this book are introductory and cursory. The material represents what I believe to be the minimum knowledge ethologists should have in order to conduct their research with insight and logic. It is meant to supplement, not substitute for, consultation with a biometrician.

6.5 BASIC EXPERIMENTAL DESIGNS

Experimental designs are protocols for measuring and/or manipulating independent variables in such a way that their singular or combined effects on the behavior being studied (dependent variable) can be determined. Choice of the proper experimental design is based on knowledge of the: 1. objectives of the research; 2. hypothesis to be tested; 3. feasibility of gathering various types of data; 4. types of experimental designs available and their relative attributes of power and efficiency. The selection of an appropriate statistical test (Chapters 12–17) is based, in large part, on the experimental design.

6.5.1 Variables, samples, treatments, blocks, individuals and measurements

There are six terms that are necessary to understand in order to build appropriate experimental designs and apply valid statistical tests. The notation and definitions I use are easy to understand and remember, but they may differ from other texts. The terms are defined below in the hierarchial manner in which they are applied to experimental designs (refer to Table 6.5) and statistical tests.

Independent variable(s) (A,B. . .)

An *independent variable* is an environmental factor that has different discrete or continuous values; we study the effect that different values of one, or more, independent variables (*A,B* ...) have on one, or more, *dependent variables* (behavior units). The independent variable in Table 6.5 is the natural *time of day*, and two values of time of day were selected for determining their effects on house mice activity: noon (*A*1) and midnight (*A*2). Table 6.5 illustrates a mensurative experiment in which the two values of the independent variable were measured.

In manipulative experiments where we artificially change the independent variables, the manipulations are called *treatments*. For example, if the experiment in Table 6.5 had been conducted in the laboratory and the photoperiod had been manipulated to create artificial noon and midnight, those two values of time of day would be considered treatments.

Sample(s) (S_{A1}, S_{A2} . . .; S_{B1}, S_{B2} . . .)

A *sample* is a set of *measurements* of the *dependent variable* (*behavior unit*) on a number of individuals. Each value of the independent variable, whose effects we are studying, is indicated by a number (1,2 . . . *n*). For example, in Table 6.5, one sample (S_{A1}) is taken at noon (*A*1), and one sample (S_{A2}) is taken at midnight (*A*2). Each Sample contains eight *measurements*.

In experimental designs where we take two, or more, samples from the same value of the independent variable, those multiple samples are called *replications*. In the majority of ethological studies we take only one sample from each value of the independent variable; therefore, we commonly refer to the number of samples, rather than the number of values of the independent variable, when discussing experimental designs and statistical tests.

Block(s) (B_1, . . . B_n)

Blocks are used in a *randomized block design* to control for the effects of a nuisance variable; the various levels or categories of that variable are denoted by $B_{1,2 . . . n}$. In Table 6.5, the individual mice are blocked by sex: females (B_1) and males B_2).

Individuals (I_1 . . . I_n)

When the effect of each value (or treatment) of the independent variable is measured on the same individuals (*repeated measures design*; discussed later in this chapter), the individuals are indicated by $I_{1,2 . . . n}$. In the repeated measures

Table 6.5. *A hypothetical example of a repeated measures experimental design used to determine the effect of time of day on activity of house mice. Four male and four female house mice were randomly selected and placed outdoors in cages with running wheels in order to measure the number of minutes each individual mouse spent using their running wheel during one hour at noon (1200–1300) and one hour at midnight (2400–0100). The activity data in the table are from the first day's measurements. (See text for further explanation)*

		Time of day $(A)^1$	
Blocks by sex of individuals	Individuals	Noon $(A1)^2$ $S_{A1}{}^3$	Midnight $(A2)^2$ $S_{A2}{}^3$
Females (B_1)	I_1	11.3 (x_{11})	52.2 (x_{21})
	I_2	8.6 (x_{12})	49.4 (x_{22})
	I_3	10.5	50.5
	I_4	9.3	51.7
Males (B_2)	I_1	9.6	50.0
	I_2	10.3	51.7
	I_3	10.9	52.6
	I_4	9.1	50.3

Notes:

[1] Independent variable.

[2] Two values of the independent variable.

[3] One sample of measurements from each Value of the Independent Variable.

design illustrated in Table 6.5, there are four female mice (I_{1-4}) and four male mice (I_{1-4}).

Dependent variable(s) (Y)

The dependent variable is the behavior unit on which *measurements* are made. That behavior unit will have one or more properties in which observations will differ in some measureable way (e.g. occurrence, frequency, duration, latency). The dependent variable in Table 6.5 is total time spent wheel running during each one-hour sample period.

Measurement(s) (x_{ij})

A measurement is the property (category or quantity) of the dependent variable (behavior unit) assigned to an observation. In all experimental designs, except the repeated measures design, each measurement is made on a different individual. In the repeated measures design illustrated in Table 6.5, the measurement of minutes of wheel running is made on each individual (I) under each value (noon and midnight) of the independent variable (time of day). When the effect of only one independent variable is being measured, the subscripts of x denote the sample (i) and the number of the measurement (j), which is also the number of the individual in a repeated measures design (Table 6.5).

Another example of the use of some of these terms is provided by Topoff and Zimmerli's (1993) manipulation experiment on the takeover of colonies of slave ant species (*Formica* sp.) by queens of a parasitic ant species (*Polyergus breviceps*). They introduced each of ten *Polyergus* queens who had attacked a dead *F. gnava* queen into five colonies each of *F. gnava* and *F. occulata*. They determined whether *Polyergus* queens would be accepted by colonies of the same species as the dead queen (*F. gnava*) and colonies of another *Formica* species (*F. occulata*) (Table 6.6).

- *Independent variable*: Slave ant species (*Formica* sp.)
- *Treatments*: They introduced the parasitic queens into colonies of two species: *F. gnava* and *F. occulata*
- *Samples*: One sample from each treatment (*F. gnava* and *F. occulata*). Each sample consisted of measurements from five different colonies.
- *Dependent variable*: Takeover of colony.
- *Measurement*: Two measurements of the dependent variable were made for each of five colonies: Category data: Takeover? (Yes or No) Quantity data: Takeover time (min.)

6.5.2 Experimental designs

Only five experimental designs will be discussed in this book. For a more extensive discussion of experimental designs applicable to animal behavior research you should consult one or more of the texts devoted to that topic (e.g. Kirk, 1968; Edwards, 1985). For good discussions of time-series, single-case (single subject) designs, consult Kratochwill (1978) and Kazdin (1982).

The five experimental designs discussed below can be divided into two categories according to whether the researcher will be studying the effects of one or two independent variables.

Table 6.6. Polyergus *queen takeover of unrelated* F. gnava *colonies and of*
F. occulata *colonies*

	Foreign F. *gnava* colony		F. *occulata* colony	
Colony	Successful takeover?	Takeover time (min.)	Successful takeover?	Takeover time (min.)
1	+	15	−	NA
2	+	70	−	NA
3	+	1	+	2 days
4	+	60	−	NA
5	+	20	+	3 days

Notes:

NA: Not applicable. +: Successful takeover; −: unsuccessful takeover.

Source: From Topoff & Zimmerli (1993). Copyrighted by Academic Press.

- *One independent variable*
  Completely randomized design
  Randomized block (repeated measures; matched pairs) design
  Incomplete block design
  Latin square
- *Two independent variables*
  Completely randomized two-factor design

Almost all the more complex designs, not discussed here, can be built by combining two or more completely randomized or randomized block designs (Kirk, 1968). Statistical tests (Chapters 12–17) will be provided for analysis of data from completely randomized, randomized block, repeated measures, matched-pairs, and completely randomized two-factor designs.

Most of the examples provided for the experimental designs described below are based on Vives' (1988) laboratory study of parental choice by larval convict cichlids (*Cichlasoma nigrofasciatum*). He released the larval cichlids into the center of a 76 litre aquarium where they could choose to stay in close proximity to individual ciclids held in screened enclosures at either end of the aquarium (i.e. within 3 cm of enclosure continuosly for 15 min.). The larval cichlids were raised under one of two conditions: 1. in the presence of predators of fry; or 2. not in the presence of predators of fry. Some of the examples provided are actual experiments that Vives conducted; however, most of the data, and some of the experiments, are hypothetical.

6.5.2a *Completely randomized design*

This is one of the simplest designs, but it can be used to compare any number of samples (or treatments) of a single independent variable. The samples are the qualitative or quantitative differences in the independent variable which you have hypothesized have a measurable effect on the behavior you are studying. If the experiment is *mensurative*, the measurements taken in each sample are from individuals randomly selected from the population; the experiment may consist of a single sample. If the samples are treatments in a *manipulation* experiment, there must be a minimum of two treatments (or treatment plus control), and individual animals are randomly assigned to each treatment. A *control* is often considered a 'zero' treatment in which everything is the same except that the experimental variable has not been imposed. However, a control can also be: 1. a 'procedural' treatment (e.g. mice injected only with saline versus saline plus drug treatment); or 2. any treatment against which other treatments are compared (Hurlbert, 1984).

One sample
Tabular form   One sample (or treatment) of the independent variable

$$\frac{S_{A1}}{}$$
$$x_{11}$$
$$x_{12}$$
$$x_{13}$$
$$\vdots$$
$$\frac{x_{1n}}{}$$
Sample Mean: $\bar{x}_{1.}$

S_{A1} = the one sample (or one treatment) of the independent variable

x_{12} = Measurement of the dependent variable on the second individual in the sample.

$\bar{x}_{1.}$ = a bar over a total indicates the mean for that total. The dot in the subscript denotes the variable over which the summation occurred; in this case, this is the mean for the sum of the measurements in this one sample.

Example   Vives (1988) put a parent cichlid in each of the screened enclosures at the ends of the aquarium and determined whether their larvae chose to stay in proximity to one or the other of the parents, or chose neither.

$\underline{S_{A1}}$ (One parent at each end of the aquarium)

$+$ (x_{11}; the first individual in the sample is indicated as having chosen a parent)

$-$ (x_{12}; the second individual in the sample is indicated as having not chosen a parent)

$\vdots$

$+$ (x_{169}; the 69th individual in the sample is indicated as having chosen a parent)

___

Of the 69 individuals tested, 47 chose to stay in proximity to one of their parents, and 22 did not choose either parent.

Two samples

Tabular Form

$\underline{S_{A1}}$	$\underline{S_{A2}}$
x_{11}	x_{21}
x_{12}	x_{22}
x_{13}	x_{23}
$\vdots$	$\vdots$
$\underline{x_{1n}}$	$\underline{x_{2n}}$

Sample means: $\bar{x}_{1.}$ $\bar{x}_{2.}$

Grand mean: $\bar{x}_{..}$

where: S_{A2} = Sample 2 (or Treatment 2) of the independent variable

x_{22} = Measurement of the dependent variable on the second individual in sample 2.

$\bar{x}_{2.}$ = a bar over a total indicates the mean for that total. The dot in the subscript denotes the variable over which the summation occurred; in this case, this is the mean for the sum of the measurements in Sample 2.

Example   Vives (1988) found that the length of time until a choice was made (reaction time=latency to choice) tended to be shorter for young from broods reared in the presence of predators of young versus those raised with no predators of young present (see Table 6.7).

There were 33 individuals in Sample $A1$ and 14 individuals in Sample $A2$. The data show that the first individual (x_{11}) from a brood raised with predators (Sample S_{A1}) took 123 s to make a choice. The last individual (x_{214}) from a brood raised without predators present (Sample S_{A2}) took 356 s. to make a choice.

Table 6.7. *An example of data in a completely randomized design with two samples (see text for explanation)*

Groups from broods raised with predators S_{A1}	Groups from broods raised without predators S_{A2}
123 (x_{11})	305 (x_{21})
207 (x_{12})	215 (x_{22})
186 (x_{13})	311 (x_{23})
$\vdots$	$\vdots$
421 (x_{133})	356 (x_{214})

Source: Based on Vives (1988).

Three or more samples
Tabular form

S_{A1}	S_{A2}	S_{A3}	$\cdots$	S_{An}
x_{11}	x_{21}	x_{31}	$\cdots$	x_{n1}
x_{12}	x_{22}	x_{32}	$\cdots$	x_{n2}
x_{13}	x_{23}	x_{33}	$\cdots$	x_{n3}
$\vdots$	$\vdots$	$\vdots$	$\vdots\vdots\vdots$	$\vdots$
x_{1n}	x_{2n}	x_{3n}	$\cdots$	x_{nn}

Sample means: $\bar{x}_{1.}$ $\bar{x}_{2.}$ $\bar{x}_{3.}$ $\cdots$ $\bar{x}_{n.}$

Grand mean: $\bar{x}_{..}$

S_{A1} = Sample 1 (or Treatment 1) of the independent variable

x_{12} = Measurement of the dependent variable on the second individual in Sample 1.

$\bar{x}_{1.}$ = a bar over a total indicates the mean for that total. The dot in the subscript denotes the variable over which the summation occurred; in this case, this is the mean for the sum of the measurements in Sample 1.

Example   A third sample could be added to the example above by adding a sample of young raised in the presence of predators of adult cichlids. Reaction times would then be taken from randomly selected individuals from each group (see Table 6.8).

In this hypothetical example we have used 21 individuals from broods raised in the presence of predators of adult cichlids (Sample A3). The last individual measured in this group (x_{321}) had a reaction time of 323 seconds.

Table 6.8. *An example of a completely randomized design with three samples illustrated with hypothetical data from larval cichlids (see text for explanation)*

Raised with predators of young S_{A1}	Raised without predators of young S_{A2}	Raised with predators of adults S_{A3}
123 (x_{11})	305 (x_{21})	216 (x_{31})
207 (x_{12})	215 (x_{22})	275 (x_{32})
186 (x_{13})	311 (x_{23})	195 (x_{33})
$\vdots$	$\vdots$	$\vdots$
421 (x_{133})	356 (x_{214})	323 (x_{321})

Linear model   A simple equation can be used to show all the sources of variation that affect the individual measurements in the three completely randomized block designs described above.

$$x_{ij} = \mu + \alpha_i + \epsilon_{ij}$$

The model (equation) states that each individual measurement (x_{ij}) is equal to the population mean (μ) plus the sample (or treatment) effect (α_i) plus an error effect (ϵ_{ij}) which is unique for each individual subject.

As in most experiments, population parameters are not known, but the samples are used to estimate those parameters. In the examples of experiments above, μ, α_i and ϵ_{ij} are unknown; but they are estimated by the following:

$$\hat{\mu} = \bar{x}_{..} \text{ (estimates } \mu)$$
$$\hat{\alpha}_i = (\bar{x}_{i.} - \bar{x}_{..}) \text{ (estimates } \alpha_i)$$
$$\hat{\epsilon}_{ij} = (x_{ij} - \bar{x}_{i.}) \text{ (estimates } \epsilon_{ij})$$

The error effect is the summed effect of all the uncontrolled *nuisance variables*; that is, all of the effects not attributable to a particular sample (or treatment). We can rearrange the linear model to show that the error effect (ϵ_{ij}) is what remains of an individual measurement (x_{ij}) after the sample (or treatment) effect and population mean are substracted from it.

$$\hat{\epsilon}_{ij} = x_{ij} - (\hat{\alpha}_i + \hat{\mu})$$

6.5.2b *Randomized block, repeated measures and matched pairs designs*

Randomized block design

This design controls for additional variability (expressed in the error effect) by assigning subjects that are similar in one or more characteristics (e.g. same sex, age, litter) to blocks. That is, subjects within each block should be more homogeneous than subjects between blocks. Blocks can also be treated as a second independent variable (e.g. two-way analysis of variance, not discussed in this book).

Tabular form

Samples of independent variable A

Blocks	S_{A1}	S_{A2}	S_{A3}	...	S_{An}	Block means
B_1	x_{11}	x_{21}	x_{31}	...	x_{n1}	$x_{.1}$
B_2	x_{12}	x_{22}	x_{32}	...	x_{n2}	$x_{.2}$
⋮	⋮	⋮	⋮	⋮⋮⋮	⋮	⋮
B_n	x_{1n}	x_{2n}	x_{3n}	...	x_{nn}	$x_{.n}$
Sample means:	$\bar{x}_{1.}$	$\bar{x}_{2.}$	$\bar{x}_{3.}$	...	$\bar{x}_{n.}$	

Grand mean: $\bar{x}_{..}$

Each x_{ij} is a measurement from a different individual or a total for several individuals in that block and sample. Individuals are blocked (e.g. B_1) according to some characteristic such as sex, age, genotype, place, or time.

Example   Vives (1988) tested the larval cichlids at an age of two days free-swimming, because by that time they feed exogenously and show good mobility, and they had previously been shown to respond most readily to models when less than six days free-swimming. If it were not possible to have large numbers of young available at the age of two days free-swimming, and it was felt that age might have an effect on their choices, then a randomized block design could be used. From the individuals raised with predators of young, five would be randomly selected and tested at age 1 day, another five at age 2, etc. The same procedure would be used for the individuals raised without predators of young.

The hypothetical data in Table 6.9 show that the total time to make a choice for the five individuals raised with predators of young, and tested at age 1 day free-swimming (x_{11}) was 650 seconds. The five individual times making up that total would be designated as x_{111}, x_{112}, x_{113}, x_{114}, and x_{115}.

Table 6.9. *A randomized block design illustrated with hypothetical data for the time it took larval cichlids of different ages to choose to stay near one of their parents*

Age (days free-swimming)	Raised with predators of young S_{A1}	Raised without predators of young S_{A2}
1	650 (x_{11})	1235 (x_{21})
2	587 (x_{12})	1307 (x_{22})
3	673 (x_{13})	1214 (x_{23})
4	706 (x_{14})	1292 (x_{24})

Repeated measures design

The *repeated measures design* is the same as the *randomized block design* except that:

1 Individuals (e.g. I_1) are the blocks (Table 8.4) or subsets of blocks (Table 6.5).
2 A measurement in each sample (or treatment) is made on each individual. For example, the measurements $x_{11}, x_{21} \ldots x_{n1}$ are all made on individual I_1.

Tabular form

Samples of independent variable

Individuals	S_A	S_{A2}	S_{A3} ... S_{An}
I_1	x_{11}	x_{21}	x_{31} ... $\bar{x}_{n1}$
I_2	x_{12}	x_{22}	x_{32} ... $\bar{x}_{n2}$
⋮	⋮	⋮	⋮ ⋮⋮⋮ ⋮
I_n	x_{1n}	x_{2n}	x_{3n} ... x_{nn}

Sample means: $\bar{x}_{1.}$ $\bar{x}_{2.}$ $\bar{x}_{3.}$ ... $\bar{x}_{n.}$

Grand mean: $\bar{x}_{..}$

Example   If we decided that the greatest effect of being raised with predators of young was shown by young at age 2 days free-swimming, we might then choose to use ten individuals from that age group to test for the effect of predator odor in the test aquarium on their time to make a choice. We would randomly select five individuals to be tested in the presence of the odor first; the other five we would test first under the conditions of no predator odor.

The hypothetical data in Table 6.10 shows that individual I_1 took 203 s to make a

Table 6.10. *An example of a repeated measures design with two samples illustrated with hypothetical data from larval cichlids (see text for explanation)*

Individuals of age two days free-swimming raised with predators of young	No Odor of predator S_{A1}	Odor of predator S_{A2}
I_1	203 (x_{11})	175 (x_{21})
I_2	185 (x_{12})	168 (x_{22})
⋮	⋮	⋮
I_{10}	226 (x_{110})	213 (x_{210})

choice when no predator odor was present, and it took 175 s. when the odor was present.

Matched pairs design

This design is a subset of the randomized block and repeated measures designs. Either design is usually referred to as *matched pairs* when only two samples (or treatments) are used.

Tabular format

Samples of independent variable		
Blocks or individuals	S_{A1}	S_{A2}
B_1	x_{11}	x_{21}
B_2	x_{12}	x_{22}
⋮	⋮	⋮
B_n	x_{1n}	x_{2n}
Sample means:	$\bar{x}_{1.}$	$\bar{x}_{2.}$

Grand mean: $\bar{x}_{..}$

Example Since both of the examples for the randomized block design and the repeated measures design (above) have only two samples, they are also considered matched pairs designs.

Linear model

The equation for the randomized block, repeated measures and matched pairs designs includes β_j, the effect attributable to the *j*th block (or individual):

$$x_{ij} = \mu + \alpha_i + \beta_j + \epsilon_{ij}$$

Therefore, by assigning similar subjects to blocks (or taking a measurement from the same individual in each sample) we have partitioned an additional source of variability out of the error effect. We can show this by rearranging the linear model:

$$\hat{\epsilon}_{ij} = x_{ij} - (\hat{\alpha}_i + \hat{\beta}_i + \hat{\mu})$$

Estimates for μ and α_i remain $\bar{x}_{..}$ and $(\bar{x}_{i.} - \bar{x}_{..})$, respectively.

$$\hat{\beta}_j = (\hat{\beta}_{.j} - \bar{x}_{..}) \text{ (estimates } \beta_j)$$

therefore, the error effect is estimated by:

$$\hat{\epsilon}_{ij} = x_{ij} - [(\bar{x}_{.j} - \bar{x}_{..}) + (x_{i.} - \bar{x}_{..}) + \bar{x}_{..}]$$

If the block effect (β_j) is appreciable then we will have been successful in reducing the error effect (ϵ_{ij}) by blocking. The relative power of the statistical analysis we use will be increased by reducing the error effect as much as possible.

6.5.2c *Incomplete block design*

This design is applicable when the number of subjects available for study is not large enough to measure each sample (or treatment) effect for each block; that is, you cannot complete a normal randomized block design.

Tabular form

Samples of independent variable

Blocks	S_{A1}	S_{A2}	S_{A3}	Block means
B_1	x_{11}		x_{31}	$\bar{x}_{.1}$
B_2		x_{22}	x_{32}	$\bar{x}_{.2}$
B_3	x_{13}	x_{23}		$\bar{x}_{.3}$

Sample means: $\bar{x}_{1.}$ $\quad$ $\bar{x}_{2.}$ $\quad$ $\bar{x}_{3.}$

Grand mean: $\bar{x}_{..}$

This design is balanced, which means that each block contains the same number of subjects, each sample (or treatment) occurs the same number of times, and subjects are assigned to the samples (or treatments) so that each possible pair of treatment levels occurs within each block an equal number of times. In partially balanced designs some pairs of treatment levels occur together within the blocks more often than do other pairs.

Table 6.11. *The densities of male and female woodrats used in Kinsey's (1976) study of social behavior in Allegheny woodrats*

Sexes in groups	Density (individuals per 65 m²)								
	2	3	4	5	6	7	8	9–12	14
All Male	X				X				
All Female	X			X					
Mixed	X		X		X		X		X

As an example, Kinsey (1976:181) tested his research hypothesis that '. . . Allegheny woodrats (*Neotoma floridana magister*) would exhibit territorial behavior when confined in relatively low-density populations in a large observation cage and at populations of higher density would exhibit increased agonistic interactions and a dominance hierarchy type of social organization'. He placed wild-trapped male and female woodrats together in a 65 m² enclosure at densities varying from 2 to 14. However, in this partially balanced design not all densities (treatments) were represented nor did each group (block) contain both males and females. For analysis, the densities were combined into two groups: low density (2–4) and high density (5–14). Table 6.11 shows the densities and sexes that were represented in the groups used.

At this point the question arises: If I have a limited number of subjects should I go to an incomplete block design or fall back on the completely randomized design? *You should always strive to design the data collection in such a way as to reduce the error effect as much as possible.*

Linear model

The equation for the incomplete block design is the same as that for the randomized block design:

$$x_{ij} = \mu + \alpha_j + \beta_i + \epsilon_{ij}$$

6.5.2d Latin square design

This design uses blocking to reduce the error effect from two nuisance variables. The levels of the two nuisance variables are assigned to the rows and columns of a Latin square. You must have the same number of samples of each of the three variables (the independent variable whose effect you are studying and the two nuisance variable whose effects you are partitioning out in the blocks).

Tabular form

The example below uses three samples of the independent variable of interest (A) and three samples of each of the two nuisance variables (B,C). The C variable samples are assigned to each of the cells in the Latin square in a balanced fashion; that is, each level of the C variable is equally represented in both the rows and columns. This necessitates the Latin square having the same number of rows, columns, and levels for each variable.

Independent variable A

Nuisance

variable B	A_1	A_2	A_3
B_1	C_1	C_2	C_3
	x_{111}	x_{212}	x_{113}
B_2	C_2	C_3	C_1
	x_{122}	x_{223}	x_{321}
B_3	C_3	C_1	C_2
	x_{133}	x_{231}	x_{332}

The subscripts for each measurement (x_{ijk}) denote the level of each independent variable in the order A,B,C. For example, x_{231} denotes level 2 for variable A, level 3 for B, and level 1 for C. Each measurement (x_{ijk}) is taken on a different individual. The individuals are randomly assigned to the combinations of variables.

Example

Using our example of the experiment of choice by larval cichlids we could measure the effect of three different regimes of raising the young on their time to make a choice, while at the same time blocking for color morph and age when tested. The following variables can be applied to the tabular format above:

Independent variable A (raising regime):

A_1 = raised without the presence of predators
A_2 = raised in the presence of predators of young
A_3 = raised in the presence of predators of adults

Nuisance variable B (color morph):

B_1 = gold
B_2 = green
B_3 = wild type

Nuisance variable C (age when tested):

C_1 = 1 day free-swimming
C_2 = 2 days free-swimming
C_3 = 3 days free-swimming

Linear model

The equation for this design includes the A variable effect (α_i), the B variable effect (β_j), and the C variable effect (v_k).

$$x_{ijk} = \mu + \alpha_i + \beta_j + v_k + \epsilon_{ijk}$$

By partitioning out the two nuisance variables, we have reduced the error effect. This can make the Latin square a more powerful design than either the completely randomized or randomized block designs, although it is rarely used in ethological research.

$$\epsilon_{ijk} = x_{ijk} - (\hat{\alpha}_i + \hat{\beta}_i + \hat{v}_k + \hat{\mu})$$

The following matrices show the assignment of levels (1,2,3, etc.) of nuisance variable C to the cells of a 4×4×4 and a 6×6×6 Latin square design.

4×4×4						6×6×6						
Variable A						Variable A						
B	1	2	3	4		B	1	2	3	4	5	6
1	1	2	3	4		1	1	2	3	4	5	6
2	2	3	4	1		2	2	3	4	5	6	1
3	3	4	1	2		3	3	4	5	6	1	2
4	4	1	2	3		4	4	5	6	1	2	3
						5	5	6	1	2	3	4
						6	6	1	2	3	4	5

6.5.2e *Completely randomized two-factor design*

This design is used to measure the effects of two independent variables. It can be extended to three or more variables by continued blocking of the variables so that blocks are split into additional blocks of another variable. Each measurement (x_{ij}) is taken on a different individual. The individuals are randomly assigned to the combinations of variables.

One sample

Tabular form   One sample each of independent variables A and B

S_{A1}	S_{B1}
x_{a1}	x_{b1}
x_{a2}	x_{b2}
⋮	⋮
x_{an}	x_{bn}

Table 6.12. *Example of a completely randomized two-factor design with one sample for each variable illustrted with hypothetical data for larval cichlids (see text for explanation)*

Raised in presence of:	
Predator odor S_{A1}	Predator Visual stimuli S_{B1}
213 (x_{a1})	187 (x_{b1})
162 (x_{a2})	219 (x_{b2})
⋮	⋮
235 (x_{a10})	174 (x_{b10})

Example   In the study of choice by larval cichlids, we might want to test the effect of being raised in the presence of the odor of a predator versus being raised where the predator can only be seen on the reaction time of the cichlids (time to make a choice). We would randomly select individuals to be raised under these two conditions and then randomly select 10 individuals from each group at age two days free-swimming for testing. They would be tested with neither predator odor nor visual stimuli present.

In this hypothetical example (Table 6.12), the second individual from the group raised in the presence of predator odor (x_{a2}) took 162 seconds to choose one of the parents.

Two samples of two independent variables
Tabular form

Samples of independent variable A

Samples of independent variable B	S_{A1}	S_{A2}	Variable B means
S_{B1}	x_{11}	x_{21}	$\bar{x}_{.1}$
S_{B2}	x_{12}	x_{22}	$\bar{x}_{.2}$
Variable A means:	$\bar{x}_{1.}$	$\bar{x}_{2.}$	$\bar{x}_{n.}$

Grand mean: $\bar{x}_{..}$

Example   In the experiment with choice by larval cichlids which we have been using as an example, we decide to look at the combined effect of color morph and

Table 6.13. Example of a completely randomized two-factor design with two samples illustrated with hypothetical data for whether larval cichlids of two color morphs and raised under two conditions chose to stay in close proximity to one of their parents

	Independent variable A		
	Raised with predators ($N=40$) S_{A1}	Raised without predators ($N=40$) S_{A2}	
Color morph Totals			
Gold S_{B1} ($N=40$)	16 (x_{11})	4 (x_{21})	20
Wild type S_{B2} ($N=40$)	17 (x_{12})	10 (x_{22})	27
Totals	33	14	47

being raised with predators of young on whether age two day free-swimming larval cichlids choose to stay in close proximity of a parent, or not.

In this hypothetical example (Table 6.13), 47 of the 80 cichlids tested chose to stay in close proximity of one of the parents; 33 of those were raised in the presence of predators, and 27 were of the wild type. Seventeen had both features; that is, wild type raised in the presence of predators.

Three, or more, samples of two independent variables
Tabular form

	Samples of independent variable A				
Samples of independent variable B	S_{A1}	S_{A2}	S_{A3} ... S_{An}		Variable B means
S_{B1}	x_{11}	x_{21}	x_{31} ... x_{n1}		$x_{.1}$
S_{B2}	x_{12}	x_{22}	x_{32} ... x_{n2}		$x_{.2}$
S_{B3}	x_{13}	x_{22}	x_{33} ... x_{n3}		$x_{.3}$
⋮	⋮	⋮	⋮	⋮	⋮
S_{Bn}	x_{1n}	x_{2n}	x_{3n} ... x_{nn}		$x_{.n}$
Variable A means:	$\bar{x}_{1.}$	$\bar{x}_{2.}$	$\bar{x}_{3.}$	$\bar{x}_{n.}$	

Grand mean: $\bar{x}_{..}$

Table 6.14. *Example of a completely randomized two-factor design with three samples illustrated with hypothetical data for whether larval cichlids of three color morphs and raised under three conditions chose to stay in close proximity to one of their parents*

	Independent variable A		
	Raised with predators of larvae ($N=60$)	Raised without predators ($N=60$)	Raised with predators of adults ($N=60$)
Color Morph	S_{A1}	S_{A2}	S_{A3}
Gold S_{B1} ($N=60$)	16 (x_{11})	4 (x_{21})	12 (x_{31})
Wild type S_{B2} ($N=60$)	17 (x_{12})	10 (x_{22})	14 (x_{32})
Green S_{B3} ($N=60$)	13 (x_{13})	7 (x_{23})	10 (x_{33})

Example   As an example of this design we will add to the example given above (see Table 6.14). We will add an additional color morph (green) and an additional condition under which the young are raised (with predators of adults).

Linear model   With the completely randomized design with two independent variables we have included the effect of independent variable A (α_i), independent variable B (β_j) and their interaction (γ_{ij}).

$$x_{ij} = \mu + \alpha_i + \beta_j + \gamma_{ij} + \epsilon_{ij}$$

6.5.3 Random, haphazard and opportunistic samples

In the experimental designs described above, it is assumed that individuals (or groups) selected to receive treatments, or selected for observation and behavioral measurements, are randomly selected from the population. In fact, it is usually very difficult (if not impossible) to accomplish this when conducting ethological research in the field. Instead, we attempt to sample a large number of individuals and select individuals in a way that we believe results in a *reasonable approximation* of a random sample. The researcher must attempt to avoid a biased sample from a population. For example, Sharman and Dunbar (1982) examined 24 field studies of baboons and found that the researchers tended to select the largest group available to them as their study group; they point out that this will likely introduce significant biases in analyses whenever group size is an important independent variable.

When selecting individuals for a randomized design or selecting random samples of behavior from an individual, it is important to understand the difference between *random, haphazard* and *opportunistic* samples.

- *Random sample*: An individual selected from a population, or a measurement taken from an individual, in such a way that all possible samples have the same probability of being selected. For example, all individuals in a population could be assigned numbers and then the sample of individuals to be observed could be randomly drawn from a hat.
- *Haphazard sample*: A sample taken on some arbitrary basis, generally convenience. For example, we might take measurements on the individuals in a group that are closest to us, or we might take measurements before lunch and after dinner, or when we think the animals are likely to be most active.
- *Opportunistic sample*: A sample of behavior taken when it occurs in any individual in the population being observed. The individual selected, then, is based on the behavior which it is performing. For example, we might be studying an unusual behavior in a species that forms large feeding aggregations; we watch all the individuals until we see the behavior occur, then we record our observations on that individual performing that behavior (all-animal all occurrences sampling, Chapter 8).

Some researchers use the term *random* to refer to the way in which their samples were collected, when really what they collected were *haphazard* samples; for example:

> This study is based on facts gleaned for the most part from random
> observations.   *[Lorenz, 1935:90]*

In some cases, research by early ethologists suffered from a lack of knowledge about experimental designs. For example, Tinbergen admitted that in his earlier studies with grayling butterflies, they had presented their models in a haphazard rather than random sequence.

> We varied the sequence of the models irregularly (not being
> sophisticated enough to vary them in a random way).
>   *[Tinbergen, 1958:182]*

Although true random sampling is often difficult in ethological field studies, it is an assumption of most statistical tests. Therefore, researchers should combat the temptation to take the easy way out and sample haphazardly; they should make every attempt to achieve or approximate random sampling. S. Altmann (1965a:494) '. . . tried to sample at random from among the monkeys. However, no systematic randomizing technique was used'.

Ethologists often sample *opportunistically* from a large number of animals that cannot be recognized individually, especially if their interest is in a particular behavior, or sequence of behaviors. For example, Cresswell (1993) studied escape responses by redshanks (*Tringa totanus*) in response to avian predators by making opportunistic observations of attacks by sparrowhawks, peregrines and merlins on individuals in a large population of redshanks.

Opportunistic samples are neither random, nor completely haphazard. We should randomly sample individuals, but instead we sample based on behavior, and we do not know how many different individuals are included in the sample. We can only hope that with a large enough sample of occurrences of the behavior we will also have sampled a reasonably large number of individuals. Even if we could recognize individuals and randomly select a sample of individuals for observation, the chances of observing the behavior of interest would be remote without spending an inefficiently long period of time making observations. For example, in Creswell's (1993) study of redshanks, he observed 696 raptor attacks on a population of 200–500 individual redshanks, during 2557 h of observation. If Cresswell had randomly selected an individual for observation from a population of 200 redshanks, that redshank would have had a probability of approximately 0.0014 of being attacked during a one-hour observation period, if the attacks were randomly distributed throughout the redshank population.

A respectable attempt at random sampling different individuals can sometimes be made by stratifying your sample by sex or body size and attempting to sample randomly within those groups. If you are observing a large group of individuals, such as geese feeding in a field, you can divide the group spatially using an imaginary grid or marker stakes placed in the field before the geese arrive. You can then sample from different and distant portions of the grid assuming that an individual goose is unlikely to move that distance during your sampling period.

When individuals can be recognized, it may still be difficult to follow a random sampling protocol because of the availability or observabilty of specific individuals. To make efficient use of your time, you might then attempt to *equalize* the number and temporal distribution of observations among individuals. For example, you might obtain 10 one-hour samples of feeding behavior from each individual during each of three sample periods (early morning, noon, late afternoon) over the period of a month; however, this type of sampling protocol is often difficult to accomplish.

> I attempted to distribute observations evenly throughout the daylight hours so that for each bird during each age period . . . foraging

behaviour was sampled during the morning, midday, and evening. This
was not always possible.   *[Yoerg, 1994:579]*

For a particular sample period, if you could not collect data on the randomly
selected individual, you should select another and continue that process until equal
samples had been collected for each individual. Collins (1984) used instantaneous
samples on every baboon in order to estimate the amount of time each animal spent
in each part of the troop. However, he had difficulty randomly selecting the focal
animal, so he settled on the following procedure which combined opportunistic
sampling and equalizing samples:

> Sample subjects [focal animals] were selected in sequence from a
> predetermined random-order list, but some choice had to be introduced
> to speed up sampling. To do this, the list was divided into triplets, and
> the subject chosen was the first one seen of the next triplet provided that
> it (a) was more than 25 m from the site of the previous sample, and (b)
> had not been sampled during the previous hour nor more than once that
> day. Some animals were however given priority over others in their
> triplet (before the start of the day's sampling) if their sample total to
> date had lagged behind . . .   *[Collins, 1984:539]*

6.6 RELATIVE EFFICIENCY OF EXPERIMENTAL DESIGNS

A measure of the efficiency of an experimental design should include the cost (time
and money) of collecting the data balanced against the accuracy and validity of
those data. The relative efficiency of two designs is often assessed by comparing
their respective experimental errors (error effects). Experimental error is the extraneous variation in the measurements due to all the nuisance variables. Its ultimate
effect is to mask the effect due to the independent variable(s).

Federer (1955:13) proposed the following formula to measure efficiency:

$$\text{efficiency} = \frac{\left(\dfrac{n_1 c_2}{\hat{\sigma}_2^2}\right)\left(\dfrac{df_1+1}{df_1+3}\right)}{\left(\dfrac{n_1 c_1}{\hat{\sigma}_2^2}\right)\left(\dfrac{df_2+1}{df_2+3}\right)}$$

where n = number of subjects

   c = cost of data collection per subject

   df = experimental error degrees of freedom

   For completely randomized design:

   error df = Total df − samples df

where: total df$=n-1$

samples df$=$number of samples-1

$\sigma^2=$estimate of experimental error per observation The experimental error can be estimated by calculating the mean of the deviations of each measurement from its sample mean:

$$\left[\frac{\Sigma(x_{ij}-\bar{x}_{i.})^2}{n} \right]$$

The subscripts designate the two experimental designs. If the ratio is greater than one, then the first design is more efficient than the second. Because of the limited control ethologists generally have in field studies, compromises are usually necessary.

6.7 DETERMINATION OF SAMPLE SIZE

The *sample size* is the number of measurements in a sample. Since in most experimental designs each measurement is made on a different individual, sample size generally refers to the number of individuals (subjects) in the sample. Experimental psychology and psychophysical studies are sometimes based on single-subject (time-series) designs (Kazdin, 1982; Kratochwill, 1978) in which a very few subjects are exposed to a series of treatments over an extended period of time. In contrast, most studies in ethology require the use of several subjects.

Sample size needs to be adequate to provide sufficient power for your statistical test (Chapter 11), but it should not be excessive. In order to determine the sample size required for your particular study, you should have an estimate of the variability of the data. This can be determined by gathering some preliminary data, or using data gathered during reconnaissance observations or by other researchers, and then calculating the standard deviation of the measurements (see section 11.7.6). The following formula for determining required sample size (Snedecor, 1946) for a test of two means can then be used.

$$n=\text{number of samples required}=\frac{s^2\,t^2}{d^2}$$

$$\text{where } s=\text{standard deviation}=\sqrt{\frac{(x_{ij}-\bar{x})^2}{n-1}}$$

$t=$tabular 't' value (Table A5) at the selected confidence level (section 11.5), and for the degrees of freedom (p. 00) in your sample

$d=$margin of error (mean$\times$designated accuracy)

For example, let us say we want to determine the difference in mean durations of coyote howls given nocturnally and diurnally. We refer to our field notes and find that we have measured the durations of six individual nocturnal howls as follows:

5.2 s	3.1
6.1	7.2 mean=5.4 s
4.3	6.7

We calculate the standard deviation $(s)=1.5$.

The tabular t value for five degrees of freedom $(6-1)$ at the 0.95 confidence level$=2.571$.

We decide to accept a 0.05 level of accuracy, then:

$$n=\frac{(1.5)^2\,(2.571)^2}{(5.4\times0.05)^2}=\frac{(2.25)(6.61)}{(0.07)}=212$$

Therefore, we must obtain 212 samples of nocturnal coyote howls in order to have a reasonable estimate of the true mean duration. We would have to make the same calculations based on a sample of diurnal howls.

If you are unable to obtain an estimate of the variation to be expected in the data, then you should err on the side of a sample size which may be larger than necessary. *Statistics estimate population parameters from sample measurements*; therefore, the larger the sample size, the better the probability that the sample statistics will closely approximate the population values.

You can also determine the necessary sample size by solving for N in power formulas for specific statistical tests. Power formulas for two t-tests are described in Chapter 11; others can be found in various texts (e.g. Cohen, 1988; Zar, 1984).

You should not attempt to increase the sample size by increasing the number of observations on individuals already in the sample and then pooling those 'k' observations on 'n' individuals to create a larger sample size consisting of 'nk' observations; this is 'pseudoreplication' (Hurlbert, 1984) or the 'pooling fallacy' (Machlis *et al.*, 1985) which increases the probability of committing a Type I error.

Increasing sample size is not the only way to increase the power of your experiment (see discussion of power in Chapter 11). For example, Still (1982) has proposed several alternatives to consider including: 1. selecting a better experimental design; 2. using a larger alpha level (section 11.5); and 3. sequential methods.

7 Experimental research

In the previous chapter, description and experimentation were discussed relative to designing a research project. Description is the approach which uses naturalistic observation in order to construct an ethogram for a species as it behaves normally. In experimental studies, we test hypotheses about the relationships between independent variables (individual or environmental variables) and dependent variables (behavior units). We can either allow the independent variables to change naturally, or we can manipulate them.

In addition, if we make a distinction between field and laboratory studies, we can categorize ethological research along a continuum from descriptive field studies to manipulative laboratory experiments.

The key difference between field and laboratory studies is that, if everything else is equal, descriptive studies are more valid when conducted under natural conditions (i.e. usually the field), and experimental manipulations can best be controlled under laboratory conditions. Nevertheless, it is sometimes difficult with certain species to make naturalistic observations under field conditions. The researcher should then consider whether natural behavior of the species can be expected, and unobtrusive observations can be made, in a laboratory environment.

Even though we might want to shift our field studies into the more controlled captive or laboratory setting, some species cannot be easily and properly maintained in captivity. For example, Tinbergen, after reflecting on some of the shortcomings of his field studies on gulls, concluded:

> It would seem to be more efficient to try to improve the field methods than to try to keep a large colony of gulls under laboratory conditions.
>
> *[Tinbergen, 1958:251]*

Not only is it sometimes difficult to move field studies into the laboratory, but conditions in the field often make manipulations very difficult. After reviewing ethological and behavioral ecology studies of African ungulates, Leuthold (1977) concluded that,

> Experiments have rarely been carried out so far, partly because much descriptive work was required at first, and partly because of the physical difficulties of manipulating wild ungulates in experimental situations.
>
> *[Leuthold, 1973:13]*

Research questions may require that measurements and manipulations be made in the field in order to be valid. This sometimes becomes evident when the same experiments are conducted in both the field and the laboratory. As an example, McPherson (1988) tested fruit preferences of cedar waxwings (*Bombycilla cedrorum*) (categorized by species, color and size) in both the field and laboratory; the differences were explained as follows: 'The lack of complete agreement between preferences for fruits in the field and in the laboratory suggests that factors important in the field but controlled in the laboratory (e.g. abundance, location) override preferences for certain fruits' (McPherson, 1988:961).

As a general rule, all ethological studies should be conducted in the field when feasible; this is especially true for descriptive studies and mensurative experiments. Changing the emphasis of your research from description in the field to experimentation in the field is a natural progression.

> The observational work has to be followed up by experimental study. This can often be done in the field. The change from observation to experiment has to be a gradual one. The investigation of causal relationships has to begin with the utilization of 'natural experiments.' The conditions under which things occur in nature vary to such a degree that comparison of the circumstances in which a certain thing happens often has the value of an experiment, which has only to be refined in the crucial tests.   *[Tinbergen, 1953:136]*

Descriptive studies and mensurative experiments in captivity, or the laboratory, will likely be more efficient, but less valid than field studies. Likewise, manipulative experiments should be carried out in the species' normal habitat when they can be conducted in a valid manner. Nevertheless, the controlled laboratory setting often plays an important role in ethological studies; several examples are given in sections later in this chapter.

Peeke and Petrinovich (1984) can be consulted for another discussion of the relative advantages and disadvantages of field and laboratory studies in animal behavior. Also, Mertz and McCauley (1980) present a similar discussion for ecological studies which can also be applied to ethological research.

7.1 THE VARYING VARIABLES

The objective of an experiment is to test one, or more, hypotheses about the relationships between variables. These may be cause-and-effect relationships between independent and dependent variables, or correlational relationships between independent variables. The relationships may be long term (evolutionary and ontogenetic) or short term (proximal, Chapter 2).

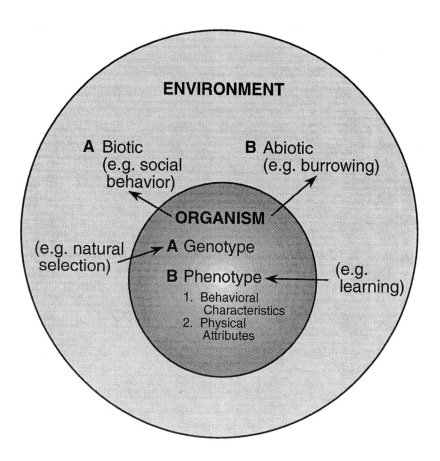

Fig. 7.1 Interactions between an organism and its biotic and abiotic external
environment. Arrows indicate factors from the environment acting on the
organism and vice versa, (see text for further explanation; drawing by Lori
Miyasato).

To understand which variables may affect an organism's behavior, we must first
recognize that an individual exists in time and space in a dynamic state, continually
under the influence of its environment, continually imposing its own effects upon
the environment, and behaving as a result, in part, of its evolutionary and ontogenetic history. Figure 7.1 is a simplified diagram of the relationship between an
organism and the environment (also see Model, Chapter 2). The environment has
both biotic (biological) and abiotic (physical) features, some of which will affect the
organism. As examples, biotic features may include vegetation type in habitat selection, interspecific predator effects on prey behavior and intraspecific courtship
signals on mate selection. Abiotic features may be important in regulating a species'

temporal activity patterns, such as circadian rhythms and seasonal cycles. That is, temperature, humidity, wind and light levels may all be important in determining the activity patterns shown by an animal or group of animals.

As discussed in Chapter 2, the organism's physical and behavioral phenotype bears the mark of natural selection on the genotype over many previous generations. The phenotype includes the action and interaction of the genotype and environment (including experience) and the animal's anatomy and physiology. The organism can exert forces on both the biotic environment (e.g. intra- and interspecific social behavior) and the abiotic environment (e.g. a badger burrowing into a hillside).

Since all the factors in Figure 7.1 are variables that have a potential effect on all behavior, the researcher must be judicious when selecting the one (or few) variable to study at any one time. The complexity of the interactions that can result from two, or more, variables must be recognized and dealt with as skillfully as possible within the limits available to the researcher (e.g. experimental designs in Chapter 6). For example, Prinz and Wiltschko (1992) studied the effect of the interaction of stellar and magnetic information during ontogeny on the migratory orientation of pied flycatchers (*Ficedula hypoleuca*).

The first step is to list all the variables that are known to affect the behavior in question or are suspected of having some effect. Some examples of the different types of variable are listed below:

I Environmental Variables
   A Biotic
      1 Members of social group
      2 Predator–prey relationships
      3 Vegetative characteristics of habitat
   B Abiotic
      1 Temperature
      2 Wind
      3 Humidity
      4 Cloud cover
      5 Topography
      6 Time–circadian and seasonal

II Organismal variables
   A Genotype
      1 Sex
      2 Parent stock
   B Phenotype
      1 Behavioral characteristics

    a Description of behavior (general categories, types, patterns and
       behavioral acts)
    b Frequency
    c Rate
    d Duration
    e Temporal patterning–circadian and circannual
    f Spatial characteristics
  2 Physical attributes
    a Morphological characteristics; e.g. shapes, color patterns
    b Physiological characteristics

This procedure of attempting to list all the important variables is useful not only when you are thinking about the potential causation of a particular behavior in anticipation of designing a study, but also when you have already decided on the variable(s) you want to measure or manipulate and want to account for other potential sources of variation (e.g. eliminate or measure them). The number of variables that could potentially affect a behavior is extremely large; therefore, the researcher should be willing to spend time compiling the list. Important variables that you overlook can often be identified by other ethologists; therefore, you should enlist the help of your colleagues in identifying other variables, as well as concurring on the variables of most concern for measurement, manipulation or control. The importance of consulting colleagues is illustrated by Wehner and Rossel's (1985) account of how Karl von Frisch came to make his dramatic discovery that bees use polarized light from the sky as a celestial compass.

> . . . two decades earlier . . . FELIX SANTSCHI had already observed that ants could find their way even when they could see nothing but a small patch of blue sky. In an interesting but unfortunately neglected work . . .SANTSCHI (1923) literally asked the question 'What is it in this small patch of sky that guides the ants back home?' SANTSCHI . . . could not tell. In one experiment he had even used a ground glass disk (which depolarized the light from the sky) and put it above a homing ant. The ant instantly stopped and searched around at random, but SANTSCHI did not draw the right inferences from this important observation. . . . After a quarter of a century had passed, in 1947, VON FRISCH did an experiment with bees almost identical to the experiment SANTSCHI had performed with ants. He got the same result, asked the same question, and – horrible dictu – could not tell either. However, VON FRISCH, then Head of the Department of Zoology at the University of Graz, was in a better position to answer such questions than SANTSCHI had ever been. At one of the next Faculty Meetings he told the story to his colleague of the Physics Department, HANS

> BENNDORF. The physicist advised him to check for polarized light.
> Next summer, in 1948, VON FRISCH did the crucial experiment: He
> placed a polarizer above a bee which performed its recruitment dances
> on a horizontal comb, and as he rotated the polarizer, the direction of
> the bee's dances changed correspondingly (VON FRISCH 1949). This
> was the first demonstration that an animal used skylight polarization
> for adjusting the direction of its course.   *[Wehner and Rossel, 1985:13]*

After identifying the variable of interest, the usual procedure is either to manipulate that variable systematically, as von Frisch did with light polarization, or to
follow it through natural changes, measuring both the variable and the behavior of
interest. The other variables you have identified as potentially having an effect must
remain constant or vary randomly, so that they can be considered to have no systematic effect on the behavior being studied. The variable being manipulated is the
independent variable (e.g. light polarization), and the behavior being measured is
the dependent variable (e.g. orientation of the bee's waggle dance; see Chapter 6 for
a further discussion of manipulating variables).

Several variables may be manipulated and/or measured simultaneously in order
to determine both individual effects and interactions. Selected analyses of this type
are discussed under multivariate analyses in Chapter 16.

When planning your experiments, always keep in mind how the various results
will (could) be interpreted.

> Bourbon on the rocks, scotch on the rocks, vodka on the rocks, gin on
> the rocks all can make you drunk – must be the ice cubes.
>   *[V. DeGhett pers. commun.]*

7.1.1 Natural variation

The first step in experimentation is to obtain clear descriptions and definitions of
the behaviors to be measured (Chapter 4). This requires obtaining those descriptions from secondary sources (Chapter 4) and/or making your own observations of
the behaviors under conditions of naturally occurring changes in the biotic and
abiotic environment. Even if you use descriptions and definitions from other
researchers, you should still gain experience in observing the behavior before conducting experiments.

Mensurative experiments make use of natural changes in the environment to
study their effects on the behavior of selected species. For example, Pengelley and
Asmundson (1971) demonstrated that the yearly activities of golden-mantled
ground squirrels (*Spermophillus lateralis*) fluctuated in synchrony with climatological variables in the environment. Foraging activity of the nocturnal bee
(*Sphecodogastra texana*) was shown by Kerfoot (1967) to be based on the lunar

cycle. Sunrise and sunset apparently trigger the onset and cessation of activity in cottontail rabbits (*Sylvilagus floridanus*) and snowshoe hares (*Lepus americanus*) (Mech *et al.*, 1966).

Some environmental factors fluctuate within seasonal ranges but vary somewhat irregularly from day to day. For example, decreasing light levels near sunset apparently trigger the initial departure towards the roost of foraging starlings (*Sturnus vulgaris*; Davis and Lussenhop 1970). Nisbet and Drury (1968) compared measurements of the density of songbird and waterbird migration to 19 weather variables in the area of takeoff. They found that migration densities were significantly correlated with high and rising temperature, low and falling pressure, low but rising humidity, and the onshore component of wind velocity.

The response of animals to simultaneous variations in the environment can also be studied. For example, Heinrich (1971) examined the feeding pattern of the caterpillar (*Manduca sexta*) and found that it was consistent for given leaf shapes and sizes. Simultaneous variation has also been the basis for many field studies of habitat selection. MacArthur (1958), for example, studied the distribution of five congeneric species of warblers while they fed on individual white-spruce trees. He divided the trees into 16 zones and measured the percentage of the total number of seconds of observation and the percentage of the total observations for each species in each zone. He found that the five species distributed themselves on the trees such that utilized different microhabitat variables (Figure 7.2).

The age and experience of the animals under investigation can be allowed to advance naturally and their behavior observed at various stages. Scott and Fuller (1965), for example, observed the changes in behavior of several domestic dog breeds from birth to maturity and were able to divide their development into four periods: neonatal, transition, socialization and juvenile. Development of behavior in the song sparrow (*Melospiza melodia*) was divided into six similar stages by Nice (1943). Drori and Folman (1967) showed a marked effect of experience on the copulatory behavior of male rats, and Carlier and Noirot (1965) demonstrated that experience improved pup retrieval in female rats. Stefanski (1967) showed that the average territory size of black-capped chickadees (*Parus atricapillus*) varied during six stages of the breeding season: prenesting, nest building, egg laying, incubation, nestling, and fledgling.

The use of natural variation has limitations which are both qualitative and quantitative. Waiting for the proper conditions to arise and attempting to gather a sufficient number of observations sometimes drives the ethologist to artificial manipulation:

> Systematic exploitation of such natural experiments – that is, systematic comparison of the situations which do and those which do not release a given response – can be almost as good as planned experiments; the

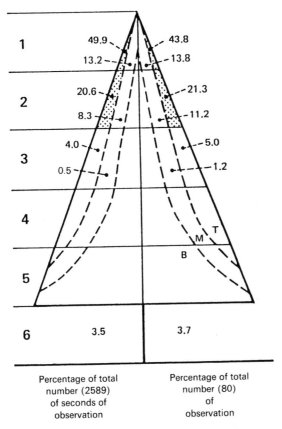

Fig. 7.2 Cape May warbler feeding positions. At least 50% of the activity is in the stippled
zones. Each branch was divided into three zones: B bare of lichen-covered base,
M, old needles; and T, new (less than 1.5 years old) needles and buds. (from
MacArthur, 1958).

important thing seems to me is not to miss the natural experiments and
yet to know when it becomes necessary to continue by planned tests.

[Tinbergen, 1958:289]

7.1.2 Artificial manipulation

Another experimental approach to the study of cause and effect of behavior is to
take control of the variables and manipulate them in the field or the laboratory.
Although the manipulation is artificial, every attempt should be made to approximate the normal stimuli and the natural changes as closely as possible.

7.1.2a *Elimination, disruption and manipulation*

When manipulating the animal or exogenous stimuli to answer questions about 'how' an animal performs a behavior there are different levels of intervention which lead to differing degrees of validity of results. For example, in determining the exogenous stimuli and corresponding sensory systems (or endogenous stimuli and corresponding hormonal/neuronal systems) involved in a behavior you can *eliminate, disrupt or manipulate* variables. These interventions (all of which are usually referred to as 'manipulations') can be made on the stimuli or the animal's anatomy (e.g. sensory system). Elimination, disruption and manipulation represent decreased levels of perturbation, respectively; in general, manipulation provides more rigorous and valid results than does elimination or disruption (see examples of bird orientation/migration studies later in this chapter). When stimuli are manipulated (e.g. von Frisch changing the plane of polarized light received by a dancing bee, p. 155), you can make predictions about the resultant behavior (e.g. orientation of the bee's dance will track with the changing plane of polarization); that is, you can invoke research and statistical hypotheses with higher resolution and greater statistical power (Chapter 11). With elimination and disruption you are only attempting to eliminate or disrupt the behavior being studied (often in an unpredictable manner). For example, in attempting to locate the circadian pacemaker it was known that surgical ablation (elimination) of the suprachiasmatic nucleus (SCN) in the brains of mammals eliminated overt behavioral rhythmicity; those experiments provided some evidence for the SCN being the pacemaker. However, conclusive evidence was provided when Ralph *et al.* (1990) conducted transplantation experiments with normal hamsters and a mutant strain with a short circadian period. They demonstrated that small neural grafts from the SCN of donor hamsters restored circadian rhythms to arhythmic hamsters whose own SCN had been ablated; the restored rhythms always exhibited the period of the donor genotype, normal or mutant (short).

If you eliminate or disrupt stimuli in order to determine their role in eliciting or orienting a behavior, the behavior may come under the control of other stimuli and sensory systems. Thus, when the behavior does not disappear, or is not disrupted, you could draw incorrect conclusions. Indeed, this is what occurred in early experiments designed to determine the environmental cues used for orientation by foraging bees and migrating birds (Gould, 1982). For example, if you were an ethologist in the early part of the century and were interested in the environmental cues that homing pigeons use to orient back to the home loft, you might have designed experiments to eliminate or disrupt potential cues. If you hypothesized that pigeons use the sun as a compass and tested them on overcast days (to *eliminate* the sun as a cue), the pigeons would still have homed using the earth's magnetic field as the cue.

Since the pigeons homed successfully, you might have concluded that they don't use the sun as a compass and therefore hypothesized that they use the earth's magnetic field. If you then attached bar magnets to their backs (to *disrupt* the magnetic field around them), but tested them on a sunny day, they would still have homed, but this time they would have been using the sun as a compass. Further, unless you recognized that you should have been controlling more than one of the variables at a time, you might conclude that the pigeons use neither the sun nor geomagnetic field as compass cues. Even if you recognized your design error and proceeded to disrupt their orientation by applying magnets and testing the pigeons on overcast days, you would not have as conclusive results as you could have obtained by *manipulating* the variables and predicting the changes in orientation (see experiments described in section 7.3).

When you eliminate or disrupt an animal's sensory system you also run the risk of affecting other anatomical and physiological systems which could be important for the behavior(s) you are measuring. The following story illustrates how attempts to eliminate a sensory system can have additional effects on the animal's behavior and the researcher's ability to interpret results:

> A zoology student had succeeded in training cockroaches, and he proudly displayed the results of his long efforts to his professor.
>
> He had his cockroaches fall in, and he gave them the command: 'Forward, march!' the cockroaches marched forward. 'Column left!' the student commanded, and all the cockroaches turned left.
>
> The professor was about to congratulate the student on this remarkable accomplishment, but the student interrupted him. 'Wait!' he said. 'I still have to show you the most important thing.'
>
> The student picked up a cockroach from the last row, pulled off its legs, and put it back in its place. Once again he commanded: 'Forward, march!'
>
> The cockroaches marched as before, except, of course, for the one without legs. 'Column left.' Again, all the cockroaches turned on command, except for the one that lay where it had been placed.
>
> The professor looked inquiringly at the student.
> The student said proudly, 'This experiment proves conclusively that cockroaches hear with their legs.'   *[Eigen and Winkler, 1981:298–299]*

This tale gives rise to three important questions: 1. Was the manipulation appropriate to obtain valid results? If so, 2. Was this severe a manipulation necessary to answer the research question? If so, 3. Was the answer to the research question worth making this severe a manipulation? Since the answer to questions 1 and 2 is 'No', you can conclude that the student was either a naive or sadistic researcher. If we give the student the benefit of assuming they were only naive and insensitive, we should recommend that they answer those questions before making any manipulation in their next experiment.

Manipulation of variables (versus elimination or disruption) is the method being employed when the researcher uses models and dummies, or conditioning (all are discussed below).

7.1.2b Models and dummies

Models constructed to mimic animals, or parts of animals, and *dummies* (stuffed skins of animals) have a long history of use in ethology. Dummies were used by Allen (1934) in his study of the courtship of ruffed grouse (*Bonasa umbellus* L.), in Chapman's (1935) study of courtship in Gould's manakin (*Manacus vetellinus vitellinus*), and by Lack (1943) in his study of aggression in robins (*Erithacus rubecula*; also see Table 8.3). Models and dummies have the advantage of allowing the experimenter to vary stimuli (e.g. visual, auditory, chemical, tactile) in a systematic way in order to measure the effect of qualitative and quantitative differences. Tinbergen was an early and exemplary proponent of the use of models (Dawkins *et al.*, 1992).

As a typical example, Tinbergen and Perdeck (1950) presented models of an adult herring gull's head to herring gull chicks. They found that the color of the spot on the bill (qualitative property) of the model had an effect on the number of pecks given by the chicks (Figure 7.3A). Tinbergen and Kuenen (1939) used simple models to demonstrate that the gaping response of nestling blackbirds (*Turdus merula merula*) and thrushes (*Turdus ericetorum ericetorum*) is oriented by the relative size of the parent's head to their body (Figure 7.3B).

Moller (1987) studied the role of badge size (extent of dark coloration on the throat and breast) on status signaling in house sparrows (*Passer domesticus*) by placing stuffed male house sparrows (dummies) near nests. Stout and Brass (1969) placed pairs of dummies, or wooden-block models with tiltable bodies and adjustable stuffed heads (Figure 7.4), in glaucous-winged gull territories; they demonstrated that the head and neck are the parts of the body that release territorial aggression displays in this species.

Some researchers have incorporated movement and/or odors and sound into their models and dummies. For example, Esch (1967) used a wooden, motor-driven model in his research on communication of food source location in honeybees. The model was the approximate size of the honeybees being studied, but it didn't closely resemble them physically; this probably had little effect since the experiments were carried out in a dark hive. The model did have the identical odor of the hive's inhabitants and performed a 'normal' waggle dance, but no bees left the hive to search for food in the direction proclaimed by the model's dance. Esch concluded that some-

Fig. 7.3   Above. A cardboard model of a herring gull head being presented to a chick
(adapted from Tinbergen 1960b by Lori Miyasato). Below. Presentation of
models of the parents' head, body and tail to study the relationships that orient
nestlings' gaping response (adapted from Tinbergen 1972 by Lori Miyasato).

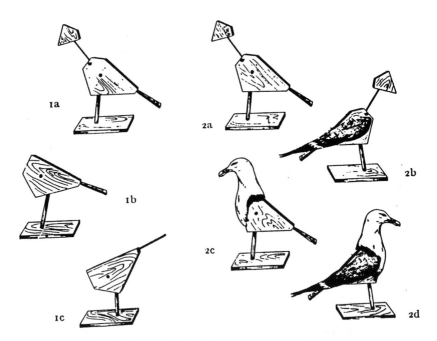

Fig. 7.4 Models and a dummy (2d) used by Stout and Brass (1969) in their study of
glaucous-winged gulls. 1a, upright, threat-postured body 1b, trumpeting-
postured body; 1c, choking-postured body; 2a, basic wooden model; 2b, upright
threat posture; 2c, model without wings; 2d, dummy showing upright threat
posture with wings.

thing more than the dance was necessary to elicit foraging. More recent research
used a motor-driven model bee which not only danced, but also vibrated artificial
wings and exuded sugar-water samples; this dummy bee was much more successful
in recruiting foragers (Moffett ,1990).

Hunsaker (1962) used a head-bobbing machine to move the model heads of
lizards (*Sceloporus* sp.) in different species-typical patterns. He found that females
selected those models which head-bobbed in the pattern typical of their own species.
Jarvi and Bakken (1984) used three dummy great tits to study the function of the
variation in the breast stripe. Their dummies could be turned 360°, by radio control,
to keep them always oriented in the direction from which the live birds approached.

Models should contain the important features of the live animal, and they
should be used in a normal context (see Curio 1975 for an excellent example of
extensive and proper use of models). In other words, 'an underlying assumption of
the method is that response to the model depends on much the same causal system
as response to the natural stimulus' (Losey, 1977:224). Unfortunately, this assumption is rarely validated. However, in Losey's experiments on the response of host fish

(*Chaeton aurign*) to a cleaner (*Labroides phthrirophagus*), he demonstrated the validity of his use of a cleaner model through three indicators: pose duration, pose-to-inspect ratio, and approach behavior of the host fish to both live cleaners and his models.

Not only must the use of models and dummies be carefully planned, but the results of such experiments must be carefully interpreted. As an example, in another aspect of Tinbergen and Perdeck's (1950) experiments on the begging response in neonatal herring-gull chicks, they changed the position of the red spot from the model's bill to its forehead. The chicks delivered significantly more pecks to the model with the spot on the bill than they did to the model with the spot on the forehead. They concluded that it was the position of the red spot on the head that caused the decrease in the chicks' responses. Hailman (1969) re-investigated this phenomenon by placing the models at different distances from the pivot point of the rod holding the model. Further, he adjusted the height of the chick so that it was always at eye level with the red spot. He had created three models: a 'normal model' with the spot on the bill, a model with the spot on the forehead and the pivot point the same ('slow model'), and a model with the spot as on the bill-spot model ('fast model'). The fast forehead-spot model received more pecks than the slow forehead model, although fewer than the 'normal model,' revealing the effect of speed of the red spot on the chicks' responses. Therefore, Tinbergen and Perdeck (1950) were correct in concluding that position of the spot is important; but Hailman demonstrated that speed of the spot is also a contributing factor.

Models and dummies should be used with appropriate caution. They may be either too simple with the important stimuli absent, or too complex with extraneous stimuli confounding the experiment. As with any tool, however, in the hands of a skilled researcher, models and dummies can be an important means of manipulation in the field.

7.1.2c *Instrumental and classical conditioning*

An important technique for manipulating variables in the laboratory is through the use of instrumental and classical conditioning. Conditioning is a powerful method for studying 'causation' and answering 'how' questions; the basic paradigms for instrumental and classical conditioning were discussed in Chapter 2.

Conditioning is the basis for many psychophysical studies designed to determine 'how' a species discriminates between various stimuli. For example, May *et al.* (1988) studied how Japanese macaques (*Macaca fuscata*) discriminate between different coo vocalizations by using *instrumental conditioning* to train individual macaques to discriminate 'smooth early high' and 'smooth late high' coo sounds. The macaques were trained to make hand contact with a metal cylinder in response to one type of vocalization and release contact in response to the other vocalization. First, generalization tests showed that the macaques responded appropriately to both natural and computer-synthesized coo sounds. Then acoustic features were systematically removed from the computer-synthesized sounds to determine the minimal elements necessary for the macaques to recognize them as distinct coo sounds. Pietrewicz and Kamil (1977) studied the ability of blue jays (*Cyanocitta cristata*) to detect cryptic moths by *instrumentally conditioning* them to respond differentially to the presence and absence of moths in projected images (slides). If the projected slide contained a moth, 10 pecks on the stimulus key resulted in the blue jay being positively reinforced with half a mealworm. The jays were able to detect the moths, but their ability was affected by the background upon which the moth was placed and the moth's body orientation. In a later study, Pietrewicz and Kamil (1979) used the same *instrumental conditioning* procedure to study search image formation in blue jays.

Often questions about 'how' an animal uses environmental cues begins with studies of what a species 'can' perceive (Miller 1985); that is, what stimuli they are capable of perceiving and responding to. For example, Lehner and Dennis (1971) hypothesized that waterfowl might use atmospheric pressure changes as a cue for orientation during migration. They used *instrumental conditioning* to train mallard ducks, in a barometric pressure chamber, to peck one microswitch when the pressure increased and another microswitch when the pressure decreased. They then exposed the ducks to sequentially smaller changes in pressure and demonstrated that the ducks could perceive atmospheric pressure changes as small as 0.4 psi. Kreithen and Keeton (1974a) used *classical conditioning* to test the capabilities of homing pigeons to detect atmospheric pressure changes. The procedure was to place the pigeons individually in an airtight chamber, change the pressure over a 5 second interval (neutral stimulus), hold the pressure steady for the next 5 seconds, and then deliver electric shock (unconditioned stimulus) to the pigeon, causing the heart rate to increase (unconditioned response). After a few presentations, the pressure change became a conditioned stimulus that caused the heart rate to increase (conditioned response) without the electrical shock being administered. Then, the pigeons' perception of different amounts of pressure change was determined by observing changes in their heart rate. They determined that the homing pigeon is able to detect atmospheric pressure changes of 10 mm of H_2O, or lower. Kreithen and Keeton (1974b) used the same *classical conditioning* procedure to determine the ability of homing pigeons to detect polarized light, a cue used by bees in orientation.

As part of their research on coyote predation, Horn and Lehner (1974) wanted to determine the lowest environmental light levels that coyotes, which hunt primarily at night, could perceive. Coyotes were individually trained to stand in a dark test chamber (Figure 7.5) and face a stimulus light projected on an opaque plastic disk

Fig. 7.5 Coyote in test chamber used by Horn and Lehner (1975) to determine the coyote's scotopic (dark adapted) light sensitivity. A stimulus patch is at the coyote's eye level at the center of the right wall; it is not illuminated in this photo. Two foot treadles are on the floor, separated by a plexiglas partition.

at the coyote's eye level. They were then instrumentally conditioned to step on a foot treadle to their right when the light was on, and a treadle to their left when it was off. Once they consistently performed this discrimination task, then intensity of the light stimulus was put under the control of the coyotes. When they stepped on the right foot treadle (indicating they could perceive the light stimulus), the light automatically decreased in intensity; conversely, stepping on the left treadle (indicating they could not perceive the light) automatically increased the light intensity. The intensity of the light stimulus was continuously recorded resulting in a graph of the coyotes' psychophysical threshold for vision at night.

Instrumental conditioning has been an important technique in studies of foraging strategies. As an example, Ha *et al.* (1990) *instrumentally conditioned* caged gray jays (*Perisoreus canadensis*) to alternate hops on two perches in order to receive food pellets. The jays could forage in two 'food patches', each of which had two perches and a pellet dispenser (Figure 7.6). The food pellets were delivered on variable ratio schedules (VR; see Chapter 2) in both patches. Both VR schedules had the same mean (e.g. mean of 40 perch hops=VR40), but one patch had a high variance about the mean and the other a low variance. The gray jays chose to forage preferentially in the high variance food patch.

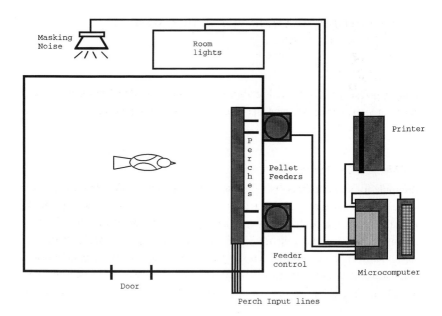

Fig. 7.6 Diagram of the instrumental conditioning apparatus used to study gray jay foraging strategies. The two patches each consisted of two perches attached to microswitches, a hole through which dispensed pellets were reached, and an automatic pellet dispenser. A microcomputer recorded perch hops and operated the feeders, as well as controlling lights and backgound noise (from Ha *et al.*, 1990). Copyrighted by Academic Press.

Laboratory research using classical or operant conditioning provides the opportunity to manipulate variables very precisely and measure the animal's resultant behavior very accurately. The drawback is that we don't always know how to translate these laboratory results into what the animal actually does in its normal environment. We only determine what the animal 'can' do; we are not sure whether that is 'how' they normally do it.

If you are interested in more detailed information on specific conditioning methods, you should consult the primary literature for papers reporting on research similar to what you are planning. Also, there are several good text books on learning and experimental psychology that present the basic methodology (e.g. Davey, 1981; Iverson and Lattal, 1991).

7.2 FURTHER EXAMPLES OF EXPERIMENTAL MANIPULATION

7.2.1 In the field

Many experiments arise from descriptive studies in the field and progress through mensurative experiments to artificial manipulation of the animal and/or its environment.

7.2.1a Manipulation of the animal

Manipulation of the animal involves altering the anatomy and/or physiology of the animal (A and P in the model in Chapter 2). For example, the role of sensory receptors and physiological state can be studied by manipulation of the animal *per se*. Layne (1967) studied the role of vision in diurnal orientation of bats (*Myotis austroriparius*) by releasing normal, earplugged, and blinded bats (two types of sensory elimination) at various distances from the home cave. None of the eye-covered bats homed, suggesting that vision is an important in homing behavior. Ehrenfeld and Carr (1967) measured the role of vision in the sea-finding behavior of female green turtles (*Chelonia mydas*) by blindfolding them or fitting them with spectacles containing different filters (elimination and disruption of the visual sense). Blindfolded turtles and those wearing red, blue, and 0.4 neutral density filters had significantly reduced orientation scores.

Morphological changes are occasionally made on animals in the field, and the effect on the animal's ability to obtain and/or retain a mate, social status or a territory is then measured. In these studies, it is the change in behavior of other individuals that engage in interactions with the altered individual which is usually being measured; but an effect can also often be found by observing the altered individual. As an example, Bouissou (1972) showed that dehorning and reduced weight decreased the ability of domestic cattle to obtain and maintain high social rank in the herd. Harris sparrows (*Zonotrichia quereula*) signal their dominance status by variations in the amount of black feathering on their crowns and throat. Rohwer (1977) ranked individuals into 14 'studliness' categories (Figure 7.7) and then altered the amount of black feathering on selected individuals to determine the effect on their status. Subordinates dyed to mimic the highest ranking birds were still persecuted by legitimate 'studlies,' and bleached birds eventually exerted their normally high-ranking dominance. The data suggested that 'cheating' (i.e. lower-ranking birds being elevated in status simply by having a darker crown and throat) is socially controlled. Moller (1987) used similar manipulations and demonstrated a 'status signaling' function for badge size (dark coloration on throat and breast) in house sparrows (*Passer domesticus*).

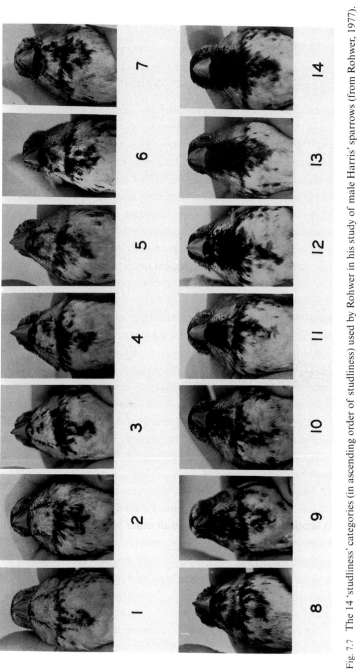

Fig. 7.7 The 14 'studliness' categories (in ascending order of studliness) used by Rohwer in his study of male Harris' sparrows (from Rohwer, 1977).

The role of the red epaulets of male red-winged blackbirds (*Agelaius phoeniceus*) was studied by D. G. Smith (1972) by dying the epaulets black on selected territorial males. He found that the epaulets were important in maintenance of territories against rival males, but they had little effect on the males' ability to obtain mates. N. G. Smith (1967) changed the eye-ring color of one member of mated pairs of sympatric glaucous gulls (*Larus hyperboreus*), Kumlien's gulls (*L. glaucoides*) and herring gulls. In all cases where the female's eye-ring color had been changed the pair broke up, but altering the male's eye-ring appeared to have no effect on the pair's behavior.

It is important in all research where animals are manipulated and the effects are studied in interactions with other individuals to observe the effects on the manipulated animal, as well as on others responding to it. This is true in both intra- and interspecific studies, such as the effects of altered males on selection by females and altered prey on selection by predators.

7.2.1b *Manipulation of the environment*

Altering the biotic or abiotic environment (see section 2.3.2b) in order to study its resultant effect on behavior ranges from gross-perturbation experiments to subtle changes in one or a few stimuli.

Stewart and Aldrich (1951) were able to get an indication of the extent of the surplus 'floating' population of unmated male birds the spruce–fir forests by drastically reducing (by shooting) a large number of territorial holders on a 40-acre tract. During nine days in June, they removed 148 territorial males, reducing the population to 19% of the original. They continued to shoot birds as they moved into the area, and by July 8 they had collected a total of 455 individuals. This is a rather drastic perturbation experiment, and as they admit 'the breeding territories were completely disrupted during the period when the original occupants were being removed and at the same time new adult males were constantly invading the area'. On a smaller scale, Krebs (1971) shot six pairs of great tits occupying territories and observed that residents expanded their territories and four new pairs took up occupancy. In contrast to these major manipulations, Tinbergen was prone to concentrate on subtle environmental changes in order to study effects without greatly disturbing the normal activities of the animals.

> The trick is, to insert experiments now and then in the normal life of the animal so that this normal life is in no way interrupted; however exciting the result of a test may be for us, it must be a matter of daily routine to the animal. A man who lacks the feeling for this kind of work will inevitably commit offenses just as some people cannot help kicking and damaging delicate furniture in a room without even noticing it.
>
> [Tinbergen, 1953:138]

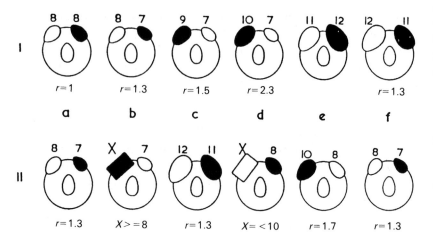

Fig. 7.8   The 'tritration' method for determining the value of an egg model. The circle
represents the nest with one egg in the nest bowl and two models on the rim. The
code numbers 7 to 12 refer to the models of the size series R. X is the model to be
measured; r is the ratio between maximal projection surfaces of the model on the
nest rim. The black model is the preferred one. I, determination of the value of
the position preference. In Ia the right side is preferred. This preference remains
when model 8 is replaced by the smaller model 7 (Ib), but can be overcome by
replacing 8 by 9 (Ic); this sequence shows that the value of the position preference
lies between r =1.3 and r = 1.5: this conclusion holds when another pair of
dummies with the same ratio is used (If). Control test Ie shows that the size
optimum for this gull exceeds size 11. II, determination of the value of model X.
Tests IIa, IIc, and IIf show that the position preference has remained unchanged.
Test IIb and IId indicate, in combination with the preceding and succeeding tests,
that the value of X is between those of models 8 and 10 of the reference size
series (from Baerends and Kruijt, 1973).

Manipulation of the environment can be conveniently divided into four types:
intraspecific, interspecific, other biotic factors, and abiotic environmental manipulations. The *intraspecific* facilitating effect of a female mallard (*Anas platyrhynchos*)
on male courtship displays was demonstrated by Weidmann and Darley (1971) by
introducing a strange female or male to resident groups of three males in the spring
and autumn. Free (1967) manipulated the *biotic environment* in honeybee hives and
showed that the amount of pollen collected increased with the amount of brood
present and decreased in the absence of a queen. He went on to isolate some of the
stimuli produced by the brood which are important in stimulating pollen collection.
    Baerends and Kruijt (1973) determined the *intra-specific* stimuli important in
releasing egg retrieval in herring gulls by presenting them with three-dimensional
model eggs placed two at a time on the edge of a nest (Figure 7.8). The relative

importance of the various configurations in releasing egg retrieval was 1. larger>smaller; 2. speckled>not speckled; 3. green>blue, red>grey; and 4. shape, other than roundness, was relatively unimportant. The *titration method* used by Baerends and Kruijt is worthy of careful consideration for other studies using models. This method allowed them to rank the models on a relative basis between and within the four categories of features.

> Our experiments with the size series showed position preference to be a quantitative phenomenon. A first choice for the smaller egg in the preferred position can always be overcome by increasing the size of the model in the non-preferred position. With our series of models gradually increasing in size it was possible to identify stepwise, in successive tests with the same bird, the minimum size of a model required to overcome position preference, when in competition with a dummy of a smaller size in the preferred position. Thus, through this 'titration,' a model was found the value of which, in combination with that of the non-preferred site, could just outweigh the combined values of the smaller model and the preferred site. Empirically it turned out that the birds were acting in accordance with the ratios between the surfaces of the maximal projections (maximal shadows when turned around in a beam of parallel light) of the models. Different pairs of models, matching each other with respect to other parameters tries (e.g., volume), or equal with regard to the difference instead of the ratio in the parameters used, proved to be unequal in counteracting position preference. The ratio between sites often remained constant for a couple of hours, and within that period the relative value of dummies with any kind of stimulus combination could be measured and expressed with reference to the standard size series. *[Baerends and Kruijt, 1973:30]*

Baerends and Kruijt's results (Figure 7.9) show how the releasing value of a model egg with respect to size is affected by the other manipulated variables, such as changing the egg shapes into a round-edged block, omitting the speckling on brown models, and adding speckling to green models. Baerends and Kruijt caution that the exactitude of the method should not be overestimated. It was limited by the step sizes in the 'titration' series, and there was considerable variability in the results of individual tests. However, it is clear that their 'titration' procedure provided increased understanding of the role of the various stimuli in the egg retrieval behavior. Also note that they conducted over 10 000 tests in their experiment.

Manipulation of eggs, although considered here as an intraspecific manipulation, might be argued to be manipulation of the biotic environment. The answer lies in the 'eyes of the beholder', the gulls; hence we'll probably never be sure.

An *interspecific manipulation* was made by Littlejohn and Martin (1969) in their

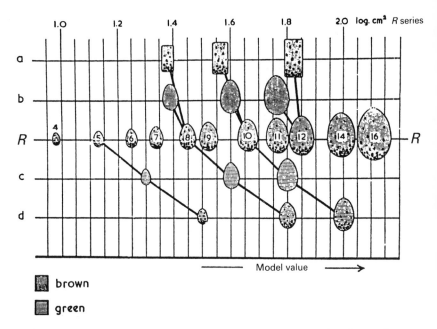

Fig. 7.9   The average values found for various models with respect to the reference size series R (standard brown, speckled models). The position of different types of model (brown, speckled, block-shaped (a); brown, unspeckled, egg-shaped (b); green, unspeckled, egg-shaped (c); green, speckled, egg-shaped (d), each in different sizes, was determined with the method described in the legend of Figure 7.8. The code numbers 4 to 16 stand for, respectively, 4/8 to 16/8 of the linear dimensions of the normal egg size (8 = 8/8). The maximal projection surfaces of the eggs of the reference series have been plotted (egg centers) along the logarithmic scale (cm^2) of the abscissa. Equal distances between points on this scale imply equal ratio values (from Baerends and Kruijt, 1973).

study of acoustic interaction between two sympatric species of frog, *Pseudophryne semimarmorata* and *Crinia victoriana*. They played a tape-recorded mating call of *C. victoriana* and synthetic signals to individual calling males of *P. semimarmorata*. The call of *C. victoriana*, if played above 80 dB, and synthetic pulsed signals with a carrier frequency of 1500 to 2500 Hz, were all effective in inhibiting *P. semimarmorata* males from calling.

Manipulation of the environment can take many forms. Two basic procedures are often used: 1. change the environment in which the animal is presently located, or 2. relocate the animal to another environment. These procedures are generally used to determine the effect of the abiotic and/or biotic (e.g. vegetation) environment; however, the confounding effects of intra- and interspecific interactions are often difficult to eliminate.

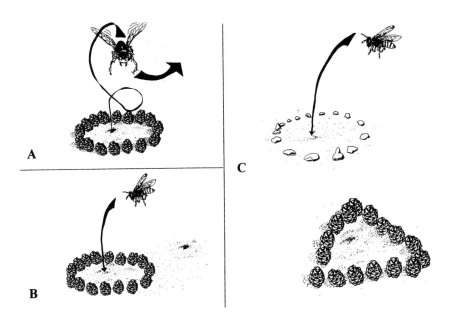

Fig. 7.10 Digger wasps memorize the landmarks around their burrows in order to find
them when they return from hunting. To test this, Tinbergen (1972) arranged a
circle of pine cones around a burrow (A), and the wasp memorized it. When the
circle of pine cones was moved a foot or two, the wasp looked for its burrow
within the pine cone ring (B). When the pine cones were arranged in a triangle,
and rocks were arranged in a circle, the wasp looked for its burrow in the circle of
rocks, demonstrating that it was the geometric configuration (circle) it was
remembering, not the objects (pine cones) (drawing by Brenda Knapp, based on
Tinbergen, 1972).

Tinbergen and Kruyt (1938) investigated the role of landmarks in the ability of
female digger wasps (*Philanthus triangulum*) to locate their burrows. They manipulated the type and geometric arrangement of objects around or near the burrow and
recorded the response of the returning wasp (Figure 7.10). This is typical of the
simple, yet cogent, type of experimentation for which Tinbergen is famous.
Remarking on Tinbergen's methods, Lorenz (1960b:xii) stated, 'He knows exactly
how to ask questions of nature in such a way that she is bound to give clear answers.'

Perdeck (1958) relocated migrating starlings (*Sturnus vulgaris*) geographically to
measure their ability to navigate to their normal winter areas. He captured adult and
juvenile starlings in the Netherlands during their southwestward fall migration and
transported them south to Switzerland where they were released. Adults were recovered northwest of the release sites in their normal winter areas along the coast of
Western Europe. The juveniles, however, were recovered west of the release sites,

indicating that they had continued to follow a westward orientation, not navigating northwest to adjust for their southerly displacement as did the adults.

7.2.2 In the laboratory

No one would argue that ethologists are found studying behavior in the field, and I have suggested that the field is the best place for ethological research when it is both feasible and valid. However, much ethological research is also conducted in captivity or the laboratory. Scientists (including ethologists) are not judged by where they work but by what they do.

7.2.2a *Manipulation of the animal*

Some species can often be more easily and accurately manipulated and observed in the laboratory than in the field; proper manipulations contribute to valid results. For example, Buchler (1976) examined the wandering shrew's (*Sorex vagrans*) use of echolocation by training six shrews to echolocate the position of a platform in order to drop to it. The shrews preferentially directed ultrasonic transmissions toward the platform before dropping. When their ears were plugged their ultrasonic transmission rate increased, but their ability to locate the platform decreased significantly. When the ear plugs were replaced with hollow tubes, they located the platform as well as when the ears were not plugged.

The effects of hormones on behavior have been investigated in numerous studies. For example, R.J.F. Smith and Hoar (1967) demonstrated that injections of prolactin failed to induce fanning behavior in male sticklebacks (*Gasterosteus aculeatus*), castration reduced the behavior, and injections of testosterone restored it. Estrogen can stimulate nest-material preparation (cutting strips of paper) in peach-faced lovebirds (*Agapornis roseicollis*) at least two weeks before it would normally occur, but only after the female is at least 98 days old (Orcutt, 1967). Lindzey *et al.* (1968) measured territorial-marking behavior in male mongolian gerbils (*Meriones unguiculatus*) which were either castrated or sham operated (control) at 30 days of age. Marking did not develop in castrates, but when injected with testosterone they began to mark earlier and reached higher frequencies than did the controls.

The effects of stimulation of various brain sites on general behavior patterns (e.g. sitting, standing, eating, crowing) in domestic chickens were studied by von Holst and von Saint Paul (1963). Dethier and Bodenstein (1958) were able to demonstrate that the recurrent nerve running from the foregut to the brain signals the brain of the blowfly when the foregut is distended and inhibits further feeding. By cutting the recurrent nerve they were able to show that the blowfly will continue to ingest until it bursts. Balaban *et al.* (1988) studied the role of brain development on species-typical behavior patterns. They created domestic chick–quail chimeras by transplanting part of the neural epithelium from a quail embryo into the developing brain of a chick embryo from which the corresponding brain region had been removed. They found that 'transplants containing the entire quail mesencephalon and diencephalon resulted in the transfer of certain aspects of species-typical crowing behavior' (Balaban *et al.*, 1988:1339).

By studying age- or genotype-dependent behavior in the laboratory, one is essentially making use of natural variation. Fuller (1967:470) focused on genotype effects and demonstrated that 'albino [house] mice otherwise cogenic with strain C57BL/6J escaped more slowly from water, were less active in an open field and made more errors on a black–white discrimination task than their pigmented congeners'. Van Abeelen (1966), also interested in genetic effects, used 30 behavioral components performed by individuals and pairs of male house mice to measure differences between strains DBA/2J, C57BL/6J, their F1 hybrids, and homozygous and heterozygous short-ear animals.

Dilger (1962) studied the behavior of hybrids between peach-faced lovebirds (*Agapornis roseicollis*), which carry nest material under feathers on their backs, and Fischer's lovebirds (*A. personata fischeri*), which carry nest material in their bills. The hybrids initially tried tucking nest material in their plumage as well as carrying it in their bills. Even though feather tucking was unsuccessful for these hybrids, it took two years before feather tucking diminished to any great extent and carrying in the bill was almost exclusive (Dilger 1962). This demonstrated the interaction between genotype and experience.

7.2.2b *Manipulation of the environment*

Examples of environmental manipulation in the laboratory are widespread in the ethological literature. As in the field, manipulation of the environment in the laboratory can consist of altering intraspecific factors, interspecific factors, other biotic factors and physical-environment variables. For example, providing domestic hens with experience in an *intraspecific* flock can change the dominant–subordinate relationships seen in later paired encounters (King, 1965). Marsden (1968) artificially induced changes in rank in young rhesus monkeys by introducing a 'strange' adult male at a time when the second-ranking female was in estrus or by removing and reintroducing the currently top-ranking female.

Numerous and various *interspecific* manipulations have been made in laboratory experiments. As an example, Kalinoski (1975) observed agonistic behavior between house finches (*Carpodacus mexicanus*) and house sparrows (*Passer domesticus*) simply by maintaining mixed groups in laboratory cages. He systematically controlled the species and sex composition as follows: Group I – four male house finches and four male house sparrows; Group II – four male finches and four female sparrows; Group III – four female finches and four male sparrows; Group IV – four female finches and three female sparrows (one female sparrow died prior to the experiment).

Turner (1964) investigated social feeding in house sparrows and chaffinches (*Fringilla coelebs*) by allowing a caged 'reactor' (either species) to observe simultaneously two individually caged 'actors' (both of the same species, either chaffinch or sparrow), one which was feeding and the other not feeding. He found that individuals of each species were attracted to feeding and nonfeeding conspecifics. Also, chaffinches were attracted more by feeding than nonfeeding sparrows, but this was not true for sparrows observing chaffinches.

The responses of a caged chaffinch to a stuffed owl located at various distances were measured by Hinde (1954). He found that at distances closer than 17 feet the chaffinch moved away, while at greater distances it moved predominantly towards the stuffed owl.

As another example of manipulation of interspecific stimuli, Wells and Lehner (1978) were able to differentially affect the ability of coyotes (*Canis latrans*) to find a rabbit by manipulating the sensory stimuli available to them. Visual, auditory and olfactory stimuli were eliminated, respectively, by testing the coyotes in the dark, with dead rabbits and with an intense masking odor of rabbit feces and urine. Also, Metzgar (1967) exposed pairs of mice to a screech owl (*Otus asio*) in a laboratory test area for 2–30 minutes. One was a 'resident mouse' (had spent several days in the test area), and the other was a 'transient mouse' (had no prior experience in the area). The owl captured 'transient mice' significantly more frequently.

Potash (1972) measured changes in the *abiotic environment* in his study of the effects of environmental noise on separation crowing by Japanese quail (*Coturnix coturnix japonica*). He found that ambient noise increased the frequency of separation crowing and the number of crows per bout, both of which should increase the detectibility of the signal and the localizability of the sender. Bradbury and Nottebohm (1969) varied the amount of light in a flight chamber and measured the ability of auditorily impaired and untreated little brown bats (*Myotic lucifugus*) to negotiate a vertical string maze. Dim light and high contrast enhanced the bats' ability and was interpreted as evidence that they use pattern vision while in flight. Reynierse (1968) investigated the effect of refrigeration and non-refrigeration during daily maintenance, the intertrial interval, and the experimental session on earthworm locomotion. He found that refrigeration before an experimental session at room temperature inhibited their locomotion, but had no effect if the session was also run under refrigeration.

Klopfer (1963) used manipulation of the *biotic environment* in the laboratory to study the role of early experience on habitat selection in the chipping sparrow

(*Spizella passerina*). He released individuals into a room in which he had placed pine boughs on one side and oak branches and leaves on the other. He found that wild caught adults and hand-reared isolated adults preferred the pine, but that hand-reared individuals, which had been previously exposed to oak, preferred the oak. Emlen *et al.* (1976) tested the orienting capabilities of indigo buntings (*Passerina cyanea*) in cages with minimal exposure to visual cues but with an artificial geomagnetic field provided by Hemlholtz coils surrounding the cage. When the horizontal component of the magnetic field was deflected clockwise by 120°, the orientation of the buntings shifted accordingly (clockwise to geographic east–southeast).

The effect of different types of feedback (see model, Chapter 2) provided by differently treated seeds was tested in black-capped chickadees (*Parus atricapillus*) by Alcock (1970). He presented the birds with striped seeds which were empty, filled with mealworm to which salt had been added, or contained mealworms treated with quinine sulphate (an emetic). There was a rapid and stable avoidance of the empty and emetic seeds, but they continued to attack the salted mealworms, perhaps because the food reward outweighed the punishment (salty taste).

7.3 FIELD TO LABORATORY: A CONTINUUM

Field and laboratory studies represent the extremes along one of the conceptual dimensions of ethological research (Chapter 1). However, in practice they complement each other in a cyclical continuum called a 'research cycle' by Kelly (1967, 1969). Beck (1977) emphasized the value of captive studies in conjunction with field research. For example, for the six years following their 13 year field study on vervet monkeys, Cheney and Seyfarth (1990:ix) 'supplemented [their] research on the Amboseli vervets with studies of captive primates'. Likewise, Kummer (1984) conducted a long-term field study and complementary captive research on the social system of Hamadryas baboons (*Papio hamadryas*).

At the two extremes we have the mensurative (non-manipulative) experiments in the field and the highly manipulative studies in the laboratory. The middle of the continuum is illustrated by studies conducted in enclosures in the field. For example, Wecker (1964) set up an instrumented enclosure that was half in an oak–hickory woodlot and half in a field, in order to investigate habitat selection in prairie deer mice (*Peromyscus maniculatus*). Wells (1977) used a large outdoor enclosure to investigate the relative priority of the coyote's distance senses in predation on rabbits. At what point does the field become the laboratory and vice versa? Hinde and Spencer-Booth (1967:169), in their study of rhesus monkey behavior, attempted to reach a compromise in an enclosure with 'a moderately complex environment under conditions which permit a moderate degree of experimental control and moderately precise recording.'

Hoffman and Ratner (1973:541) suggested 'that laboratory investigations complement and explain the frequently puzzling data obtained in a natural setting'. For example, Cheney and Seyfarth (1990) conducted a 13-year field study of social interactions, including communication, in vervet monkeys. In order to test the hypothesis that vervets recognize members of other groups, they used playbacks of vocalizations (Chapter 9) employing the same technique that had been used to study neighbor recognition in songbirds (Brooks and Falls, 1975). Nevertheless, when interpreting some of their results on vervet 'concepts' Cheney and Seyfarth (1990: 94–95) concluded that 'Definitive proof that monkeys are capable of solving social analogies, and that language training is not a necessary prerequisite, can only come from laboratory tests'.

Ideally, however, research should undergo the field–laboratory cycle several times, utilizing to best advantage the important attributes of each. Menzel (1969) considers this process analogous to 'zooming in' and 'zooming out' with a lens. Avian orientation/migration studies provide a good example. Matthews (1951) found that homing pigeons released in unfamiliar territory under clear skies were able to fly off directly toward home, but if the sky was overcast they became disoriented. That the sun was a cue used in orientation was given further support by Kramer (1952) who placed starlings in a circular cage with six windows giving a view of the sky only. The starlings showed migratory restlessness (Zugunruhe), fluttering in the proper migratory direction when the sky was clear but in random directions when it was overcast. Kramer altered the apparent position of the sun with mirrors and was able to reorient the starlings in a predictable manner. Schmidt-Koenig (1961) kept pigeons under artificial day–night conditions six hours out of phase with the normal day. When the pigeons were released they oriented 90° from the current direction, showing that they were using a biological clock and the sun's position as a cue. Kramer trained starlings to find food in particular trays in a circular cage using only the sun as a cue to direction. When the cage was covered and the starlings were presented with a stationary light, they used the light as if it were the moving sun and changed their direction at the rate of 15°/hour. Meyer (1964) used instrumental conditioning, discrimination tests in the laboratory to show that pigeons could indeed detect movement of 15°/hour.

Night-migrating warblers (*Sylvia atricapilla, S. borin* and *S. curruca*) were tested in a planetarium by Sauer (1957) and were shown to use stellar cues in orientation. Emlen (1967) measured the nocturnal orientation of caged indigo buntings outdoors under the natural night sky and then took them into the planetarium, where they continued to orient themselves correctly when the planetarium sky was set for local conditions. They reversed themselves when the north–south axis of the planetarium sky was reversed, and they were disoriented when the planetarium sky was diffusely illuminated or darkened. Emlen (1970) also used manipulation of star pattern movement in a planetarium to demonstrate that the axis of celestial rotation was important in the development of migratory orientation by young indigo buntings.

Both field and laboratory studies have provided convincing evidence that geomagnetic fields are an orientation cue sometimes used by birds. Moore (1977) showed that nocturnal free-flying passerine migrants responded to natural fluctuations in the earth's magnetic field. Electromagnetic fields produced by large antennas were shown to alter the path of free-flying migrants (Larkin and Sutherland, 1977) and gulls held in an orientation cage (Southern, 1975). Homing pigeons become disoriented when released under an overcast sky with a bar magnet attached to their backs (Keeton, 1974) or with Helmholz coils on their heads (Walcott and Green, 1974). In carefully controlled laboratory investigations with a cage surrounded by Helmholz coils, use of the inclination of the axial direction of the magnetic field (increased downward dip as the magnetic north pole is approached) for orientation was demonstrated in the European robin (*Erithacus rubecula*) (Wiltschko and Wiltschko 1972) and indigo bunting (Emlen *et al.* 1976).

The research described above represents a very few examples of the multitude of studies that have been conducted in the field and laboratory using both mensurative experiments and various degrees of manipulation. Literature reviews of species, concepts and behavior types will generally provide studies representing all approaches from description in the field to manipulation in the laboratory. Examples can be found in D.E. Davis's (1964) review of the relative contribution of field and laboratory research to our understanding of aggression and the role of hormones in aggressive behavior. Above all, astute researchers recognize the value of both description and experimentation (mensurative and manipulative) and how the various approaches can be applied in both the field and laboratory best to answer their research questions (e.g. Holldobler and Wilson 1990).

> I see neither halos nor horns on either a real experiment or on accurate observations. Any method is a special case of human experience, and it cannot surpass the limitations of its human interpreters.
>
> *[Menzel, 1969:80]*

II Collecting the data

8 Data collection methods

We are now at the point where: 1. the research question has been asked; 2. the subjects chosen; 3. reconnaissance observations made; 4. the objectives formulated; 5. a descriptive or experimental approach determined, and, if experimental; 6. the research hypotheses stated; 7. the behavioral units to be measured determined; and 8. the experimental design established. Now it is necessary to decide on the procedures to be used to collect the data. You should also select the statistical tests to be used in the analyses (see Chapters 12–17) before beginning to collect data.

8.1 RESEARCH DESIGN AND DATA COLLECTION

Research design and data collection are mutual dictators. The research design chosen will dictate the data to be collected; likewise, a knowledge of the type and amount of data that can be collected will partially dictate the research design to be used. Research design and data collection are in harness together, and the pushing and pulling that each does to the other will depend on the individual study and the experience of the researcher. Experienced observers will use a knowledge of the animal's behavior and types of data that they can expect to collect to push for the best research design. On the other hand, neophyte researchers may allow a research design, selected for statistical attributes, to pull them around in the field, attempting to collect nearly impossible (and sometimes behaviorally meaningless) data.

Research design and data collection must complement each other for the study to be efficient and the results valid. Even seemingly well-planned research can sometimes benefit from redesign and additional (or modified) data collection. Do not be afraid to evaluate carefully your research design and data collection methods while you are conducting your study. However, *do not redesign your research until you have carefully assessed your present and future losses in time, money and data.*

8.2 SCALES OF MEASUREMENT

Data collection involves the assignment of numbers to observations and observations to categories. This process is referred to as *measurement.*

Scales of measurement are levels of resolution (or accuracy) of measurement. The four scales (Stevens, 1946), represent points along a continuum of resolution. That is, some types of data will appear to fall between two scales and may be difficult to categorize.

Table 8.1. *Frequencies of behaviors observed during the 10 min. following an aggressive interaction and the 10 min. control period*

Behavior	Post-aggression	Control
Return/stay within 1 m	160	0
Hindquarters present	70	1
Proximity	28	5
Groom	13	1
Contact	8	0
Embrace	9	0
Redirect*	60	0
Tongue flick*	26	2

Notes:
More than one of the above behavioral categories could occur per 10-min. post-conflict period.
* Not included in the analysis of non-aggressive post-conflict behavior.
Source: From Gust and Gordon (1993) Copyrighted by Academic Press.

The scales of measurement relate to the type of variable being measured: nominal scales measure *attributes*; ordinal scales measure *ranked variables*; interval and ratio scales measure *continuous variables* (see Sokal and Rohlf, 1981a, for a discussion of these types of variables). The four scales are listed in order (below) from the lowest resolution (nominal) to the highest resolution and most restricted measure (ratio; Stevens, 1946).

Nominal Scale: Observations (measurements) are classified into predetermined, qualitatively different categories (e.g. behaviors: sitting, walking, feeding). The data are in the form of counts (i.e. frequencies of occurrence). For example, Gust and Gordon (1993) studied conflict resolution in sooty mangabeys (*Cerocebus torquatus atys*) by recording the occurrences of selected behaviors for the 10 min. following an aggressive interaction and a 10 min. control period (the same time during the next observation day; Table 8.1).

Ordinal Scale: This scale is the same as nominal scale with the addition that the categories are ordered with respect to a qualitative or quantitative property. The ordering must be stable, and it must hold throughout the entire scale. See Table 8.2 for examples.

The more intervals the researcher uses in the scale (while maintaining accuracy), the higher the resolution of the measurements will be, which will make the statistical analysis more powerful (Chapter 11). For example, Ekman and Sklepkovych (1994)

Table 8.2. *Examples of ordinal scales of measurement*

Score	Distance from model[1]	Score	Activity level[2]	Score	Vocal response level[3]	Score	Aggression level[4]
0	>30.0	1	Sleeping	0	None	1	Intruder antennated (as in 2), but if mobile, is not followed, if intruder is stationary, resident ant does not stop
1	18.0–29.9	2	Lying alert	1	Few, intermittent		
2	10.5–17.9	3	Sitting	2	Few, close	2	Intruder antennated for less than 2 s; if mobile, intruder is followed slowly for several cm; if intruder is stationary, resident stops
3	6.2–10.4	4	Standing	3	Many, intermittent		
4	3.7–6.1	5	Walking (slow pace)	4	Many, close		
5	2.2–3.6	6	Walking (steady pace)	5	Full or continuous	3	Rapid antennation of intruder, antennae extended for greater than 2 s
6	1.3–2.1						
7	0.7–1.2	7	Gentle play (mouthing, etc.)			4	Mandible gaping, rapid antennation; 'sidling' (maintaining a lateral orientation to and slowly circling intruder)
8	0.4–0.6	8	Excited pacing				
9	0.2–0.3	9	Chase-play (rough-and-tumble)			5	Alarm (running, abdomen elevation and vibration) and recruitment
10	<0.2					6	Intruder 'held', but released; biting; no abdomen-curling
						7	Intruder 'held' (as in 8), but released; abdomen-curling (stinging posture) by residents, but no stinging; biting
						8	Intruder surrounded and 'held' in mandibles by petiole and appendages; appendages pulled bitten off; eventual stinging
						9	Immediate lunge, grab and stinging

Notes:
[1,3] From Studd and Robertson (1985).
[2] From Bekoff and Corcoran (1975).
[4] Examples for ants, from Obin and Vander Meer (1988).
Source: Copyrighted by Bailliere Tindall.

scored level of aggression during feeding bouts in Siberian jays (*Perisoreus infaustus*) on a scale of 0 to 3; Studd and Robertson (1985) used a scale of 0 to 5 to score movement of yellow warblers; Riechert (1984) scored agonistic encounters in spiders on a scale of 1 to 35. Note that the number of intervals in the examples in Table 8.2 vary from 6 to 9.

These have all been examples of using an ordinal scale on the *dependent variable* (behavior units), but ordinal scales can also be used to rank values of the *independent variable*. For example, Moller (1987) used an ordinal scale of 1 to 5 to rank the variation in badge size (amount of black coloration on the breast and throat) in house sparrows; Rohwer (1977) used 14 intervals to rank badge size ('studliness' categories) in Harris' sparrows (Figure 7.7). Alatalo *et al.* (1990), in their study of female preference, ranked the percentage of brown and/or grey feathers on the backs of male pied flycatchers (*Ficedula hypoleuca*) using an ordinal scale of 1 to 7. Ward (1988) measured sexual dichromatism in 24 fish species using a four-point ordinal scale of skin color on three areas: the head, and the dorsal and ventral surfaces; the sum of the three ranks was used as the measure of sexual dichromatism. Another example of an ordinal scale of measurement for an independent variable is the Beaufort Scale of wind velocity measurement described in Chapter 4.

Measurements (scores) based on an ordinal scale are sometimes weighted to reflect relative differences in the *intensity* of behaviors, and they are occasionally combined with other measurements to create a *composite score*. Example of both of these are provided by Lightbody and Weatherhead's (1987) study of female choice versus male competition in yellow-headed blackbirds (*Xanthocephalus xanthocephalus*) (Table 8.3); note that the ranks jump from 6 to 8, then to 10 and 13. In addition, the ranks were multiplied by the durations of the male's behavior and then summed for a composite score.

Also, all the behaviors do not have to have different scores. If several behaviors reflect the same intensity, they can be assigned the same rank. As an example, Cigliano (1993) studied dominance and den use in *Octopus bimaculatus* by recording the occurrence of eight attack and withdrawl behaviors. These behaviors were weighted on an 'intensity of response scale' of 0 to 4, but four of the attack behaviors were all given the score of 2. Likewise, Barki *et al.* (1992) determined the effect of size and morphotype on dominance in prawns (*Macrobrachium rosenbergii*) by recording 18 agonistic behavior acts which were scored on an ordinal scale of -3 to $+3$, with some behavioral acts receiving the same score.

Interval scale: This scale is the same as the ordinal scale except that the amount of the differences between respective categories is the same and is known; this necessitates a unit of measurement which permits additivity (see Table 8.6). The zero point is not known, or is arbitrarily defined, for measurements on an interval scale.

A common example is temperature measurement; the zero point is arbitrary, to

Table 8.3. *Interval scale for measuring the behavior of 11 male yellow-headed blackbirds (*Xanthocephalus xanthocephalus*) towards a taxidermic mount of a female in a copulatory position placed in their territory for 5 minutes*

Score	Behavior
1	Distance (>5 m) and non-attentive
2	Close (<5 m) and non-attentive
3	Distant, silent observation
4	Close, silent observation
5	Distant, agitation
6	Distant, agitation and vocalization
8	Close, agitation
10	Close, agitation and vocalization
13	Direct attack/mount

Source: From Lightbody and Weatherhead (1987).

wit the different zero points on the centrigrade and fahrenheit scales. In neither scale is the zero point the absence of temperature; likewise, 80 °F is not twice as warm as 40 °F. Compass directions and time divisions (e.g. time of day, weeks, years) are also divisions on an interval scale (Zar, 1984). The length of time it takes individual birds to fly after an alarm call is given (latencies) would generally be considered to be on an interval scale of measurement. These time-to-fly latencies can be compared to each other, but the zero point for flight is not really known, although we would probably use our hearing of the alarm call as an arbitrary zero point.

Interval scales of measurement are uncommon in ethological studies except when measuring spatial or temporal characteristics. However, Maxim (1976) constructed an interval scale of 17 behavior categories, including 'attack', 'stare', 'lips-mack' and 'grimace', for use in studies of social relations in pairs of rhesus monkeys. The scale was based on observations of 120 pairs of monkeys; a scale value for each behavior category was then established by the relative frequency distribution of responses between categories. The theory behind Maxim's interval scale was 'The Law of Categorical Judgement', that 'is a set of equations which, using frequency distributions of responses to a set of stimuli, establishes the parameters and mean scale values of those stimuli' (Maxim, 1976:125).

Ratio scale: This scale is the same as the interval scale except the zero point is known. This scale is commonly used with continuous variables, such as duration and distance. For example, Mesce (1993) studied shell selection behavior in two

Table 8.4. *Time required for five individual hermit crabs (* Pagurus samuelis*) to locate a shell in the light and in the dark*

Individual	Time (s)	
	Light	Dark
1	3	323
2	8	241
3	37	216
4	3	118
5	4	57

Source: Abridged from Mesce (1993). Copyrighted by Academic Press.

Table 8.5. *Examples of data with different scales of measurement*

Behavior	Data recorded Behavior code for measurement	Scale of measurement
A occurred	*A*	Nominal
A occurred at intensity level 2	A2	Ordinal
A occurred at intensity level 2, at 1:00 pm	A2/1300	Interval
A occurred at intensity level 2, at 1:00 pm, at a distance of 8 m from the stimulus model	A2/1300–8	Ratio

hermit crab species; one measure of their use of vision was the time required to locate a shell in light and in dark (Table 8.4).

To determine the scale of measurement used in a study, examine the data *as it was recorded* (i.e. the 'raw data'), as shown in Table 8.5.

Note that data with a scale of measurement of less resolution can be extracted from those with higher resolution (e.g. ordinal from interval or ratio: A2 from A2/1300 or A2/1300–8). This is sometimes done for data analysis with nonparametric tests (Chapter 13). However, it is best to collect data with as high a resolution scale of measurement as is feasible and valid, and then sacrifice the resolution later (i.e. convert ratio to ordinal data) if you are not convinced it was collected accurately and/or you find that the data do not meet the criteria for parametric statistical tests (see below and Chapter 12). Remember to keep the original data, for you may want to come back to it later, and at that time you may need the higher-resolution scale of measurement.

The scale of measurement chosen for collecting data will, in part, determine the experimental design (Chapter 6), as well as the statistical tests that can be used (Chapter 13 and 14). Only nonparametric statistical tests should be used with nominal and ordinal data, whereas either parametric or nonparametric tests can be used on interval and ratio data (Chapter 12). These restrictions are based on the operations which are permissible on data from the different scales of measurement (Table 8.6). For additional information on scales of measurement consult one of the many, enlightening discussions available (e.g. Drew and Hardman, 1985; Ghent 1979; Walker, 1985).

8.3 SAMPLING METHODS

The discussion of sampling methods which follows is based almost entirely on J. Altmann's (1974) excellent review. The sampling method you select for your research will be based on: 1. your research question(s); 2. your experimental design; 3. the number and types of behavioral units you have selected to measure (states and/or events); 4. the scale of measurement; and 5. a multitude of practical considerations, such as observability, experience, and availability of equipment.

Experimental designs (Chapter 6) and statistical tests (Chapter 13 and 14) assume that the sampling methods described below will be used on randomly selected individuals and/or on behaviors randomly selected from individuals. You might want to review section 6.5, on random, haphazard and opportunistic samples.

In the discussion to follow, an example of how each method is used will be based on Figure 8.1, a hypothetical behavior record for six mule deer (*Odocoileus hemionus*) showing only two events (standing-up and lying-down) and one state (feeding). In order to be sure that you understand the diagram, confirm the following statements:

1 All individuals fed during the 90–minute period.
2 All individuals fed during at least two of the three 30–minute periods.
3 None of the individuals fed between the events of lying down and standing up.
4 Individual IV was the only one that did not feed every time it was standing.

8.3.1 Focal-animal (pair, group) versus all-animal sampling

When selecting a sampling method, the first question you must ask is which of the two categories below describes your research. That choice will help you decide whether to use *focal-animal* or *all-animal* sampling (Figure 8.2).

Table 8.6. *Relationships between scales of measurement and admissible operations*

Admissible operations[1]	Scales of measurement			
	Nominal	Ordinal	Interval	Ratio
Classifying (e.g. trotting or walking)	X	X	X	X
Ordering (e.g. trotting is faster than walking but slower than galloping)		X	X	X
Subtraction to determine relative differences (e.g. Individual A is found on a rock at 10 °C, B on a rock at 20 °C, and C on a rock at 30 °C; C's rock is the same number of degrees hotter than B's, as B's rock is to A's; *but* B's rock is not twice as hot as A's rock)			X	X
Division to determine ratios (e.g. A's rock is 5 m from the stimulus, and B's and C's rocks are 10 and 20 m from the stimulus, respectively. B's rock is twice as far from the stimulus as A's rock)				X

Notes:
[1] Each admissible operation also includes all those listed above it.
Source: Based in part on Denenburg (1976), Howell (1992), and Siegel (1956).

Does your research question(s) involve?:

1 Several *behaviors* and/or few *individuals* (e.g. food gathering, territorial defense, and courtship; behavior of the alpha male in a group) – *Focal-animal sampling*
   or
2 Few *behaviors* and/or several *individuals* (e.g. vigilant behavior; percentage of time spent feeding; synchrony of feeding in a group) – *All-animal sampling*

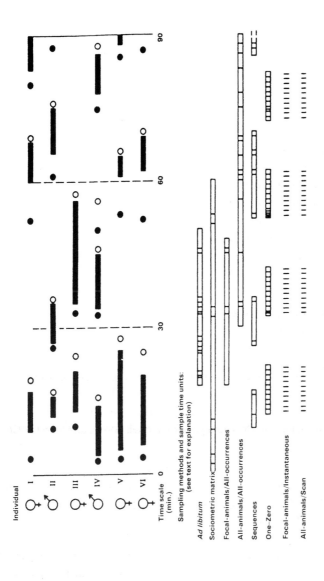

Fig. 8.1 Hypothetical record of occurrence of three behaviors in six penned adult mule deer: ●=standing up (event); ○=lying down (event). ■=feeding (state). The time period recorded by each sampling technique is shown in the lower part of the figure.

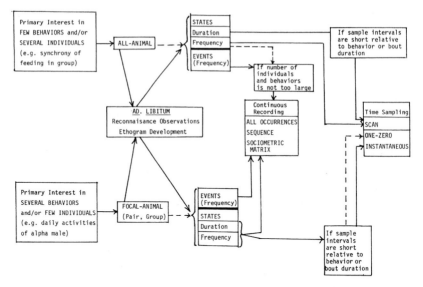

Fig. 8.2   Flow chart for selecting animal behavior sampling methods. The dashed lines
denote alternatives (see text for explanation).

Focal-animal and all-animal sampling are *always paired* with one of the other sampling methods discussed below, and each is best suited to be paired with only some
of those other methods (Figure 8.2)

8.3.1a   *Focal-animal sampling*

With *focal-animal* sampling, one individual (pair, group) is the focus of observations during a particular sample period. With this method you can accurately
measure *several behaviors in selected individuals* by observing only one (focal-
animal) at a time. For example, Goss-Custard and Sutherland (1984) observed focal
animals for periods of five minutes in their study of three different feeding methods
in oystercatchers (*Haematopus ostralegus*). Since you are concentrating all your
effort on a single individual (pair, group) and one, or a few, behaviors, focal-animal
sampling can provide accurate data on frequencies and durations of behavior when
combined with All Occurrences sampling (see below; Figure 8.2).

For our example of focal-animal sampling (Figure 8.1), we focus on male IV and
record all occurrences of standing up, feeding, and lying down. At the beginning of
the 30 minute observation period shown in the lower part of the diagram we note
that he is lying down. He stands up during the 15th minute, begins feeding one
minute later, feeds for 11 minutes, lies down during the next minute, and is still lying
down when our sample period ends. In this very simplified example, it would proba-

bly have been feasible to also record those behaviors for other individuals (e.g. the other males).

Although a specific individual receives highest priority when focal-animal sampling, this method does not necessarily restrict us to only that individual. For example, when social behavior is recorded, a focal-animal sample on an individual provides a record of all acts in which that animal is either the actor or receiver (J. Altmann, 1974). Some research requires that observations be made on two individuals simultaneously (e.g. mating, aggression/submission). For example, Huxley (1968) made his observations on one *focal-pair* at a time in his study of courtship in great crested grebes. Accurately recording the behavior of two individuals simultaneously can often be difficult so Smuts used two observers in her study of the formation of relationships in olive baboons (*Papio cyanocephalus*); one observer focused on the female and the other on the male (Ackerman, 1988).

In order to answer some research questions, observation of *focal-groups* may be required. As an example, Fujioka (1985) observed nine focal-families, one at a time, in his study of sibling competition in cattle egrets (*Bubulus ibis*).

8.3.1b *Out-of-sight time*

Focal-animals in the field and captivity often disappear from view of the observer. Therefore, it is necessary to record the time intervals in each sample period that the individual being observed is out of sight; the result is 'missing data' for those time intervals. The sampling protocol/data problem created by an animal under observation temporarily disappearing from view (an error of apprehending, see section 8.4) has not been successfully resolved. That is, there is no truly valid procedure for determining what behavior(s) occurred while that animal was out-of-sight. However, there is a general relationship between the predictability of what behavior occurred, the duration of behaviors most commonly observed, and the time the animal is out of sight. Four methods for dealing with this problem are discussed below relative to the duration of behaviors and time out of sight. Although *none of these methods is truly valid*, the hazards in using them can be reduced by following these guidelines:

For out-of-sight periods of long duration and when the durations of common behaviors are short (relative to the out-of-sight periods) do one of the following:

1 Delete the time out-of-sight from the sample; duration of the sample period is reduced accordingly.
2 Delete the time out-of-sight from the sample, but increase observation time until the time the animal was actually observed equals the time required for the predetermined sample period.

These two methods are valid only when the observer believes that the probability of performing any of the behaviors while the animal is in view is the same as when it is out of sight.

For out-of-sight periods of short duration and when the durations of common behaviors are long (relative to the out-of-sight periods) do one of the following:

1   Assign the behavior seen when the animal goes out-of-sight to the out-of-sight period.

2   Assign the behavior seen when the animal comes back into view to the out-of-sight period.

If the behavior is the same for 1 and 2, then the behavior recorded is more likely to be valid.

Behaviors occupying the largest percentage of the animal's time budget are those that are most likely to be interrupted. Two factors which should, perhaps, override or dictate use of the above methods are experience and common sense. Probably no one knows the animal better than you do; therefore, follow the course of action which you consider to be the most appropriate. Also, it is often wise to deal with data using two or more methods for comparison.

Losito et al. (1988) developed a 'focal-switch' sampling method to determine time budgets of mourning doves in view-restricted habitats. They used a 'standard wait period' to decide when to end sampling or pursue flocks lost from sight. This method reduced the out-of-sight data, increased the observers' efficiency by 12%, and saved 24% of the samples from premature termination.

Focal animals may be selected because of a research objective (e.g. mothers: duration of maternal behavior) or an experimental design restraint. For example, the design might call for a random sampling of individuals within a population, age class or treatment group.

Focal-animal sampling may strain the observer's ability to record data accurately, especially in dense aggregations or highly social species (e.g. monkeys). However, focal-animal all-occurrences sampling does provide for a rigorous examination of the behavior of individuals, and it is this type of sampling which will provide the most accurate and valid data to test hypotheses. For this reason, J. Altmann (1974) concluded that with the proper choice of behavior units, sample periods and focal individuals this method will generally be the best to use.

8.3.1c   *All-animals sampling*

With this method you sample a *few behaviors in a relatively large number of individuals*. For example, you could sample All Occurrences (see below) of urination in a herd of pronghorn antelope to determine its frequency at different times of day. Or you could score for each individual pronghorn whether it was feeding at selected points in time to determine the synchrony of the herd's foraging behavior (see example of *all-animals scan sampling* below).

All-animals sampling also allows you to collect data opportunistically on a specific behavior as it is sporadically performed by individuals in a large population. For example, you could observe the same herd of pronghorn antelope and opportunistically collect data on courtship sequences as they occur (see section 8.3.3a).

As in focal-animal sampling, individuals disappearing from view can be a problem for all-animal and *ad libitum* sampling (discussed below). Since it is rare that all individuals are equally visible, some researchers have attempted to measure individual *observability* (Chalmers, 1968; Sade, 1966). At some regularly scheduled time period (e.g. half-hour intervals) censuses are taken of the individuals which are visible; these are called *observability samples*. J. Altmann (1974:239) suggests that these adjustments are of limited compensatory value: 'Observability samples provide an accurate correction only to the extent that the probability of a behavior being recorded if any individual performs that behavior is directly proportional to the percent of time that the individual is visible.' Altmann goes on to point out that there are, at least, the following three potential sources of failure to obtain consistently proportional samples:

1　Individual or class-specific differences in observability may vary with different behaviors.
2　A specific behavior may affect observability.
3　Observers' preferences (decisions) in sampling specific behaviors introduce biases, as well as their attempts to compensate for these biases (see observer bias in section 8.4).

8.3.2 *Ad libitum* sampling

As *ad libitum* implies, no restraints are employed in sampling behavior with this method; these are opportunistic observations that result in 'typical field notes'. What is generally recorded are the behaviors of those individuals (or groups) that are most easily observed.

Ad libitum sampling is used during early reconnaissance observations and when you are developing an ethogram (Chapter 4). This type of sampling is also commonly used in descriptive research. J. Altmann (1974:235) stated that, 'In field studies of behavior [*ad libitum* sampling is] perhaps the most common form of behavior record', an unfortunate and regretable fact. That is not to say that 'typical field notes' are not important, but rather that the relative abundance of observations made in this way, even for descriptive studies, could have been improved by using some form of random or systematic sampling.

During periods in the field when you are not involved in collecting data for your experimental research, *ad libitum* field notes can often yield insights. The value of these unplanned, *ad libitum* field notes was heralded by Tinbergen:

> Scientific examination naturally requires concentration, a narrowing of interest, and the knowledge we gained through this has meant a great deal to us. But it has become increasingly clear to me how equally valuable have been the long periods of relaxed, unspecified, uncommitted interest . . . an extremely valuable store of factual knowledge is picked up by a young naturalist during his seemingly aimless wanderings in the fields. Nor are the preliminary, unplanned observations one does while relaxed and uncommitted without value to the strict experimental analysis.   *[Tinbergen, 1958:287]*

Since *ad libitum* sampling is most often used when ethologists are recording as much as they can during an unplanned encounter with a species, during reconnaissance observations, or when they are developing an ethogram, the value of those field notes as data for hypothesis testing is very limited. When the observations are treated as data and quantitative comparisons made, the following assumption generally must also be made: the true probability of observing the different sexes, age classes and behaviors is reflected in your notes. Comparisons cannot be made across time since the number of samples is probably small and the samples were not taken randomly.

For example, we might encounter the six deer represented in Figure 8.1 and observe them for the 32 minutes indicated in the lower part of the figure. During that period we would record that: 1. all but female I fed; 2. all individuals laid down; and 3. three of the six individuals stood up. That is, we could have recorded those behaviors or we could have been temporarily focusing on other behaviors and missed portions of the complete record. Let us say that when female VI laid down at minute 25, she laid close to female III, and they began head-butting intentions and making threats to each other. We might have focused our attention on them for the next 15 minutes and missed seeing male II stand up, feed, and lie down again. That is, unless we had decided previously to record all occurrences of the behaviors diagrammed in Figure 8.1, it is unlikely that we could have duplicated that portion of the diagram in our 32–minute sample from our field notes.

Besides providing information about the feasibility of a planned study (reconnaissance observations) and development of an ethogram, *ad libitum* sampling provides questions, ideas and hypotheses for future research and often reveals rare, but significant, behavioral events.

8.3.3 Continuous recording sampling methods

With continuous recording sampling methods, the researcher records a complete account of all behavior units of interest; that is, we would obtain data on: occurrence, duration and sequences of both states and events. These sampling methods provide the most complete and accurate data.

8.3.3a *All-occurrences sampling*

It may be desirable to concentrate on one, or a limited number of, behaviors and record all occurrences (called 'event-sampling' by Hutt and Hutt, (1974, and 'complete record' by Slater, 1978). All-occurrence sampling is often combined with focal-animal sampling since it is difficult to record all occurrences accurately on several individuals simultaneously. All-occurrence sampling of selected behaviors is possible if the following conditions exist:

1 The animals and the behaviors are easily observed.
2 The behaviors have been carefully defined so that they are easily recognized.
3 The behaviors do not occur more often (or more rapidly) than the observer can record them.

This method of sampling can provide accurate data on the following:

1 Frequency and rate of occurrence (and temporal changes in rate) of the selected behavior(s).
2 Restricted sequencing (see example below).
3 Behavioral synchrony (see example below).

For example, assume we are interested in the sequence of the initiation of feeding (event) and the synchrony of feeding (state) in the six mule deer in Figure 8.1. We selected the 60-minute sample period shown in the lower part of diagram and recorded the initiation and termination of feeding for each individual (events indicated in the diagram). We can then examine the data for the sequence of initiation of feeding for the six individuals, as well as how many and which individuals fed at the same time (behavioral synchrony). In this case, we can gain information about a state (feeding) by recording all occurrences of two events (initiation and termination of feeding).

Fig. 8.3 Courtship of mallard duck, a species of surface-feeding duck, involves the following sequence of behaviors: (3) tail-shake, (2) stretch-shake, (3) tail-shake, initial posture, (1) head-flick, (4) grunt-whistle, (3) tail-shake, initial posture, (5) head-up-tail-up, (6) looking towards the female, (7) nod-swimming, and (8) showing the back of the head. Also shown are down-up (9) and bridling (10). Bridling is a postcopulatory display. (Drawing by Hermann Kacher in collaboration with Konrad Lorenz; permission granted by Hermann Kacher.)

8.3.3b *Sequence sampling*

In sequence sampling the focus is on a chain of behaviors. These may be performed by a single individual (e.g. courtship displays in male ducks; Figure 8.3) or they may be behaviors alternating between two (or more) individuals (e.g. courtship in the queen butterfly; Figure 8.4).

The initiation of a sample period is usually determined by the beginning of a sequence. An experienced observer can often anticipate the initiation of a sequence in an individual and an impending interaction between two or more individuals. The sample period ends when the observed sequence terminates.

There may be difficulty in specifying the beginning and end of a sequence, as well

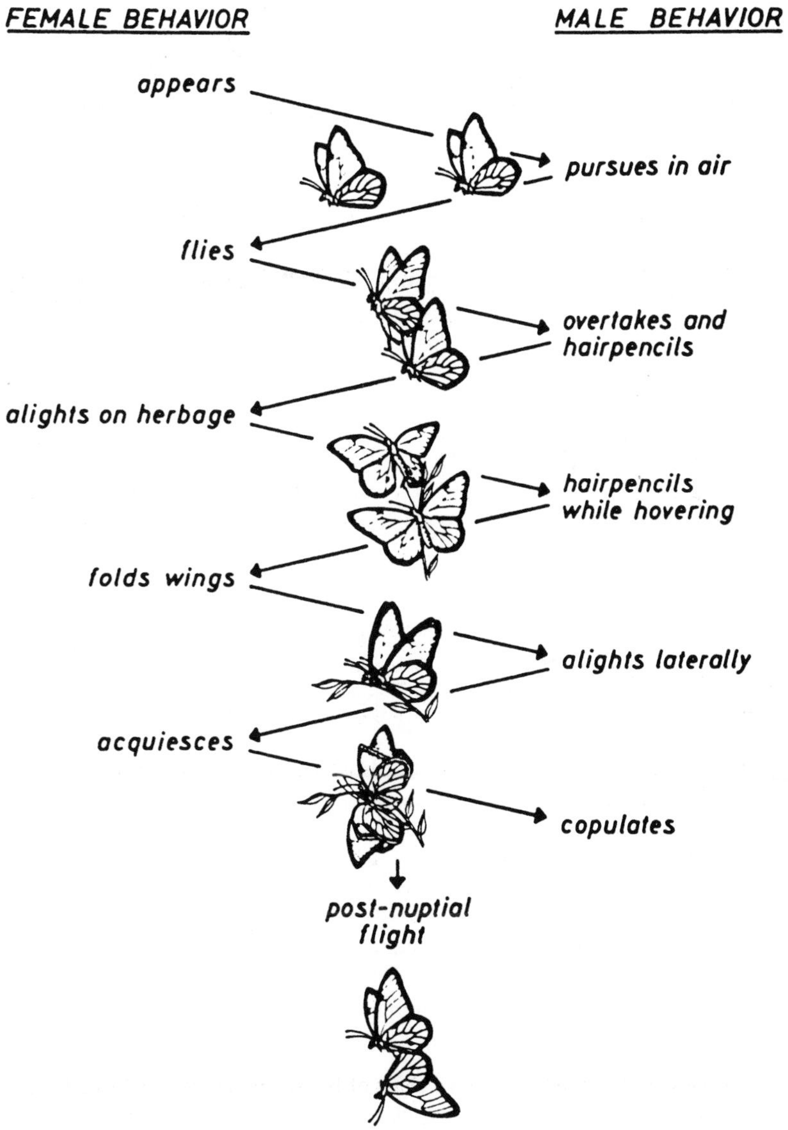

Fig. 8.4 Courtship behavior of the queen butterfly. Note that the female permits copulation to occur only after the male completes a series of courtship actions, including extruding the hairpencils and releasing pheromones (from Brower *et al.*, 1965).

as in selecting individual sequences or social interactions at random. J. Altmann (1974) discussed the sampling bias caused by differing lengths of individual sequences (or social interactions):

> If . . . the observer always begins sampling at the onset of a sequence and chooses the next sequence to sample at random among sequence onsets or in any other way that samples sequences of each length in proportion to their *frequency of occurrence*, the resulting data will be unbiased with respect to sequence length: the total time spent with sequences of, say, duration d_i, will be proportional to d_i times f_i, where f_i is the frequency of sequences of length d_i. Then the time spent with sequences of different lengths, not the probability of choosing such sequences, will be in proportion to the total time taken up by sequences of that length.   *[J. Altmann, 1974:250]*

Sequence sampling of social interactions has some other potential problems. Interactions may branch or converge; that is, an interacting group may break up (branch) into subgroups or other individuals may join an interacting group which is under observation (converge).

It is up to the observer to record as clearly as possible the branching and converging of interactions. These are recognizable events that are often important parts of social interactions among large groups. They may force the observer to develop additional observational skills, including peripheral vision; but they should be included in the observational record or they should be eliminated from the experimental design. For example, Hazlett and Bossert (1965) restricted their observations to interactions between two individual crabs (dyads).

As an example, we might be interested in the sequence of behaviors that the mule deer in Figure 8.1 go through from the time they stand up to feed until they lie down. The first two bars (sample periods) after sequence sampling show our observations of male II. We begin our sampling as he stands up and terminate it when he has laid down. We might, of course, be interested in additional behaviors that are not included in the diagram, such as limb movements associated with lying down and standing up. The third and fourth bars illustrate a sample of interactions in which we are looking at a relationship between the behavior of females V and VI. We record the events which occur for each of them, beginning with the first one to stand up and terminating when the other has laid down; the next sample period begins when one of them stands up. The data would then be analyzed for correlations between the two individuals' behaviors; if significant correlations existed, then we would design experiments to test for cause and effect.

Note that in the previous example we used all-occurrences sampling to determine the sequence of initiation of feeding in the six deer.

8.3.3c *Sociometric matrix*

A *sociometric matrix* is really an experimental design or a way of tabulating data. Collection of data for a sociometric matrix can be considered a special type of *all-occurrences* sampling in which the observer records interactions between pairs of individuals (e.g. transmitter–receiver, groomer–groomee) or records social interactions of an individual (*focal animal*) during a specified sampling period.

For example, we might want to measure the synchrony between individuals in standing up in the group of six mule deer in Figure 8.1. We suspect that the behavior of certain individuals, when standing up, stimulates others to stand up, so we construct the sociometric matrix below. We record the initiator and follower, if the follower stands up in less than 60 seconds after the initiator has stood up. The data in the matrix below are from the 60–minute sample period in Figure 8.1.

Follower

		I	II	III	IV	V	VI
	I					1	
	II			1			
Initiator	III						
	IV	1		1		1	1
	V	1					1
	VI					1	

The above data *suggest* that individual IV is more of a leader than a follower; but we would, of course, need to record a large number of these interactions in order to demonstrate significant correlations.

In most instances the researcher uses a sociometric matrix to test for one-sidedness in dyadic interactions. Therefore, an attempt is made to record as many interactions as possible without regard to random or systematic sampling. Hence, the data cannot be compared between cells, and the matrix cannot be treated as a true contingency table, but more as a form for tabulating data.

8.3.4 Time sampling

When using a Time sampling method, the observer records either: 1. the behavior state that the animal(s) is performing at points in time (instantaneous/scan sampling; point sampling, Dunbar 1976); or 2. whether a behavior state or event occurred during a sample interval delineated by points in time (one–zero sampling). These methods are often used under the following conditions:

1 We want to gather data on a few behaviors while simultaneously sampling a relatively large group of individuals (e.g. studies of behavioral synchrony; daily activity patterns; percentage of time spent in behaviors of specific interest).
2 We want to gather data on a larger number of behaviors on a few individuals (e.g. juvenile females), than we can with continuous sampling methods (e.g. time budgets for an exhaustive list of mutually exclusive behaviors).
3 We want to maintain high inter-observer reliabilty (see below) when several observers with varying levels of ability and experience are by necessity involved gathering data.

8.3.4a *One–zero sampling*

With one–zero sampling, the observer scores whether a behavior occurs (one), or not (zero), during a short interval of time (sample period). It is suitable for recording states and/or events. This method has been referred to as 'time-sampling' (Hutt and Hutt, 1974) or the 'Hansen system' (Fienberg, 1972). This method has the following features:

1 In each sample period the occurrence or non-occurrence (not frequency of occurrence) is scored.
2 Behaviors of one or more individuals can be recorded in each sample period.
3 Occurrence refers to either an event or a state (ongoing at some point during the sample period).
4 The sample periods are generally short (e.g. 15 seconds), and several (e.g. 50) are used in succession.
5 It is easy to record data using this method; it is not as demanding of the observer's total concentration, hence you can often use longer observation periods.
6 Inexperienced observers can quickly learn to use this method with a level of accuracy that results in high measures of inter-observer reliability (see below).

As an example, in Figure 8.1, there are four groups of one–zero samples of 10 sample periods each. The sample periods are of one minute duration, although they would normally be much shorter. If we are interested in one–zero scoring of the event of standing up (for any individual in the group), then the first set of 10 sample periods would all contain '0' scores, the second set two '1' scores, the third set three '1' scores, and the fourth set two '1' scores.

Although, by definition, one–zero sampling requires that a behavior be scored only once per sample period regardless of the number of times it occurs, some researchers have recorded all occurrences but analyzed the data as one–zero scores (e.g. Kummer, 1968). Slater (1978) suggested that one–zero data may be useful as a first-approximation look at associations between behaviors, by determining how frequently they occur together in the sample intervals.

J. Altmann (1974:253) pointed out a common fallacy in interpreting one–zero data: 'It is too easy for both author and reader to forget that a one–zero score is not the frequency of *behavior* but is the frequency of *intervals* that included any amount of time spent in that behavior'. However, the data from one–zero samples are sometimes presented as *Hansen frequencies* (the number of sample periods in which the behavior occurred/total number of sample periods, based on Hansen, 1966), when the sample periods are short in duration but large in number. S. Altmann and Wagner (1970) described a method for using Hansen frequencies to estimate the mean rate of occurrence of events when the events approximate a Poisson distribution (Chapter 11); that is, ' that the behavior occurs randomly at a constant rate, that the chance of two or more simultaneous occurrences of the behavior is negligible, and that the chance that a particular behavior will occur during an interval is independent of the time that has elapsed since the last occurrence of that behavior' (Altmann and Wagner, 1970:182); see Chapter 15 for a discussion of calculating rates of behavior.

Caution should also be used when converting one–zero scores to percentage of time spent in a behavior (Simpson and Simpson 1977). This would be accurate only if the behavior lasted for the complete sample periods in which it was scored. If researchers desire to use one–zero data for 'time-spent' estimates, they must determine how closely their data approximate the above condition.

Another potential problem arises when recording a state (ongoing behavior) that continues through several sample periods and is scored for each one. In this case there is no close relationship between the number of scores (i.e. intervals in which the behavior occurred) and actual frequency of occurrence (Dunbar, 1976). However, if the sample period is sufficiently short relative to the behavior's duration and the interval between successive occurrences, then the observer can obtain (with reasonable accuracy) both frequency and duration through careful data analysis. For this to be fairly accurate, the probability of both a termination and an onset occurring in one sample must be negligible. Martin and Bateson (1986, 1993) illustrate a method of determining the length of sample intervals when using any time-sampling method. It is based on obtaining true frequencies and durations through continuous sampling and then determining the amount of error that would have been introduced if the same behavior had been time sampled at different intervals. In addition, Adams and Markley (1978), using computer-generated 'behaviors' with a variety of known mean durations and frequencies, found instantaneous sampling (see next section) to be superior to one–zero sampling for estimating duration of behaviors; their results support the conclusions of several similar studies. However, there are also studies that report one–zero sampling as being reasonably accurate, especially when the frequency of occurrence of the behavior is high and the sample interval is short relative to behavior duration and bout length (e.g. Griffin and Adams 1983; Rhine and Ender 1983). Rhine and Flanigan (1978) compared how well One-Zero, All Occurrences, and Instantaneous sampling methods reflected the actual rates and durations of primate social behaviors. Besides concluding the One-Zero sampling can be an accurate method, they suggested five situations in which this method would be especially useful:

1   To provide a single measure of social relatedness that combines the percentage of time individuals spend together and their actual frequency of interaction.
2   To provide high inter-observer reliability.
3   To obtain data to be compared with data previously collected using one–zero sampling,
4   To avoid arbitrary definitions of a behavior's start and end times.
5   To obtain information about social relatedness for less time, effort and expense than other methods.

In contrast, Kraemer (1979) summarized the results of her analysis of the use of one–zero sampling of primate behavior as follows:

> . . . in order to maximize the possibility of comparing results of studies done in different research milieus, to render the type of changes which are seen to occur clearly interpretable in terms of the characteristics of the behavior, and to eliminate the possibility of obscuring a real change in behavior by an unfortunate choice of sampling interval, methods other than One–zero sampling are preferred   *[Kraemer, 1979:243]*

The major disadvantage of this sampling method is that accurate information about actual frequency and actual duration may be lost, especially if the observer does not use appropriate sample intervals. The researcher has to weigh this disadvantage against the ease of recording observations and the high inter-observer reliability which this method provides. J. Altmann (1974:258) concluded that 'In short, neither ease of use nor observer agreement *per se* provides an adequate justification for the use of this technique.'

8.3.4b *Instantaneous and scan sampling*

Instantaneous sampling

Instantaneous sampling is a special type of time sampling in which the observer scores an animal's behavior at predetermined 'points' in time. This method has been called 'time-sampling' by Hutt and Hutt (1974), 'point sampling' by Dunbar (1976), and 'on-the-dot sampling' by Slater (1978). The major benefit of instantaneous sampling is the relative ease of recording data versus all-occurrences sampling.

This method works well with behavioral states, but it is not recommended for use with events. Behavioral events and the sampling points are both instantaneous; hence the probability of them occuring together is remote.

Instantaneous sampling is often used to obtain data on the time distribution of behavioral states in an individual; that is, to determine time budgets. This method provides reliable estimates of true time use if the sampling interval is short relative to the mean duration of the states being measured, and if average time use is calculated from data collected from several individuals. It is often used to sample states, since the probability of scoring events with this method is remote. This method can be used to obtain data on the time distribution of behavioral states for an individual. For example, Dunbar (1976) found that sampling intervals of 5, 10, 15, 30 and 60 seconds all gave reliable estimates of time use for states that varied in mean duration from 14.0 to 124.6 s. Likewise, Tyler (1979) found that the same sampling intervals, used by Dunbar, gave reliable estimates of time budgets for behavioral states that ranged in mean duration from 40 to 50 s. Poysa (1991) cautions that, although averages from several individuals may provide reliable estimates of time use, data from individual instantaneous samples may differ greatly from true time use; this can be a serious problem if you try to relate this individual variation in time use data to an independent variable.

As an example, suppose we wanted to record the time distribution of feeding in female III in Figure 8.1. We set our sampling points in the middle of the periods we used for one–zero sampling. Occurrences of feeding would be recorded at sample points 2 through 9 in the first set, 4 through 10 in the second set, 1 through 4 in the third set, and at no sample points in the fourth set. The frequency of occurrence recorded decreased as our sample periods progressed; you can see how closely this approximates the real situation in the diagram.

Scan sampling

Scan sampling is simply a form of instantaneous sampling in which several individuals are 'scanned' at predetermined points in time and their behavioral states are scored; that is, instantaneous samples are taken on several individuals at the same time. This allows the observer to record relatively accurate data on a few behaviors for a relatively large number of animals. For example, we could use the same sample points in Figure 8.1 for scan sampling all six deer that were used in the previous example to instantaneous sample female III.

For example, in a study of locust feeding behavior, five locusts, in individual containers, were scan sampled at 10 second intervals, and one of four behaviors was scored (Simpson *et al.*, 1988). Six mother–pup pairs of harp seals were scan sampled at 30 second intervals, and one of 18 behaviors were scored in a study of maternal behavior (Kovacs, 1987). As many as 80 female grey seals were scan sampled at 10 min. intervals to obtain time budgets for 11 behaviors (Anderson and Harwood, 1985). From these examples, it can be seen that sampling interval is determined by the number of individuals being scanned and the number of behaviors being recorded.

The number of individuals scanned and the number of behaviors cannot be so large, nor the sampling interval so short, that all individuals cannot all be easily sampled, with time to spare, before it becomes time to take the next scan sample. An observer should attempt to be as instantaneous as possible, for the longer they linger on one individual, the more the sample approximates a series of short focal-animal samples of unknown durations. Estimates of time spent scanning individuals which are difficult to observe, as well the time spent scanning the entire group, should be made.

One important use of instantaneous and scan sampling is to estimate the percentage of time that individuals spend in various activities (i.e. time budgets). One–zero sampling is effective in determining time spent in different behaviors only under limited conditions (see above; Simpson and Simpson, 1977), hence it is not recommended. Caution should be used when estimating rates and relative frequencies for instantaneous and scan samples, since they tell us nothing about the frequencies of the behaviors or when each state actually began and ended. The caution is the same as that expressed for one–zero samples. 'In the special case where the interval between instantaneous samples is short enough that no more than one transition can occur between consecutive samples, the resulting data are essentially equivalent to that of focal-animal sampling for rate and relative frequency estimates, but have a greater margin of error for duration estimates' (J. Altmann, 1974:261).

8.3.5 Summary

Each sampling method described above has recommended uses. By properly combining focal-animal or all-animal sampling with one of the other methods, a researcher is able to maximize accuracy, reliability and efficiency of data collection and ensure that valid data are collected for testing the research hypothesis.

Table 8.7. *Sampling methods and recommended uses*

Sampling method	State or event sampling	Recommended uses
1. *Ad libitum*	Either	Primarily of heuristic value; suggestive; records of rare but significant events
2. Sociometric matrix completion	Event	Asymmetry within dyads
3. Focal-animal	Either	Sequential constraints; percentage of time; rates; durations; nearest neighbor relationships
4. All occurrences of selected behaviors	Usually event	Synchrony; rates
5. Sequence	Either	Sequential constraints
6. One–zero	Usually state	None
7. Instantaneous and scan	State	Percentage of time; synchrony; subgroups

Source: From J. Altmann (1974)

Figure 8.2 provides a flow chart that will assist in deciding which sampling method is appropriate for your study. J. Altmann (1974) has also provided a table to assist in the selection of the proper sampling method (Table 8.7).

Different sampling methods produce different types of data, including different scales of measurement. Hence, the validity of the research and the statistical tests that can be used will be affected by the sampling method employed.

Dunbar (1976) presented an excellent demonstration that the use of various behavioral parameters and sampling methods *to answer the same question* can lead to different conclusions. His demonstration challenged the internal validity of several methodologies. He estimated 'social relationships' among individual gelada baboons (*Theropithecus gelada*) divided into 11 age–sex classes, using three different behavioral units that showed the extent to which individuals: 1. *interacted* with each other; 2. *groomed* each other; and 3. were in a particular *spatial relationship*.

Data were collected in the following seven combinations of behavioral units and sampling methods:

1 The number of social contacts between individuals (i.e. the number of times they interacted, irrespective of the number of social acts exchanged in each interaction) (focal-animal and all-occurrences sampling).

2 The number of social acts exchanged between individuals (all-occurrences sampling).

3 The number of 30-second time intervals in which individuals interacted (one–zero sampling).

4 The number of 60-second instantaneous or point samples in which individuals were interacting.

5 The number of grooming bouts exchanged between individuals (all-occurrences sampling).

6 The number of 60-second instantaneous or point samples in which individuals were grooming.

7 The number of 30-second time intervals in which individuals were within arm's reach of each other (one–zero sampling).

The data collected by the seven methods are shown in Table 8.8. There are obvious differences in the data among the seven sampling method–behavioral unit measurement techniques. When each pair of measurements is compared to determine if they rank the interactees in the same order, the correlations are quite high. Therefore, any of the methods would appear to be suitable if we were interested in only an ordinal scaling of the interactees (e.g. adult females interacted more than three-year males, who interacted more than adult males).

However, if we try to make direct comparisons among relative probabilities of interaction (interval scaling) (Table 8.8), we find larger discrepancies are due in part to the following:

1 By using seven different measurement techniques a 'social interaction' has been operationally defined seven different ways.

2 Social interactions between different sex–age classes are often expressed in different ways.

3 Each type of sampling method has inherent biases including the problems of frequency and duration (discussed previously; see also Dunbar, 1976).

Dunbar's exercise exemplifies the need to state clearly the hypothesis to be tested and to define carefully the behaviors to be measured. Make certain that the sampling method you choose is a valid measure of the behavior you want to measure. This is essentially a subjective decision dictated by your definition of the behavior you want to measure; but it is strengthened by a sound knowledge of the biases inherent in the various sampling methods. If possible, compare different sampling methods on the same data, as did Dunbar.

Other reviews of animal behavior sampling methods can be found in Altmann (1984), Crockett (1995), Lehner (1992), Martin and Bateson (1986, 1993), and

Table 8.8. *Frequencies with which subjects 'interacted' with the various age–sex classes as determined in different ways (all subjects combined) and, based on these data relative strength of relationship with each age–sex class as determined by the different estimators given as a probability*

| | Age–sex class of interactee | | | | | | | | | | | |
Estimator	Adult male	6 y male	5 y male	4 y male	Adult female	3 y male	3 y female	2 y male	2 y female	Yearling	Infant	Total
Frequency of interaction												
1. Number of contacts	3	3	1	2	8	11	3	3	2	10	1	47
2. Number of acts	19	10	3	7	47	34	8	15	8	17	1	169
3. One–zero interacting	34	33	3	17	99	49	11	39	9	10	1	305
4. Point interacting	15	15	1	7	44	19	3	16	5	1	—	126
5. Grooming bouts	16	9	1	5	38	18	7	15	2	3	—	114
6. Point grooming	15	15	1	7	44	17	3	16	4	—	—	122
7. Spatial measure	45	37	4	19	111	51	17	40	21	67	3	415
Relative strength of relationship												
1. Number of contacts	0.064	0.064	0.021	0.043	0.170	0.234	0.064	0.064	0.043	0.213	0.021	
2. Number of acts	0.112	0.059	0.018	0.041	0.278	0.210	0.047	0.089	0.047	0.101	0.006	
3. One–zero interacting	0.111	0.108	0.010	0.056	0.325	0.161	0.036	0.128	0.030	0.033	0.003	
4. Point interacting	0.119	0.119	0.008	0.056	0.349	0.151	0.024	0.127	0.040	0.008	0.000	
5. Grooming bouts	0.140	0.079	0.009	0.044	0.333	0.158	0.061	0.132	0.018	0.026	0.000	
6. Point grooming	0.123	0.123	0.008	0.057	0.361	0.139	0.025	0.131	0.033	0.000	0.000	
7. Spatial measure	0.108	0.089	0.010	0.046	0.267	0.123	0.041	0.096	0.051	0.161	0.007	

Notes:

These date were pooled and treated as though they came from the same individual or sex-age class for the purpose of comparing methodologies. No conclusions about gelada-baboon behavior should be drawn from the data.

Source: From Dunbar (1976).

Sackett (1978). Also, an overview of behavior sampling methods and examples of them being used in a zoo is provided by the videotape program created by Steven Hage and Jill Mellen: *Research Methods for Studying Animal Behavior in a Zoo Setting* (available from the Minnesota Zoological Garden, Apple Valley, MN 55124).

8.4 OBSERVER EFFECTS

As methods have biases, so do observers, but bias is only one factor that contributes to observer error (Rosenthal, 1976). Observer errors can contribute to a decrease in both reliability and validity; therefore the results are only as good as the observer (Kazdin, 1982).

Good data mean that they are an *accurate* measurement of the true situation. Observers may be *precise*; that is, their data will not vary greatly, but because of biases they might not be accurate. For example, consider two riflemen firing at separate targets at a rifle range. Rifleman A groups his five shots closely (good precision) and in the bull's-eye (good accuracy). Rifleman B groups his five shots closely (good precision) but to the right of the bull's-eye (poor accuracy). Rifleman B's accuracy is biased by his habit of pulling his rifle slightly to the right while squeezing the trigger.

To carry the analogy to completion, we may say that each man's shots were measurements of the location of the bull's-eye. Rifleman A's measurements (shot holes in target) were both reliable (precise) and valid (accurate). Rifleman B's measurements were reliable, but inaccurate.

Several potential observer errors, which may have severe effects on the results of ethological studies, have been discussed by Rosenthal (1976) and are depicted in Figure 8.5. *Observer effect* is due to the visual presence of the observer, or other stimuli (e.g. odor) from the observer, and results in a change in the animal's behavior. In psychology and sociology, this change in the subject's behavior is known as the *Hawthorn effect* (Martin and Bateson, 1986). An indirect observer effect was measured by Gotmark and Ahlund (1984). They determined whether observers attracted predators (crows and gulls) to the nests of common eiders (a result that would affect both the reproduction and behavior of the eiders); they found that human disturbance did increase foraging effort and success for gulls, but not for crows. Examples of direct observer effects on several species, in both field and laboratory studies, are described in Davis and Balfour (1992).

> Of course, it is sheer metaphysical conceit to claim that we ever do achieve a strictly naturalistic picture of behavior. To do so would require access to phenomena 'as they are.' The best we can do is to try for samples of events that are representative and valid, and hope that our analysis of what happened is reasonably accurate.   *[Menzel, 1969:80]*

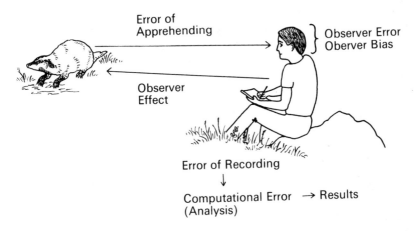

Fig. 8.5   Types of observer effects encountered in ethological research (drawing by Dan Thompson).

Error of apprehending is due to the physical arrangement of the animal and/or the observer making it difficult to observe the behavior. For example, in a study of mockingbird behavior, Breitwisch (1988) stated his avoidance of error of apprehending as follows:

> I did not record wing-flashing, a display sometimes given in the presence of predators and other contexts . . . because I could not reliably observe birds wing-flashing in the crowns of nest trees.   *[Breitwisch, 1988:63]*

Observer error is caused by many factors, including inexperience and poorly defined behavioral units. Observer error also includes observer 'drift' and 'decay' defined by Hollenbeck (1978:96) as follows: 'Drift refers to the *movement of an observer* in time from some base point either in a positive or negative direction. Decay, on the other hand, implies *that the instrument*, which includes the observer and the scoring categories, is drifting beyond the bounds of acceptable measurement error.'

Observer Bias is principally related to expectancies of the observer. Biases may be consistent or inconsistent and either high or low. Rosenthal and Rubin (1978) provide a thorough review of this bias, and Arber (1985) addresses it from the standpoint of the philosophy of science. Following his study of mountain sheep, Geist (1971) stated his concern about this bias as follows:

> How could I, the same person, separate the process of data gathering from any thought I might have about those data? As the lone investigator I had to record observations and simultaneously think about the observed events in order to recognize relationships. This is a

> very important problem, for if data gathering is not clearly separated
> out, it can be biased by some subconscious preference.   *[Geist, 1971:xii]*

In truth, research hypotheses are really personal prophesies, and to strengthen confidence in our ethological insight most of us would like to see them come true. In addition, Bell (1979) has argued that humans are inherently more 'subjective' than 'objective'. We must guard against these basic human traits if we are to reduce observer bias to an acceptable level. This bias can be controlled in manipulation experiments by using a *blind* procedure. The manipulations are made by one researcher, and a different person takes the measurements 'blind' to the treatment to which each animal has been exposed. Martin and Bateson (1986, 1993) provide a good overview of the effectiveness of blind experiments.

Error of recording may be due to poor techniques and equipment, mental lapses in the observer, and inexperience. Some researchers have chosen to use a separate observer and recorder (e.g. Simpson *et al.*, 1988); depending on the abilities and experience of both, this procedure could either increase or decrease the error of recording.

Computational error is due to incorrect manipulation and analysis of data. It is the error that results if an inappropriate statistical test is used (Chapters 12–17).

All of these observer effects can have a severe effect on the results of the research. The results may not only be invalid for this study, but they may also serve as the basis for another ethologist's research and compound the problem.

8.5 RELIABILITY

Selection of the proper behavior unit, scale of measurement, states or events, and sampling method will all affect your study's reliability and validity (Chapter 6).

Reliability refers to the reproducibility of the measurements. That is, how much can we rely on our own ability to obtain very similar data again (*intra-observer reliability*)? Other observers, in this study or future studies, should also be able to replicate our measurements; that means we should have good *inter-observer reliability*. This is, of course, often difficult, since skill in observation develops through practice. Accurate and precise observation is not only a general skill (good ethologist), but also a specific skill (e.g. good butterfly ethologist or good acoustical-communication ethologist).

Hollenbeck (1978) concluded that reliability consists of both stability and accuracy. However, this is true only of inter-observer reliability. As was discussed earlier (p. 210), as long as *precision* (stability) is maintained, an observer will be *reliable* but might still be *inaccurate*. Therefore, intra-observer reliability is solely a measure of stability (or precision), whereas accuracy affects validity (Chapter 6). However, accuracy will almost certainly affect *inter-observer reliability* (Kazdin, 1982), since few observers are likely to have the same biases. An accuracy criterion can be established by using an 'expert' observer (e.g. the principal investigator), or a 'calibrating' observer (Kratochwill and Wetzel, 1977), or the consensus of several observers. Measurements of reliability are discussed below.

8.5.1 Intra-observer or self-reliability

Observers do not see and hear all they would like to during each observation. Many times in the field, after having observed an unusual behavior, I have muttered 'I wish I could see that again'. We rarely have that opportunity. Only if we have captured it on videotape or film are we afforded the privilege of seeing it again (then only in two dimensions), and few observers are able to videotape all their observations. Therefore, we must be as efficient and accurate as possible in recording behavior as it occurs.

Experience is the most important factor in increasing efficiency and accuracy and is necessary with the following aspects of all ethological studies:

1 Observing animals, in general
2 Observing individuals of the species to be studied
3 The behavioral units to be measured
4 The sampling methods to be employed
5 The data recording methods and equipment to be used

Intra-observer reliability can be measured by videotaping or filming a segment of the animal's behavior which is under investigation and then observing the same segment several different times at varying intervals. Your perception of the behavior, and the data you record, should not differ whether you see the segment three times in succession or at two-day intervals. We all question the accuracy of our own observations and are prone to attempt improvement. However, strive for objectivity and stability; do not redefine behavior units as you proceed, unless you are willing to start the research over again. Be careful not to overlay interpretations on your observations as you record data. This is often very difficult; but every observer must check the tendency to see the behavior of animals in light of the observer's own experiences (see Chapter 3). Continually assess the validity of your data and the behavioral units you have selected. However, *do not redesign your research in midstream without careful consideration of the consequences*. Search for and acknowledge weaknesses in your study, such as an inability to identify individuals.

Many types of study, particularly of social behavior, necessitate the *recognition of individual animals*. Any longitudinal study of a group of animals is strengthened by recognition of individuals. This can be assisted by marking individuals (discussed later in this chapter) or by relying on individual variation in morphology (e.g. antler configuration) and/or behavior (e.g. walks with a limp). D.K. Scott's reliability in identifying individuals in her study of the social behavior of Bewick's swans (*Cygnus columbianus bewickii*) was measured by Bateson (1977). More than 100 individual swans were identified by Scott as they were photographed in color by Bateson. The 30 clearest slides were selected and shown to Scott 14 days later, at which time she correctly identified 29 of the 30. She then correctly identified 23 swans from 30 inferior photographs. This test demonstrated that Scott could reliably identify a large number of individual swans and provided increased credibility to her claim that she could identify some 450 individual swans.

8.5.2 Inter-observer reliability

... the greatest source of errors in using observations made by others is that no two people who look at the same thing see the same thing.

[Lorenz, 1935:93]

Lorenz was referring to basic differences in observers' perceptions of the same behavior that are a result of both *inherited* and *learned* traits. For example, poor eyesight and hearing may be inherited, but training will provide an observer with the ability to use his senses with increased effectiveness.

Several factors mentioned above as contributing to observer effects also affect inter-observer reliability:

* *Error of apprehending*: For example, can each observer see and hear the animal(s) equally well?
* *Observer error*: For example, are the behavior units clearly defined for all observers? Poor intra-observer reliability will be reflected in poor inter-observer reliability.
* *Observer bias*: For example, do one, or all, observers have preconceptions about what behavior the animal is likely to perform in different situations?
* *Error of recording*: For example, is each observer recording data in the same manner?

Poor inter-observer reliability is a frequent problem in field studies, where errors of apprehending are difficult to overcome. The other observer effects can usually be reduced to a minimum through training and experience.

Inter-observer reliability should be measured periodically in order to insure accuracy and validity in the study as a whole. A simple, first approximation measure of inter-observer reliability is *percentage agreement*. This entails taking a sample of observations made by all observers and determining the percentage which are the same. This can also assist you in detecting observer effects as a source of aberrant data.

For example, suppose that four observers are collecting instantaneous samples (every 20 seconds) of behavior of zebra stallions living in small family groups in central Africa. They are interested in developing a time budget for the following behaviors: feeding (F), resting (R), grooming (G), walking (W), agonistic (intragroup: A_1, intergroup: A_2). In order to measure inter-observer reliability they decide to all collect data for the same 10-minute period (30 samples) on the same zebra stallion (focal animal). The hypothetical results are presented in Table 8.9.

The four observers recorded the same behaviors only 60% (18/30) of the time. However, Observers 1, 2 and 4 recorded the same behaviors 93% (28/30) of the time, and Observers 1 and 4 had 100% reliability. Therefore, the team as a whole was not very reliable (60%), and Observer 3 was the most unreliable. Observer 3 should not be allowed to collect data, and efforts should be made to increase his or her reliability to >95%. The reliability of 93% of Observers 1, 2 and 4 is probably acceptable, but attempts should be made to improve it. However, reliability >90% may be the most we can hope for in field studies, due primarily to error of apprehending.

Hailman (1971) tested the effect of stimulus preference on the begging response of newly hatched laughing-gull (*Larus atricilla*) chicks. Most data were collected simultaneously by two observers, and inter-observer reliability was checked using 'percentage agreement'. In another experiment, Hailman tested all the chicks, and then another observer tested the same chicks while Hailman was not present. The results showed perfect rank-order agreement with regard to stimulus orientation; but the repeated testing had suppressed the response rates.

Successive tests of inter-observer reliability can be used to measure bias in inter-experimenter reliability, as was Hailman's objective (above). However, to measure inter-observer reliability best the observers must see the *same* behavior, either simultaneously in real-time or by videotape or film. If videotapes or films are used, observers viewing the behavior in these media should not be tested against those viewing the live behavior. Simply put, inter-observer reliability can be accurately measured only if everything is the same except the observer.

Several authors have cautioned that 'percentage agreement' does not provide an accurate measure of inter-reliability since it includes agreement that could have occurred by chance alone (e.g. Sackett *et al.*, 1978; Hollenbeck, 1978; Kratochwill and Wetzel, 1977; Adams and MacDonald, 1987). In addition, Hollenbeck listed four further weaknesses of 'percentage agreement' (provided by D.P. Hartmann, 1972) and concluded that it should not be used at all.

More accurate measures of inter-observer reliability are correlation coefficients, which reflect the degree of agreement between observers which is *above chance agreement*.

Table 8.9. *Hypothetical results of four observers collecting instantaneous samples of feeding (F), resting (R), grooming (G), walking (W) and agonistic behavior (intragroup A_1, intergroup A_2) from a single zebra stallion (focal animal)*

Instantaneous sample	Observer				Same observations observers		
	1	2	3	4	1,2,3,4,	1,2,4	1,4
1	F	F	F	F	x	x	x
2	F	F	F	F	x	x	x
3	F	F	W	F		x	x
4	G	G	G	G	x	x	x
5	G	G	G	G	x	x	x
6	G	G	R	G		x	x
7	F	F	F	F	x	x	x
8	F	F	W	F		x	x
9	F	W	W	F			x
10	F	F	W	F		x	x
11	G	G	G	G	x	x	x
12	G	G	G	G	x	x	x
13	W	W	W	W	x	x	x
14	A_1	A_2	A_2	A_1			x
15	A_1	A_1	A_2	A_1		x	x
16	A_1	A_1	A_2	A_1		x	x
17	G	G	G	G	x	x	x
18	A_1	A_1	A_1	A_1	x	x	x
19	F	F	F	F	x	x	x
20	W	W	F	W		x	x
21	F	F	F	F	x	x	x
22	G	G	G	G	x	x	x
23	F	F	W	F		x	x
24	F	F	F	F	x	x	x
25	F	F	R	F		x	x
26	R	R	R	R	x	x	x
27	R	R	F	R		x	x
28	R	R	R	R	x	x	x
29	F	F	F	F	x	x	x
30	F	F	F	F	x	x	x
Totals					18	28	30

8.5.2a *Phi coefficient*

The phi coefficient can be used to measure correlations between two sets of two nominal variables, which are cast into a 2×2 matrix in the format below. It varies from 0 when the observers' scores are statistically independent to 1 when they are in complete agreement.

A	B	A+B
C	D	C+D
A+C	B+D	

$$Phi = \frac{AD - BC}{\sqrt{[(A+B)(C+D)(A+C)(B+D)]}}$$

If there is complete agreement, the B and C cells are zero so phi becomes 1.

$$Phi = \frac{AD}{\sqrt{(A \times D \times A \times D)}} = 1$$

For example, from the data in Table 8.9, we might suspect that the major problem Observer 3 is having is in determining whether the zebra is feeding (F) or walking (W). We use Observer 1 as our calibration observer and compile their data from each instantaneous sample into the matrix below. It shows the number of times they both scored the behavior as F or both scored it as W, and the number of times Observer 3 scored W when Observer 1 scored F and vice versa.

Observer 3

	F	W	
Observer 1 F	8 (A)	5 (B)	A+B=13
Observer 1 W	1 (C)	1 (D)	C+D=2
	A+C=9	B+D=6	

$$Phi = \frac{8 - 5}{\sqrt{[(13)(2)(9)(6)]}} = \frac{3}{37.5} = 0.08$$

This very low Phi coefficient demonstrates that Observer 3 had a difficult time determining whether the zebra stallion was walking or feeding.

Table 8.10. *An agreement matrix for Observers 1 and 2 in Table 8.9*

	Behavior codes[1]	Observer 1						Proportion of total for observer 2 (P_2)
		F	R	G	W	A_1	A_2	
	F	13	0	0	0	0	0	13/30 = 0.43
	R	0	3	0	0	0	0	3/30 = 0.10
	G	0	0	7	0	0	0	7/30 = 0.23
Observer 2	W	1	0	0	2	0	0	3/30 = 0.10
	A_1	0	0	0	0	3	0	3/30 = 0.10
	A_2	0	0	0	0	1	0	1/30 = 0.03
Proportion of total for observer 1 ($P1$)		0.47	0.10	0.23	0.07	0.13	0.00	

Notes:
[1] See Table 8.9 for details of codes.

8.5.2b *Kappa*

Kappa is a statistic developed by Cohen (1960) which is the percentage agreement corrected for chance agreement (Adams and MacDonald, 1987).

$$Kappa = \frac{(P_o - P_c)}{(1 - P_c)}$$

where: P_o = observed proportion of agreements
P_c = chance of proportion of agreements

Kappa is derived from an agreement matrix which shows agreement on the diagonal. As an example, we will measure the inter-observer reliability between Observers 1 and 2 in Table 8.9; their agreement matrix is shown in Table 8.10. The diagonal in the table is the number of occurrences of agreement between the two observers for each behavior.

$$P_o = \frac{\text{sum of diagonal entries}}{\text{total number of entries}} = \frac{28}{30} = 0.93$$

$$P_c = (P_1 \times P_2)$$

where: P_1 = for each behavior (e.g. F) recorded by Observer 1, the proportion of the total number of behavioral acts recorded that were behavior F.
P_2 = the same as above but for Observer 2

$$P_c = (0.43 \times 0.47) + (0.10 \times 0.10) + (0.23 \times 0.23) +$$
$$(0.10 \times 0.07) + (0.10 \times 0.13) + (0.02 \times 0.00)$$
$$= (0.20) + (0.01) + (0.05) + (0.01) + (0.01) + (0.00)$$
$$= 0.28$$

$$\text{Kappa} = \frac{(0.93 - 0.28)}{(1 - 0.28)} = \frac{0.65}{0.72} = 0.90$$

The kappa value of 0.90 is very close to the simple 'percentage agreement' which is 0.93 (see above); however, note that by taking chance agreement into account kappa is lower. What is an acceptable kappa coefficient? Fleiss (1981) considers a kappa of 0.60–0.75 as good and >0.75 as excellent, but Bakeman and Gottman (1986) are inclined to be concerned with a kappa less than 0.7. Bakeman and Gottman (1986) illustrate how to calculate whether a kappa coefficient differs significantly from zero (based on Fleiss *et al.*, 1969).

8.5.2c *Pearson product moment correlation coefficient*

Another measure of inter-observer reliability that has been used is the Pearson product moment correlation coefficient (p. 389). This will measure the correlation between two sets (e.g. two observers) of interval or ratio data. However, Hoffman (1987) used this correlation coefficient to measure the inter-observer reliability of two observers for recording number of lunges (*nominal data*) and duration of encounters (*ratio data*) in fruit flies (*Drosophila melanogaster*) (Table 8.11).

8.5.2d *Kendall's coefficient of concordance*

Kendall's coefficient of concordance (W) was developed by Kendall (1948), and is used to measure the correlation between ranked variables (ordinal data). It is discussed in Chapter 14 as a statistical test for data analysis, but it is also described here for use in measuring inter-observer reliability. It can be used when several observers have ranked behaviors with regard to some parameter, such as frequency, duration or intensity.

$$W = \frac{\Sigma(R_j - \bar{R})^2}{\left(\frac{1}{12}\right)K^2(N^3 - N)}$$

where: $\bar{R}$ = mean rank = $\dfrac{\Sigma R_j}{N}$

R_j = sum of the ranks for each behavior category across observers
N = number of behavior categories
K = number of observers

Table 8.11. *Reliabilities for duration of encounter and number of lunges: correlation coefficents between encounters in a* D. melanogaster *trial scored twice by two observers*

	Second time scored	
First time scored	Observer 1	Observer 2
Duration of sequence		
Observer 1	1.000	0.997
Observer 2	0.997	0.999
Number of lunges		
Observer 1	0.992	0.992
Observer 2	0.993	0.994

Notes:
Coefficients are based on scores for 116 encounters.
Source: From Hoffman (1987). Copyrighted by Bailliere Tindall.

As an example, the data from all four observers in Table 8.7 have been treated in ordinal format by ranking the behaviors in order of frequency of occurrence as recorded by each observer (Table 8.12)

$$N=6$$

$$\bar{R}=\frac{\Sigma R_j}{N}=\frac{84}{6}=14$$

$$(R_j-\bar{R})^2=(-10)^2+(2)^2+(-5.5)^2+(2.5)^2+(2)^2+(9)^2$$
$$=(100)+(4)+(30.2)+(6.25)+(4)+(81)$$
$$=225.7$$

$$W=\frac{225.7}{(1/12)(4)^2(216-6)}=\frac{225.7}{280}=0.80$$

The calculated coefficient ($W=0.80$) indicates relatively low inter-observer reliablity between the four observers when their data are rank ordered according to frequency of occurrence.

Reliability (intra- and inter-observer) can be measured for agreement of behavior recorded for each instantaneous sample, frequencies, durations, latencies and sequences, using one or more of the measures described above. If your research involves more than one observer, remember that your major objective is to answer

Table 8.12. *Data from Table 8.9 organized in an ordinal format for the calculation of Kendall's* W *statistic*

Observer	Behavior code					
	F	R	G	W	A_1	A_2
1	1	4	2	5	3	6
2	1	4*	2	4	4	6
3	1	4	2.5	2.5	6	5
4	1	4	2	5	3	6
$R_j =$	4	16	8.5	16.5	16	23

Note:

* The value 4 is given to behaviors R, W and $A1$ for observer 2, since he observed each three times. This is the average for ranks 3, 4 and 5.

the research questions as validly as possible and transmit that information to others as accurately as possible. These objectives can be met only if your data are reliable.

8.6 IDENTIFICATION AND NAMING OF INDIVIDUALS

A necessary prerequisite for many ethological studies, beyond the initial reconnaissance observation stage, is that the observer be able to recognize individuals. In particular, longitudinal studies (i.e. studies of individuals or groups over long periods of time) make individual recognition necessary. As more is learned about the behavior of animals, it becomes increasingly clear that generalizations are difficult to make. Early naturalists talked about the behavior of species. We now know that there are often major differences between populations, social units and individuals. Hence, ethological research is building from knowledge gained from studies of various individuals and groups and attempting to produce limited generalizations.

8.6.1 Natural marks

The best situation an ethologist can have with regard to individual identification is to be studying a species in which morphological differences are sufficiently great to provide easy identification.

In some cases differences are obvious to even the casual observer. These can be the result of natural mutilations or mutations (e.g. broken horns, melanistic ground squirrels). However, these types of markings are generally not sufficiently wide-

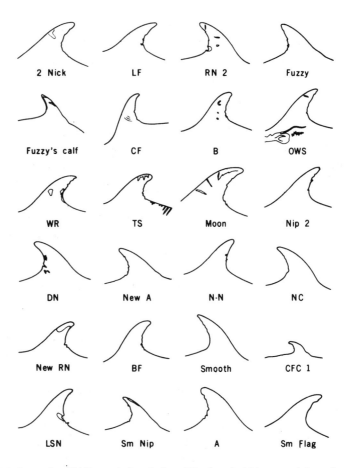

Fig. 8.6 A sample of 24 fine variations in dorsal fins found within a population of bottlenose dolphins. Lines within the fin boundaries represent light pigment spots or scar tissue (from Wursig and Wursig, 1977).

spread in a population to allow large numbers of individuals to be recognized. Many species provide sufficient individual variation for easy recognition, e.g., facial patterns in oryx (Saiz ,1975); stripe patterns in zebras (*Eguus granti*) (Klingel, 1965); pelage patterns and scars in grey seals (*Halichoerus gryus*) (Anderson and Harwood, 1985); wing-tip patterns of kittiwakes (Tinbergen, 1974); spots on each side of the body of bonnethead sharks (*Sphyrna tiburo*) (Myrberg and Gruber, 1974). Variations in the dorsal fins of bottlenose porpoises (*Tursiops truncatus*) (Figure 8.6) allowed Wursig and Wursig (1977) to identify 53 individuals during their 21–month study of group composition; the same individual characteristics were used by Connor and Smolker (1985).

It is often necessary for the ethologist to spend considerable time in order to solve the problem of individual identification. Rudnai (1973) developed a method of individual identification for African lions (*Panthera leo*) based on variations in the pattern of vibrissa spots, which lie in four to five parallel rows between the upper lip and nose (Figure 8.7A).

The individual vibrissa-spot patterns were recorded in the field on schematic profile sheets. Figure 8.7B shows the schematic profile of an individual with its pattern of vibrissa spots and nick in the left ear. Note also that the vibrissa-spot patterns are not the same on both sides. Photos of each side of the lion's head were blown up and the vibrissa-spot patterns were checked against the schematic (Figure 8.7B) drawn in the field.

The basis for individual identification was the number and position of the vibrissa spots in Rows A and B (Figure 8.7A). Decisions were made whether the spots in Row A were above or between spots in Row B. This pattern was then transferred to a schematic version (Figure 8.7C). The spots in Row B were numbered consecutively from anterior to posterior. All the possible positions for spots in Row A (above and between spots in Row B) were provided in the diagram. With continued practice Rudnai was able to use this method of individual identification with ease and confidence. This same method has been used by Patty Moehlman (pers. commun.) as an aid in identifying individual black-backed jackals (*Canis mesomelas*).

Pennycuick and Rudnai (1970) also developed a method to test the reliability of identifying individual lions in this manner. Their method was based on an estimate of the probability that two individuals would have markings that could not be distinguished. They began by determining the probability of occurrence of the different patterns in the population studied. Spots in Row A have never been observed in positions other than 1 through 13.

Regardless of the physical characteristics selected for use in identifying individuals, photographs of each individual are valuable as a continuing 'field guide' (Figure 8.8). Pennycuick (1978) provides an overview of identification using natural markings.

8.6.2 Capture and marking

When natural markings are not available or when individuals are observed only rarely (e.g. homing and migration studies), the researcher must mark the animals in some way. This generally necessitates capture of the animal, although some techniques have been developed for marking at a distance (e.g. dye darts) and for self-marking by the animal.

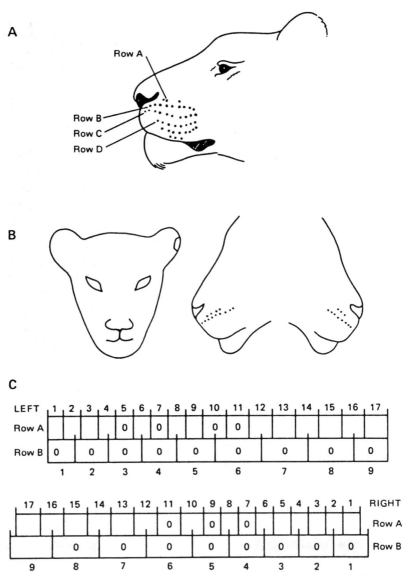

Fig. 8.7 A Profile of a lion's face, showing the vibrissa pattern. Row B is used as a scale for recording the positions of the spots in Row A. B. Schematic profiles, on the form used for recording in the field. The full-face outline is used to record such imperfections as nicks in the ears. The lion's sex, approximate age, and pride (if any) are recorded on the same sheet. C. Schematic representation of the spot pattern shown in A. The positions in Row B are by definition consecutively filled, starting with No. 1, although the total number is variable (adapted from Pennycuick and Rudnai, 1970).

Fig. 8.8   Carlos Mejia with photos of individual giraffes for identification in the field
(from Moss, 1975).

8.6.2a *Capture*

The overiding factors when considering various capture techniques is humaneness and efficiency; efficiency reflects effectiveness relative to expenditure in time and money.

The various capture techniques available for the numerous species studied by ethologists are too numerous to discuss here, but excellent reviews are available. Young (1975) provides a general guide to capturing wild animals. The Canadian Wildlife Service and US Fish and Wildlife Service published a manual which includes capture methods for birds (Anonymous, 1977). Techniques for capturing birds and bats are reviewed by Bub *et al.* (1991), and techniques for birds and mammals are reviewed by Day *et al.* (1980). The use of drugs in the capture and restraint of animals was reviewed by Harthoorn (1976).

Some marking techniques require that animals be recaptured for identification or remarking. Recapturing using the same technique can be difficult for some species, especially so for 'trap shy' individuals. For example, our experience trapping gray jays (*Perisorius canadensis*) has demonstrated that individuals are difficult to capture a second time in Potter traps, but they are relatively easily recaptured using mist nets. Mech *et al.* (1984) developed a radio-transmitter collar for wolves that contained a radio-triggered anesthetic dart which allowed the animal to be easily recaptured for measurements of the animal or repair to the collar.

Capture, recapture, handling and marking have ethical and humane implications (e.g. Cuthill. 1991; Laurenson and Caro, 1994; Appendices C and D), as well as practical implications, such as changes in behavior (see below and Chapter 9).

8.6.2b *Marking*

Various types of markers for animals have been developed for different species and purposes (e.g. dyes, leg bands, ear tags, collars, and radio transmitters, see p. 313). The following are a few examples of the diversity of techniques employed: Milinski (1984) marked individual stickleback fish by tying white and/or black conical, plastic cylinders to their first or second dorsal spine. Kovacs (1987) individually marked 201 harp seal pups (*Phoca groenlandica*) with Roto-tags through the hind flipper, and 167 of these pups were also dye-marked with individually distinct color patterns. Howard (1988) glued numbered tags to the heads of American toads. Fagerstone and Johns (1987) describe a microchip transponder which provided a 10 digit code for identifying ferrets (*Mustela putorius furo, M. nigripes*), both in-hand and remotely. Microchip implants for individual identification of animals as small as mice are available from Bio Medic Data Systems, 255 West Spring Valley Ave., Maywood, NJ.

Marking individual, nocturnal animals for observations after dark may pose special problems. Some animals will habituate to visible lights; natural marks or marks designed for daylight observation can be used (Hill and Clayton, 1985). Often, however, observations of nocturnal animals require that night-vision devices be used (Chapter 4), or special markers be attached to the animals.

A marker that allows the observer to follow visually the nocturnal movements of individual small rodents is the betalight. These are sealed glass capsules coated internally with phosphor and filled with tritium gas, which emits low-energy beta particles causing the phosphor to glow. They can be obtained in disc or tube shapes ranging in length from approximately 6.5 to 76 mm (Figure 8.9A). Available in a wide variety of colors, they are being used to a limited extent on small mammals, especially in conjunction with binoculars or starlight scopes (O.J. Reichman, pers. commun.). Davey *et al.* (1980) used betalights for individual identification of rabbits. Betalights are available through Saunders-Roe, Ltd., Middlesex, United Kingdom, but can only be used under a permit from the Nuclear Regulatory Commission in Washington, DC. Also, Wolcott (1977) describes a miniaturized optical telemetry transmitter that he field tested on nocturnal crabs; it produces accurately timed flashes that can be seen for several hundred meters with a night vision scope. Batchelor and McMillan (1980) developed a collar with light-emitting diodes (LEDs) for their study of wallabies (*Petrogale penicillata penicillata*); the LEDs could be programmed to give individually identifiable flashes, and the flasher switched off during the day to conserve battery power. Lemen and Freeman (1985) tested the efficacy of using low toxicity fluorescent pigments (red, orange and green) on small mammals. They found that when illuminated with ultraviolet light the animals could be easily located and their movements traced up to 900 m. A wide variety of markers designed for nocturnal observations are reviewed by Hill and Clayton (1985).

Coulson and Wooller (1984) individually marked kittiwakes with different amounts of [60]cobalt (radioactive) sealed into aluminum leg bands; they used Geiger tubes connected to a ratemeter to measure the amount of radiation, which then identified the individual kittiwake. Radioactive marking techniques for small mammals are discussed by Linn (1978).

The type of marking to be used should be determined after considering the following factors:

1 Number of individuals to be identified.
2 Distance over which identification is necessary; marking should be conspicuous enough to make individual identification easy for the researchers but not so eye-catching and garish as to be resented by the public (e.g. Petko-Seus *et al.*, 1985).

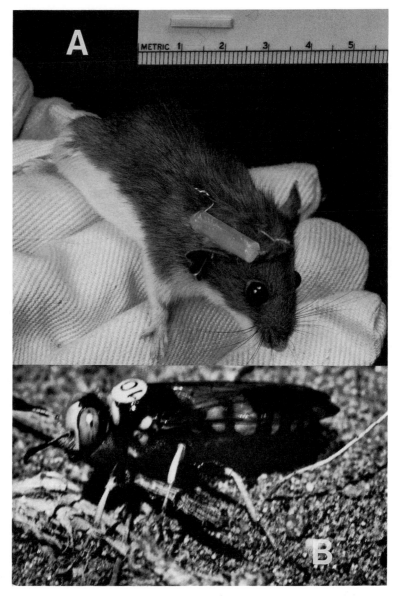

Fig. 8.9 A. Betalight attached to mouse (*Peromyscus leucopus*) (photo by Mark
Stromberg). B. Numbered-tag 10 attached to the back of a wasp (*Philanthus
bicinetus*) (photo by Darryl Gwynne).

3 Length of time identification is necessary.
4 Effects the marker might have on resultant behavior, even though some studies have found no effects; for example, water voles (Leuze, 1980), lions (Orford *et al.*, 1989), cheetahs (Laurenson and Caro, 1994), and polar bears (Ramsay and Stirling, 1986).
5 Ease of capture.
6 Effect capture might have on the animal or group.

A volume on animal marking, edited by Stonehouse (1978), contains chapters by different authors on marking a variety of animal groups including: zoo animals (Ashton), laboratory animals (Lane-Petter), mammals (Twigg), birds (Patterson), reptiles (Swingland), snakes (Spellerberg and Prestt), seals (Summers and Witthames), whales (Brown), bats (Stebbings), fish (Laird), and invertebrates (Southwood). Marking techniques for mammals, birds, and reptiles are reviewed by Day *et al.* (1980). Bird banding techniques are provided by Anonymous (1991) and Bub *et al.* (1991), and an annotated bibliography of bird marking techniques has been prepared by Marion and Shamis (1977). Emery and Wydoski (1987) provide an indexed bibliography of marking techniques for aquatic animals. Walker and Wineriter (1981) present an excellent summary of marking techniques for insects. In addition, technical reports and papers on particular species often contain methods of marking developed for that species.

The number of unique combinations of individual marks available with various schemes can be calculated using the following formula (also see Walker and Wineriter, 1981):

$$N_{n,k} = \frac{n!}{k! \, (n-k)!}$$

where: N=number of unique combinations available; that is, the number of individuals that can be uniquely marked.

n=number of different positions on the animal where a mark will be placed
k=number of positions that will be marked on each individual

For example, we select six positions where we can put a black mark on a mouse (three positions along each side), and we decide to mark two positions each time. The number of mice we can individually mark is:

$$N_{6,2} = \frac{6!}{2! \, (6-2)!} = 15$$

Buckley and Hancock (1968) devised a computer program for generating individual combinations of color and aluminum bands for birds.

Ethologists must weigh the relative advantages and disadvantages of utilizing

Table 8.13. *Relative advantages and disadvantages of natural marks and artificial markers*

Type of mark	Advantages	Disadvantages
Natural	Unnecessary to capture or handle the animal	Possible ambiguity (lack of reliability) in the markings: 1. Change of markings over time 2. Often less inter-observer reliability
Artificial	Positive identification	Markers beings lost (Royall *et al.*, 1974) Capturing, handling, or marking affecting behavior or survival (Herzog, 1979; Southern and Southern, 1983) Markers themselves affecting behavior (Boag, 1972; Perry, 1981) Public attitudes towards marked animals (Petko-Seus *et al.*, 1985)

natural marks or artificial markers before launching a marking program. Some factors to consider are provided in Table 8.13. Two of the most important potential effects of artificial markers is on the behavior of the marked individual and other individuals interacting with the marked individual. For example, Ramakka (1972) found that radio-transmitters resulted in atypical breeding behavior in male wood-cock. Burley (1981, 1988) reported that zebra finches selected mates on the basis of preferences for certain leg band colors, and Swaddle and Cuthill (1994) found that female zebra finches selectively chose symmetrically leg-banded males over asymmetrically banded ones.

8.6.3 Assignment of numbers and names

Ethologists have devised many rationales and clever schemes for assigning numbers or names to animals. For example, Cheney and Seyfarth's (1990:313) system 'for naming vervets was originally devised to enliven the many hours we and our colleagues spent watching sleeping or solitary monkeys. . . . At the beginning of our study we assigned all females, males, and juveniles in each group to a particular theme, such as scandals, prisons, enigmas, 'countries that aren't countries' and so on.' Since the infants were difficult to sex, they attempted to avoid bias by assigning

Table 8.14. *Potential biases associated with assigning identifiers to animals*

Type of identifier	Example	Potential biases
A. Number or letter	No. 117; RG	Only those biases we superstitiously associate with numbers like 13, our birth date, higher numbers being 'better' or 'worse', or our initials
B. Named for object or physical characteristic	'White face' 'Gibraltar'	Only those characteristics reminding us of other individuals towards whom we are biased
C. Named for behavioral characteristic	'Limpy'	Researcher unconsciously carrying over behavior attributes which actually change
D. Named for person	'Leakey'	Naming in honor of a person or because animal possesses particular characteristics similar to that person's; the researcher might not want to see those characteristics change or that honor tarnished (e.g. alpha individual becoming most subordinate)

names that were not specific to either sex. It is this potential bias in making objective observations which can become a problem when numbers or names which are particularly meaningful to the observers are assigned to individual animals.

> The naming of a wild animal should not be done casually, for a name colors one's thinking about it forever afterwards.   *[Schaller, 1973:45]*

We all would like our subjects to become well known like Jane Goodall's chimps 'Fifi', 'Leakey', and 'Graybeard' and the Gardners' 'Washoe'. Certainly, the individual animal given a name will be remembered longer than the animal which is assigned a number. However, names carry certain connotations or overtones for each of us. This is obvious in the naming of our children and our hearing a name and forming a mental image of the person. Tinbergen has been acutely aware of these potential biases, but appreciates the intimacy one gains through assigning names to individuals under study. He reflects on a female kittiwake that was in the colony studied by Dr. Esther Cullen:

> . . . another bird, a female, was too shy to mate; although she kept visiting males through season after season, she was always too nervous

to stay with any of them. (This bird was inadvertently named Cleopatra before her character was known.) *[Tinbergen, 1958:210]*

Although less likely than names, numbers can also create potential biases. We may have a favorite, or lucky, number which when assigned to an animal will instill a desire in us to see that animal do well in competition for food or agonistic encounters over a territory or dominance.

I am not suggesting that ethologists must necessarily randomly assign only numbers to their research animals and remove all romanticism, empathy and levity from their observations and discussions with colleagues. Rather, we must be aware of the potential biases which arise when naming and numbering animals and guard against letting them influence our observations, data recording and interpretation of results.

The biases inherent in the types of identifiers applied to animals are listed in increasing order in Table 8.14.

9 Data-collection equipment

Data-collection equipment varies from a notebook and pencil to computers and automated data-collection devices. Most of us are mesmerized by technologically advanced electronic equipment; therefore, we commonly strive to use the most sophisticated electronic methods available for collecting data, but there can be inherent problems. For the increased speed and ability to handle quantities of data provided by a data logger or computer, there might be a decrease in reliability, although Whiten and Barton (1988:146) 'are confident that these worries are unfounded'. Nevertheless, you can see what you have written in a notebook or on a data form, but you cannot be sure that what you have spoken into a microphone is being recorded on tape, or what you have typed on a keyboard (even though it appears on the screen) is stored properly in memory. When data loggers and computers are working properly they are extremely powerful tools for data collection and storage. Although the reliability of audio-tape recorders and computer hardware is now very high, problems can still arise.

The data-collection equipment necessary for your study will be dictated by your experimental design (Chapter 6) and sampling methods (Chapter 8). A good rule-of-thumb is: *use only equipment with a level of technology* necessary *to collect a complete and accurate record of the behavior(s) of interest in a format for efficient data analysis.* For example, collecting limited samples of a few behaviors from a small number of animals, such as Sordahl's (1986) test of whether avocet (*Recurvirostra americana*) and stilt (*Himantopus mexicanus*) distraction displays were directional (see Chapter 17), requires nothing more than pencil and notebook; the chi-square test (Chapter 14) used by Sordahl could also be calculated on a piece of paper using the data from his field notebook. Likewise, in their field study on the function of copulation calls in female baboons (*Papio cynocephalus*), O'Connell and Cowlishaw (1994) collected their data using a dictaphone and checksheets. However, collection of large quantities of data on several, rapidly occurring behaviors from several individuals can require a data logger or computer; that data should be collected and stored in a format that can be analyzed by computer. Several other factors to consider when selecting data-collection equipment are illustrated in Table 9.1; the reader will undoubtedly be able to add several factors of their own (see also Table 3 in Holm, 1978, and Martin and Bateson. 1993).

Binoculars and spotting scopes were discussed in Chapter 4.

Table 9.1. *Characteristics of two data-collection methods*

Characteristic	Notebook	Computer keyboard
Reliability		
Equipment	High	?
Inter-observer	Less training	More training
Intra-observer	High	High
Speed	Low	High
Quantity of data	Medium	High
Data feedback	Can be reviewed in the field	Must be computer-crunched; remote computer access possible
Power	Limits of the observer	Battery restricted when in field
Altering data-collection data	Easy, which is not always a good characteristic	More difficult

Several other types of data-collection equipment are described below. Some of the older, simpler equipment used to streamline data collection are presented in part for their historic value and as a demonstration of the ingenuity of ethologists. Also, a description of that equipment might serve as encouragement for those researchers who have more time to locate or construct (or modify) simple, suitable equipment than they have money to spend on technologically advanced data loggers and computers. However, as Martin and Bateson (1993) point out, the advent of relatively inexpensive, portable computers has all but eliminated the need for custom-built equipment. Nevertheless, it is not difficult (nor is it wise) to use technological overkill when collecting data.

If you need, and have the opportunity to use, a data logger or computer for data collection, it is always wise to have a simpler method (e.g. check sheets) available as a backup. Although portable computers have become very reliable (Whiten and Barton, 1988) remember Murphy's Law: 'If anything can go wrong, it will'. Even before you conduct reconnaissance observations (Chapter 4), you can sometimes anticipate problems by conducting 'imaginary data collection' from your couch; this is similar to the 'thought experiments' described by Sorensen (1991). Lay back, close your eyes, envision the animal, and envision the process of data collection. Focus on problems such as behaviors occuring too rapidly to record accurately on a check sheet or difficulty finding the correct keys on your computer. Some of these problems can be overcome with different equipment and others can be overcome with practice. If you fall asleep during this 'thought data collection' you might consider a different research project, or perhaps a different profession.

9.1 NOTEBOOK AND PENCIL

At the first level of data collection equipment is the notebook in which *ad libitum* notes are recorded. *Ad libitum* notes are the written descriptions found in typical field notes (Chapter 4) and are recommended only for reconnaissance-type observations (Chapter 4). However, some sampling methods lend themselves to data collection *formats* that can easily be accomodated with a notebook and pencil. For example, you might take typical field notes in a descriptive study of social organization of species X but have a format for recording dominant-subordinate interactions between individuals and groups. That format might be,

> 4→7   (Group 4 dominates Group 7)
> WO→YR   (Individual White/Orange dominates Individual
>   Yellow/Red)

and these notations would be embedded in your notes where you would more fully describe the interactions. Data collection formats are a step below *data forms* (discussed below) which also can be included in your field notebook.

Trotter (1959) added a time base to his note-taking by constructing a motorized device in which a strip of paper was slowly unwound from a roll and passed behind a slit in small steps. The steps were of equal size and occurred at a set time interval. This method can be viewed as either *ad libitum*, sampling with an automatic time base, or a type of event recorder with flexibility in the type of behavioral observation to be recorded at each interval. The size of the slit must be designed for the amount of data to be recorded, the speed at which the paper steps must be in concert with the data being collected. Too small a slit and steps occurring too rapidly could be very frustrating.

Handwritten notes can also be made directly into a portable computer for storage and later retrieval. Input occurs when the observer writes with a special stylus directly onto the computer screen. With some systems, such as Apple's Newton and Write-Top (Linus Technologies, Inc. 1889 Preston White Dr. Reston, VA 22091), software enables the computer to learn the user's handwriting. Handwritten input is then digitized and a character-recognition algorithm converts the written symbols into ASCII. Other systems (e.g. Casio's ScreenWriter Digital Diary) records and stores the writing in a graphic format and displays it just as it was written. Use of these computer systems probably does not offer any significant advantage over paper and pencil.

Recording your verbal descriptions of behavior into an audio-tape recorder (or other device) can be substituted for written notes as well as some of the data forms discussed below (see section on audio-tape recorders below). Also, brief notes can be recorded on small, tapeless recorders which store the verbal descriptions on a computer chip. For example, 'Speak!' (Voice Recognition Technologies) is a battery-powered, hand-held voice data entry device with built in voice recognition capabilities. It also has a 50–key keyboard and bar code reader. However, the capacity of these pocket-sized recorders varies. 'Voice It' has only a two minute capacity size (Voice It Technologies, Inc. Fort Collins, CO 80525); 'Voice Organizer' uses 512 KB of memory to store 99 timed notes (Voice Powered Technology International, Inc., 15260 Ventura Blvd., Suite 2200, Sherman Oaks, CA 91403); but 'Flashback' (Norris) uses interchangeable 30 minute or 1 hour clips.

9.2 DATA FORMS

Data forms, or check sheets, are the next level of organization above *ad libitum* data collection and Trotter's method. They are the minimum data-collection method for any well-designed study. Hinde (1973:393) cautions that 'every check sheet [data form] must be designed with an eye both to the problem in hand and to the idiosyncrasies of the observer.'

Kleiman (1974) described two types of check sheets (data forms) used at the National Zoological Park. One is a 'time-sample check sheet' which is used for focal-animal samples of 4–5 minutes taken every hour. The second form is used for all-occurrence sampling, in which the animal(s) is observed for 30–60 minutes at a prescribed time each day and the occurrence and duration of selected behaviors recorded.

The steps discussed below for designing data forms are the same as those that should be used before employing one of the more complex data-collection methods. Those methods are merely more sophisticated ways of collecting the same data you would get from a data form. A computer does not do anything (indeed it cannot) that cannot be done by hand.

9.2.1 Characteristics of behavior units

An essential part of the design of research is defining the units of behavior to be measured (Chapter 6). The units selected will not only affect the validity of the study but will also determine the efficiency with which the data are collected.

> The degree of selection [of behavioral units] depends in part on the problem, but also on the fact that there is an inverse relation between how much is recorded and the precision with which it is taken. For

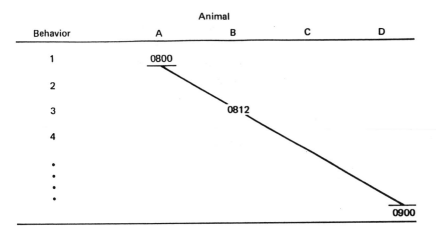

Fig. 9.1 Example of a two-dimensional data form representing a three-dimensional
format. Animal A engaged in Behavior 3 at Time 0812 during a sample period
which ran from 0800 to 0900.

quantitative treatment, it is as important to know that an item of
behavior did not occur as to know that it did, so items selected for
recording must not be missed.   *[Hinde, 1973:396]*

Be aware of whether your behavioral units are *mutually exclusive, partially over-
lapping* or *completely redundant*. The more exclusion there is between behavioral
units, the more easily the data can be recorded.

9.2.2 Columns and rows

Most data forms are two-dimensional representations of three (or four-) dimensional (or variable) data. Figure 9.1, for example, illustrates three variables (individual, behavior and time) of the four commonly recorded in behavior studies. The
fourth, spatial relationships, can be treated separately or incorporated on a standard data form. It could be incorporated on the form in Figure 9.1 by merely including the code NW with the 0812 to indicate that it occurred in the northwest quadrat
of the enclosure.

Columns should be *grouped* and clearly delineated to reduce the time it takes the
observer to find the appropriate column. Often behaviors can be grouped according
to larger categories or animals can be grouped according to social units. The
emphasis, however, should be on ease of recording with an eye to organization for
data transfer later. For example, if the data are to be entered into a computer
spreadsheet or database, it is helpful to have the data form organized so that you

can read across rows entering data without having to skip over columns. This, however, should be of lower priority than ease of getting data accurately onto the forms.

Leave both *blank columns and rows* and a large column on the right for *comments*. Notes written in the comments column are sometimes the key to later interpretation of enigmatic results. Blank columns and rows provide flexibility for adding important behavioral units or animals to your data collection. These columns are probably used more often than not.

9.2.3 Coding

Coding data increases speed of recording. The shorter the code the more rapidly it can be recorded and the smaller the space that is necessary for entering the data. However, accuracy and reliability should not be sacrificed for speed; so every effort should be made to devise a code that is simple and easy to recall. Some examples are given in Table 9.2:

Specific coding schemes are often designed for particular studies. For example, Sackett *et al.* (1973) designed a three-digit three-dimensional coding system for recording the behavior of a single monkey (Table 9.3) and a four-digit four-dimensional coding system for recording the social interactions of a focal monkey tested (5 min. trials) in a group of four monkeys (Table 9.4).

Schleidt *et al.* (1984) designed a generalized letter/symbol format for coding visual behavior patterns and tested and refined it in a study of bluebreasted quail (*Coturnix chinensis*) behavior. They suggested that the format had the potential to serve 'as a prototype for a generally usable standard for behavioral coding systems in birds, or for the members of other taxa' (Schleidt *et al.*, 1984:193). I recommend examining their coding system for possible adoption or revision before designing your own, although Bakeman and Gottman (1986) argue against adopting someone else's coding scheme. I also recommend perusing Bakeman and Gottman's (1986) chapter on developing a coding scheme.

Whatever coding system you devise or adopt, it should make data collection easier, not more difficult. Also, it should be obvious, not cryptic (e.g. Yotsumato 1976).

9.2.4 Data-form examples

The following data-form examples are simplified to serve as a basis from which individual researchers can construct their own forms to meet their individual needs. The forms are designed to correspond to the sampling methods discussed in Chapter 8.

Table 9.2 *Examples of data codes for behaviors, animals, time and spatial locations*

Behavior	Preening	P
	Flying	F
	Preening	Pr
	Pecking	Pe
Animal	Edward	E
	Mary	M
	Flash	F
	Red	R
	Green	G
	Blue	B
	Left blue	LB (colored leg bands,
	Right blue	RB ear tags, or marks)
	Left green	LG
	Right green	RG
Time	First sample period/first minute	1/1
	Third sample period/fifth minute	3/5
Spatial	Quadrat 7	7
	Southeast quadrat	SE
	In pond	P
	On hillside	H
	Next to Edward	/E

9.2.4a Sociometric matrix

Social-relationship data are generally collected as *all-occurrences* (e.g., agonistic) or in *instantaneous samples* (e.g. nearest neighbor at a particular time). However, other methods can also be used (Dunbar, 1976).

The forms shown in Figure 9.2 can be used in the sampling of *all occurrences*.

The forms in Figure 9.3 may be used when collecting sociometric data involving *instantaneous samples*. We can add zones of concentric circles around individuals and record the individuals in each zone (Figure 9.4).

Additional data concerning the exact location of individuals can be gained by using a data sheet with the spatial features represented directly. For example, we might be using a combination of focal-animal and instantaneous *sampling* to determine the social relationships between animal A and the other three members of the

Table 9.3. *A three-digit, three-dimensional scoring system for recording the behavior of a single monkey*[1]

Major category (exclusive behaviors)	Qualifier (nonexclusive behaviors)	Object direction (nonexclusive)
1 Sleep–passive	0 None	0 None
2 Stereotypy	1 Visual (Only)	1 Self
3 Explore	2 Physical–manipulation	2 Diaper
4 Play	3 Oral	3 Cage
5 Disturbance–fear	4 Physical+Oral	4 Feeding site
6 Aggress–threat		5 Food
		6 Feces
		7 Genitals
		8 Outside cage
		9 Special code
	Vocalization coding (nested code)	
0 Vocalization	1 Coo	0 Single sound
	2 Screech	1 Repetitive sound
	3 Other	

Note:
[1] Digital codes and category labels are shown for each scoring dimension. Codes are entered in a keyboard in the sequence major category–qualifier–object, followed by pressing an enter key.
Source: From Sackett *et al.* (1973).

group. Our data form has a series of groups of concentric circles, each one centered on animal A. Hinde (1973) refers to these as 'bull's-eye' check sheets (see also Figure 18.9). We use one for each instantaneous sample (Figure 9.5).

Eventually, sociometric data are generally cast into some form of a matrix in order to illustrate the relationships.

9.2.4b Focal-animal sampling

We could head each data form with the animal's designation (e.g. *A*21) and concentrate on two or more of the variables (e.g. behavior, time, spatial relationships of other individuals, or the environment) (Figure 9.6).

Table 9.4. *A four-digit, four-dimensional coding system for recording social interactions of a focal individual tested in a four-monkey group*

| | 1 | 2 | Digit position | |
| | | | 3 | 4 |
Code	Role of focal individual	Focal individual behavior	Interactor behavior	S ID-direction
0	Nonsocial	Passive	Passive	Nonsocial
1	Initiate – with physical contact	Explore	Explore	Monkey 1
2	Initiate – no physical contact	Withdraw	Withdraw	Monkey 2
3	Reciprocate – with physical contact	Disturbance–fear	Disturbance–fear	Monkey 3
4	Reciprocate – no physical contact	Rock–huddle–self–clasp	Rock–huddle–self–clasp	Monkey 4
5	Ignore – with physical contact	Stereotypy	Stereotypy	Self
6	Ignore – no contact	Play	Play	Toy
7		Sex	Sex	Ladder–shelf
8		Threat–aggression	Threat–aggression	Window
9			No response[2]	

Note:

[1] The four monkeys in the group are arbitrarily labeled from 1 to 4 as subject ID (SID) codes.

[2] The no response category for the third-digit interactor behavior Code 9 refers to social interactions initiated by the focal subject that produce no change in behavior of the potential interactor from that occurring before the initiation.

Source: From Sackett *et al.* (1973).

Supplantee

	A	B	C	D
A		✓ ✓ ✓	✓	✓ ✓ ✓ ✓ ✓
B			✓ ✓ ✓	✓ ✓ ✓ ✓
C	✓ ✓ ✓ ✓	✓ ✓ ✓		✓ ✓ ✓ ✓ ✓ ✓
D			✓	

(row label: **Supplanter**)

A check is made in the appropriate box every time one individual supplants another.

Supplanter   **Supplantee**

A B C D	B C D C	B C D C	C C D	B C D	C D B	C C D	D D	C A	C D	B C	D C
A B C D	B C D C	B C D C	C C D	B C D	C D B	C C D	D D	C A	C D	B C	D C

The code for the individual supplanted is entered sequentially in the boxes to the right of the supplanter; note that *sequential* data are obtained.

Fig. 9.2   Examples of sociometric matrix data collection forms.

Sample period	Nearest neighbor			
	A	B	C	D
1	B	A	D	C
2	D	C	B	C
3	B	C	D	C
4	B	A	D	C

The code for the nearest neighbor is entered under each individual at each instantaneous sample; note that reciprocal relationships (e.g., sample period 1) are not obligatory.

Fig. 9.3   Example of an instantaneous sampling data collection form.

Sample period	Individuals and zones											
	A			**B**			**C**			**D**		
	1	2	3	1	2	3	1	2	3	1	2	3
1	C	B			A		A	D				C
2		B		A				D				C
3		B/C			C/A		A/B					

Note that here reciprocal relationships are obligatory, so that we can collect data on only half the animals.

Fig. 9.4 Example of an instantaneous sampling data collection form with a higher resolution of spatial information than in Figure 9.3.

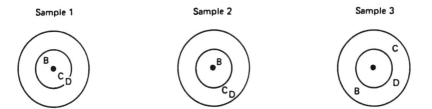

Fig. 9.5 An example of a 'bull's-eye' data form which combines focal-animal and instantaneous sampling. This form provides data on proximity of individuals.

Animal:

Behaviour	Time of occurrence				
Feeding	0915	0918	0935	0937_1	0943
Grooming	0917	0937_2	0956		
Resting	0856	0900	0951		
Agonistic	1002				

Note that we also obtain frequency of occurrence and gross sequences; finer sequences can be obtained by showing the order of occurrence of events indicated at simultaneous times (e.g., $A\ 0937_1$, $B\ 0936_2$).

Fig. 9.6 An example of a focal-animal sampling data collection form.

9.2.4c *All-occurrences sampling*

We are often interested in the who, when, and where of a particular behavior (or behaviors). A typical data form would be as follows in Figure 9.7.

If we are using quadrats (NW, NE, SE, SW) as indicated in Figure 9.7, we can use a data form that resembles a quadrat layout (see Figure 9.8).

9.2.4d **Instantaneous and scan sampling**

In contrast to all-occurrences sampling, here we are interested in the who, what, and where of behavior at particular points in time (see Figure 9.9).

9.2.4e **One–zero sampling**

In this type of sampling we might ask: During sequential one-minute samples, what individuals are engaged in a particular behavior? (see Figure 9.10).

We could reverse the 'Behavior' and 'Animal' labels and ask: During sequential one-minute samples, what behaviors did the focal animal engage in?

9.2.4f **Sequence sampling**

With this type of sampling we are interested in either intra- or inter-individual sequences of behavior. For example, we may be interested in all grooming sequences of male mallard ducks that are initiated by a tail-shake. We decide to discount the individual performing the behavior but begin sampling whenever a tail-shake (TS) occurs, then recording head-shakes (HS) and wing-flaps (WF) as they occur (Table 9.5). We could also list the possible movements and number them according to the order of occurrence, as in Table 9.6.

We might also be interested in the sequence of occurrences of events in a group of individuals engaged in some type of behavior (e.g. encountering predator or prey or leaving a sleeping site). Here we can record the times of occurrence of significant events (Table 9.7). This is essentially a scan sample with the sample period being designated by the occurrence of significant changes in events.

Hinde (1973) also suggests the use of a data sheet with a precalibrated time scale (Figure 9.11). This form is essentially the same as that provided by an event recorder, which is highly recommended if available.

Behavior: ——————————

Animal	When and where the behavior occurred				
A	0830/SE	0900/SE			
B	0840/NW				
C	0702/NW	0715/SW			
D	0702/NW	0717/SW			

Fig. 9.7 An example of an all-occurrences sampling data collection form.

Animal   A

	0830 0900

B

0840	

C

0702	
0715	

D

0702	
0717	

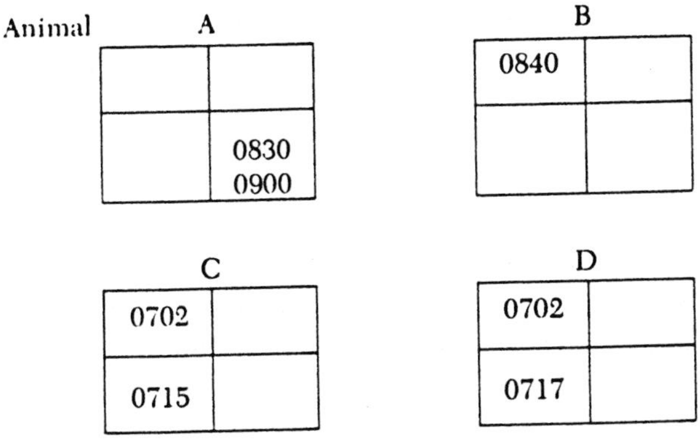

The times of occurrences are then recorded in the
appropriate quadrats.

Fig. 9.8 An all-occurrences sampling data collection form for recording the same data as
in Figure 9.7. in a different format.

Animals, behavior and location

Sample times	A beh.	A loc.	B beh.	B loc.	C beh.	C loc.	D beh.	D loc.
0800	Pr	SW	R	NW	R	NW	R	NW
0801	F	SW	R	NW	F	SE	F	SE
0802								
0803								
0804								

Fig. 9.9   An example of a scan sampling data collection form.

Behavior: _____

Sample times	A	B	C	D
0800	x	x		
0801		x		
0802	x		x	
0803				
0804		x		

We merely make a check mark if an individual animal is performing the behavior during each sample period.

Fig. 9.10   An example of a one-zero sampling data collection form.

Table 9.5. *A hypothetical example of sequence sampling the grooming behavior of male mallard ducks*

Sample	Sequence initiated by TS			
1	TS	HS	WF	HS
2	TS	HS	WF	
3	TS	HS	WF	HS
4	TS	HS	WF	TS
5	TS	HS	TS	

Table 9.6 *A hypothetical example of sequence sampling based on the data in Table 9.5*

| Sample | Movement | | |
	HS	WF	TS
1	2/4	3	
2	2	3	
3	2/4	3	
4	2	3	4
5	2		3

Table 9.7 *Individuals and behavior*

Time	A	B	C	D
0520	St	Y	St	S
0521	W	Y	Y	S
0526	W	W	W	St
0528	D	D	W	W
0531	L	L	L	L

9.3 CLOCKS AND COUNTERS

For many studies in the field, counting and timing behaviors can be accomplished using a hand-held mechanical counter and stopwatch (e.g. Carpenter and Grubitz 1961; also see below). For example, Anderson and Harwood (1985) recorded the behavior of each bull and cow grey seal (*Halichoerus grypus*) on a multi-key counter during 10 minute scan samples. When several different behaviors are being measured, other types of multi-channel event recorders can also be used (see below).

Pushbutton switches connected to electromechanical clocks and counters have been used for many years in laboratory studies (e.g. Mitchell and Clark, 1968). Pressing a switch advances a counter and activates the respective clock which runs until the switch is released. The result is summary data in the form of number of occurrences and total duration for each behavior for the sample period. In addition to the clocks and counters this equipment necessitates a 28–volt DC power supply

BABY Whisky   DATE 6/5/71   NO. 1   TEMP. 11°C   WIND 15-20   CLOUD 7/10Se

Time	On Mother			Off Mother		Grooming			Leaves Appr				Play	Remarks
	Eyes shut	On nip	Off nip	Under 60cm	Over 60cm	Other monkey	by M	by B	M	B	M	B		
00	✓	✓												
	✓	✓												
01	✓	✓												
	✓	✓												
02		✓	✓											
			✓	✓	✓					✓				
03			R₃		✓					✓✓				
				✓	✓									
04					✓									
					✓									
05					✓				✓			✓		
				✓	✓									
06					✓									

Time	On Mother			Off Mother		Grooming			Leaves		Appr		Play	Remarks
	Eyes shut	On nip	Off nip	Under 60cm	Over 60cm	Other monkey	by M	by B	M	B	M	B		
27					✓									
				✓						✓	✓		2 Whoos	
28					✓									
					✓									
29		A₁✓	✓	✓	✓							✓		
		✓			✓									
Total	4	7	3	14	53				5	6	1	10		

Missed ____ Total on & off ____ Total <60 cms only ____ Total >60 cms only ____ Total off ____
R1 (mother moves) ____
R2 (mother rejects passively) ____
R3 (mother pushes, etc.) ____
A1 (mother puts arm round) ____
A2 (mother accepts passively) ____
M (mother's initiative) ____
Whoos (on) ____
Whoos (off) ____

Fig. 9.11 Check sheet for mother–infant relations in captive rhesus monkeys. Each row represents one half minute. Ticks are placed in the columns if the activity in question occurred during the half minute except that 'leaving' and 'approaching' (that is, distance between mother and infant increases from less than 60 cm to more, or vice versa) are recorded each time they happen. R_1, R_2, and R_3 are categories of rejection of the infant by the mother, A_1, A_2, and M are categories of acceptance. C, R, and T mean that animal C initiated rough-and-tumble play with infant by the mother. Whisky. <60 cm only' refers to number of half mins. in which infant was off mother and not more than 60 cm from her (from Hinde, 1973, after Hinde and Spencer-Booth, 1967).

and pulse formers, all of which were a mainstay in most experimental-psychology laboratories but were later replaced by integrated circuitry (e.g. McPartland *et al.*, 1976) and more recently by computers (Flowers and Leger 1982). Nevertheless, the old electromechanical equipment is still available in many laboratories and is suitable for many observational studies. However, this equipment is not readily adaptable to field work since it is bulky and uses 110–volt AC or a 28–volt DC power supply which is normally inverted and transformed from 110–volt AC.

9.4 CALCULATORS

An inexpensive hand-held calculator can be used as a counter in the field, but without a printout capability (see below), one calculator is required for each behavior. Where automatic counting of a single type of event (e.g. hops on a perch) is required, an inexpensive calculator can be modified to accept inputs from photoelectric cells, solenoids, microswitches and sound-activated switches (Knight *et al.*, 1985). Schemnitz and Giles (1980) review several activity-recording instruments for wildlife studies.

Ely (1987) used an inexpensive, hand-held, portable, printing calculator (Casio HR12) to record the behavior of white-fronted geese (*Anser albifrons*). Each of 10 different behaviors was assigned a number (0–9) on the keyboard. He first printed out baseline information for each sampling session, then entered a behavior for each 5 second scan sample (Chapter 8). A metronome (see below) signaled when to record a behavior and when to press the return key, which resulted in the row of numbers being printed onto a paper tape (Figure 9.12). Each row of numbers could represent a time, individual or group, for either scan or instantaneous sampling (Chapter 8). The printouts were sorted and stored in the field and later entered into a computer in the laboratory.

9.5 STRIP-CHART EVENT RECORDERS

Strip-chart event recorders can be used to record the occurrence and duration of events and states. With the time-honored Esterline-Angus 20 channel event recorder, switch closures activate solenoids that move a pen to one side as it is tracking along a piece of moving chart paper. The microswitches can be assembled into a keyboard (Figure 9.13) so that the observer can record the occurrence of a behavior by depressing the assigned microswitch and holding it down for as long as the behavior continues (Staddon, 1972). Eisenberg (1963) used switches that locked closed when lifted up, could be depressed into the neutral open position, or could be pushed down and remained closed as long as they were depressed. Switches can also be triggered by the animal (e.g. Wecker, 1964), or by timers, etc., in the experimenter's absence.

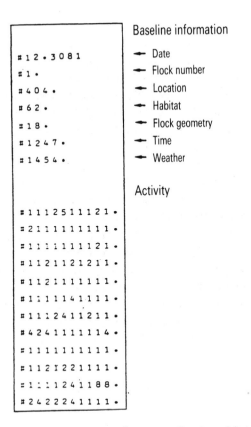

Fig. 9.12 Portion of a calculator printout recording the activity of a white-fronted goose. Activities were recorded every 5 s (10 different 5 s intervals/line). Activities: 0=loser of 'social', 1=feed, 2=alert, 3=extreme alert, 4=walk/swim, 5=social, 6=sleep, 7=rest, 8=preen, 9=fly (from Ely 1987).

The output from a typical event recorder (e.g. Esterline-Angus) consists of a series of tracings which travel at a speed set by the observer. The tracings are displaced when a switch is closed, indicating the occurrence of a behavior (Figure 9.13). These traces then provide a continuous record of the frequency, duration and temporal patterning of behaviors (Mason 1960).

Hutt and Hutt (1974) describe the 60–channel Peissler event recorder with a built-in keyboard (see Fig. 8.4 in Lehner, 1979). It was originally designed for use in studies of the social behavior of squirrel monkeys (at the Max Planck Institute fur Psychiatrie in Munich) and has some limitations for widespread adoption. However, if still available, it may be suitable and sufficient for data collection in some studies.

Older *cumulative recorders* (see Fig. 8.5 in Lehner 1979) are a special type of

Fig. 9.13   Recording behavior using a 20–key microswitch keyboard and a 20–channel
Esterline-Angus event recorder.

electro-mechanical event recorder which have been replaced by electronic printers. These recorders have been a standard method for recording animal responses (e.g. bar presses) per unit time in operant conditioning studies. They can be useful in any study where the researcher wants to illustrate the rate of some behavior since the recorder's output is essentially a graph of cumulative animal responses across time. Wolach *et al.* (1975) modified a cassette tape recorder to record responses, converted the output into a relay closure, and then played the tape back to operate a cumulative recorder.

The event recorders described above are inconvenient to use since the researcher usually must transcribe the tracings of the chart output into numerical data, and then enter the data into a calculator or computer for analysis. Data loggers and computers (discussed below) eliminate these steps, making data collection not only more efficient but often more accurate; every additional step which the researcher must perform by hand, from the raw data to the analysis, introduces additional probability for human errors.

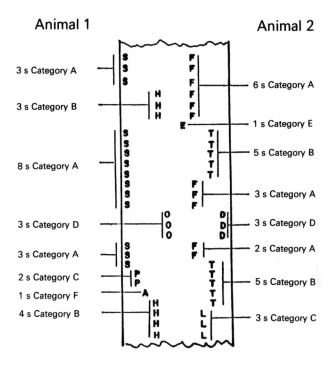

Fig. 9.14 Sample printout from a stenograph machine used to record behavior from two rats simultaneously. The behaviour has been separated into nine categories, A–I (from Heimstra and Davis, 1962).

9.6 STENOGRAPH

Heimstra and Davis (1962) described their use of a stenograph machine for recording behaviors simultaneously from two animals. However, the stenograph is adaptable to several additional formats of recording data. It prints several letters as well as numbers from 0 to 9. One or more numbers and/or letters can be depressed simultaneously. On older stenographs, the record is printed on a 2.5-inch-wide paper tape which steps one line after a key (or keys) is/are depressed (Figure 9.14). Newer machines store the transcription in a computer-compatible format, and some can be integrated with a computer to provide an instant translation of shorthand and display it on a video monitor. The stenograph is relatively compact, light, and very quiet.

Heimstra and Davis (1962) studied the effect of various drugs on the behavior of rats paired in a small wooden box. They separated the behavior into nine categories (A–I) and recorded the occurrence of the behaviors of the two rats simultaneously. They recorded durations of behavioral states by depressing the same key at one-second intervals, cued by an electronic metronome (discussed later in this chapter).

Data recorded on a stenograph are limited to the number of different keys available, but conversely are expanded by the user's capacity to depress any number of keys simultaneously. The stenograph could probably be modified to step one line automatically at set intervals through a motor drive rather than through the depression of the keys. Data recorded on a stenograph can provide information on: 1. the frequency and rate of occurrences, 2. duration; and 3. sequences.

The Microwriter (Microwriter, 251 East 61 St. New York 10021;) is a hand-held, portable, electronic device, similar to the stenograph in that letters and numerals are encoded using combinations of only five keys. A small screen displays the input as it occurs. However, like stenography, 'microwriting' is a skill that must be learned.

9.7 COMPUTER-COMPATIBLE DATA LOGGERS

Data loggers are used to collect (encode) and temporarily store data which is then transferred to a microcomputer (or mainframe) for reduction, long-term storage, organization and analysis (e.g. Morgan and Cordiner, 1994). These processes, illustrated for the Digitorg system in Figure 9.15, are essentially the same for all data logger systems, whether the data is recorded on magnetic tape or on computer chips.

In 1973, Sackett et al. summarized the use and availability of behavioral data acquistion systems for recording data in the laboratory or field. Data logger hardware and software systems available prior to 1978 were reviewed by Lehner (1979). Some of those systems are still worth investigating as possible low cost, yet effective alterantives to newer data logger systems discussed below. Many of those earlier data loggers are no longer being manufactured, but they are not extinct. For example, the Datamyte 900 (Conger and McLeod, 1977; Scott and Masi, 1977; Torgerson, 1977; Smith and Begeman, 1980), 801 and 1000 (Gerth et al., 1982) are still used by some ethologists. They have been replaced by the DataMyte 3055 (Allen-Bradley Co., Minnetonka, MN), which was designed for business applications such as factory production monitoring and quality control, but which may have application in behavioral data collection. Likewise, Microprocessor Operated Recording Equipment (MORE) Co., built a data logger which was designed for collection of behavioral data (e.g. Kodric-Brown, 1988). It was replaced by the OS-3 data logger (Observational Systems, Seattle, WA) which is still being used to collect behavioral data but is no longer being manufactured. Other data loggers which may still be available are the SSR System 7 (Stephenson and Roberts, 1977; Semiotic Systems Corp., Madison, WI), The Assistant (Human Technologies, Inc., St. Petersburg, FL), and the Polycorder (Omnidata International Inc. Logan, Utah) evaluated for fieldwork by MacCracken et al. (1984).

When selecting a data logger system the researcher should answer several questions related to the data logger's different 'abilities':

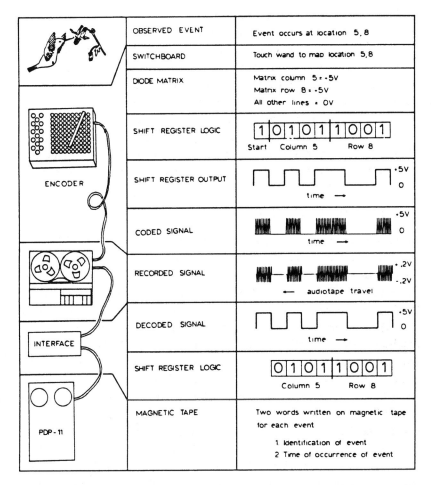

	OBSERVED EVENT	Event occurs at location 5, 8
	SWITCHBOARD	Touch wand to map location 5,8
	DIODE MATRIX	Matrix column 5 = -5V Matrix row 8 = -5V All other lines = 0V
ENCODER	SHIFT REGISTER LOGIC	1 0 1 0 1 1 0 0 1 Start Column 5 Row 8
	SHIFT REGISTER OUTPUT	+5V / 0 time →
	CODED SIGNAL	+5V / 0 time →
	RECORDED SIGNAL	+.2V / -.2V ← audiotape travel
INTERFACE	DECODED SIGNAL	+5V / 0 time →
	SHIFT REGISTER LOGIC	0 1 0 1 1 0 0 1 Column 5 Row 8
PDP-11	MAGNETIC TAPE	Two words written on magnetic tape for each event 1. Identification of event 2. Time of occurrence of event

Fig. 9.15 Information processing by the Digitorg system. Stages of information processing are shown from top to bottom (from Gass, 1977).

- *Capability*: What are your objectives?
- Will it do what you want?
- Will it do more than you need?
- How efficient is it in logging the data you need?
- *Reliability*: Is it a proven system?
- What is its on-the-job record?
- Do you or your organization have the capabilities for repair?
- *Compatibility*: Is the system compatible with your present data collection and computing system?

- How difficult will it be to integrate it?
- *Portability*: Can it be easily carried into the field?
- What is the battery life?
- How much trade-off has there been in capability for portability?
- *Accountability*: Is it really worth the cost?
- Can it be quickly utilized or will large amounts of time be lost in adapting to the system?

Remember that data loggers are only a faster and more efficient way of collecting and storing data. They will not substitute for a poorly designed study by magically changing useless data into useful data. They can be no more accurate than the researchers who use them.

9.8 MICROCOMPUTERS

Microcomputers can be used to collect (observer or animal input), organize, store and analyze data. Microcomputers, such as the IBM PC (and compatibles or clones) and the Apple Macintosh, are powered by 110 VAC and are designed for desktop use. Over the last 15 years microcomputers have become increasingly popular for collecting data in enclosure and laboratory settings (Flowers and Leger, 1982). A discussion of the types and specifications of microcomputers, including an overview of their use in ethology is found in Appendix B.

Computers have been shrinking in size and weight from desk top microcomputers to portable laptop, notebook and hand-held computers. One laptop computer that is designed to be used under harsh conditions in the field is the Bison Explorer (Bison Instruments, Inc. 5708 W. 36th St. Minneapolis, Mn 55416;). Notebook and hand-held computers have become increasingly popular for the collection and short-term storage of behavioral data since they are very compact, lightweight and have a relatively long battery life and large data storage capacity (Noldus *et al.*, 1989). This large data capacity has enabled many to also be used for limited data analysis. Competition has caused prices to fall and capabilities to increase. Some manufacturers of popular notebook computers are: Acer, Apple, AST, Canon, Compaq, IBM, NEC, Tandy and Toshiba. Hand-held computers that have been used for behavioral data collection include: Hewlett-Packard HP41,71 and 95LX, Husky Hunter, Psion Organiser and Series 3, Sharp Wizard, and Tandy 102. You should select a computer that can be easily programmed or used with available data collection software and interfaced with your microcomputer or mainframe [see examples below; also Kieras (1981) offers additional advice].

9.8.1 Data Collection

In ethology, microcomputer-based data collection can occur in at least the following four ways: 1. recording of visual and/or acoustical observation data via observer input; 2. the animal automatically recording its own behavior; 3. recording of spatial data, usually from film or videotapes (see discussions of digital photography and videotape analysis later in this chapter); and 4. recording of radiotelemetry data (see section 9.15.2).

9.8.1a *Data collection via observer input*

A microcomputer (or laptop, notebook or hand-held computer) and appropriate software constitute a 'system' that can be used by an observer to record each occurrence of a behavior in either live animals or from a video recording. For example, Mendl (1988) used the Madingley Interactive Computer for Recording Observations ('MICRO'; Styles, 1980) to collect data on play behavior in the domestic cat. The system allowed him to record when a behavior was performed, who performed it, and to whom or what it was directed. Godwin (1994) used the Behavior Events Acquisition and Analysis System (BEAST; WindWard Technology, 45–415 Akimala St. Kaneohe, Hawaii 96744) to record the behavior patterns of anemonefish (*Amphiprion melanopus*).

Hensler *et al.* (1986) used a portable computer (TRS-80 Model 100) to record the foraging behavior of several color-marked European starlings (*Sturnus vulgaris*). They wrote a program which allowed the observer to record the bird identifier, the location and the bird's activity (arriving or departing the nest box). To facilitate data entry, they relabeled the microcomputer's keys with adhesive stickers. Giraldeau *et al.* (1994) also used a TRS-80 programmed as an event recorder in their study of foraging behavior in chipmunks. Unwin and Martin, (1987) designed a behavioral data collection system also based on a portable computer (Epson PX-8) and specially designed software. Their system ('computer event recorder') was based on the criteria listed below (Unwin and Martin 1987:88–89), which are appropriate when the researcher intends to assemble a portable microcomputer based data collection system by purchasing the computer and writing the software, in contrast to using commercially available software – also see Martin and Bateson's (1993: 110–112) listings of desirable and essential features for an event recorder:

> a) The event recorder should be based on a standard, commercially produced microcomputer that is currently available. In comparison with custom-built hardware, this should make it less expensive, available for immediate use, reliable and independent of specialist support.

b) The hardware should be small, light, portable and suitable for recording under field conditions.

c) Memory should be protected by battery back-up, to prevent the loss of data and software in the event of a momentary power failure.

d) The event recording software should be relatively simple and written in a way that can be understood and modified by non-expert users. This means that it must be written in a high level language such as BASIC.

e) The software should run on other computers without extensive modification, so that the user is not committed to one particular machine. This means that the programs must be written using commands or functions that are available in most dialects of BASIC.

f) The event recorder should have similar capabilities to check sheets, including the ability to record social interactions involving two or more identified individuals and comments written in specialized notation or plain English.

g) It should be possible to obtain a 'hard copy' print-out of the data immediately [after] the recording ends, both in the form of a literal record of each key-press and in summary form.

h) It should be possible to transfer data to another computer for permanent storage and further analysis.

As another example, Whiten and Barton (1988) used lightweight, hand-held computers (HP41, HP71) to record the behavior of baboons, which they followed from dawn-to-dusk over difficult terrain in the climatic extremes of Africa. The computers were programmed in BASIC to provide accurate real times, durations and latencies of multiple behaviors. Tones of various pitches and durations could be programmed to signal that the correct key had been pressed or signal the time for a scan sample.

Software for data collection can either be: 1. written by the researcher; 2. obtained from other researchers who wrote the programs; or 3. purchased commercially.

The basic information recorded by data collection software is the occurrence and time of a behavior for each input by the observer. From this data, the program can then derive frequencies, durations, latencies and sequences of different behaviors. As an example, a flow chart of Whiten and Barton's (1988) focal-animal all-occurrences and scan sample programs is shown in Figure 9.16.

Many programs also allow the observer to record additional input for each behavior, such as who performed it, where it was performed and to whom or what it was directed (e.g. Mendl, 1988; also see overview of commercial software below).

Researchers who have developed their own data-collection programs are often willing to share them with colleagues. In addition, several data-collection programs, which can be used directly or modified for specific research needs, have been published in *Behavior Research Methods and Instrumentation*. For example, somewhat specialized programs were published by Bernstein and Livingston (1982) and Hargrove and Martin (1982), and two more generalized programs were published by Flowers (1982).

Data-collection software is commercially available through several sources; it differs widely in capability and price. Features to look for when contemplating the purchase and/or use of data collection software (other than compatibility with available hardware, availability of technical support, and price) include: 1. clock resolution (a slow program will 'miss' rapid key presses), 2. the maximum number of different behaviors (subjects, locations, etc.) that can be recorded, 3. simultaneous recording of 2 or more behaviors; 4. appropriate summarization of data; 5. the ability to calculate descriptive statistics; and 6. the storage of data in data files compatible with your spreadsheets, databases and statistical packages.

Box 9.1 gives brief overviews of four software packages available at the time this book was written. These overviews are not exhaustive and will be inaccurate for updates. Contact the supplier of each software package for current capabilities, compatabilities and costs. All four programs described below use on-screen menus to assist the researcher in configuring the data collection format and selecting options for data summarizing, display, storage, analysis and transfer.

Whether you write, borrow or purchase data-collection software, it need only have the capabilities necessary for your research. However, especially when purchasing software, consider that your research may expand into more complex experimental designs and sampling methods in the future.

Besides using the keyboard or mouse, behavioral data can also be input to a computer using a bar-code reader. For example, Line *et al.* (1987) developed a computer/bar-code system for recording behavioral observations in their studies of rhesus macaques. They recorded the frequency and duration of 51 behaviors which were printed with a unique bar code on a plasticized sheet. Behaviors were entered into the computer by scanning the bar code for the behavior with a light pen; it took approximately 1second to obtain a correct reading.

9.8.1b Data collection via animal input

Another type of data collection involves having the animal record its own behavior. For example, in an enclosure or laboratory operant arena, a microcomputer can be programmed to record bar or key presses (basically any type of switch closure) on a 24 h basis (e.g. Ha *et al.*, 1990). In turn, the microcomputer can trigger rewards on virtually any type of reinforcement schedule imaginable (Rayfield, 1982; Gordon *et al.*, 1983; Kallman, 1986; Matthews and Ladewig, 1994; O'Dell and Jackson, 1986). This type of system removes the requirement of an observer continually being

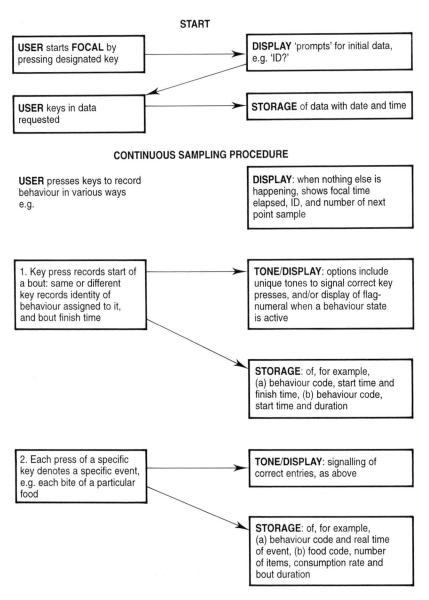

Fig. 9.16 Generalized outline of a routine used to record a 30-minute focal observation on a foraging baboon illustrates several options likely to be relevant to most users. An alarm tone signals the end of the focal at 30 minutes, and the display prompts the user to confirm the end (at which point an end-of-focal code is stored) or postpone it, permitting completion of a final data entry (from Whiten and Barton, 1988).

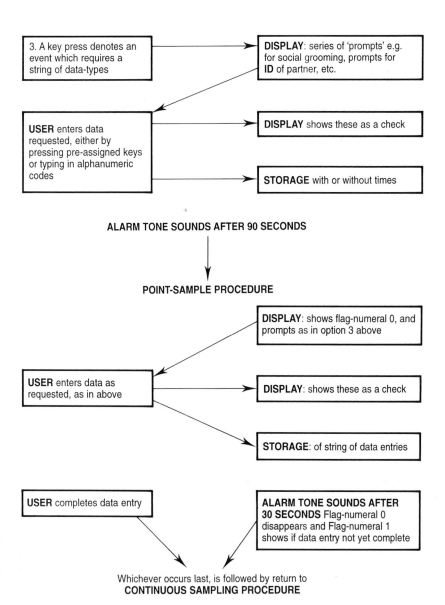

3. A key press denotes an event which requires a string of data-types

DISPLAY: series of 'prompts' e.g. for social grooming, prompts for ID of partner, etc.

USER enters data requested, either by pressing pre-assigned keys or typing in alphanumeric codes

DISPLAY shows these as a check

STORAGE with or without times

ALARM TONE SOUNDS AFTER 90 SECONDS

POINT-SAMPLE PROCEDURE

DISPLAY: shows flag-numeral 0, and prompts as in option 3 above

USER enters data as requested, as in above

DISPLAY: shows these as a check

STORAGE: of string of data entries

USER completes data entry

ALARM TONE SOUNDS AFTER 30 SECONDS Flag-numeral 0 disappears and Flag-numeral 1 shows if data entry not yet complete

Whichever occurs last, is followed by return to
CONTINUOUS SAMPLING PROCEDURE

Box 9.1 Behavioral data collection software

BEHAVIOR CHRONICLES

Hardware Requirements:

IBM-PC or compatible; at least 286 processor with 2 megabytes RAM, VGA monitor and at least a 40 megabyte hard drive; this program runs in a *Windows 3.1* environment.

Input

Mouse; clicked on icons or labels.

The Setup menu is used to configure the subject and behavior files and record the name of the observer. Observations can be divided into initiator, behavior and recipient, and they can be recorded as a group or single entry (date, time, initiator, behavior, recipient). Behaviors are classified as events or states which dictates whether the computer records the behavior immediately in the data file (event) or waits for you to click on the finished icon (state).

The Mode menu is used to select live observations, observations from videotape (the program controls the VCR)or analysis.

Clicking on the Edit icon allows the observer to add comments or delete observations in a file.

Output:

To screen, printer or disk; data format is compatible with most statistical packages.

Analyses:

Summary and descriptive statistics; time-line arrays for autocorrelation and other analyses using ABOUT TIME utility program.

Summary:

Somewhat more difficult to learn because of its increased capabilities, but the manual is very helpful. Researchers may find this program difficult to use for recording rapidly occuring interactions since you must use the mouse to first click on the initiator's name, then click on the behavior, and then click on the recipient's name. A key-driven version is being written expressly for data collection. Price range is medium for the main program and its associated utility programs. The program is free to members of the American Zoo and Aquarium Association.

Source:

Crispen R. Wilson
Chesterfield, MO

Box 9.1 (*cont.*)

EVENTLOG

Hardware Requirements:

IBM-PC (or compatible)

Input:

Keyboard

User determines which keys to activate and assigns labels to those keys.

Keys are pressed and held down while a behavior is occurring; several keys can be depressed simultaneously.

User can sets timers with auditory or visual signals to cue intervals for one–zero and instantaneous/scan sampling.

Data recording can be interrupted to add typed notes to the data file.

Output:

To screen, printer or disk.

Analyses:

Files are compatible with most major statistical packages.

Summary:

Easy to learn and use.

Source:

Conduit

The University of Iowa, Oakdale Campus Iowa City, Iowa 52242
Henderson, R.W. (1988).

EVENT-PC 3.0

Hardware Requirements:

IBM-PC (or compatible), Apple Macintosh, Commodore 64

Input:

Keyboard

Two sets of 20 keys (selectable through the shift key) can be assigned to behaviors, sexes, individuals, etc. and given labels.

Two input formats:

1. Press-and-hold key for one behavior at a time.
2. Press key once at start and once at end of each behavior to record multiple simultaneous behaviors.`

Box 9.1 (*cont.*)

Output:

> To screen, printer or disk; disk storage format compatible with most spreadsheets and statistical packages.

Analyses:

> Graphic 'strip-chart recorder' output for visual inspection of interval relationships (mimics an Esterline-Angus event recorder output).

> Summary statistics (N, mean, standard deviation) for each behavior category.

> Sequential analysis (SEQ) program available which uses EVENT-PC files to calculate: 1. monad, dyad and triad frequencies, 2. uncertainty and stereotypy indices; and 3. chi-square expected frequencies, goodness-of-fit values, and degrees of freedom for dyad frequencies.

Summary:

> Easy to learn and use; will meet the needs of many researchers conducting rather 'simple', straightforward studies, especially when the behaviors are defined as mutually exclusive (i.e. only one behavior can occur at a time); inexpensive.

Source:

> Dr James C. Ha
> 9402 224th SW
> Edmonds, WA 98020
> email: jcha@u.washington.edu

THE OBSERVER 3.0

Hardware Requirements:

> Base package: PC (versions available for DOS and Windows) or Apple Macintosh

> Support package for hand-held computers: several models of hand-held computers manufacured by Psion, Hewlett-Packard, Husky and others.

> Support package for videotape analysis: video cassette recorder, video time-code generator and reader. Optional: tape controller, character generator, frame grabber.

Box 9.1 (*cont.*)

Input:

Keyboard

Can be configured to record up to 100 subjects and 1000 behaviors using the following sampling methods: *ad libitum*, focal animal, instantaneous/scan, all occurrences, and one–zero.

The researcher designates behaviors as events or states and which are mutually exclusive. Behavioral elements can be grouped in up to 16 classes. Each class contains up to 99 elements.

Two input formats:

    1. Key press for start/end. A single key press signals the start of a behavior and the end of the previous behavior if it has been designated mutually exclusive.

    2. Press and hold key for the full duration of the behavioral state. Behaviors can overlap, but the number of keys that can be depressed simultaneously depends on the type of keyboard used.

You can select to record from single (focal animal) or multiple actors. Modifiers can be used to code the receiver, object, intensity, or direction of the behavior.

You can interrupt data collection to edit the data file. When interfacing The Observer with a video cassette recorder (see 'Hardware requirements' above), you can change the playback speed of VCR (slow-motion, pause, reverse, etc.) while the software keeps the behavioral data stream in synchrony with the video tape.

Output:

To screen, printer or disk. Data format is compatible with most spreadsheets, databases and statistical packages.

Analyses:

Will calculate statistics across a complete sampling period, for single observations, or for event- or time-based windows within an observation.

Analysis options include: intra- and inter-observer reliability; time-event tables and plots (mimics Esterline-Angus event recorder); descriptive statistics on frequency of occurrence and duration of events or states; nested analyses (e.g. behaviors and locations);

Box 9.1 (*cont.*)

---

correlations; lag sequential analysis; integrated analysis of continuous signals (e.g. heart rate) and observed behavior.

Summary:

This is the most comprehensive data-collection software available. It is relatively easy to learn, considering the choices you must make between its various configurations for recording data. Its support packages for numerous hand-held computers make it very useful for field studies. Since it is the most comprehensive package, it is also the most expensive.

Source:

Earlier version of The Observer (v. 2.0): Noldus (1991).

Software review of The Observer 2.0: Boccia, (1992) (in Multiple Authors 1992)

Noldus Information Technology bv

Costerweg 5

6702 AA Wageningen

The Netherlands

e-mail: info @ noldus.nl

---

present, and is generally more accurate and provides more information (in a computer-compatible form) than the electro-mechanical devices which have been used. Wildhaber *et al.* (1994) used a microcomputer to control and monitor continuously their experiment on foraging behavior in bluegills. A passive type of data logging is possible through the use of treadles or light beams, which are recorded by the microcomputer. Systems such as these, in which the microcomputer must detect on-or-off switch closures, are quite inexpensive to design and simple to program (Symonds and Unwin, 1982).

9.8.2 Data storage and manipulation

The storage, editing and manipulation of ethological data is no different than that of any other kind of data. Most commonly, spreadsheet programs (such as Lotus *1–2–3* or Microsoft *Excel*) or database managers (Ashton-Tate's *Dbase*, Microrim's *Rbase*, or Microsoft's *File*) are used. If data-collection software is used, it is important to remember that the collection software must produce data files on disk which are compatible with the disk files used by the storage/manipulation software. If data is not recorded in a computer-compatible form, then these storage and manipulation programs provide a convenient data entry option for later analysis. A short discussion of software packages for statistical analyses can be found in Chapter 16.

9.9 AUDIO-TAPE RECORDERS

Ethologists use audio-tape recorders for three different purposes: 1. to record observations verbally described by the researcher 2. to record sounds produced by animals under study; and 3. to store data in a format compatible for later transfer to a computer for storage and analysis (also see section 9.7 on data loggers).

9.9.1 Data collection on audio-tape recorders

This method of data collection has several advantages and disadvantages. The most noteworthy advantages are: 1. being able to observe continuously while recording data; and 2. flexibility of input; additional observations and comments can be easily recorded. The greatest disadvantages are: 1. recorders stopping and tape running out; 2. real time being difficult to measure accurately if the recorder changes speeds, slightly, at different temperatures; 3. transfer of data from the tape often being difficult since most observers do not adhere to a strict format when recording the data; and 4. speaking into the microphone may disturb the animals being observed.

At the beginning of each data-collection period you should record the same preliminary data that you record in written field notes (Chapter 4). The format for recording observations can be the same as written *ad libitum* field notes or any of the other sampling methods (Chapter 8). As with check sheets, if the behavior is complex or occurs rapidly, some type of coding should be used (discussed earlier in this chapter). The code must be clearly defined and the sounds of the code words must be easily discriminable for future transcription.

Data transfer from audio-tapes can take many forms. You can transfer the data to a check sheet (e.g. Tacha, 1988; also see section 9.2) or, for some studies, a complete sequential transcript may be advisable, such as that used by Hutt and Hutt (1974) in their study of 'free field' behavior in children (Box 9.2). Data can also be transfered directly into computer data files using standard data collection software designed for direct computer input (see above). Also, some programs are designed specifically for recording data from audio-tapes. For example, Noldus Information Technologies offers an Audio-Tape Analysis Kit for use with The Observer 3.0 software which provides accurate coding and timing of behavioral data from audio-tape.

Audio-tape recorders available for note-taking vary in size from small pocket-sized microcassette recorders (field studies) to large reel-to-reel decks (enclosure

---

Box 9.2   Transcript of the tape-recorded commentary made during a three-minute session of observation on a child's 'free field' behavior. Numbers designate location in the room; strokes designate termination of an activity; numbers above the strokes are the duration of the activity (from Hutt and Hutt 1974)

$9\frac{1}{2}$ $2\frac{1}{2}$

Standing 4 looking bricks, holding wire / / walks 7/8 twirling / / looking bricks

$2\frac{1}{2}$ 4 $3\frac{1}{2}$

/ / walks to screen 15 / / turns to 0's call, walks 10/11 / / twirling, looking bricks

7   2   2   4

twirling 11 / / walks 10/11 to 2 / / bangs wall / / looking bricks, twirling 5/6 / /

2

walks 7 picks brick / / runs screen 15, puts brick in mouth and bangs on screen

6   3 $2\frac{1}{2}$

/ / rubbing screen walks 13 to 9 / / puts brick window / / standing 5 bangs brick

2   4

on window / / turns throws brick 15 and goes after it / / picks up brick 15 throws

4   6 $3\frac{1}{2}$

at 0 / / walks 8, climbs on chair / / sitting chair, looks screen to 0 to window / /

$5\frac{1}{2}$ $2\frac{1}{2}$ $4\frac{1}{2}$

looks at brick / / gets off chair walks 7 / / picks up brick throws at screen / /

3   2 $4\frac{1}{2}$

banging screen walks 13/14 / / throws brick at 0 / / turns, runs 8, climbs chair / /

8                        $3\frac{1}{2}$

stands arm of chair holding door frame / / jumps on seat, turns / / bangs

9                                $3\frac{1}{2}$

window holding on to chair / / turns to 0's signal, reaches for 0's brick / / sits

13   2

chair looking over side at floor / / looks window / / looks ceiling, leaning over

5   2   4

back of chair, hand in mouth / / looks door / / gets off chair looking 12 / /

3   8

walks 13 / / stands 13 biting jumper looking corner / / turns, walks 13 to 9 to 5,

10   3 $3\frac{1}{2}$

biting jumper while walking to 6 to 7 / / throws brick to 16 / / twirls / / walks 16

$4\frac{1}{2}$ $2\frac{1}{2}$ 5

to 15 / / picks brick throws it to 1 / / walks to 8 climbs on to chair / /

---

and laboratory studies). Since there is no need for high fidelity in this type of use, microcassette recorders can be run at half speed (1.2 cm/s) to get 90 minutes of recording time per side of tape. Voice activation of the recorder or an on–off switch on the microphone is almost a necessity, and a rechargeable unit is especially useful to the researcher who makes long, daily forays into the field. Keep the recorders clean and in good working order; check from time to time while you are recording (VU meter or indicator light) to be sure your voice is being recorded.

Examples of the use of audio-tape recorders for data collection are provided by Brockway (1964), Burley *et al.* (1994), Eisenberg (1967), Kaufman and Rosenblum (1966), Polak (1994), Poysa (1994), Rosenblum *et al.* (1964), and Sorenson (1994).

9.9.2 Recording animal sounds

When recording animal sounds on audio-tape you will find it important to identify the recordist(s), animal(s), habitat, and behavioral context, geographic location, date, time and climatological conditions. This may be given as commentary on the tape immediately before or after the recording, or simultaneously on another track of the tape (see below). Additional written records of the recording should also be kept in your field notebook.

Recorders used for animal sounds, in contrast to recorders used for note-taking, should be of the highest quality and fidelity. The following characteristics should be checked when selecting an audio-tape recorder for this purpose:

- *Frequency response*: range from highest to lowest in hertz (Hz5cycles/s). The number of decibels (dB) from a flat curve is also usually indicated (e.g. ±dB).
- *Signal-to-noise ratio*: the ratio of background noise from the recorder to the signal put on tape. A good ratio should be about 55–60 dB.
- *Tape speed*: represents a trade-off between quality of recording (high speed) and economy (low speed). Good recorders should have the capacity for 38 cm/sec.

Monophonic recorders come in models to record on the *full track* or *half track*. The full-track recordings are probably of the highest quality; but the half-track models allow you to turn the tape over and record on the second side. There is some loss of quality since only one half of the ¼–inch tape is being recorded on each side.

Stereophonic recorders (reel-to-reel or cassette) split the tracks and allow you to record from two sources simultaneously. That is, you can record animal sound on one track and a verbal description of the other track simultaneously (two-track). Also, two observers, each responsible for one animal, can simultaneously record their observations, one on each track (Grant and Mackintosh, 1963). Four-track recorders allow the recording of two simultaneous tracks on each side.

For many years ethologists used reel-to-reel recorders exclusively since they generally made higher-quality recordings than cassette recorders, although suitable recordings could be made with the best cassette models (Bradley, 1977). Now, cassette recorders are gaining increasing use because their capability for making high-quality recordings is coupled with their compact size. The two portable reel-to-reel recorders most often used by ethologists for recording animal sounds (including footdrumming by kangaroo rats; Randall, 1994) are the Uher (Figure 9.17) and Nagra (Figure 9.18), although other high-quality recorders are also used. Sony, Marantz and Uher all make high-quality cassette recorders.

Some of the reel-to-reel and cassette recorders being used for recording animal sounds are:

- *Reel-to-reel*:
  Nagra IV-S (Brown and Waser, 1984)
  Uher 4000 (Eales, 1985; Randall, 1994)
  Uher 4200 (Randall, 1994)
- *Cassette*:
  Marantz CP430 (Adhikerana and Slater, 1993)
  Marantz PMD 221 (Elowson and Snowdon, 1994)
  Marantz PMD 360 (Given, 1993)
  Sony TCD-5M (Brenowitz and Rose, 1994)
  Sony WM-D6C (Rothstein *et al.*, 1988)

The above list is by no means exhaustive as to the makes or models of recorders that are suitable for recording animal sounds. You should check the literature and consult other ethologists who have made recordings of the same, or similar, sounds and species you will be recording. Also consult with reliable suppliers of animal sound recording equipment, such as Saul Minneroff Electronics, Inc. (574 Meacham Ave. Elmont, NY 11003).

Audio tapes have three important characteristics: 1. *thickness*; 2. *backing*; and 3. *signal-to-noise ratio*.

Tapes are generally available in three thicknesses: 0.5, 1.0, and 1.5 mm. Trade-offs are involved when you choose the thickness of tape for your recordings. The thinner the tape, the more stretching that may occur, the more tape you will get per reel, and the more chance for print-through, i.e., the tendency for a recorded signal to magnetize the adjacent wound tape (Bradley 1977). A 1.0 mm tape is usually a reasonable compromise.

Backings are generally either the newer polyester plastic (Mylar) or the older cellulose acetate. Polyester plastic is preferred, since cellulose acetate tends to be weaker and is more prone to stretching, warping and wrinkling.

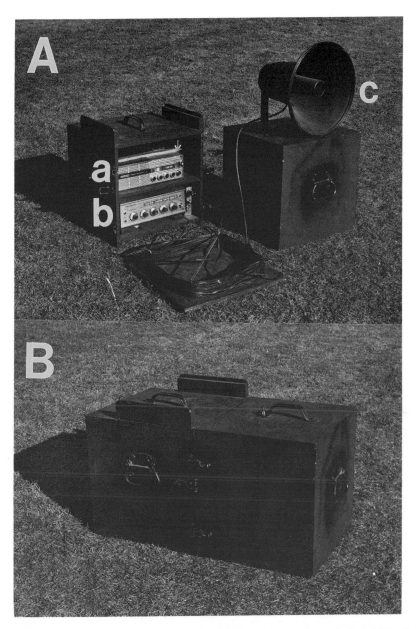

Fig. 9.17 Audio playback equipment: a. Uher 4000 Report-L recorder; b. Realistic MPA-20 amplifier; c. Realistic PA-12 trumpet speaker. B. Equipment in A assembled into carrying case.

Fig. 9.18 Parabolic reflector covered with camouflage netting and wind-shielded
microphone wired to a Nagra IV audiotape recorder.

Low-noise tapes are superior to normal tapes in their ability to improve the
signal-to-noise ratio. Some recorders have a separate setting for low-noise tapes.

Microphones (mikes) come in two basic designs: dynamic (moving-coil) or condenser. The condenser mike may be superior, but it is more complicated and often
requires additional repair. For field recording, dynamic mikes are probably the best
choice. Look carefully at the *impedance* and *output*. The microphone's impedance

must match the recorder and the output should be approximately –57 to –53 dB (Bradley, 1977).

Differential directional characteristics are provided by *omni-directional, cardioid,* and *supercardioid mikes*. Omnidirectional mikes are essentially sensitive to sounds in all directions, while cardioid mikes are most sensitive to sounds in front of them. Supercardioid mikes (shotgun mikes) are highly directional and increase the relative intensity of the sound at which they are directed (within a small angle in front of the microphone) by remaining insensitive to unwanted sounds (noise) outside of that angle. In essence, they increase the signal-to-noise ratio. Walbott (1982) provides additional information on microphones.

Cardioid microphones may be used with a *parabolic reflector* that focuses sound received over the width of the reflector onto the microphone, which is set at the focal point of the parabola (Figure 9.18). Parabolic reflectors should be wider than the wavelength of sound that you are recording. For example, many songbird vocalizations are in the frequency range of 2–8 kHz, with wavelengths of 0.03 m to 0.15 m. Therefore, a parabolic reflector with a width of 0.46 m is sufficient. However, coyote vocalizations are often around 500 Hz, with a wavelength of 0.61 m; therefore the parabolic reflector should be at least 0.61 m in diameter in order to make high-quality recordings. Parabolic reflectors are more effective than shotgun microphones when recording over distances that exceed 10–25 m. Several parabolic microphones are available, such as PBR-330 (Saul Minneroff Electronics, Inc. – see address above) and Dan Gibson E.P.M. (R.D. Systems, 290 Larkin St. Beffalo, NY 14210).

Many of the microphones used by ethologists are models manufactured by Sennheiser, although other excellent microphones are also available (e.g. Saul Minnerof Electronics; Shure). Once again check the literature (e.g. the references listed with the recorders, above), other ethologists, and suppliers.

Acoustic biotelemetry has been used to transmit animal sounds from animals equipped with microphone/transmitters to receivers and recorders a short distance away (e.g. Gaultier, 1980; Montgomery and Sunquist, 1974). Alkon *et al.* (1989) developed and tested an acoustic biotelemetry system which transmitted usable sounds from Indian crested porcupines (*Husrix indica*) for a distance up to 1 km. They tested the ability of seven briefly trained observers to identify correctly behaviors from recordings of seven different behaviors (feeding, drinking, sniffing, walking, digging, moving, threat huffs). Overall, they correctly identified 82 ± 12% (SE) of the behaviors from the recordings and were 93% correct for recordings of feeding and walking.

9.9.3 Playback of sounds

Sounds are played to animals in an attempt to determine their effectiveness in stimulating or inhibiting behavior. In this way the function and/or effect of biotic or abiotic environmental sounds can sometimes be determined. For an overview of the use of playback in ethological research see McGregor (1992).

Normal (unmodified) animal sounds are often played to conspecifics and their resultant effect observed. For example, Waser (1975b) demonstrated that playback of the gray-cheeked mangabey's (*Cercocebus albigena*) 'whoopgobble' vocalization mediated intergroup avoidance, and Lehner (1982) demonstrated that coyotes differentiated between 'group howl' and 'group yip-howl' vocalizations by their different vocal responses to playback of the two vocalizations.

Animal sounds are sometimes modified to determine the functions of their different components. For example, Emlen (1972) modified the recorded songs of indigo buntings (*Passerina cyanea*) and through playback demonstrated that: 1. species recognition is coded in the note structure, inter-note interval, and note length; 2. individual recognition is coded in the details of note structure; and 3. motivation cues are reflected in song length and singing rate. Using a unique approach, Simmons (1971) picked up bat cries in two condenser microphones and played them back to the bat with different delay times, simulating 'phantom targets' at different distances.

Playback can also be used to reveal the significance of interspecific sounds. For example, Cade (1975) showed that female parasitoid flies (*Euphasiopteryx ochracea*) containing living larvae were attracted to dead crickets attached to speakers, through which cricket songs were played.

The effects of natural (or synthesized) abiotic environmental sounds also can be studied by playing them to animals. For example, Larkin (1977) showed that tape recordings of thunder, bird calls, and artificial sounds played to migrating birds through a loudspeaker system slaved to a tracking radar often caused the birds to turn away from the sound. Heppner (1965) found that high-intensity noise had no effect on the ability of captive robins (*Turdus migratorius*) to find earthworms, further supporting the hypothesis that the robins were primarily using visual cues.

Several types of equipment and method can be used in playback studies. Several older techniques can now be replaced by computer technology. An example of older technology (that may still be suitable for some research) is the *pattern playback* that was designed to synthesize human speech for research on the recognition of consonants (Denes and Pinson, 1973). It is essentially the opposite of a sound spectrograph. A sound spectrogram pattern is drawn on a piece of paper that is then run through the Pattern Playback, which converts the images drawn on the paper into sound played through a loudspeaker. The Pattern Playback has been used to segregate the relative importance of components of sound in transmitting information. Likewise, Emlen (1972), in his study of indigo bunting song, used the time-consuming technique of cutting and splicing audio-tape in order to rearrange the order of the indigo bunting's notes for playback. Sound on audio-tape can now be digitized and stored in a computer where it can be stored, modified and manipulated and then converted back to an analog signal (normal sound) for playback. All of the newer sound analysis software programs have these capabilities (see section 9.10.3).

For playback of either natural, modified or synthesized sounds, the recorder, amplifier and speaker should be of sufficient quality and fidelity to broadcast a good reproduction of the sound. Lehner (1976) used the playback equipment in Figure 9.17 in his study of coyote vocalizations; the study required the broadcast of relatively low frequency, often 'noisy' (few pure tones) vocalizations at high intensity (volume).

The audio-tape recorders discussed above (section 9.2.2) are also suitable for playback of most animal vocalizations and mechanical sounds. Some of the reel-to-reel and cassette recorders being used by ethologists to playback animal sounds are:

- *Reel-to-reel*:
  Nagra IV-S (Dyson *et al.*, 1994)
  Revox A700 (Eiriksson, 1994)
  Tandberg Series 14 (Chaiken *et al.*, 1993)
  Uher 4000 (Brown and Waser, 1984)
- *Cassette*:
  Marantz CP430 (Adhikerana and Slater, 1993)
  Marantz PMD 200 (Marzluff, 1988)
  Marantz PMD 430 (Given, 1993; Allan and Simmons, 1994)
  Marantz PMD 3340 (Rothstein *et al.*, 1988)
  Sony TCD-5M (Brenowitz and Rose, 1994)

Like the list of recorders given earlier, this list is not exhaustive as to the makes or models of recorders that are suitable for playback of animal sounds. You should check the literature and consult other ethologists who have conducted playback of the same, or similar, sounds to those you will be broadcasting. Once again you should consult with reliable suppliers of playback equipment, such as Saul Minneroff Electronics, Inc.

9.9.4 Ultrasonic detectors

There are two types of commercially available detectors which have been used to detect the ultrasonic sounds of animals (Fenton *et al.*, 1973). These devices have been used in research on bats (Fenton, 1970; Kunz and Brock, 1975) and insects (Klein, 1955). See Sales and Pye (1974) for a review.

One ultrasonic detector is manufactured by Holgates of Totton, Southampton, United Kingdom. It uses a capacitance microphone capable of responding to frequencies between 10 and 180 kHz as well as electronic tuning to limit the input band width. Another that uses a crystal microphone adjusted for maximum sensitivity at 40 kHz is manufactured by Alton Electronics Co., Gainesville, Florida. Paige *et al.* (1985) provide a schematic diagram for constructing an inexpensive, hand-held ultrasonic detector.

9.10 ANALYSIS OF ANIMAL SOUNDS

Several types of sound spectrographs and computer/software systems are available for the analysis of animal sounds. Some of these tools are briefly described below.

9.10.1 Equipment

9.10.1a Kay Sona-Graph Model 7029A

The Kay Sona-Graph Model 7029A (Figure 9.19A) is an electronic device for converting tape recorded sound to a visual display. It records sound input from an audio-tape recorder onto a metallic drum. The recorded sound is then reproduced by a stylus which scans the various frequencies across time and electrically burns a sheet of paper to produce a 'picture' of the sound with frequency on the vertical axis and time on the horizontal axis. You have the option of representing frequency on a linear or logarithmic scale; it is usually reproduced on a linear scale, although Marshall (1977) argued for use of the log scale. The frequency range and duration of the sound spectrogram produced depends on the speed at which you set the metallic recording drum to spin while recording from the tape. For example, a sound spectrogram which displays 20–2000 Hz on the vertical axis will have a duration of 9.6 s, and an 80–8000 Hz display will have a duration of 2.4 s. It will produce several types of display, all of which are useful in the analysis of vocalizations or mechanical sounds (e.g. grasshopper calls).

The *normal display* of the Sona-Graph is the sonagram shown in Figure 9.20A. Time is represented on the horizontal axis (1 s/mark), frequency (pitch) is represented on the vertical axis (1 kHz/mark), and amplitude (intensity) is represented by the blackness of the mark. Sound spectrograms are produced on paper 14.5 cm × 32.4 cm, only a portion of which is shown in Figure 9.20.

From the sonagrams above we can see that the coyote's howl began with three bursts of energy over several frequencies, with none of the frequencies being clearly defined; these are essentially introductory 'barks'. The howl portion began at a relatively low frequency and then rose to approximately 1.6 kHz, where it remained for approximately two seconds, at which point the frequency dropped off sharply.

Since the difference between intensities is one of degree (relative blackness), we can use the *contour display* to delineate areas of equal intensity in seven gradations

(Figure 9.20B). This provides only relative measurements and says nothing about the actual intensity of the sound.

Another feature of the Sona-Graph that provides a relative measure of intensity versus frequency is the *section display* (Figure 9.20C). This display samples the recording at six or fewer predetermined points and presents relative amplitude as a horizontal mark at each frequency, inverted from the normal sonagraph (frequency increasing from the top to the bottom of the paper). Note that the section through the bark shows a much wider range of frequencies than that through the howl. Marler and Isaac (1960) describe a device for modifying the sound spectrograph to make frequency-versus-amplitude sections serially through a syllable at intervals down to 2.5 ms.

Sections are useful for determining the relative amplitude of frequencies in a particular syllable or sonagram. However, they cannot be used to make absolute measurements (e.g. number of decibels) without considerable difficulty, and they should not be used for comparisons between sonagrams since they are affected by the investigator's choice of settings on the sound spectrograph.

Vocalization terminology has been rather inconsistently applied, with few authors using similar terms. Kroodsma (1977) used the terms in Figure 9.21A to detail song development in the song sparrow. These terms are similar to those used by Rice and Thompson (1968) for indigo bunting vocalizations. Although not totally satisfactory (Kroodsma, pers. commun.), these terms are applicable to vocalizations of numerous other species and are useful in sonagram analysis. Temporal patterns are extremely important in insect sounds. Bentley and Hoy (1972) developed the terminology in Figure 9.21B for their study of the genetic control of cricket (*Teleogryllus gryllus*) song patterns.

The Kay Sona-Graph Model 7029A has been used for more than two decades in ethology, but it is no longer being manufactured (although limited parts are available; Kay Elemetrics Corp. – address below). Although more sophisticated equipment is now being marketed (see below), used 7029A machines may still be available, and they are satisfactory for analyzing sounds in many studies (e.g. Miller 1994, Payne and Payne 1993).

9.10.1b Kay DSP Sona-Graph Model 5500

The Kay DSP Sona-Graph (Figure 9.19B) is a workstation that combines a real-time sound spectrograph, a computer-based data-acquisition system and a dual channel FFT analyzer. Sound input is stored digitally for display and analysis, and it can be downloaded to another computer for storage or analysis by other computer programs (see below). It is a menu-driven system that displays oscillograms (wave forms), contour power spectrums and spectrograms on the video monitor that can

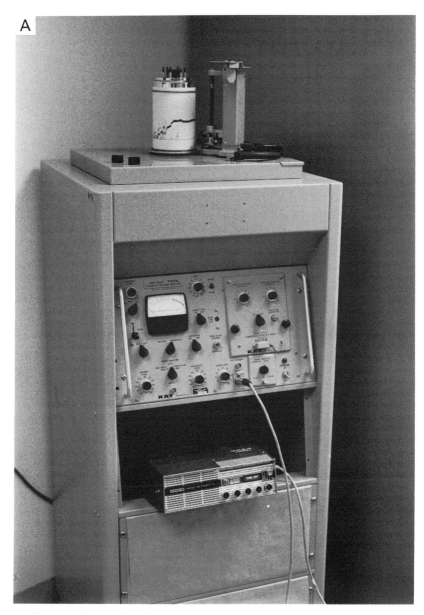

Fig. 9.19A Kay Elemetrics 7029A Sona-Graph and Uher 4000 audio-tape recorder.

Fig. 9.19B   Kay Elemetrics DSP Sona-Graph Model 5500 interface with a microcomputer.

then be printed. It has a history of use in ethological studies (e.g. Brenowitz and Rose, 1994) and is available with several hardware and software options. The newer Model CFL 4300 is a completely computer-based system that may replace the DSP 5500 for animal sound analysis (Kay Elemetrics Corp., 12 Maple Ave. Pine Brook, NJ 07058).

9.10.1c   *Uniscan II and Ubiquitous Spectrum Analyzer*

Two sound-analysis machines that are no longer being manufactured, but may still be available for use, are the Uniscan II and the Ubiquitous Spectrum Analyzer. Both can be used for analyzing animal sounds.

The Uniscan II (Multigon Industries Inc.) system includes a keyboard, processor, monitor and printer. It produces a real-time display of a sound spectrogram to the monitor or printer in several selectable frequency ranges. Any 1.6 second segment can be frozen on the display for measurements of frequency and duration.

Ubiquitous is the trade name for the Federal Scientific Spectrum Analyzer. It is a real-time, time-compression scanning analyzer which can be used for analysis of animal vocalizations with the addition of a display system (Hopkins *et al.*, 1974). A digital system is used to speed up the signal, and an analog system sweeps the time-

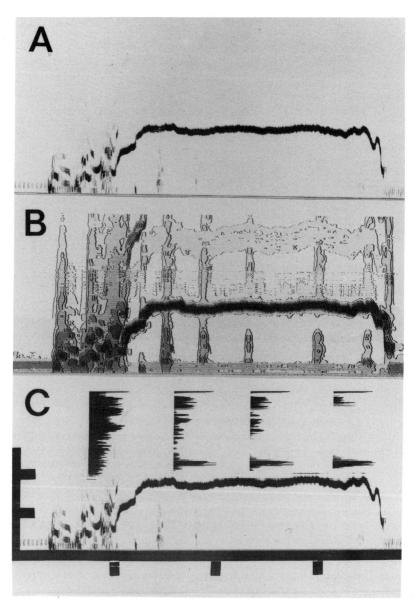

Fig. 9.20 A. Normal display sound spectrogram (sonagram) of a coyote howl; B. Contour display of the same howl; C. Section display above the normal display. Time is marked on the horizontal axis in one-second intervals. Frequency is marked on the vertical axis in one kHZ intervals. (See text for an explanation of the types of sound spectrograms.)

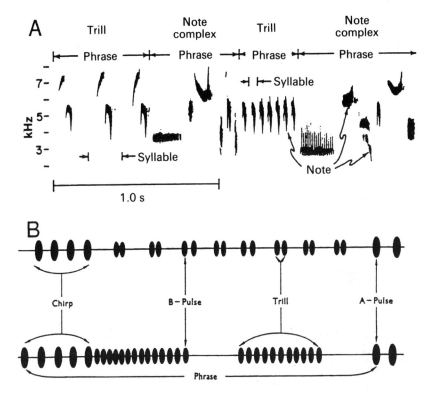

Fig. 9.21   A. Terminology used in a study of song development in the song sparrow (from
Kroodsma, 1977); B. Diagram of structural components and terminology of
Telegryllus songs: upper line=*T. oceanicus*; lower line=*T. commodus*. Interchirp
interval=interval between onset of A-pulses. Intrachirp interval=interval
between onset of B-pulses. Intertrill interval=interval between onset of last B-
pulse in one trill and the first pulse of the next trill (from Bentley and Hoy, 1972).

compressed signal through a filter. Spectrograms can be displayed on a storage
oscilloscope or photographed for permanent copy. Narrow and wide bandwidth
analyses are possible, and section displays can be made at intervals as short as 3.125
ms.

One advantage of this system is the speed at which spectrograms can be produced. Hopkins *et al.* (1974) report that a 2.4–second-long spectrogram takes
approximately 1.3 minutes to analyze on the Kay Sonagraph 7029A and only 9.6
seconds on the Ubiquitous. Another advantage is the relative ease with which
section displays can be produced. Spectrograms produced by the Ubiquitous are
grainier than those produced by the Sonagraph; however, this apparently does not
affect interpretation (Hopkins *et al.*, 1974).

9.10.1d Desktop computers

Desktop computers, commonly the IBM-PC (and compatibles) and the Apple Macintosh, are frequently used with specially designed software to store and analyze animal sounds digitally.

Digitizing tape recordings of animal sounds is accomplished with an analog-to-digital (A/D) converter, often a circuit board inserted into the computer where the sound will be stored. For example, *Drosophila* courtship songs were digitized using a Campbridge Electronics Design 1401 (Ritchie and Kyriacou, 1994) and a Canopus Sound Master (Tomaru and Oguma, 1994), and Gerhardt *et al.* (1994) used a Soundfx interface board to digitize tree frog calls.

9.10.2 Analysis of sound spectrograms

Sound spectrograms (hardcopy) are generally used to measure: 1. frequencies (Hz), both dominant frequencies and harmonics; 2. durations of sounds and time intervals between them; and 3. relative intensities of portions of the sound. They can also be used to compare components of the sounds between and within individual animals. Hall-Craggs (1979) provides several useful suggestions for basic sound spectrogram analysis, and Thompson (1979) offers suggestions for preparing sound spectrograms for publication. The techniques described below involve using hand-operated mechanical devices (e.g. rulers, calipers, computer stylus) and human judgement. Although time consuming and generally less accurate than computer analysis, they may be suitable for some studies.

Frequencies are measured from either a *narrow band filter* display on a *normal* spectrogram or from a *section* display. Transparent overlay grids are useful in the making of these measurements. Frequency measurement is more accurate when lower-frequency spreads are used for display (e.g. 20–2000 versus. 160–16000 Hz). Contour displays are often useful to determine more accurately the dominant frequencies when large areas are burned. Horii (1974) described a method for producing digital sound spectrograms with simultaneous plotting of intensity and fundamental frequency. A digitizer system (X, Y cursor, teletype and computer) was used by Field (1976) to analyze sound spectrograms of wolf vocalizations. The cursor is moved along a selected frequency band (e.g., dominant frequency). The X, Y axes of the cursor's plane of movement represent time and frequency, respectively. The operator depresses a button at predetermined points along the trace, and the X, Y coordinates are transmitted to the computer for storage and printed out on the teletype. Field (1976) used an overlaying grid to locate coordinate sample points every 0.05 second. The greater the variation in frequency of the sound, the more often the coordinates should be sampled. The coordinate sets can then be used to redraw the trace and compare for its accuracy in representing the original spectrogram.

Duration measurements are generally made from *wide band pass* spectrograms. However, the mark intensity can affect the measurements if they are either under- or over-burned.

All the measurements described above (and many more) can be made more quickly and accurately with a desktop computer and special software.

9.10.3 Computer analysis of recorded sounds

A specialized field of data analysis has developed around the use of microcomputers in bioacoustics. Software currently exists to permit input of animal sounds from a tape recorder or microphone to a microcomputer where it is digitized and stored, using hardware such as the Unisonic (for IBM-PC and compatibiles), MacRecorder digitizer, or various analog-to-digital interface boards (see section 9.10.1d); the maximum length of sound that can be digitized and stored at one time is limited to the computer's available random access memory (RAM). Once the sound is digitally stored, the software can quickly produce spectrographs of frequency, time and intensity, and oscillographs of time, amplitude and frequency. These spectrographs and oscillographs can then be printed out or analyzed further. Some software can make matches between sound segments (e.g. MATCH; Payne and Payne, 1993) and produce three-dimensional visualizations of sound measurements.

Another important aspect of this bioacoustical software is the ability to manipulate sounds which have been digitized and stored in the microcomputer. Portions of the sound can be deleted, duplicated, moved, reversed or frequency-altered at the touch of a key. The modified sound can then be played directly to an animal or fed back into a tape recorder with a built-in (or peripheral) digital-to-analog converter (e.g. Allan and Simmons, 1994; Randall, 1994). Sounds can be created from scratch using these programs, or even more easily and inexpensively, using any of the large number of music programs on the market.

Davis (1986) describes the Personal Acoustics Lab (PAL), which is a microcomputer based system for digital signal acquisition, analysis and synthesis. Some of the commercially available sound analysis software packages are listed below:

- *Canary* (Apple Macintosh)
  Bioacoustics Research Program
  Cornell Laboratory of Ornithology
  159 Sapsucker Woods Rd.
  Ithaca, NY 14850
- *MacSpeech Lab* (Apple Macintosh)

GW Instruments
P.O. Box 2145
264 Msgr O'Brien Hwy #8
Cambridge, MA 02141
- *SIGNAL* (IBM-PC)
  Engineering Design
  43 Newton St.
  Belmont, MA 02178
- *SoundEdit v.2.0.3* (Apple Macintosh; can edit frequencies only from 0–11 KHz; not designed for computer analyses)
  Farallon Computing
  2150 Kittredge St.
  Berkeley, CA 94704
- *SoundEdit Pro* (Apple Macintosh)
  MacroMind Paracomp, Inc.
  600 Townsend St.
  San Francisco, CA 94103

Although most sound-analysis programs will perform the functions described in previous paragraphs, some are easier to use, have faster sampling rates and greater dynamic ranges, and have additional graphic and analysis capabilities; for example, Weary and Weisman (1993) state that MacSpeech Lab and SIGNAL are more 'sophisticated' software packages than SoundEdit v.2.0.3. You should obtain additional information from researchers who have used the software, as well as the distributors of the software, before choosing a software package to use or purchase; sometimes demonstration programs (shortened, simplified versions) are available for you to try.

9.11 PHOTOGRAPHY

Ethologists should make a photographic record of their research. Pictures should be sharp, well composed and suitable for reproduction if needed. Prints are necessary for publication, while color transparencies (slides) are very useful for oral presentations.

Photos should depict the: 1. study site; 2. animals studied; 3. equipment and methodology; 4. results of data analysis (tables and figures), and 5. your interpretation of the results (e.g. models; Chapter 18). As photos are taken a log should be kept of photo number, date, time, location, subject matter, why the photo was taken (i.e. what you were trying to depict) and what in particular you should note when you see the transparency or print. In addition, to improve future photos, you may record environmental conditions, lens used, film type, shutter speed, lens opening (f stop), filters, and any exposure compensations made. This log may be kept as part of your field notebook.

The techniques of good photography are beyond the scope of this book; complete and useful discussions can be found in Blaker (1976) and Anonymous (1970).

9.11.1 Still photography

The most useful *still camera* for the ethologist is the 35–mm single lens reflex (SLR) camera. A distinct advantage of the SLR camera is that the image you see in the viewfinder is the same (93–100% accuracy) as the image that is recorded on the film. The SLR camera is also compact, lightweight and versatile. It accepts a large variety of film types (see following sections), although the most commonly used is color-slide film. Ideally, an ethologist entering the field in an unfamiliar area should be prepared with two cameras loaded with different types of film depending upon the proposed use of the photos or slides. Typically, a black-and-white negative film for black-and-white prints is kept in one camera and a color slide film for presentations is loaded in the other. Color prints can also be made from the slides, if necessary. It is helpful if both cameras will accept the same lenses so that they can be easily changed or interchanged.

There are a large number of makes and models of 35–mm SLR on the market today, most of which have their own group of ardent supporters. Nikon and Leica are excellent camera systems known for their quality and versatility; however, Leica is very expensive. Minolta, Canon, Pentax and Olympus are other camera manufacturers to consider seriously, each having within their systems the necessary equipment for simple to complex photography. These camera brands will have a complete line of accessories to cover your photographic needs, including a wide variety of lenses, motor drives (automatic film advance), and flashes. The following features should be considered necessities in any camera you use or purchase:

1 Maximum shutter speed of at least 1/1000th second.
2 Automatic and manual exposure settings with a maximum lens opening of at least f1.9; that is, the f stops to go as low as f1.9.
3 Through-the-lens light meter.
4 Black camera body to reduce reflections and glare directed to the animal.

The following are optional features to be considered:

1 Electronic cable release – enables remote firing of the camera (connects to the motor drive).
2 Depth-of-field preview – permits you to stop the lens down manually to preview depth-of-field with a given f-stop.

3   Interchangeable finder screens – allows replacement of a split-image focusing screen with a clear matt screen for easier focusing.

4   Data back – provides on frame information when photo is taken; that is, frame number, date, time, and exposure settings (information provided depends upon capability of different backs).

5   Water resistence or waterproof – some cameras are sealed with rubber gaskets that resist leakage to moderate depths underwater. A rating for a depth of only 3 meters will be worthwhile in heavy rain and if the camera is dropped in a stream.

Today, cameras have advanced far beyond the cameras of 15 years ago. Computer chips, instead of manual mechanisms, control most of the camera's functions including focus, exposure and flash photography. However, the Olympus OM-3, Canon F-1, and the Pentax K-1000, have mechanically controlled shutter speeds and therefore rely on batteries only to run the exposure meter. The Pentax K-1000, however, does not accept a motor drive or data back. Electronic cameras, through years of design and testing, have reached a high level of reliability and performance. Their ability to self-adjust the exposure in difficult and contrasting lighting conditions, auto-focus, and imprint data on a photo make them a very valuable tool in research. However, since electronic (automatic) cameras rely on batteries to function, it is recommended that batteries be replaced yearly and spare batteries be kept on hand.

The majority of auto-focus cameras have the ability to self-focus accurately in near dark conditions; they will usually have the standard manual focusing available also. Exposure metering systems in cameras give you a variety of choices from full manual exposures to fully automatic exposures controlled by the camera's electronics. The following list describes several types of exposure metering systems and defines their function:

1   Standard program – the camera sets both shutter speed and lens aperture with a bias of hand-held shutter speed of 1/125 s or above.

2   Wide program – the camera sets both shutter speed and lens aperture with a bias of smaller aperture over shutter speed for greater depth-of-field.

3   Tele program – the camera sets both shutter speed and lens aperture with bias towards higher shutter speeds to freeze action.

4   Shutter-priority auto – you set the shutter speed and the camera sets the lens aperture.

5   Aperature-priority auto – you set the lens aperture (f stop) for depth-of-field and the camera selects the correct shutter speed.

6   Manual exposure – you set both the shutter speed and the lens aperture with the guidance of the built-in light meter.

All of these exposure metering systems will, under normal lighting conditions, give you the correct exposure. However, when selecting a camera for use or purchase determine whether that camera has the system that best meets your needs.

Lenses play an important role in the quality of your photographs. The quality of a lens varies optically and in durability. A very expensive camera will take poor-quality photos if a poor-quality lens is used. The 'standard' lens that is most often supplied with a 35 mm camera is a 50 mm focal length lens. It is considered to have a normal perspective (angle-of-view). Any lens that has a focal length longer than 50 mm is a 'telephoto' lens, while a lens with a shorter focal length is a 'wide angle' lens. Wide angle lenses are used where a wider perspective is desired (e.g. habitat photos, photos in tight quarters), and telephoto lens are used to magnify subjects that are far away. The magnification of an object is directionally proportional to the focal length of the lens; that is, a lens with twice the focal length will double the magnification (e.g. a 100 mm lens produces twice the magnification of a 50 mm lens). Zoom lenses have variable focal lengths built into them, such as a 28 – 80mm zoom lens. Their advantage is that you can carry one or two zoom lenses rather than several fixed focal length lenses. The disadvantage of zoom lenses is that their quality varies greatly and is often not as good as a fixed focal length lens. Commonly used zoom lenses are 28–80 mm and 80–200 mm.

An accessory which many ethologists find useful is a *motor-drive* unit which automatically advances the film after each shot is taken. A motor drive comes built-in to some cameras. The speed of film advancement ranges from 1.5 to 5 frames-per-second. Besides allowing you to take photos very rapidly, a motor drive allows you to maintain continual observation of the animal(s) through the viewfinder without moving it to advance the film manually. You can then concentrate on the animal's behavior and photograph carefully selected behavior units, especially sequences, for later analysis or presentations.

Automatic film advances are necessary when cameras are left set up in the field and are triggered by an animal's activity. For example, Savidge and Seibert (1988) used an infrared device to trigger a camera that photographed predators when they visited artifical nests.

Electronic flashes are helpful when additional illumination is necessary and the subject is within range of the flash output. For nocturnal animals, this may be the only means to photograph them in their natural habitat; however, flashes are likely to alter their behavior. Photographing small animals (e.g. field mice) by natural light often produces unsatisfactory photos. The combination of a slower film for quality and a small aperture for depth-of-field forces you to use a slow shutter speed and a tripod for support. A flash will allow the use of a small aperture for greater depth-of-field, better detail on low light subjects, and the ability to freeze movement with a higher shutter speed. Try to use a flash that is the same brand as the camera so that it will integrate well with the camera's electronics and provide accurate autoexposures. Manually controlled flash exposure is also available with these units.

Another important piece of equipment is a *tripod*. Tripods are often left behind because of their additional weight and bulk, but their advantages outweigh their disadvantages. A good tripod will allow you to shoot at slower shutter speeds and use the aperture setting that you may need for greater depth-of-field. The rule-of-thumb for acceptable hand-held photography is to always use a shutter speed faster than the focal length of the lens you are using. For example, the slowest shutter speed you should use for a hand-held camera with a 50 mm lens is 1/60th second. Shooting at slower speeds will result in unsharp photos. When photographing from a fixed position, such as a nest site, a tripod will eliminate the need to constantly hold the camera in order to keep the subject in the viewfinder at all times. This will often result in photos of behaviors that occur too rapidly for you to bring a hand-held camera into position and shoot. There are several sources of information on tripods including a leaflet from Christophers, Ltd. (2401 Tee Circle, suites 105/106, Norman, OK 73069).

You should choose a camera that is ruggedly built and handles well. If you are unsure, talk to other ethologists and find out which cameras they have used and recommend. Before purchasing or going into the field with a camera you should become familiar with its functions. Understand the metering system (see above) and other features that you intend to use. Shoot a roll of film through the camera to verify that the camera is operating properly and you understand how to use it. The sequence of setting the exposure, aiming the camera, releasing the shutter, and advancing to the next frame should occur in one easy motion.

The most common *film* used in ethology is color slide film. Slides can be projected for oral presentations and used in analyzing behaviors (e.g. visual displays). With special processing, color or black-and-white prints can be made from color slides. The second most common film is black-and-white negative film. If the primary purpose is publication, use film that is designed for black-and-white prints (Table 9.8). Black-and-white prints made from color slides are of poorer quality. Below are several characteristics which you should consider when selecting the proper film for your intended use (Tables 9.8, 9.9).

Film speed is represented by an ISO (International Standard Organization) number (Tables 9.8, 9.9). This replaced the ASA (American Standards Association) number, even though ISO and ASA numbers are equivalent. DIN is the German standard of sensitivity to light; it corresponds to, but is not the same as, ISO numbers. These numbers are relative and comparable from one film to another, regardless of the manufacturer. For example, a film with ISO of 100 is twice as fast as one of ISO 50. The speed of a film indicates its relative sensitivity to light. Increased film speed: 1. allows the use of faster shutter speeds to freeze motion;

Table 9.8. *Selected Kodak black-and-white film for use in 35 mm still cameras*

Film	Speed ISO	Definition Graininess	Definition Resolving power	Sharpness	Degree of enlargement allowed	Suggested uses
Panatomic-X	32	Extremely fine	Extremely high	Very high	Very high	For prints. With a special reversing process it will produce slides. It should be used when the emphasis is on very-high-quality prints for publication or enlargement
Plus-X Pan	125	Very fine	High	Very high	High	For prints. A good all-around film that combines reasonable speed with high definition qualites
T-MAX 100	100	Very fine	Very fine	Very high	High	Sharper than PLUS-X
TRI-X Pan	400	Medium	Medium	Very high	Moderate	For prints. Its major quality is high speed which can be pushed to ASA 800 in some cameras. It can be used in very low light (e.g. forest) or to stop motion (e.g. running antelope)
T-MAX 400	400	Fine	Medium to high	Very high	Moderate	Sharper than TRI-X Can be pushed up to 1600 ISO
2475 Recording (Estar-Ah Base)	1000	Coarse	Low	Very high	Low	For prints. This is a poor-quality film which has only its speed to recommend it. It should be used only when very low light or high speed call for it
High Contrast Copy Film 5069	64	Fine	High	Very high	Very high	For copying printed materials (e.g. photos, charts, tables, drawings, etc.). Useful in preparing visual aids for presentations and field trips
T-MAX P3200	3200 multi-speed	Medium coarse	Medium	High	Low to Moderate	Has an ISO Range of 1000–6400. Allows photography in very low light at high speeds with good results

Table 9.9. *Selected color-reversal film for use in 35 mm still cameras*

Film	Daylight speed ISO	Definition			Type of picture and degree of enlargement allowed	Suggested uses
		Graininess	Resolving power	Sharpness		
Kodachrome 25	25	Extremely fine	High	High	Slides[2] High	Has high color quality and a wide exposure latitude. It should be used under most daylight conditions when sufficient light is available and fast motion does not need to be stopped
Kodachrome 64	64	Extremely fine	High	High	Slides[2] High	Combines good definition with relatively high speed. It does not have the color quality of Kodachrome 25, so it should be used only when extra speed is necessary
Ektachrome 100	100	Extremely fine	Medium	Medium	Slides Moderate	Should be used as a substitute for Kodachrome-64 when you expect to do your own processing
Ektachrome 200	200[1]	Extremely fine	Medium	Medium	Slides Moderate	For use in dim light, shade, or to stop rapid movement; also with telephotos lacking large lens openings, in order to increase depth of field
Fujichrome 50	50	Extremely fine	High	High	Slides[2] High	Same as Kodachrome 25. Excellent color rendition
Fujichrome 100	100	Extremely fine	High	High	Slides[2] High	Good definition with higher speeds. Excellent blues and greens. Can be pushed to ISO 200 with good results

Notes:

[1] This ISO can be pushed to 400 with special processing.

[2] High-quality prints can also be obtained through an additional process.

2. allows for a smaller lens opening (aperture=f stop), which provides greater depth-of-field (range of distance in front of and behind the subject that will be in focus); and 3. generally produces a grainier photograph (see next section).

You may want to modify the film speed rating recommended by the manufacturer based on the results obtained with your camera. If your photos are consistently overexposed or underexposed you may want to adjust your setting of the film speed as follows:

Overexposed: recommended film speed $\times$ 2=1 f stop less exposure

Underexposed: $\dfrac{\text{recommended film speed}}{2}$ =1 f stop more exposure

If you are still getting poor-quality photos after making these compensations, then your camera may be in need of adjustments by the manufacturer or a local camera repair person.

Definition is the clarity of detail in the photograph; that is, how closely it represents the actual scene as viewed by the human eye (aside from color rendition). Definition is the result of several factors, the three most important for our purposes being graininess, resolving power and sharpness.

Graininess refers to a dot-like or grainy appearance of a photograph when viewed closely. A photo that is very coarse-grained will appear like a series of dots rather than a continual gradation of colors and shades. Graininess is particulary important if a photo is going to be enlarged or used in publication, since graininess magnifies with enlarging. Generally, graininess increases with increased film speed (higher ISO number) and overexposure.

Resolving power refers to a film's capability to reproduce fine details, such as two lines very close together. This is generally not important in most ethological work except in photographs of items such as equipment, charts and maps.

Sharpness refers to the definition of edge in a photograph, for example, the definition of the side of an elk's antler against a background. If sharpness is poor, the antler will appear soft or 'fuzzy'. Sharpness is affected by the speed and type of film, lens quality, and shutter speed relative to how steady the camera is held.

Exposure latitude is the degree of overexposure and underexposure that the film can be shot at and still produce acceptable photographs. Most continuous-tone negative films have an exposure latitude of 5 f stops; 3 f stops overexposed and 2 f stops underexposed. Slide films have less latitude; most range from 1½ stops overexposed to 2 stops underexposed.

Color sensitivity is the range of wavelengths of light to which the film is sensitive. All the films listed in Tables 9.8 and 9.9 are *panchromatic*; they are sensitive to all the visible colors as well as ultraviolet radiation.

Table 9.10 *Reciprocity and recommended* f *stop corrections for 35 mm color films*

Film type	1 s	10 s
Kodachrome 25	+½ stop	+2 stops
Kodachrome 64	+½ stop	N/A
Ektachrome 100	+1 stop	+1½ stops
Ektachrome 200	+½ stop	N/A
Ektachrome 400	+½ stop	+1½ stops
Fujichrome 50	No change	+½ stop
Fujichrome 100	No change	+½ stop
Fujichrome 400	No chage	+1 stop

Note: One stop=doubling the exposure

Reciprocity failure is the loss of a film's light sensitivity during long exposures, normally longer than one second. This can be corrected by doubling the exposure time for shots of one second, or more. Color films may show a color-shift (differing sensitivity to different wavelengths of light) during exposures over two seconds. Table 9.10 lists the reciprocity and recommended f stop corrections for several common films: Most negative films require an increase of one stop with exposures longer than one second.

In addition, *infrared films* are available for special uses. Kodak High Speed Infrared film is available in 36–exposure rolls for 35–mm cameras. It is fine-grained with moderately high contrast, and medium resolving power and sharpness. It can be used to photograph through haze or to record behavior of nocturnal animals lighted by infrared bulbs. The speed of the film is highly variable, depending on the ratio of visible to infrared light available.

Storage is an important consideration for all types of film. All films are damaged by high temperature and high humidity. Films can be obtained in vapor-tight packaging if you anticipate working in areas of high humidity. Refrigerating film extends its useful life well beyond the expiration date printed on the box. Before using film that has been refrigerated, allow 2–3 hours for the film to reach ambient temperature in it's plastic container (condensation may form on the film if removed). Kodak makes the following storage recommendations for black and white film:

For storage periods of up to:	2	6	12	months
Keep film below:	75°F	60°F	50°F	

Keep film away from industrial gases, motor exhaust, and vapors of mothballs, formaldehyde, solvents, cleaners, and mildew or fungus preventatives.

Static electricity caused by rewinding film too rapidly in cold weather will cause streaks or dots on the film. Also, X-ray inspection units in airports, despite their claims, may damage film whether it is new or exposed. X-ray damage is cumulative and may not show up until after additional exposure. Check film through by hand or protect it in special, lead-lined bags available at camera stores.

Prints, negatives and slides should be kept in a cool, dry area where they will be safe from damage, but where they can be easily retrieved. Store negatives and slides in archival, plastic pages which can be put into three–ring binders or stored in a file cabinet. This will protect your original photos and provide for easy access and viewing. A cataloging system will organize your photos and may be based on: 1. separate research projects; 2. behavior types; 3. species; or 4. field seasons. Computer programs are available that will catalog by numbers and captions, search for specific slides, and print out labels for slides. Ethologists must develop a system which they find most useful. In addition, attempt to reduce possible losses or damage in the mail by sending photos properly packaged in separate packages; if possible, send duplicates instead of original slides and prints instead of negatives.

Another storage medium is provided by *digital photography*. As examples, Kodak's models DCS 420 and DCS 460 (high resolution) combine digital imaging with a Nikon N90 SLR camera body. They are available in monochrome, color and infrared models which store the images on removable 170 MB RAM cards. One card will store from 30 high-resolution images (6 million pixels; DCS 460) to 100 images (DCS 420); by changing storage cards, 300 images can be captured on a single 1 hour battery charge. Using appropriate interfaces, the images can be downloaded to Apple Macintosh II, Powerbook, Quadra and IBM-PC (and compatible) computers. They can then be used in computer displays, made into prints or slides, or stored in portfolio CDs. Additional information can be obtained from: Digital & Applied Imaging, L&MS, MC 00532; Eastman Kodak Co., PO Box 92894, Rochester, NY 14692–9939.

9.11.2 Motion-picture photography

The obvious advantage of both motion pictures and videotape is that they allow you to record a two-dimensional visual representation of entire behavior patterns. The two-dimensional restriction can be overcome, in part, by the use of two or more strategically located cameras. In addition, it provides the capacity synchronously to record sound (produced by the animals, or the environs, or dictated by the observer).

Hutt and Hutt (1974) list five situations in which motion pictures and videotape are especially useful: 1. swift action; 2. complex action; 3. subtle behavioral changes; 4. complex behavioral sequences; and 5. the need for precise measurements of parameters.

The first choice you have to make is the film-size format you want to use. The two

Table 9.11 *Relative advantage of Super 8 mm and 16 mm filming*

Super 8 mm	16 mm
1. Cheaper cameras, film and processing	1. Pictures with greater sharpness, resolution, and definition
2. Lighter equipment	2. Pictures brighter when projected to same size
3. Convenience of cartridge film	3. Cameras often more durable
	4. Film often easier to handle for editing and analysis
	5. Larger film capacity
	6. Better for sync. sound

basic choices are 16 mm and Super 8 mm; standard 8–mm film has essentially disappeared. The relative advantages of each are listed in Table 9.11; however, it is basically a choice between lower cost (Super 8 mm) and higher quality (16 mm). If you don't intend to do much filming, borrow or rent a Super 8 mm camera. If you intend to make filming an integral part of your studies and can afford it, use 16 mm (Figure 9.22).

In selecting a camera you will be confronted with a trade-off between cost and certain features (e.g. lenses, built-in exposure meter, filming speeds, durability, etc.). Some of these features are discussed below. Remember to purchase what you need, but not more than you need. Also, if possible, try before you buy.

Lenses should be selected with an eye toward the uses you intend for your equipment. If the camera has a lens turret, then you might select three lenses, such as 10 mm (wide-angle), 26 mm (standard), and 75 mm (telephoto). It may be necessary to use a telephoto lens as large as 600 mm (Dan and Van der Kloot 1964) or 1000 mm. If the camera will handle only one lens, then a zoom (26 mm to 75 or 100 mm) is very useful. If you are working with insects, a close-up lens and extension tubes are often desirable. Select high-quality lenses with large apertures approaching $f1.1$.

The diversity of *films* available for movie-making is so great as to confuse the neophyte. Selection of the proper film is generally a trade-off between film speed (amount of light necessary for proper exposure) and picture quality. Black-and-white film is cheaper to purchase but more expensive to process, while color film provides an additional dimension which is not only esthetically pleasing but often necessary in some ethological studies. Table 9.12 provides a list of Kodak films that are useful for filming animal behavior. Additional, more specialized films can be found in Eastman Kodak's publication *R-31, Kodak Photographic Materials Guide*, as well as from other manufacturers. For example, infrared film can be used in con-

Fig. 9.22   Bolex H-16 16 mm movie camera with 75 mm telephoto lens.

junction with infrared lighting (e.g photofloods and infrared filters) to obtain motion pictures under nocturnal conditions. (Delgado and Delgado 1964).

The *filming speed* you choose will depend on the purpose of the filming. Normal projection speeds for Super 8 mm and 16 mm are 18 and 24 frames/second, respectively. The effect of accelerated motion is produced by filming at slower speeds (e.g. 2–10 frames/second), and slow motion is produced with greater filming speeds (e.g. 32–64 frames/second). If you are interested in frame-by-frame analysis (see below) then the faster you film, the smaller the change in the animal's position from frame to frame. Faster filming speeds also allow for unsteadiness by the cameraman; but it means changing film more often and increased costs in purchasing and processing the larger amount of film.

Various filming speeds and the authors' rationale for their use can be found in the literature. For example, in terms of frames per second, 2–7 or 48 (Eibl-Eibesfeldt, 1972), 16 (Clayton, 1976; Havkin and Fentress, 1985), 18 (Fleishman, 1988), 22 (Diakow, 1975), 24 (Kruijt, 1964; Dane and Van der Kloot, 1964), 32 (Duncan and Wood-Gush, 1972), 64 (Bekoff, 1977a), 128 (Hildebrand, 1965), and 800 and 1000 (Grobecker and Pietsch, 1979) have all been used. Time-lapse photography can be used to obtain instantaneous samples of behavior over extended periods of time (also see section 9.11.3, on film analysis). For example, Capen (1978) used an 8 mm movie camera set to take a frame at either one-, 1.5 or two-minute intervals in his study of nesting behavior in white ibises. When the cameras were set at two-minute intervals, five days of photos could be obtained from one film

Table 9.12. *Selected Kodak reversal motion-picture films*

Film	Daylight Speed (ISO)	16 mm/ Super 8	Characteristics	Suggested uses
Black-and-white				
Plus-X	50	16/8	High degree of sharpness, good contrast, and excellent tonal gradation	General outdoor photography
Tri-X	200	16/8	Excellent tonal gradation	Under adverse lighting conditions
Color				
Kodachrome 40	40	8	Good color rendition	General outdoor photography
Ektachrome 160	160	8	Higher speed	Adverse lighting
Ektachrome 7239	160	16	Good color rendition and sharp images	General outdoor photography
Ektachrome high speed daylight	400	16	High speed	Adverse lighting and high-speed photography

cartridge. Borgia (1986) used an infrared system to trigger a super 8 motion picture camera (Figure 9.23) in his study of bowerbird behavior.

Both Super 8 mm and 16 mm films and cameras are available for simultaneous recording on a *sound track*. The sound reproduction is generally not of high quality, but can be useful for recording the observer's commentary during filming. Good-quality sound recordings are best made with 16 mm cameras (e.g., Bolex H-16, Figure 9.22) that will synchronize with a high-quality tape recorder, such as the Nagra IV-L.

9.11.3 Film analysis

Ethologists take motion pictures for basically two purposes: 1. to have a visual record of the behavior for illustrative purposes (presentations and publications, e.g. J.M. Davis 1975); and/or 2. for analysis of (a) specific individual behaviors

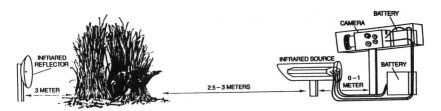

Fig. 9.23 The monitoring system used in a study of the satin bowerbird. An infrared
beam, invisible to the bowerbird, is projected through the avenue of the bower
to a reflector. When the beam is interrupted, a super-8 motion-picture camera
exposes one frame every two seconds. Birds were also observed from blinds. The
system enabled the researcher to monitor the behavior and identity of bower
owners and visitors at 33 bowers for the 50–day mating season. The researchers
are now using a more sophisticated system based on videocameras. From Borgia
(1986). Copyright © 1986 by Scientific American, Inc. All rights reserved.

(Milinski, 1984); (b) individual movements (Hailman, 1967; Havkin and Fentress,
1985), including locomotion (Hildebrand, 1965; see also Chapter 10) and social displays (Barlow, 1977; Bekoff, 1977 a,b); c) intra-individual sequences (Tinbergen,
1960a); Balgooyen, 1976); (d) inter-individual sequences (Diakow 1975); and (e)
spatial relationships (Dane and Van der Kloot, 1964).

Analysis of film is conducted either frame-by-frame or by sampling frames at
regular intervals, e.g. every 24th frame (Golani, 1973). If frames are to be selected at
intervals for analysis, an intervalometer can be coupled with the camera to expose
frames at set intervals (Figure 9.24). This provides a more efficient use of film.

Steele and Partridge (1988) projected Super-8 film of courting male *Drosophila*
onto the underside of a glass table and copied the males' movements onto tracing
paper; from these tracings they measured each male's maximum angular lag and top
speed during their courtship dance. Analyses are generally conducted with either
film editors that have a built-in projection screen (Hutt and Hutt, 1974), an optical
data analyzer (e.g. LW International; Milinski, 1984; Havkin and Fentress, 1985), or
an analyzer-projector (e.g. Lafayette Analyzer; Lafayette Instrument Co.,
Lafayette, Indiana). The latter projects the film onto a large screen (Dane and Van
der Kloot, 1964). Whichever system is used, it should have a reverse and a frame
counter. The frame counter, coupled with the filming speed, provides a time base for
measuring the latencies, durations and inter-act periods of behaviors.

A digitizing tablet can be used to record data on the spatial position of an animal
or part of its body directly into a computer. For example, Fleishman (1988) digitized the head and dewlap position of displaying *Anolis* lizards from Super-8 movie
frames. (These same techniques are described for videotapes in a later section.)

Frame-by-frame analysis has been used to measure the movements of the tongue

Fig. 9.24   An 8 mm sequence camera and intervalometer inside a weatherproof housing.

of boas (*Constrictor constrictor*) (Ulinski, 1972) and the foot of a mollusc (*Cardium echinatum*) (Ansell, 1967). Illustrations of the results of their analyses are shown in Figure 9.25. Head movements relative to particular behaviors have been analyzed frame by frame for the Burmese red jungle fowl (*Gallus gallus spadiceous*) (Kruijt, 1964), laughing gull chick (*Larus atricilla*) (Hailman, 1967), and domestic duck (*Anas platyrhynchos*) (Clayton, 1976; Figure 9.26).

Spatial relationships between courting goldeneyes were measured by Dane and Van der Kloot (1964) by projecting film frame by frame onto a screen that they had divided with 16 equally spaced vertical lines. Distances perpendicular to the camera's line of sight are relatively easy to measure; but the perspective of depth is lost in measurements parallel to the line of sight. Dane and Van der Kloot list other complications and restrictions which are common to similar types of film analysis:

> (1) Birds are often passing in and out of the field of view of the camera. When the final analysis is undertaken, there is always the chance that an action given by a bird outside of the field is affecting those recorded on film. This problem was minimized by analyzing discrete groups. (2) Computing the distance between birds, and thus the relative position of each individual flock, is difficult when using a telephoto lens. (3) When testing for a relationship between the actions of two birds, there is always the possibility that one is not distinguishing the pair which is

actively interacting. If this were the case, correlations which actually exist might be overlooked. (4) Finally, though unlikely, a movement which was too subtle to be detected on the film, might be a stimulus for another individual. *[Dane and Van der Kloot, 1964:285]*

A computer system for frame-by-frame analysis of film, FIDAC (Ledley, 1965), has been described by Watt (1966). The system consists of a cathode-ray-tube generator which projects an ordered array of rows and columns of spots of light through the film frame, where the intensity of the light transmitted is measured by a photocell as one of seven different levels of gray. This information is then transmitted to a digital computer. The computer can be programmed to control the location of the array of spots of light, their density in the array, and the area covered. The system has both high speed and high resolution. This system, or a similar one, may find useful application in ethological studies of movement where the animal is filmed against a light background.

In summary, I have not mentioned the vast array of additional equipment (e.g. light meters, filters, tripods) that may be necessary for proper filming. These items should be discussed with your local camera dealer. Likewise, the various techniques which will improve your motion pictures and their analysis can best be gathered through discussions, experience and literature (Dewsbury, 1975; Matzkin, 1975; Wildi, 1973). Both 8 mm and 16 mm film can be transfered to videotape for analysis or to provide additional copies for yourself and colleagues.

9.12 VIDEOTAPE RECORDING AND ANALYSIS

Behavior is often thought of as an animal doing something. Only movie and video cameras will accurately capture that activity, although computer cameras and high-speed motor-driven slide cameras are sometimes acceptable alteratives. Videotapes (or films) can be used to simply gain experience with an animal's behavior, even before reconnaissance observations are made on live animals. By viewing the same footage several times you learn to anticipate behaviors; you see subtleties in behavior which you often miss in a single observation.

Videotape has several advantages and disadvantages relative to movie film (Walbott, 1982; Table 9.13). Movie film is often used to document behavior for long-term storage, but films can also deteriorate. When you want to record behavior to be reviewed soon after it occurs, and frequently in the future, videotape is recommended.

Videotaping systems vary from separate camera and recorder (reel-to-reel or cassette) to the combination of camera and recorder into a camcorder. Camcorders, which are the most appropriate systems for field work, use cassette tapes which vary in size from the large VHS to the small 8 mm.

A

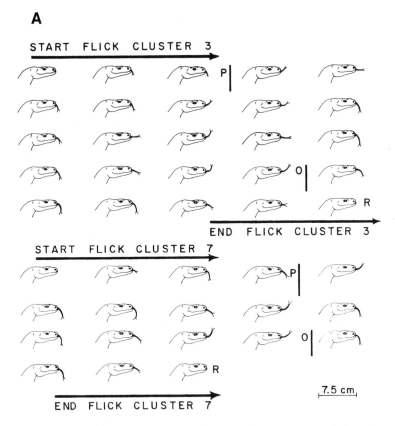

Fig. 9.25A Pattern of boa tongue movements in lateral view. Tracings of each frame in
motion pictures of two complete flick clusters are illustrated. Successive pictures
are about 42 ms apart in time. The ends of the protrusion phase (P), the
oscillation phase (O), and the retraction phase (R) are indicated by vertical lines.
The figures should be studied from left to right in each line. (from Ulinski, 1972).

The lightweight, compact, battery-powered VHS-C and 8 mm cassette palm-
corders (Figure 9.27) make videotaping in the field relatively easy. Motor driven
zoom lenses allow the researcher to obtain a broad or focused view of behavior.
Built-in microphones record environmental sounds (those from the animals are
generally not of sufficient quality for analysis) and also allow the researcher to make
verbal notations while recording. Although these popular camcorders are relatively
resistant to moisture and light impact, they are not designed for the harsh conditions to which many field ethologists might expose them. As with still cameras, you
should check the capabilities for proper use and resistance to abuse for any camcorder you are considering using. The Fieldcam is an 8 mm closed circuit video

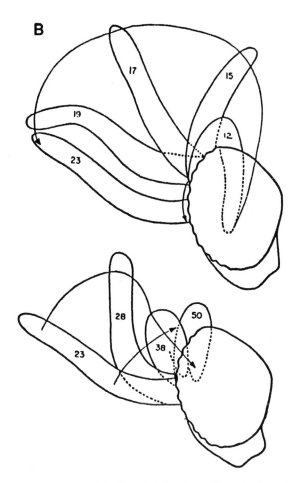

Fig. 9.25B The movements of the foot of the bivalve mollusc, *Cardium echinatum*, during a single leap, shown with reference to the shell as a fixed object. The positions were taken from motion picture of the movement, the numbers indicating the number of the frame corresponding to each position (16 frames/second). Active (frames 12 to 23) and recovery (frames 23 to 50) are shown separately (from Ansell, 1967).

system specifically designed for field use. It is an integrated, all weather, compact, real-time video monitor, remote camera (infrared and visible light sensitive) and recording system. Batteries will power the external camera system for up to 20 hours and the camcorder and monitor for up to 12 h; the camcorder will record up to 120 min. per tape. Several Fieldcam systems are available from Fuhrman Diversified, Inc. 905 South 8th St. Laporte, TX 77571.

A common use of videotaping is to document behavior for later data recording and analysis. For example, Weigensberg and Fairbairn (1994) videotaped 12–18 h

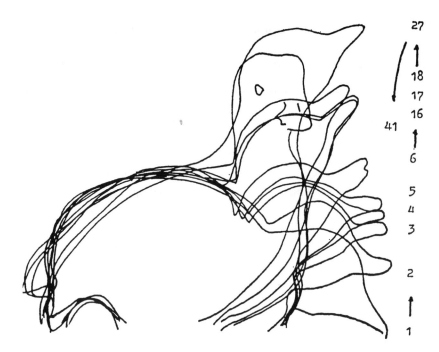

Fig. 9.26   The duckling's drinking response illustrating the bill-lift element. This composite line drawing is based on frames from a motion picture film (16 frames/second). The sequence of numbers corresponds to the frame numbers beginning as the bill leaves the water (from Clayton, 1976).

trials in their study of mating behavior in water striders; they point out that video-taping 'allows the detection and accurate quantification of short-lived behaviour patterns and continuous monitoring of behaviour of long durations' (p.895). Data from videotapes can be recorded on check sheets or input directly into a computer using standard data-collection programs (see section 9.10.1d). For example, Roberts (1994), in his research on vigilance sequences in sanderlings, used a com-puter-based event recording system to record the times of behavior events from videotape. Also, several specialized systems and programs have been designed specifically for recording data from videotapes. Krauss *et al.* (1988) describe the hardware and software of a computerized multichannel event recorder, Videologger, for analyzing videotapes. It records a starting and stopping signal on the second audio track of the videotape to mark the beginning and end of the segment being analyzed. The microcomputer uses the signals to reset its internal clock and store in memory the onset time and duration of keypresses for any number of behaviors. It was designed to run on an Apple II computer but can be converted to an IBM format. The system consists of five software programs which

Table 9.13. *Relative advantages and disadvantages of videotape and movie film for ethological studies*

Videotape	Movie Film
Advantages	
1. Immediate playback	1. Better quality
2. Reusable	2. Easily analyzable frame by frame providing an accurate time base for studies of movements
3. Tape relatively inexpensive	3. With wind-up cameras, time in the field limited only by the amount of film
4. Easily duplicated	4. Equipment generally light
Disadvantages	
1. Poorer-quality picture with less expensive video recorders	1. Time delay for developing
2. Equipment run off batteries with limited chargeable life	2. Film usable only once
3. Equipment sometimes heavy	3. Film relatively expensive
	4. Duplicating more expensive

are available gratis from the authors. The Behavior Chronicles software (see section 9.10.1d) includes a videotape analysis mode in which the computer screen clock is synchronized with the VCR and an icon on the computer screen allows the researcher to control the VCR with the computer's mouse.

Several programs designed for recording data from videotapes are available commercially. CAMERA is a system which includes software and a keyboard which the researcher interfaces with an IBM-PC; each button on the keyboard generates a sound with a unique pitch providing the researcher with immediate auditory feedback. CAMERA was reviewed by van der Vlugt *et al.* (1992) and is available from ProGAMMA, P.O. Box 841, 9700 AV Groningen, The Netherlands. PROCORDER is another program for recording behavioral data from videotape; it was reviewed by Tapp and Walden (1993) and is available from Jon Tapp and Associates c/o Jon Tapp, 306 Liberty Lane, Lavergne, TN 37086. Noldus Information Technology offers a Video Tape Analysis System for use with The Observer 3.0 software; it is available in three different option packages.

Fig. 9.27   Stephanie Bestelmeyer using a Panasonic VHS-C palmcorder, model PV-S62, to record waterfowl behavior.

Videotape can be reviewed at slower or faster speeds to enhance the researcher's ability to observe and measure behavior. For example, Grandin (1989), in her study of pigs, found that high-speed reviewing of videotape recorded at 0.9 frames/s revealed subtle nosing and rooting movements as easily seen vibrations of the snout. Frame-by-frame analysis of videotape allowed Weihs and Katzir (1994) to demonstrate the hydrodynamic function of bill sweeping in the spoonbill (Figure 9.28).

Time-lapse videotaping is often useful to obtain instantaneous/scan samples of behavior over long periods of time. For example, Grant (1987) used time-lapse video-recording to provide a 'continuous' record of the behavior of a beagle bitch and her pups over a three-week period. Time-lapse videotaping has also been used in studies of the behavior of calves (Dellmeier *et al.*, 1985), and Burmese red jungle-fowl (Hogan and Boxel, 1993).

Movements and spatial relationships of animals are frequently measured using videotapes. Earlier, researchers often traced the movement of the animal on an acetate transparency layed over the video monitor (e.g. Crawford, 1984); the animal's position on the acetate sheet could then be transferred to a computer using a digitizing tablet (e.g. LC-12, Terminal Display Systems Ltd., Fuiman and Webb, 1988). Methods are available for digitizing the animal's position across time directly from the video monitor. This technique requires the use of special software and

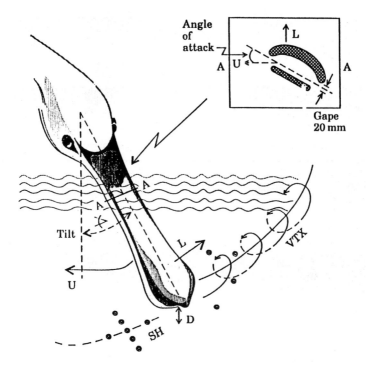

Fig. 9.28 Schematic representation of a spoonbill's sweeping, showing the various
geometric parameters. Also shown are the simulated prey items placed on the
bottom to test displacement (undisturbed pattern on the left) and the bill tip
vortex streamlines, indicating shell motion on the right. U: Sweeping velocity; L:
lift; A–A: cross section of the bill; D: distance of the tip of the bill to bottom;
VTX: induced vortex; SH: empty snail shells ('prey') (from Weihs and Katzir,
1994). Copyrighted by Academic Press.

peripheral equipment (a screen digitizer) which records (or allows the user to
record) the position of an animal on the video monitor screen and transfers that
digitized position to a computer. For example, Richardson (1994) recorded the
location of minnows from a videotape and then transferred that data to a computer
which was used to calculate all inter-fish distances. Watt and Young (1994) video-
taped daphnia (waterfleas) in a water tank illuminated with polarized light using a
Panasonic model AG 6720 time lapse video-recorder; they then recorded data by
digitizing the movement tracks of five randomly selected daphnia. Young and Getty
(1987) tracked the movements of daphnia in three dimensions using a novel two-
camera system, described below:

> The video system had two separate black-and-white video cameras
> (Link, type 109b) which produced views of the tank from different

directions. One camera needed its horizontal and vertical scan directions reversed for both cameras to produce pictures the same way round and the same way up. A single video-recorder (National NV8030) was used to store both images, which were electronically multiplexed on recording and re-separated on playback. The two images were displayed as red and green pictures on an RGB monitor, and created an anaglyphic stereo display when viewed through red/green glasses. When viewed directly, each animal was represented by a pair of dots, one red and one green, whose distance apart (disparity) increased steadily if the animal swam upwards in the tank. Digitizing the position of both red and green dots enabled us to compute x, y and z coordinates for the animal.

[Young and Getty, 1987:542–543]

Software used in conjunction with the digitizer cursor or stylus often allows you to: 1. calibrate your own coordinate system or select Cartesian or polar coordinates; and 2. choose whether to digitize single points, continuous data stream, or user-defined increments (e.g. SigmaScan Measurement System. Jandel Scientific, 65 Koch Road, Corte Madera, CA 94925).

Some commercial systems include all the equipment necessary, including the animal enclosure. Coughlin *et al.* (1992) used a commercial system (CritterSpy filming apparatus) to measure the swimming and search behavior of clownfish larvae in three dimensions. Other video-computer systems designed to measure animal activity in small enclosures are commercially available from Columbus Instruments International Corp. P.O. Box 44049, Columbus, OH 43204, MED Associates, Inc. P.O. Box 319, St. Albans, VT 05478, and Omnitech Electronics, Inc. 5090 Trabue Rd. Columbus, OH 43228. Some of these systems can also be used for gait analysis (see Chapter 10), but one specifically designed for that purpose is available from Peak Performance Technologies, Inc. 7388 S. Revere Parkway, Suite 601, Englewood, CO 80112. Software for tracking animals from frame-to-frame on videotape which runs on Apple Macintosh computers is available from James B. Hoy, USDA-ARS-MAVERL, P.O.Box 14565, Gainesville, FL 32604.

Digitizing systems can be automatic, scanning the screen at a given frequency and recording position(s) of animals by the contrast in color or gray scale between the animal and the background. These automatic systems are not sophisticated enough at this time to record positions of rapidly moving objects with a high degree of accuracy, and they are quite expensive (*Science* 1985, vol.227:1567). Manual systems, in which the observer touches the position of the animal on the screen with a wand or light pen to record automatically this position in the computer, are more accurate and much less expensive. An entire single frame (i.e. the entire picture) can be digitized by a computer either from videotape or live using a camera and a frame-grabber board (e.g. PCVISIONplus Frame Grabber; Imaging Technology, Inc.). Computer cameras are also available which produce a digital image that is then fed into the computer; these include the Fotoman Digital Camera with IBM-PC and Apple Macintosh models (Logitech Inc. 6505 Kaiser Dr. Fremont, CA 94555–9971), Apple Computer's QuickTake 100, and the IBM-PC compatible EDC-1000 (Electrim Corp. P.O. Box 2074, Princeton, NJ 08543).

Video cameras can be made sensitive to *infrared* by replacing the normal vidicon tube with an infrared-sensitive tube. For example, Davis and Hopkins (1988) video-taped the behavior of electric fish (*Gymnotus carapo*) in a near-infrared illuminated tank using an infrared-sensitive video camera (GTE 4 Te E-44 with a Newvicon tube). Wells and Lehner (1978) flooded a large room with infrared light and used a infrared-adapted video camera to study the predatory behavior of coyotes in the dark. Conner and Masters (1978) described a video system for viewing in the near infrared (700 to 1000 mm), which was used to observe the nocturnal courtship of an arctiid moth and nocturnal predatory behavior of the Florida mouse (*Peromyscus floridanus*). Grant (1987) used a low light/infrared sensitive camera (Sanyo) and time-lapse video-recorder for his 'continuous' record of the behavior of a beagle bitch and her pups over a three-week period. Video cameras can also be used to view *ultraviolet* light (Eisner *et al.*, 1988), such as the ultraviolet reflectance by gorgets of sunangel hummingbirds (Bleiweiss, 1994). Videotapes can also be made underwater. For example, McKaye *et al.* (1994) used an RCA model CMR 300 with wide angle lens in a Jaymar housing to record the parental behavior of catfish.

9.13 STOPWATCHES

Stopwatches (Figure 9.29) are a time-honored piece of equipment in ethological studies, which are still useful today (e.g. Randall, 1994; Yoerg, 1994). They are used primarily to measure durations and latencies. *Electronic digital stopwatches* have rapidly replaced the old *mechanical stopwatches* (Carpenter and Grubitz, 1961) in many ethologists' pockets.

Most electronic digital stopwatches are comparable to good mechanical stopwatches in size, but they are generally lighter, less expensive and easier to use. However, some researchers are prone to question the dependability of electronic devices (even though their accuracy is greater) and do not like to tie themselves to batteries.

Electronic digital stopwatches are easier to read, and many have several functions that are useful to the ethologist. For example, the Heathkit stopwatch (Figure 9.29) provides five functions, listed in Table 9.14, plus two programmable functions which might prove useful for laboratory work.

Kits for digital stopwatches can be purchased from several manufacturers (e.g. James Electronics, San Carlos, California). Wolach *et al.* (1975) described an economical method for converting an electronic handheld calculator into a digital stopwatch with increments in 0.10 seconds.

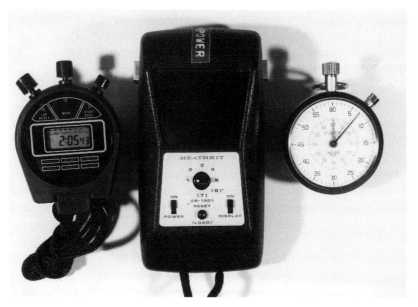

Fig. 9.29   Left to right: electronic digital stopwatch; Heathkit programmable digital
stopwatch; mechanical analog stopwatch.

9.14 METRONOMES

Metronomes provide a time base for field observations. They allow observers to
enter a time point in their notes (e.g. every 10 seconds put a slash), time instanta-
neous/scan samples (Chapter 8), and provide an electronic signal (audio or visual)
at intervals (e.g. one second) which can then be counted in order to determine durations of behaviors.

An electronic metronome designed and constructed by Jim Starkey and used in
our studies can be set to beep at 1 or 10 second intervals through a small earphone.
It is both small and lightweight, which makes it suitable for fieldwork. Lockard
(1976) described a metronome which has a pulse rate continuously adjustable from
0.5 to 20 seconds and includes a light-emitting diode providing a visual signal at the
set intervals (Figure 9.30). Wiens et al. (1970) designed an electronic metronome
which emits tone pulses through a small earphone at intervals which can be varied
from 1 to 20 seconds. Their metronome was used by Dwyer (1975) in his study of
time budgets in gadwall ducks (*Anas strepera*). Reynierse and Toeus (1973) describe
a metronome which produces pulse rates of from five per second to one per 30
seconds and can be built using the circuit diagram they provide; Martin and
Bateson (1993) also provide a schematic for a metronome. Figure 9.31 gives two
schematics for constructing your own electronic metronome.

Table 9.14. Functions provided by Heathkit Model GB-1201E digital stopwatch

Function	Description	Illustration	
1	Duration of separate behaviors plus total duration of session		+total duration 0 through C
2	Time from one event to another; latencies; plus total time		+total time from 0 to C
3	Accumulated time for several occurrences of a particular behavior; plus total time of session		+total time 0 thorugh A_3
4	Latencies for events from a single starting point; plus total latencies		+total latencies $A+B+C$
5	Duration of separate occurrences of a behavior; plus total duration of all occurrences		+total duration $A_1+A_2+A_3$

9.15 DETERMINING GEOGRAPHIC LOCATION

It is necessary to determine the geographic location of animals in order to plot home ranges and territories (Figure 17.8), and in some cases to locate them for observation. The researcher can make accurate locations based on grids marked out in the study area (e.g. Figure 17.7) and reasonably accurate determinations based on a knowledge of the immediate study area (e.g. Figure 17.6). Less accurate determinations can be made over large geographic areas if the researcher is capable of locating their position on topographic maps based on the terrain around them. The Global Positioning System has improved the accuracy of determining a researcher's geographic location, and biotelemetry has made the location of unobserved animals efficient and reasonably accurate.

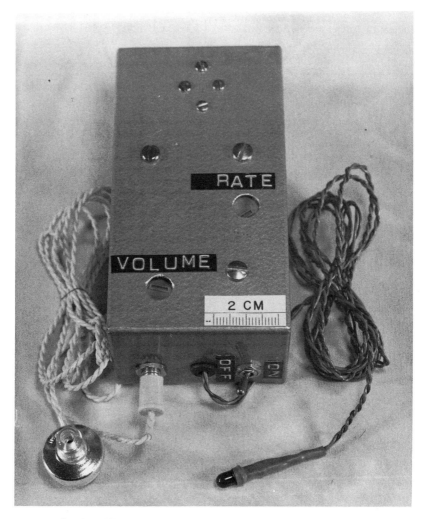

Fig. 9.30   An electronic metronome which provides an auditory signal (earphone) and
visual signal (light emitting diode) (from Lockard, 1976).

9.15.1 Global Positioning System and Argos Satellite System

The Global Positioning System (GPS) was developed for the US Department of
Defense as an accurate targeting and navigational system and has been made available for civilian use. Hand-held electronic receiving units (Figure 9.32) use the
signals from two, or more, of the 24 orbiting satellites in the system to calculate an
accurate position by triangulation. The principle of the one receiver, two (or more)

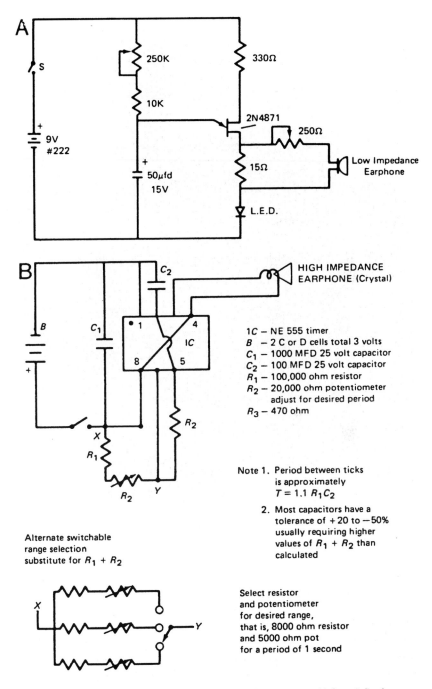

Fig. 9.31 Two schematics for constructing an electronic metronome (A from J. Starkey, pers. commun.; B from Lockard, 1976).

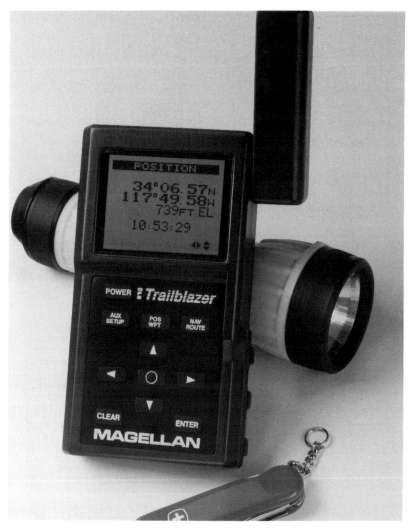

Fig. 9.32   Magellan hand-held global positioning system receiver (photo courtesy of
Magellan Systems Corp.).

transmitting satellites GPS system, is the opposite of two receiver, one transmitter
biotelemetry system illustrated in Figure 9.34.

A researcher can stand on the spot that was occupied by the animal and use the
receiver to determine their location on the earth by latitude–longitude and UTM
(Universal Transverse Mercator) coordinates with an accuracy to within 50 ft. The
receivers have a small screen which will display location coordinates, elevation, and
time, and graphic displays to assist in navigating to a pre-programmed location (e.g.

nest site, den, display ground). They can also keep you from getting lost by directing you back to the point where you started. The receivers vary in capabilities and displays, but most will track several satellites, update position every second and operate for 4 to 20 hours on batteries. They can be obtained from sporting goods/outdoor equipment stores, or you can contact manufacturers directly, such as Magellan (960 Overland Court, San Dimas, CA 91773).

An overview of the more technical aspects of the GPS is provided by Logsdon (1992), and a popular (but informative) account was written by Brogdon (1993).

Animals can be automatically tracked worldwide using the Argos Data Collection and Location System (DCLS), a joint venture between governmental agencies of the United States (NASA and NOAA) and France (CNES). Harris *et al.* (1990) review the use of the DCLS to monitor movements and activities of 10 species of large mammals in Alaska and the Rocky Mountain region; mean error of locations of captive animals was estimated to be 954 m. An overview of animal tracking by satellite is provided by Taillade (1992), and Priede and Swift (1992) contains a series of chapters on tracking various species by satellite. Animals are more commonly located and tracked using earth-based biotelemetry systems (described below).

9.15.2 Biotelemetry

Biotelemetry has been used to record remotely information from a wide variety of species for several decades. Many additional species have been radio-tracked since Brander and Cochran compiled their list in 1969, and many technological advancements have been made since the overviews of Slater (1965), Fryer *et al.* (1976), and Long (1977), and the bibliographies by Schladweiler and Ball (1968) were published; however, useful basic information on biotelemetry can still be found in these sources.

Biotelemetry has found many unique and valuable applications in animal behavior research (Macdonald and Amlaner, 1981; Priede and Swift, 1992), but it has primarily been used for the following purposes:

1 To locate an unobserved animal for plotting its movements and calculating its home range (see Chapter 17).

2 To assist in locating an animal for direct observation of its behavior (Figure 9.33).

3 To record activity and physiological data such as EEG, EKG, respiratory rate, and internal or surface body temperature.

4 To transmit animal sounds for recording and determination of behavior (see section 9.9.2).

Fig. 9.33 Andy Sandoval using a hand-held radiotelemetry receiver and yagi antenna to locate bighorn sheep for behavioral observations.

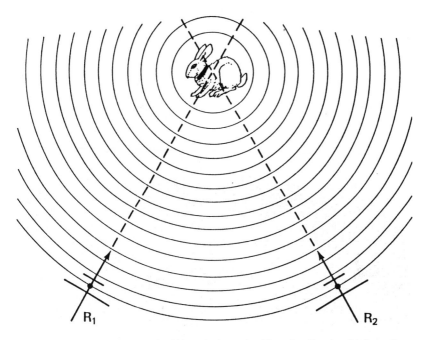

Fig. 9.34   Diagram of location of rabbit at A, determined by using directional information received from two receiving stations, R_1 and R_2.

Basically, the location of an unobserved animal is determined by triangulating from the simultaneous directional information provided by two or more antennae (Figure 9.34) or receiving stations.

Besides animal location, diverse additional information can be transmitted including physiological data (Mackay, 1993), and behavior, such as animal sounds (vocalizations or mechanical sounds which indicate specific behaviors; Alkon *et al.* 1989; also see section 9.9). Animal activity can be determined using various techniques such as motion-sensitive transmitters (e.g. Gillingham and Bunnell, 1985) or putting the receiving antennae close to a specific location (e.g. nest; Benedix 1994). Also, a change in the characteristics of the signal received can often be correlated with specific behaviors, such as walking and wing-flapping.

Given all the information and data that biotelemetry can provide the researcher, it is not without its problems as is to be expected. As examples, triangulation errors can occur due to characteristics of the signal and the environment (e.g. Laundre *et al.*, 1987), transmitters can be lost (e.g. Schulz and Ludwig 1985) or cause pathological conditions (Guynn *et al.*, 1987), and repeated captures to repair transmitters can cause mortality (King and Duvall, 1990). Likewise, the capture and handling necessary to attach transmitters, as well as the transmitters

Fig. 9.35   Radio-collared coyote, receiver, and hand-held yagi antenna (photo by A. Olsen).

themselves, can have effects on the animal's behavior (see Chapter 8; Cuthill, 1991; Laurenson and Caro, 1994). These potential effects are not necessarily to be expected, but they should be guarded against through proper procedures and monitoring. For example, research has been conducted on the effect of radio-transmitters (external and implants) on the behavior of several different duck species, including mallards (Greenwood and Sargent, 1973; Korschgen *et al.*, 1984), canvasbacks (Perry, 1981; Korschgen *et al.*, 1984), blue-winged teal (Greenwood and Sargent, 1973), African black ducks (Siegfried *et al.*, 1977), red-heads (Woakes and Butler, 1975; Korschgen *et al.*, 1984), pintails (Korschgen *et al.*, 1984), and tufted ducks and pochards (Woakes and Butler, 1975); various behavioral effects were found.

Potential problems aside, when used properly, biotelemetry is a valuable research tool that has promoted the development of an array of sophisticated equipment. Various types and sizes of radio transmitters with various battery lives and signal strengths have been developed for use on a wide variety of species. Transmitters can be attached externally to the animal using various standard (e.g. harnesses, collars; Figure 9.35) and unique techniques, such as velcro on the back of penguins (Heath, 1987); transmitters can also be implanted in the animal (e.g. Green *et al.*, 1985;

Benedix, 1994; see Anderka and Angehrn, 1992, for a review of transmitter attachment methods). Receiving systems can be as simple as a portable antenna and receiver, tuneable to the frequencies being broadcast by each animal's transmitter. The researcher obtains a directional fix on the animal by rotating the antenna until the signal strength is the greatest, indicated by a VU meter and/or an audio signal in earphones.

Decades ago, automatic radio-tracking systems were devised which transferred time and directional information into a mainframe computer where it was stored and plotted by an X–Y plotter to show an animal's movements and home range (Cochran *et al.*, 1965). Today, automated biotelemetry systems vary from large, permanent radio-tracking systems to small portable systems. As an example of a permanent, large (yet sophisticated) system, the Starkey Project in Oregon was established in a 25 000 acre study area to monitor continuously the movements of 180 radio-collared animals (elk, deer and cattle) over a 10-year period. This system makes use of the federal government's Loran-C navigational system as follows:

> A base station computer sends out a location request to a different collar every 15 seconds. Once a collar receives the computer's signal, a pager inside the collar turns on a transmitter and receiver. During the next 10 seconds, the receiver collects Loran-C signals from three out-of-state Loran towers located in George, Washington; Fallon, Nevada; and Middleton, California. The transmitter inside the animal's collar sends the signals over a microwave link to one of Starkey's field station towers, which retransmits the signals to the base station's Loran-C hardware. Positioning software translates the Loran signals into Northing and Easting coordinates, then computes locations using differential statistics.
>
> *[Anonymous, 1989]*

At the other extreme, portable microcomputers can be used in field-operated biotelemetry systems automatically to record data, such as animal location (Angerbjorn and Becker, 1992) and activity (Cooper and Charles-Dominique, 1985).

Biotelemetry equipment can be developed and/or built by researchers (or technicians) from the many published schematic diagrams, or the researcher can purchase equipment from companies specializing in the development and manufacture of radiotelemetry equipment. For additional information on equipment, applications and manufacturers/suppliers, you should consult the references below as well as papers published on the use of biotelemetry with your species of interest, or closely related species.

I have provided only a cursory overview of the potential for using biotelemetry in ethological research. For additional information, consult the following compendiums on the use of biotelemetry for various purposes, with various species, in both the field and captivity: Amlaner (1989), Amlaner and Macdonald (1980), Asa (1991), Cheesman and Mitson (1982), Cochran (1980), Kenward (1987), Mackay (1993), Mech (1983), Priede and Swift (1992) and White and Garrott (1990).

10 Selected examples of data collection and description

As stated in previous chapters, all ethological research begins with a description of *what* the animal does. In some cases a quantitative description of behavior is the objective of the study; in other cases it is the foundation for experimental research (Chapter 6). Sometimes your previous research will provide the descriptions necessary for experimentation, or the literature and other researchers can provide that information (Chapter 4). Often, however, researchers find themselves collecting and quantifying data in order to provide complete and accurate descriptions of behavior. This chapter provides a few examples of this process.

10.1 INDIVIDUAL BEHAVIOR

It is generally preferable to study individual behavior before examining interactions between two or more individuals. Knowledge of individual behavior, experience in data collection, and discipline in concentration on particular units of behavior can then be applied to the study of social behavior.

10.1.1 Terrestrial tetrapod locomotion

Studies of terrestrial tetrapod locomotion begin with descriptions of the spatial and temporal relationships of limbs. The position of the limbs is sampled at high rates of speed, generally with the use of motion pictures. Videotape generally does not provide the frame-by-frame analysis possible with movie film.

Bullock (1974) studied locomotion in the pronghorn antelope (*Antilocapra americana*). He took movies with a Pathe Professional Reflex 16 mm movie camera mounted on a modified rifle stock, using black-and-white Kodak Plus-X Reversal film exposed at 80 frames per second. He studied the film frame by frame with a Zeiss-Ikon Moviscope Viewer fitted with a 2× magnifying lens. To facilitate more in-depth study, Bullock also projected the film onto a solid screen with a 35 mm film-strip projector; he then traced the sequences in silhouette form on the screen. These analyses generated footfall formulas and gait diagrams for the pronghorn's various gaits. These descriptions derive from determining when each foot is on and off the

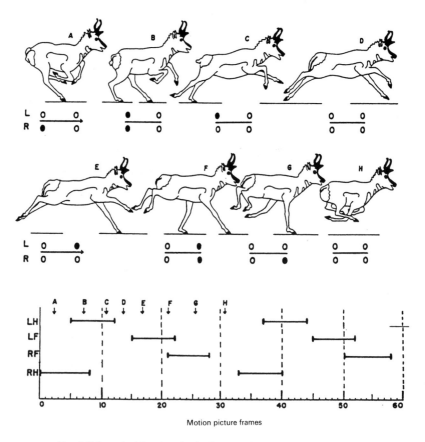

Fig. 10.1   Footfall formula (above) and gait diagram of a pronghorn antelope employing a
lateral (rotary) gallop (from Bullock, 1974).

ground. Bullock was able to do this at 1/80th-second intervals. For example, Figure
10.1 illustrates the footfall formula and gait diagram for the lateral gallop.

Since the movie was taken at 80 frames per second, only 0.75 s is depicted in the
gait diagram. From the information in Figure 10.1, it can be shown that this gait is
both rapid and asymmetrical (see Hildebrand, 1977; Muller-Schwarze, 1968).
Bullock's (1974) analysis also included: 1. support intervals of fast gaits; 2. leads; 3.
turning; 4. change of gait; 5. synchronization of gait and lead; and 6. speed.
Cocatre-Zilgien and Delcomyn (1993) have proposed the use of 'state diagrams' to
reveal more suitably gait trends and differences between gaits at a glance. Mendel
(1985) describes the use of frame-by-frame analysis of the gait patterns and step
lengths of the three-toed sloth's hands and feet during climbing and terrestrial locomotion.

10.2 SOCIAL BEHAVIOR

10.2.1 Displays

A display is 'a behavior pattern that has been modified in the course of evolution to convey information' (Wilson, 1975:528; Beer, 1977). Displays are often dramatic, eye-catching behaviors and have attracted the attention of ethologists since the inception of the discipline. The classical studies of the comparative behavior of the Anatidae (Heinroth, 1911; Lorenz, 1941) were based primarily on displays.

10.2.1a Description of displays

The first step in the study of behavior pattern is description. The components of the display must be described clearly and completely, without bias as to interpretation (Chapter 4). Many hours of observation are generally necessary before you will feel comfortable with your description.

> Only by watching, writing down, drawing, realizing how much you are not certain about, watching again, and thus completing your description step by step, can you attain a reasonable accuracy and completeness.
>
> *[Tinbergen, 1953:131]*

Tinbergen's study of the comparative behavior of gulls (*Laridae*) has long served as a model of careful description of displays. As an example the following is his description of the choking display.

> CHOKING [Figure 10.2]. In this posture, the bird squats and bends forward. The tongue bone is usually lowered, the neck is held in an S-bend, and the bill is pointing down. In this position the head makes rapid downward movements, usually however without touching the ground. The carpal joints are often raised, and the wings may even be raised and spread, and kept stationary for seconds. A muffled, rhythmical sound is given which may or may not be in time with the pecking movements. The breast is 'heaving' strongly, particularly in large gulls. Often the lateral ventral feathers are raised. The bird may be facing another bird, or face away from it, or take up an intermediate orientation.
>
> *[Tinbergen, 1959:16–17]*

10.2.1b Quantifying displays

Baerends *et al.* (1955) studied the courtship displays of the guppy (*Lebistes reticulatus*) by observing them in large aquariums. Besides describing and quantifying

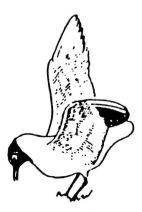

Fig. 10.2   Choking display in black-headed gull (from Tinbergen, 1959).

various postures and movements, they also studied the occurrence of the black markings on the males. For ease in description and recording data, they assigned a number to each of the markings (Figure 10.3A) and then measured the frequency of occurrence of the various patterns in different behavioral contexts (Fig. 10.3B). The relative variability of displays (e.g. duration) can be determined using the coefficient of variation (Chapter 11).

Bekoff (1977a) used movement along a single coordinate to measure the variability in the duration and form of the play bow in three canid species and one hybrid. Movies (Super 8 and 16 mm) were made of individuals at 64 frames/second and then analyzed frame by frame. Duration was determined according to the number of frames during which the bow was maintained. Form was measured as a declination of the shoulders relative to standing height on a grid system (Figure 10.4A). Similar techniques were used by Hausfater (1977) to study tail carriage in baboons (*Papio cynocephalus*) (Figure 10.4B).

In Bekoff's (1977a) study, each individual's shoulder height was divided into ten equal segments to normalize individual differences. Each of the ten segments was then divided into fourths to increase the resolution of measurement.

Tail position as a component of pelecaniform displays was measured by Van Tets (1965). He divided the vertical component into nine 30° sectors and measured the frequency of occurrence of the tail elevations during different displays (Figure 10.5). Similar techniques have been used to measure joint angles in the study of the development of coordinated limb movements in domestic chicks (Bekoff, 1976).

Havkin and Fentress (1985) delineated nine contact zones on the bodies of wolf pups which they used for scoring snout contacts during interactions. They then analyzed 16 mm film to measure the effect of body pitch, snout contact placement,

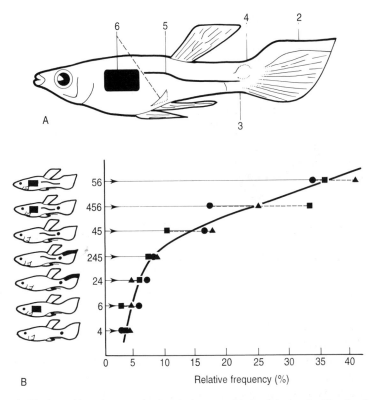

Fig. 10.3 A. Black markings that may develop during courtship in male guppies. Number 1 (not shown) denotes the overall darkness of the entire body. B. Relative frequency of occurrence of marking patterns associated with copulatory attempts (●) sigmoid postures (■), and sigmoid intention movements (▲). The mean frequency of each pattern is shown by the curve (from Baerends *et al.*, 1955).

mutual orientation and combative strategies on the unbalancing (falling) of one of the pups.

Golani (1976) used several limb and body axes (Figure 10.6) in applying the Eshkol–Wachmann (E–W) movement notation to a description of the motor sequences in the interactions of golden jackals (*Canis aureus*) and Tasmanian devils (*Sarcophilus harrisii*). The E–W notation system uses a coordinate system (Figure 10.7) from which to describe a movement of any part of the body. The coordinate system can be applied relative to: 1. the individual's body; 2. the environment; or 3. a social partner. Figure 10.8 shows the coordinate system centered on an owl's head in order to describe head movements.

Each movement is noted on a 'score page' in terms of: 1. point of beginning; 2. point of ending; 3. spatial and 4. temporal units of movement. For example,

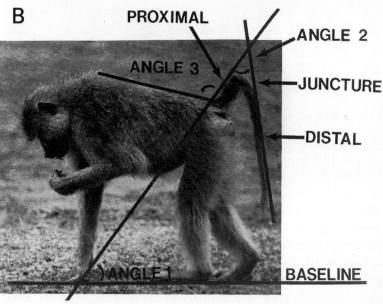

Fig. 10.4   A. Coordinates for measuring form of the canid play bow (from Bekoff, 1977a).
B. Coordinates for measuring tail carriage in baboons (courtesy of G. Hausfater).

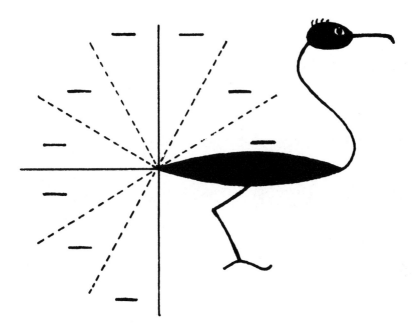

Fig. 10.5 Diagram showing the nine 30 degree sectors that were used for frequency distributions of tail elevations.

the owl's head movements illustrated in Figure 10.8 would be noted as in Figure 10.9:

In this case we are using the beak position as an index of head position. The beak started at point 2/0 and stopped at point 2/6. This represents two ($\xrightarrow{2}$) units of measurement since the coordinates are set at 45° angles. Each block represents 0.2 second; the movement thus took 1.2 seconds.

What has been presented is a simplified version of Golani's movement notation method, and his paper should be consulted for more detailed descriptions of additional notations and more sophisticated uses. However, be cautioned that many studies will not require this intense an analysis in order to answer the research questions. Weigh the increased resolution obtained with the method against the fact that unfamiliarity will make it initially tedious and perhaps introduce errors of recording.

Trochim (1976) described a three-dimensional method for quantifying body position. Height, width, and depth of various body points are taken from videotape. These coordinates can then be graphed by a computer (Figure 10.10) and analyzed for various parameters, including inter-individual distance and body activity. Trochim describes the collection and analysis of data, as well as the computer programs which he developed. Commercial software is now available to analyze movements.

Fig. 10.6   A pair of golden jackals during precopulatory behavior. The superimposed bars indicate the body parts that were considered as separate limb segments during a study of displays (from Golani, 1976).

10.2.2   Dominant–subordinate relationships

Another example of data collection, organization and description is illustrated here for dominant–subordinate relationships. Studies of dominant–subordinate relationships have been made on individuals in the same group (e.g. sooty mangabeys; Gust and Gordon, 1994), different groups of the same species (e.g. vervet monkeys; Cheney and Seyfarth, 1990), and groups of different species (e.g. horses, deer, bighorn and pronghorn; Berger, 1985). The concept of one, or more, individuals (groups) dominating one, or more, other individuals (groups) has been studied for many years by ethologists using several procedures.

10.2.2a   Measuring dominant–subordinate relationships

Typically, determination of dominant–subordinant relationships involves the following (Boyd and Silk, 1983):

1   Identifying a behavior, or behaviors, associated with dominance.
2   Establishing a criterion by which winners and losers of interactions can be distinguished unambiguously.

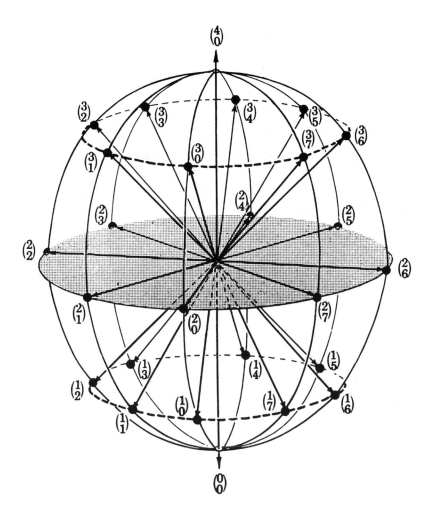

Fig. 10.7 The Eskol–Wachmann coordinate system for measuring displays. For each pair
of numerals, the lower indicates the horizontal, and the upper the vertical,
coordinate. In Golani's (1976) study, one unit of displacement equals 45°. (By
permission of the Movement Notation Society, Israel.)

3 Collecting data on interactions between individuals.
4 Assessing the temporal consistency of the outcomes of interactions (e.g.
   individual A is consistently dominant over individual B).
5 Constructing a dyadic interaction matrix in which individuals are ordered
   in accordance with some convention (see below).
6 Diagramming the dominance hierarchy which results from the rank order
   in the interaction matrix (see below).

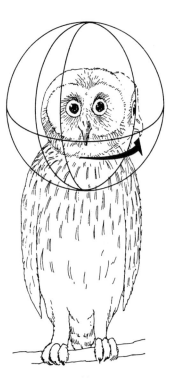

Fig. 10.8 The Eskol–Wachmann coordinate system superimposed on a owl's head in order to denote the movement. In this example the head has moved 90° to the left from point $\frac{2}{0}$ to point $\frac{2}{6}$. (See text for explanation.)

2	2				2
0	→				6

10.9   Eshkol-Wachmann notation for the owl's head movement in Figure 10.8.

Measurements of dominant–subordinant relationships are generally conducted using one, or both, of the following methods (contexts): 1. the researcher stages an equal number of dyadic interactions between all individuals in a group; that is, each individual is matched with every other individual in the group an equal number of times (e.g. Smith and Hale, 1969); or 2. the researcher records naturally occuring interactions in wild or captive groups (e.g. Cheney and Seyfarth, 1990), sometimes manipulating the environment to encourage conflict (e.g. introducing a limited amount of preferred food). This second method generally results in different numbers of interactions between different dyadic combinations of individuals (e.g.

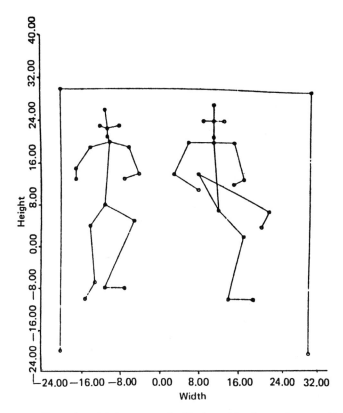

Fig. 10.10 Front-view of Cal-comp graph of body position for two human subjects (from Trochim, 1976).

Horrocks and Hunte, 1983). The two methods can provide conflicting data. For example, in two of six flocks of chickens studied by Guhl (1953), he found correlations between number of contests won (method 1) and number of individuals dominated in the flock (method 2) were less than 0.50. Data collection in the second (natural, group) context is more time consuming, but it will generally allow you to construct a more valid hierarchy.

The behavioral units selected for measurement (e.g. Hausfater, 1975), the context(s) in which interactions are observed (described above), and the criteria of dominance have varied widely (Bekoff, 1977b; Kaufmann, 1983). Ivan Chase (pers. commun.) has suggested that there are at least three methods of deciding when one animal has dominated another during an aggressive interaction. First, you can use an arbitrary, yet intuitive, criterion based upon observations of the animals' behavior. For example, Chase (1982) used the following criteria in his study of hierarchy formation in hen chickens:

> One animal was considered to dominate another if she: (1) delivered any combination of three strong aggressive contact actions (pecks, jump ons, and claws) to the other and (2) there was a 30 minute period following the third action during which the receiver of the actions did not attack the initiator.
>
> *[Chase, 1982:220]*

Secondly, you can use a binomial approach; that is, in each aggressive interaction an individual is scored as either a winner or loser (based on submission or fleeing; e.g. Brown, 1963). An animal is considered dominant over another individual if it wins significantly more (e.g. binomial test; see Chapter 14) than it loses in encounters with the other individual. Thirdly, Chase has suggested that you could use a combination of the first two methods.

Another common criterion for the expression of a dominance relationship is priority of access to a limited resource (e.g. food, shelter, space, estrus female, etc.). Priority of access can be demonstrated through the supplanting of one individual by another without overt aggression being displayed (see Richards' 10 measures below). Dominance does not always provide priority of access to all valued and limited resources; hence, there may be different dominant–subordinate relationships for different resources (Huntingford and Turner, 1987).

When individuals are ranked by different criteria, the rankings are often not comparable (Bekoff, 1977b; Bernstein, 1970; Syme, 1974). However, Richards (1974) used the ten factors listed below to assess dominance rank in six groups of captive rhesus monkeys and found that they produced comparable results.

1 Priority to food incentives
   a Order to daily food ration
   b Order to milk bottle
   c Interactions at milk bottle
   d Order to approach experimenter during food offers
2 Agonistic encounters
3 Displays
4 Gestures for fear-submission
   a Yielding ground/avoidances
   b Cautious approaches
   c Nonsexual presentations/mountings
   d Fear-grins

10.2.2b *Dominance hierarchies*

The data collected on aggressive interactions, based on the criteria for determining winners and losers, are entered into a matrix. Each individual in the group is listed along each axis of the matrix; one axis is labeled winners, the other losers (see Table

Table 10.1. *An example of a dyadic interaction matrix (see text for explanation)*

		Loser				
		D	E	A	C	B
	D		24	3	0	0
	E	0		13	0	0
Winner	A	21	11		0	0
	C	12	16	17		14
	B	37	31	41	0	

10.1). The dyadic interaction matrix that results provides the basis for generating a dominance hierarchy.

Brown (1975) provided the following list of steps to follow in the construction of a dyadic interaction matrix (dominance matrix):

1 *Observations*: B>D, C>A, B>A, C>B, B>D, etc.: B>D means B won an encounter with D. In most cases, these encounters take the forms of supplanting rather than fighting.

2 *Starting order*: Choose an arbitrary order, e.g. DEACB.

3 *Starting matrix*: Enter the number of wins and losses observed in the matrix, as illustrated in Table 10.1.

4 *Treatment of reversals*: A win by one individual over another that has won the majority of encounters with the first is termed a reversal. Rearrange the order so that only reversals fall below the diagonal, so far as possible; that is, change the above order to CBDAE or CBEDA or CBAED.

5 *Treatment of intransivity*: An order in which an individual dominates another (wins the majority of encounters) that dominates the first is termed intransitive or circular. Rearrange to minimize the inevitable ambiguity. From the circular relationship shown in Figure 10.11 there are three main alternatives, as shown. In the three alternatives not shown, the

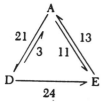

Fig. 10.11 Diagram of the intransitive dominance relationships for individuals A, D and E from the matrix in Table 10.1.

departure from linearity involves two individuals rather than one. Place the individuals that are in the least ambiguous relationships (lowest proportion of reversals) in linear order. This procedure tends to minimize the total number of encounters entered below the diagonal.

6 *Final matrix*: The one order that best reflects the order of dominance within the group is then CBADE. A matrix may then be constructed.

Best

	A	D	E			E	A	D			D	E	A
A		21	11		E		13	0		D		24	3
D	3		24		A	11		21		E	0		13
E	13	0			D	24	3			A	21	11	

	C	B	A	D	E	Wins	Losses
C		14	17	12	59	59	0
B	0		41	37	31	109	14
A	0	0		21	11	32	74
D	0	0	3		24	27	70
E	0	0	13	0		13	125

The dominance hierarchies described above rank individuals on an ordinal scale (see Chapter 8); that is, C ranks above B, and B ranks above A. Boyd and Silk (1983) describe a more complex, statistical method for generating a cardinal index of dominance rank (versus the ordinal heirarchy above) based upon paired comparisons. It incorporates information on interactions that result in wins, losses and ties. They 'then describe how to evaluate: 1. whether the rank differences between individuals are significant; and 2. whether differences in the cardinal hierarchies based on different behaviours or the same behaviour at different times are significant' (Boyd and Silk, 1983:45). Also see the critique of Boyd and Silk's method by Rushen (1984) and their reply (Silk and Boyd, 1984).

10.2.2c *Analysis of linearity*

Dominance hierarchies are generally divided into two types: *linear* and *nonlinear*. Linear hierarchies are transitive; that is, one top-ranking individual (alpha) dominates all other group members, the second-ranking individual (beta) dominates all individuals except the alpha, the third-ranking individual (gamma) dominates all individuals except alpha and beta, and so on down the hierarchy. In nonlinear hierarchies, there are one or more intransitive (circular) relationships, such as individual A dominating B, and B dominating C, but C dominating A. Rankings in a hierarchy and type of hierarchy can change. For example, Murchison (1935) found that the ranking in a group of six domestic fowl roosters changed from a nonlinear to a linear hierarchy as they matured (Figure 10.12).

Perfectly linear hierarchies are relatively rare, making most hierarchies technically nonlinear. Perfectly linear hierarchies are unidirectional. They can contain reversals (i.e., a subordinate wins an occasional encounter with a dominant individual), but they cannot contain any individuals of equal status or have any circularity such as: A→B→C. The nonlinearity is, however, of varying degrees, and some so closely approximate perfectly linear hierarchies that they should be considered linear.

Landau's index of linearity has been discussed by Bekoff (1977b) and Chase (1974). The index (h) is calculated according to the following formula:

$$h = \left(\frac{12}{n^3 - n}\right) \sum_{a=1}^{n} \left[v_a - \frac{(n-1)}{2} \right]^2$$

where:

n = number of animals in the group

V_a = number of animals that individual 'a' dominates

The portion of the formula $(V_a - [(n-1)/2])^2$ is calculated for each individual in the group, and these are then summed.

The term $12/(n^3 - n)$ normalizes the index so that it ranges from 0 (nonlinear) to 1 (perfectly linear). Bekoff (1977b) agreed with Chase (1974) that $h \geq 0.9$ would be a reasonable (although arbitrary) cutoff criterion for 'strong', nearly linear hierarchies.

As an example, we will calculate the Landau index of linearity for the dominance hierarchy of 16-week-old domestic-fowl roosters in Figure 10.12A.

$n = 6$

YY	dominated	Blue, G, R, W, Y	;	$V_{YY} = 5$
Blue	dominated	W, R, Y	;	$V_{Blue} = 3$
G	dominated	Blue, R, Y	;	$V_G = 3$
R	dominated	W, Y	;	$V_R = 2$
W	dominated	R, G	;	$V_W = 2$
Y	dominated	none	;	$V_Y = 0$

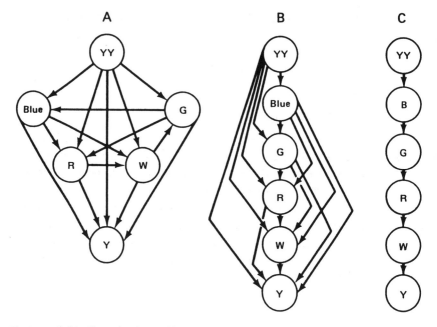

Fig. 10.12 A. Nonlinear dominance hierarchy in a group of 16–week-old domestic fowl roosters. B. Linear hierarchy in the same group of roosters at 32 weeks of age. (A and B after Murchison, 1935). C. Shorthand diagram of linear hierarchy in B; all individuals above are assumed to dominate all those below.

$$h=\left(\frac{12}{n^3-n}\right)\sum_{a=1}^{n}\left[v_a-\frac{(n-1)}{2}\right]^2$$

$$h=\frac{12}{n^3-n}\{6.25+0.25+0.25+0.25+0.25+6.25\}$$

$$h=\frac{12}{n^3-n}(13.5)=\frac{12}{210}(13.5)=0.057(13.5)=0.77$$

The low h value of 0.77 reflects the high degree of nonlinearity in the 16–week-old roosters.

As another example, we will calculate the Landau index of linearity for the dominance hierarchy of 32–week-old domestic-fowl roosters in Figure 10.12B. By comparing the two diagrams (A,B) in Figure 10.12 you would expect the heirarchy of the 32–week-old roosters to have a higher index of linearity.

$$n\ \ =6$$

$$V_{YY}=5$$

$V_{Blue}=4$

$V_G = 3$

$V_R = 2$

$V_W = 1$

$V_Y = 0$

$h=(0.057)(6.25+2.25+0.25+0.25+2.25+6.25)$

$h=(0.057)(17.5)=1.0$

The h value of 1 is just as we would expect for this perfectly linear hierarchy. Bekoff (1978b) demonstrated an h value of 1 in litters of coyote pups (*Canis latrans*) at various ages. Calculating the index of linearity for the hierarchy in Brown's example (above) will generate an h of less than 1.

When individuals are close in rank, that is they supplant each other approximately equally, then assigning clearcut dominance status to one may indicate more linearity to the hierarchy than truly exists. However, Landau's index can still be used when individuals are of equal rank by applying the following rule:

For individuals of equal rank:

$V_a=1$ for each individual dominated plus 0.5 for each individual of equal rank.

For example:

Individuals D and E of equal rank

A	$V_A=5$
↓	$V_B=4$
B	$V_C=3$
↓	$V_D=1.5$
C	$V_E=1.5$
↙ ↘	$V_F=0$
D = E	
↘ ↙	
F	

$h=(0.057)(6.25+2.25+0.25$
$+1+1+6.25)$
$=(0.057)(17)=0.97$

An example where individuals B, C, D and E are of equal rank:

$V_A=5$
$V_B=2.5$
$V_C=2.5$
$V_D=2.5$
$V_E=2.5$
$V_F=0$

$h=(0.057)(6.25+6.25)$
$=(0.057)(12.50)=0.71$

10.2.2d *Dominance indices*

Dominance indices provide a measure of how dominant an individual is in a group, rather than their rank relative to other individuals, as in the hierarchies discussed above.

Barki *et al.* (1992) measured the dominance of individual freshwater prawns involved in agonistic interactions. They assigned each of 18 agonistic behavioral acts a weight on an ordinal scale of aggressiveness of from -3 to $+3$. For each individual a Dominance Index (DI) was calculated using the following formula:

$$DI = \frac{\Sigma(\text{act's frequency} \times \text{act's weight})}{\Sigma(\text{act's frequency})}$$

The winner of an interaction was assigned to the individual with the higher dominance index.

Most dominance indices are some form of the ratio of an individual animal's wins (or other indications of dominance) to the individual's total number of interactions, as shown below.

$$\text{Dominance Index (DI)} = \frac{W}{T}$$

where: $W =$ no. of wins

$T =$ total no. of dominant–subordinant interactions with other individuals

This simple index varies from 0 to 1. It does not take into account the individual's success with different opponents, only the total for all opponents. Eden (1987) used a dominance index (below) in his study of magpies which incorporates the W/T ratio for encounters with each opponent.

$$DI = \frac{W_i/T_i}{N}$$

where: $N =$ total no. of opponents

$W_i =$ no. of wins in interactions with opponent 'i'.

$T_i =$ total no. interactions with opponent 'i'.

This index varies from 0 to 1. If the individual has the same number of interactions with each opponent, this yields the same index as the simple ratio (W/T). Eden's index is essentially the same as that used by Crook and Butterfield (1970) except that they used ($N-1$) in the denominator.

Berger (1977) devised a 'dominance coefficient' to measure the relative dominance of feral horses within their respective bands. He incorporated the number of interactions initiated in his ratio since he assumed that the horses with the highest frequency of successful bouts and the lowest number of bouts initiated by them were dominant within the band. He then calculated the intra-band dominance as follows:

$$\text{Dominance Coefficient} = \frac{(a/b \times 100)}{(i+1)}$$

where: a = no. successful bouts

b = total no. bouts

i = no. interactions initiated

This index varies from 0 to 100. Anyone considering the use of Berger's dominance coefficient for the species they are studying should carefully consider the effect of using the number of interactions initiated (i) in the denominator. For example, an animal that wins only 5 out of 10 interactions (50%) and initiates none has the same dominance coefficient (50) as an animal that wins 10 out of 10 interactions (100%) and initiates one.

An index of relative dominance between two individuals was used by Beilharz and Cox (1967) in their study of swine. It is based on the proportions of wins by each individual pig in interactions between them.

$$\text{Relative dominance between individuals '}i\text{' and '}j\text{'} = d_{ij} = \frac{p}{n} - \frac{n-p}{n}$$

where: p = no. of wins by pig 'i'

n = total no. interactions between pig 'i' and pig 'j'.

This index ranges from 0 to ± 1.

Festa-Bianchet (1991) used a dominance index (based on Clutton-Brock *et al.* 1986) in her study of bighorn sheep which removes the effects of age. She calculated the dominance index for each of the different age cohorts of ewes. First, she calculated the following ratio:

$$\frac{OS_D + O_U + 1}{YS_D + Y_U + 1}$$

where: OS_D = no. of ewes of the same age or older dominated by the subject ewe.

O_U = no. of older ewes with whom the subject ewe interacted with no clear outcome.

YS_D = no. of ewes of the same age or younger that dominated the subject ewe.

Y_U = no. of younger ewes with whom the subject ewe interacted with no clear outcome.

Festa-Bianchet included the unclear interactions since the older ewe won 92% of the interactions; therefore, she assumed that normally the older ewe should have won an interaction with a younger ewe. The ratio (above) was used to rank the ewes in each age cohort, then the ranks were divided by the number of ewes in the cohort. This provided dominance indices from 0.11 to 1.00.

Changes in dominance between age cohorts of cock red grouse hatched from year to year was measured by Moss and Watson (1980) using the following index:

$$C = \frac{N_1}{N+N_1} \times 100 - 50$$

where: C=change in dominance between years.

N=no. of wins by cocks hatched in year t.

N_1=no. of wins by cocks hatched in year $t+1$.

With this index: when there is no change, $C=0$; when $N=0.5N_1$, $C=+17$; when $N=2N_1$, $C=-17$. Moss and Watson added the Cs together from year to year to provide a cumulative index.

10.2.3 Social organization

> Society: a group of individuals belonging to the same species and organized in a cooperative manner.   *[Wilson, 1975:7]*

Social organization is the behavioral organization (type, temporal, and spatial) of the society's members. Analysis of social organization is one of the most complex endeavors an ethologist can undertake, because it necessitates the integrative analysis of social behavior both within and between group members. This includes the development of social behavior in the individual (socialization) and the interaction of group members over time (social phases).

Interactions between individuals serve as the basis for social relationships, which are then integrated into the society's social organization. Hinde and Stevenson-Hinde (1976) have presented these relationships in diagrammatic, but rather complex, form (Figure 10.13).

In order to understand fully the social-behavior matrix which is the structural basis for social organization, the researcher should begin the analysis of social organization at the level of individual behavior, then dyadic interactions. However, a superficial knowledge of social organization can be gained by looking at relationships and perhaps even structure.

S.A. Altmann (1968) determined the relationships among and within sex–age classes of monkeys by observing interactions among individuals. This allowed him to describe the social organization of rhesus monkeys at the relationship level in the diagrammatic form of a sociogram (Figure 10.14).

Interactions, relationships, and structure are specific to individual group social organizations and should not be generalized to other species or even other popula-

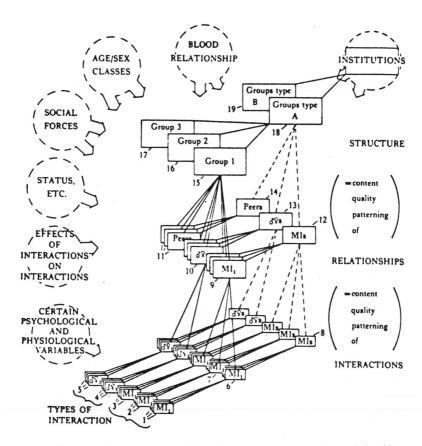

Fig. 10.13 Diagrammatic representation of the relations between interactions, relationships, and social structure, shown as rectangles on three levels, with successive stages of abstraction from left to right. The discontinuous circles represent independent or intervening variables operating at each level. Institutions, having a dual role, are shown in both a rectangle and a circle.

tions of the same species without confirming evidence. Differences in group composition, group size, and habitat can all affect social organization. For example, at the level of interactions, Farr and Herrnkind (1974) measured the frequency of occurrence of courtship displays in the guppy (*Peocilia reticulata*) at different densities of pairs. The correlation between pair density and courtship interactions is shown in Figure 10.15.

The types of structure found in animal societies have been classified in diverse ways by several authors. One of the most useful overall classifications of social organizations is that proposed by Brown (1975) in Table 10.2. Other classification schemes have been proposed for specific groups of animals, such as Jarman's (1974)

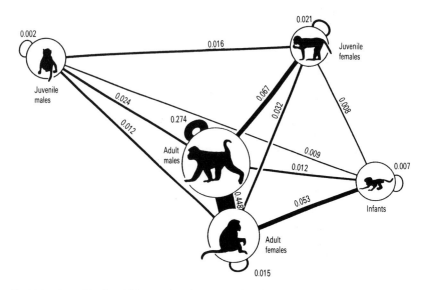

Fig. 10.14   A model of social interactions between and within classes of rhesus monkeys. Probabilities of various intractions, shown by the figures and the thickness of the lines, are calculated from a hypothetical population with equal representation of age–sex classes (from S. Altmann, 1968).

categorization of African antelope social organization into five classes. Likewise, mating systems have been classified for several animal groups including vertebrates (Emlen and Oring, 1977; Wittenberger, 1979, 1981), mammals (Clutton-Brock, 1989), birds and mammals (Davies, 1991), insects (Thornhill and Alcock, 1983), and males and females (Alcock, 1993).

Within the conceptual framework discussed above, how do you go about study-ing social organization? Using the methods and equipment discussed previously, you should begin at the level of individual behavior and interactions (such as those discussed in the section 10.2.2 on dominant–subordinate relationships) and build through relationships to the level of structure. This can be accomplished only through intensive and extensive observation spread over several seasons. One should focus on certain aspects of social interactions at each level in Figure 10.13. This can be accomplished by following the questionnaire compiled by McBride (1976) following a committee's discussion of the information necessary to describe adequately the social organization of a species. Numerous established ethologists contributed to the questionnaire prepared by McBride.

In addition, Owen-Smith's (1974) description of the social organization of the white rhinoceros (*Ceratotherium simum*) is a concise example of a 'typical' social

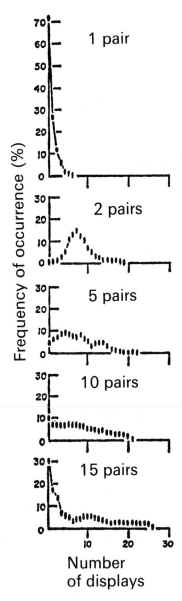

Fig. 10.15 A running average of frequency of ocurrence of male guppy courtship displays in an observation period (from Farr and Herrnkind, 1974).

Table 10.2. *Types of intraspecific animal groups*

Types of groups	Examples
1. Kin groups	
Clones – groups formed by asexual reproduction of sessile colonial invertebrates, typically in permanent physical contact	Colonial coelenterates
Families – groups formed by one or two parents and their most recent offspring	Goose and swan families Coyote (*canis latrans*)
Extended families – groups formed from families by failure of many offspring to leave parents	Prairiedogs (*Cynomys*) Some primate groups Mexican jay (*Aphelocoma ultramarina*) Gray wolf (*Canis lupus*)
2. Mating groups	
Pairs – monogamous groups of two	Scrub jay (*Aphelocoma coerulescens*) Lar Gibbon (*Hylobates lar*)
Harems – groups in which a male attempts to keep females together and away from other males, with or without cooperation of females	Red deer (*Cervus elephas*)
Leks – groups formed by attraction of males (and subsequently females) to a communal mating ground; eggs or young produced elsewhere	Lek birds and mammals Hilltopping butterflies Hawaiian *Drosophila*
Spawning groups – groups of both sexes formed at localized spawning grounds; no provisioning of young	Many fishes and amphibians
3. Colonial groups	
Groups formed by colonial nesting of pairs or one-male harems; young provisioned at nest	Tricolored blackbird (*Agelaius tricolor*) Many sea birds Some bats and seals
4. Survival groups	
Groups formed by aggregation of randomly related, usually nonbreeding individuals who are mutually attracted by each other	Foraging flocks Night roosts of New World blackbirds Ducks and geese Herding mammals Fish schools Bachelor groups of wapiti

Table 10.2. (*cont.*)

Types of groups	Examples
5. *Aggregations* (coincidental groups)	
Groups formed by physical factors acting on migrating or moving animals	Hawks migrating along a mountain ridge
	Land birds migrating through a mountain pass
	Whelks on a sheltered ocean rock
	Stream-surface insects on a calm eddy
Groups formed by attraction to a common resource, such as food or water	Bears at a garbage dump

Source: Adapted from Brown (1975).

organization study. Although it is lacking in some respects, it can serve as a model for studies of this type. Also, consult Kummer's excellent study (1968) of the social organization of hamadryas baboons, Crook *et al.*'s excellent conceptual model of the structure and function of mammalian social systems (1976), and Eisenberg's discussion of social organization in mammals (1966).

III Analyzing the results

11 Introduction to statistical analyses

Alpha-coefficient: Equivalent of an Italian sports car.
Type I error: Making one misteak.
Type II error: Making two misteakz.   *[Norman and Streiner 1986]*

Analysis is the ordering, breaking down, and manipulation of data in order to obtain answers to research questions. What we will be primarily concerned with in this chapter and Chapters 12–17 are: 1. initial ways to look at data; 2. first approximation sample statistics; and 3. parametric and nonparametric statistical tests.

11.1 STATISTICAL PRINCIPLES

Statistics are measures computed from observations in a sample. *Statistical tests* are procedures whereby hypotheses are tested. Kerlinger has defined statistics in the way the term is most commonly used:

> Statistics is the theory, discipline, and method of studying quantitative data gathered from samples of observations in order to study and compare sources of variance of phenomena, to help make decisions to accept or reject hypothesized relations between the phenomena so studied, and to aid in making reliable inferences from observations.
>
> *[Kerlinger, 1964:148]*

In order to make statements concerning the results of their experimental research, ethologists must support their conclusions with statistical tests. Like other biological scientists, the ethologist assumes that there is some order to animal behavior and it is, therefore, amenable to statistical testing.

> In biology most phenomena are affected by many casual factors, uncontrollable in their variation and often unidentifiable. Statistics is needed to measure such variable phenomena with a predictable error and to ascertain the reality of minute but important differences. Whether biological phenomena are in fact fundamentally deterministic and only the variety of causal variables and our inability to control

these make these phenomena appear probabilistic, or whether biological processes are truly probabilistic, as postulated in quantum mechanics for elementary particles, is a deep philosophical question.

[Sokal and Rohlf, 1969:5]

Needless to say, and as intriguing as it is, our purpose here is not to deal with the philosophical question, but rather to justify the assertion that ethologists employ statistics.

11.2 STATISTICAL HYPOTHESES

When I discussed the scientific method in Chapter 1, I stated that the scientific method is basically a matter of hypothesis testing. Also, when we examined the design of ethological research I stated that a *research hypothesis* is our best guess as to the answer to our *research question*. Research hypotheses refer to the phenomena of nature; that is, tentative predictions about the causation, function, ontogeny, or evolution of some aspect of behavior. Karl Popper, a contemporary philosopher of science, has said that in order to be scientific, a theory (I assume he would include research hypotheses) must be testable and falsifiable (Maynard Smith, 1984). This problem has plagued many of the hypotheses concerning animal awareness (e.g. Griffin, 1976, 1984a, b); to many ethologists, these hypotheses are very intriguing but not falsifiable (e.g. Campbell and Blake, 1977; Davis, 1984; Krebs, 1977).

Statistical hypotheses are statements about population parameters that are amenable to evaluation by statistical tests. A statistical hypothesis is either a *null hypothesis* (H_0) or an *alternative hypothesis* (H_1). Both need to be stated in order to conduct a statistical test. The statistical test is a procedure whereby a researcher chooses which one of the dichotomous set of mutually exclusive and exhaustive hypotheses (H_0 and H_1) is to be rejected and which one is to be accepted. This is done at some predetermined risk of making an incorrect decision (Type I and Type II errors, to be discussed later).

Statistical hypotheses are either exact, directional, or inexact, nondirectional. For example, the hypothesis that the mean number (μ) of vocalizations (j) given by a population of bobwhite quail each day is 100 is expressed by

$$H_0 : \mu_j = 100$$

and is an exact null hypothesis, and the hypothesis that the number is equal to, or greater than, 100

$$H_0 : \mu_j \geqq 100$$

is a directional null hypothesis. The alternative hypothesis for the first H_0 is

$H_1 : \mu_j \neq 100$

and is a nondirectional alternative hypothesis. The alternative hypothesis for the second H_0 is

$H_1 : \mu_j < 100$

and is a directional alternative hypothesis. Hypothesis testing in ethological research usually involves an exact H_0 and a nondirectional H_1. Whether the null hypothesis is exact or directional determines whether a one-tailed or two-tailed statistical test is applicable (see below).

11.3 HYPOTHESIS TESTING

> The main purpose of inferential statistics is to test research hypotheses by testing statistical hypotheses. *[Kerlinger, 1964:173]*

Statistical tests are designed to determine whether you can reject the null hypothesis and thus accept the alternative hypothesis. Therefore, the alternative hypothesis should closely approximate the research hypothesis (i.e. what you believe the true situation to be). That is, if you are attempting to demonstrate statistically what you believe to be the case from observations, your H_1 should state what you have observed. For example, if you believe that a population of quail rarely if ever call more than 100 times a day, your statistical hypotheses should read as follows:

$H_0 : \mu \geq 100$

$H_1 : \mu < 100$

If our statistical test is significant and we reject the H_0, we then infer that the H_1 is correct. In this particular example we are interested in testing whether the population mean for the number of quail calls per day is equal to or greater than 100. That is, if it is significantly less than 100, we will reject the H_0 and accept the H_1. Since we are interested in only one side of the distribution, in this case the left-hand side of the distribution (see Figure 11.1), our statistical test will be one-tailed. One-tailed tests are associated with exact (directional) null hypotheses. At the 0.05 level of significance we would have that probability (0.05) of committing a Type I error (see next section).

If we were simply testing whether the population mean for number of quail calls per day was significantly different from 100 ($H_0 : \mu = 100$; $H_1 : \mu \neq 100$), then half of the 0.05 probability of a Type I error would be associated with each tail of the curve (see Figure 11.2). Therefore we would be conducting a two-tailed test which is associated with inexact (non-directional) null hypotheses.

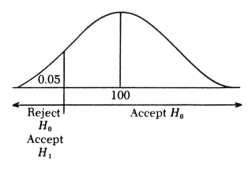

Fig. 11.1   Regions under the normal curve for rejection, and failure to reject, the null hypothesis that the number of quail calls per day is equal to or greater than 100 (one-tailed test; see text).

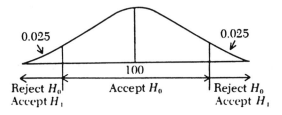

Fig. 11.2   Regions under the normal curve for rejection, and failure to reject, the null hypothesis that the number of quail calls per day is significantly different from 100 (two-tailed test; see text).

It is important to know which type of statistical test (one- or two-tailed) you are conducting in order to obtain the correct values from the statistical tables in Appendix A. Values in a statistical table for two-tailed tests are the same as for one-tailed tests with twice the level of significance (i.e. one-half the alpha value); that is, if the level of significance for the two-tailed test is 0.05 then the tabular values are the same as for a one-tailed test with a 0.025 level of significance, and vice versa.

Hypothesis-testing procedures are important to the ethologist in designing, analyzing, and interpreting research. However, they should not be allowed to blind the careful observer or overshadow common sense.

> Hypothesis-testing procedures should be viewed as tools that aid an experimenter in interpreting the outcome of research. Such procedures should not be permitted to replace the judicial use of logic by an alert analytic experimenter.   *[Kirk, 1968:33]*

A statistical test is used to compare the null and alternative hypotheses and make a choice between them. The null hypothesis (H_0) is essentially a prediction of a sampling distribution of anticipated values for a sample statistic (e.g. mean). If the sample statistic, generated by the data collected, falls within the sampling distribution of anticipated values, then we fail to reject the null hypothesis.

- Sample statistic: statistic generated from data that are used to estimate population parameters (e.g. mean, standard deviation)
- Test statistic: statistic computed from data; used to test a statistical hypothesis (e.g. χ^2, t, F)

The following is the stepwise procedure used in hypothesis testing:

1 From your research question and research hypothesis (Chapter 5) formulate a null hypothesis (H_0) and the alternative hypothesis (H_1).
2 Select an appropriate sample statistic and test statistic.
3 Select a level of significance (alpha level; see below) and a sample size (N).
4 Collect the data (Chapters 8,9).
5 Compute the sample statistic and the test statistic. If the test statistic's value falls in the region of rejection, the H_0 is rejected and the H_1 is accepted. Failure to reject the H_0 is not the same as accepting it, although this is often done.

Failure to reject the H_0, can result from two conditions:

1 The H_0 is true.
2 The experiment was not a valid test of the hypothesis; that is, the experiment could have been improperly designed or conducted.

When our test fails to reject the H_0, we don't know which of the two conditions (or both) produced the failure to reject the H_0. Generally, we will first scrutinize our research design (e.g. behavior units measured, sample size) and our data collection (e.g. potential observer errors). If we are convinced that our experiment was valid, then we are more willing to accept the H_0 is true and take the chance of committing a Type II error (see below).

This concept is very similar to the judicial concept of 'guilt'. That is, if a jury fails to find a defendant 'guilty' they are considered 'not guilty', but they are not necessarily considered 'innocent'. There are several possible reasons for failure to find guilt; only one reason is that the defendant is innocent.

11.4 TYPE I AND TYPE II ERRORS

Researchers take a chance in hypothesis testing. They can commit basically two types of errors in making a decision to accept or reject a null hypothesis.

True State of The World

Decision:	H_0 True	H_0 False
Reject H_0	Type I error $p = \alpha$	Correct decision $p = 1 - \beta$ = power
Fail to reject H_0	Correct decision $p = 1 - \alpha$	Type II error $p = \beta$

Fig. 11.3 Possible outcomes from making decisions about the results of statistical tests (from Howell, 1992). Copyrighted by Wadsworth Publishing Company.

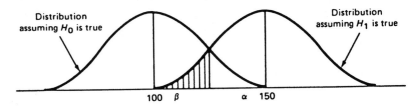

Fig. 11.4 Example of regions of sampling distribution represented by Type I (α) and Type II (β) errors.

- Type I error: Rejection of the H_0 when it is true. The probability α is the risk of making *A* Type I error.
- Type II error: Failure to reject the H_0 when it is false. The risk of making a Type II error is designated as β.

These errors are a reflection of our decisions to accept or reject hypotheses relative to the true state of the world (Figure 11.3). The probability of making Type I and Type II errors is illustrated in Figure 11.4.

11.5 SELECTING THE ALPHA LEVEL

Selecting the probability level for a Type I error (alpha level) is generally done by convention. This is also called the level of significance for a particular statistical test.

Papers presented at professional meetings and published in journals are reports on acquired facts (data) and the researcher's conclusions drawn from those facts. First, *facts* are not necessarily *truths*. Secondly, *conclusions* are *opinions* which others may or may not share, even though opinions are also facts. Specific alpha levels are generally accepted criteria for a researcher to reject a null hypothesis and for others to accept that researcher's conclusion (e.g. song duration of individuals in population X is different from song duration of individuals in population Y). That is, if the research procedures and statistical test were valid, and the statistical test was significant at the criterion alpha level we accept as a truth that the song duration of individuals in population X is significantly different from the song duration of individuals in population Y.

The criterion alpha level for most researchers is 0.05 (significant), but for others it is 0.01 (highly significant). These alpha levels are used for no other reason than that it is generally accepted that they represent a reasonable risk of making a Type I error. Generally, values that are greater than 0.05 are not considered to be statistically significant (Sokal and Rohlf, 1969:161).

The alpha level selected is for a single, independent, statistical test of a hypothesis. When multiple non-independent tests are conducted, such as when the effect of a variable is measured on several subgroups, then the probability of a type one error increases. When n tests are performed, the probability becomes:

$$\text{Probability of a Type I error} = 1 - 0.95^n$$

If a researcher is conducting multiple tests, then *Bonferroni's Correction* should be used to determine the alpha level to be applied to each test (Bakeman and Gottman, 1986).

$$\text{Alpha level for each test} = \frac{\text{Alpha level for overall study}}{\text{Number of tests performed}}$$

For example, if the researcher wants to maintain an alpha level of 0.05 for the overall study, but he or she is conducting ten tests, then:

$$\text{Alpha level for each test} = \frac{0.05}{10} = 0.005$$

11.6 POWER OF A TEST

The power of a test is the probability of rejecting the null hypothesis when it is false and the alternative hypothesis is true; that is, the probability that you will make a correct decision in your favor while avoiding a Type II error (β). Therefore, power $= 1 - \beta$. Remember that you should state your hypotheses in such a way that in order to support your research hypothesis you must reject the null hypothesis. Increased power of a test increases the probability of your rejecting the null hypothesis if your research hypothesis is correct. For example, if you believe that male goldfish spend different amounts of time in shaded and sunlit areas (research hypothesis), you then state your hypotheses in such a way that you expect to reject the null hypothesis.

$$H_0 : \hat{\mu}_{\text{shade}} = \hat{\mu}_{\text{sunlight}}$$

$$H_1 : \hat{\mu}_{\text{shade}} \neq \mu_{\text{sunlight}}$$

You then collect data by making numerous sample observations on several randomly selected male golfish by recording the length of time that the each goldfish is in the shaded and in the sunlit area (see Chapters 6,7,8). From this you can compute mean times (sample statistics) for each male goldfish in each area ($\hat{\mu}_{\text{shade}}$, $\hat{\mu}_{\text{sun}}$) and generate a test statistic with which you can test your null hypothesis. The power of your statistical test will determine whether you will reject the null hypothesis, if in fact it is false; that is, these male goldfish actually spent different amounts of time in the shaded and sunlit areas.

The power of a specific statistical test is a function of three interacting factors (Glantz, 1992; Howell, 1992):

1. The alpha level you have chosen (i.e. the probability of a Type I error).
2. The sample size (N).
3. The size of the difference between the two populations you want to detect relative to the amount of the variability in the populations. This is called 'effect size' (ES).

Below is a description of the general procedure for determining power and examples for two t-tests (see Chapter 13) based on Howell (1992) and Welkowitz *et al.* (1976). Besides these references, additional descriptions of the calculation of power for various statistical tests can be found in numerous texts, including Cohen (1988), Glantz (1992), Kraemer and Thiemann (1987) and Zar (1984).

General procedure for determining power:

1. Specify the alpha level.
2. Calculate, or set, the *effect size* (ES). Effect size can be calculated using the appropriate formulas (see below). When the data are not readily available, ES can be set using the conventional values (Table 11.1) which Welkowitz *et al.* (1976) describe as arbitrary but reasonable.
3. Calculate delta (see below).
4. Use Table 11.2 to determine the power of the test as a function of delta and alpha.

Calculations of E.S. and delta for:

Student's t-test for independent means

$$ES = \frac{\mu_1 - \mu_0}{\sigma} \qquad \text{Delta} = ES \sqrt{\left(\frac{N}{2}\right)}$$

where: μ_1 = mean of population 1

μ_2 = mean of population 2

σ = population standard deviation

Matched-sample t-test

$$ES = \frac{U_1 - U_0}{\sigma_{x_1 - x_2}} \qquad \text{where: } \sigma_{x_1 - x_2} = \sigma \sqrt{2(1 - \rho)}$$

where: ρ = correlation between the two variables; ρ equals zero if they are independent (Howell, 1992)

$\text{Delta} = ES\sqrt{N}$

Table 11.1 *Conventional values of effect size for use in determining power of a statistical test*

Size of the difference to be detected	Effect size (ES)
Small	0.20
Medium	0.50
Large	0.80

Source: From Cohen (1988).

The power of a statistical test, for any given level of significance, can be increased in basically five ways:

1 Increase the sample size. Still (1982) cautions that increasing the sample size may be done too hastily resulting in more animals being used than is necessary for reasonable power. In order to reduce the amount of animal suffering, Still recommends that researchers carefully consider other methods of increasing power (below).

2 Use a more powerful statistical test, such as a parametric test (if valid) rather than a nonparametric test (see power-efficiency in Chapter 12).

3 Select an experimental design that more precisely measures treatment effects and has a smaller error effect (Chapter 6). Still (1982) suggests using repeated measures rather than randomized groups but cautions about carryover effects, learning and boredom affecting the measured behaviors. Also, McConway (1992) notes that exposing individual animals to multiple treatments may result in more suffering than exposing more individuals to single treatments.

Table 11.2. *Power as a function of delta and significance criterion* (α)

	One-tailed test (α)			
	0.05	0.25	0.01	0.005
	Two-tailed test (α)			
Delta	0.10	0.05	0.02	0.01
0.0	0.10[1]	0.05[1]	0.02	0.01
0.1	0.10[1]	0.05[1]	0.02	0.01
0.2	0.11[1]	0.05	0.02	0.01
0.3	0.12[1]	0.06	0.03	0.01
0.4	0.13[1]	0.07	0.03	0.01
0.5	0.14	0.08	0.03	0.02
0.6	0.16	0.09	0.04	0.02
0.7	0.18	0.11	0.05	0.03
0.8	0.21	0.13	0.06	0.04
0.9	0.23	0.15	0.08	0.05
1.0	0.26	0.17	0.09	0.06
1.1	0.30	0.20	0.11	0.07
1.2	0.33	0.22	0.13	0.08
1.3	0.37	0.26	0.15	0.10
1.4	0.40	0.29	0.18	0.12
1.5	0.44	0.32	0.20	0.14
1.6	0.48	0.36	0.23	0.16
1.7	0.52	0.40	0.27	0.19
1.8	0.56	0.44	0.30	0.22
1.9	0.60	0.48	0.33	0.25
2.0	0.64	0.52	0.37	0.28
2.1	0.68	0.56	0.41	0.32
2.2	0.71	0.59	0.45	0.35
2.3	0.74	0.63	0.49	0.39
2.4	0.77	0.67	0.53	0.43
2.5	0.80	0.71	0.57	0.47
2.6	0.83	0.74	0.61	0.51
2.7	0.85	0.77	0.65	0.55
2.8	0.88	0.80	0.68	0.59
2.9	0.90	0.83	0.72	0.63
3.0	0.91	0.85	0.75	0.66
3.1	0.93	0.87	0.78	0.70

Table 11.2. (*cont.*)

	One-tailed test (α)			
	0.05	0.25	0.01	0.005
	Two-tailed test (α)			
Delta	0.10	0.05	0.02	0.01
3.2	0.94	0.89	0.81	0.73
3.3	0.96	0.91	0.83	0.77
3.4	0.96	0.93	0.86	0.80
3.5	0.97	0.94	0.88	0.82
3.6	0.97	0.95	0.90	0.85
3.7	0.98	0.96	0.92	0.87
3.8	0.98	0.97	0.93	0.89
3.9	0.99	0.97	0.94	0.91
4.0	0.99	0.98	0.95	0.92
4.1	0.99	0.98	0.96	0.94
4.2	0.99	0.99	0.97	0.95
4.3	[2]	0.99	0.98	0.96
4.4		0.99	0.98	0.97
4.5		0.99	0.99	0.97
4.6		[2]	0.99	0.98
4.7			0.99	0.98
4.8			0.99	0.99
4.9			0.99	0.99
5.0			[2]	0.99
5.1				0.99
5.2				[2]

Notes:

[1] Values inaccurate for *one-tailed* test by more than 0.01.

[2] The power at and below this point is greater than 0.995.

Source: Adapted from Welkowitz *et al.* (1976).

4 Increase the precision of measurements during data collection. This might entail increasing the resolution of measurements (Chapter 8) from a nominal or ordinal scale to an interval or ratio scale so that a nonparametric statistical test could be replaced with a more powerful parametric statistical test (if the other criteria are met; Chapter 12).

5 Select a larger alpha level, such as 0.05 rather than 0.01.

11.7 SAMPLE STATISTICS

Sample statistics (often called 'descriptive statistics') are used to define the nature and distribution of the data. They should be calculated immediately to give the researcher a first approximation look at his or her results. Sample statistics will often provide insight into whether statistical tests will show significance, or not. Many pocket calculators have provisions for calculating sample statistics and, in some cases, are pre-programmed to conduct selected statistical tests.

11.7.1 Sample distributions

Sample data from a population show characteristics that reflect both the population's properties and the sampling methods used. Proper sampling methods (Chapters 6 and 8) must be selected so that a valid measure of the populations can be made.

The choice of appropriate statistical tests will be based to a large extent on the distribution of the sample data.

EXAMPLE: We measure the duration of fighting a mirror image in 25 male siamese fighting fish (*Betta splendens*).

Duration in seconds:

3.7	4.2	3.8	7.7	5.6
9.5	3.7	8.6	3.3	2.5
10.8	4.5	5.9	4.7	4.1
2.4	6.9	4.4	4.4	4.5
5.8	1.6	11.7	6.6	7.6

We can now calculate sample statistics from the above data.

11.7.2 Sample mean, mode and median

We compute the *sample mean* ($\bar{X}$) for the data above by summing (Σ) the sample measurements (x_i) and dividing by the sample size (N).

$$\bar{X} = \frac{\Sigma x_i}{N} = \frac{142.5}{25} = 5.7$$

The sample mean should not be considered the *norm* since the mean might rarely, if ever, occur. For example, we might measure the duration of fighting a mirror image in each of the 25 male siamese fighting fish 100 times and have none of the durations equal 5.7 seconds. A more representative measure of the norm is the *mode*, the measurement that occurs most often in the data. For our small sample of data on the male fighting fish, there is no clear mode since 3.7, 4.4 and 4.5 all occurred twice (see listing below). But if we had made 100 measurements on each fish, a mode probably would be evident.

The *sample median* is the measurement with an equal number of measurements on either side of it. It can be determined by arranging the measurements in order. For example, the measurements for the 25 male fighting fish above would be arranged as follows:

1	1.6	10	4.4	18	6.6
2	2.4	11	4.4	19	6.9
3	2.5	12	4.5	20	7.6
4	3.3	*13	4.5 *Median*	21	7.7
5	3.7	14	4.7	22	8.6
6	3.7	15	5.6	23	9.5
7	3.8	16	5.8	24	10.8
8	4.1	17	5.9	25	11.7
9	4.2				

With 25 measurements, the median value will be the 13th measurement; in this case it is 4.5.

In order to plot the frequency distribution, the measurements are placed into equal intervals. For example, we can place the 25 measurements into one-second intervals (Table 11.3).

These measurements are then plotted in a histogram (see Figure 11.5).

The difference between the sample median and sample mean in the above frequency distribution (see Figure 11.5) demonstrates that the sample data are not normally distributed. That is, sample data are normally distributed when their frequency distribution is the same on either side of the mean (see below).

11.7.3 Skewness

When sample data are not normally distributed, they are skewed, either positively (the curve tailing off to the right toward higher values) or negatively (the curve tailing off toward lower values) (see Figure 11.6). See Chapter 12 for a further discussion and a description of a test for skewness.

11.7.4 Location

Data sample distributions may be alike in form but may differ in location. For example, the two curves in Figure 11.7 are both skewed positively and have the same variability but differ in their location on the scale of measurements.

Table 11.3 *The 25 durations of mirror fighting by the male siamese fighting fish organized according to the number of occurrences of durations within one second intervals from 0 to 12 s*

Interval (s)	No. occurrences
0–1	0
1–2	1
2–3	2
3–4	4
4–5	7
5–6	3
6–7	2
7–8	2
8–9	1
9–10	1
10–11	1
11–12	1

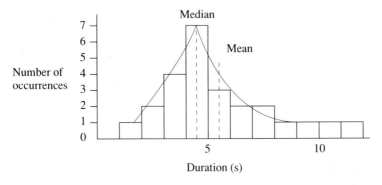

Fig. 11.5   Histogram plot of durations of fighting a mirror image for 25 male siamese fighting fish broken into one second intervals (data from Table 11.3).

The location of a sample distribution is specified by quantities such as the mean and the median. These are referred to as *location parameters*.

Sample distributions are often worth plotting in order to obtain a visual image of their skewness + variability (= form) and locations.

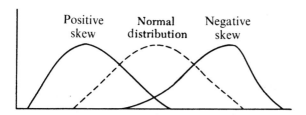

Fig. 11.6 Illustration of normal and skewed distributions of data.

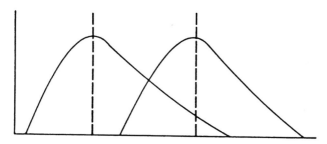

Fig. 11.7 Illustration of two data distributions that have the same form (i.e. same positive skewed distribution and same variability) but differ in location (i.e. their means differ).

11.7.5 Variability

Data samples from two, or more, populations may both be normally distributed and have the same location, but may differ in variability. That is, the frequency distribution of the data on either side of the mean may be the same within each population, but be different between the populations. For example, curves A and B (see Figure 11.8) are two data samples which are both normally distributed. However, curve A represents much more variable data spread over a larger range (10–50).

Skewness and variability are characteristics which combine to determine the *form* of a sample distribution.

11.7.6 Standard deviation

When means are compared, it is also important to know how much variability there is in the original measurements (x_i) from which those means were derived. The *standard deviation* (s) is a measure of that variability about the mean and is represented by the formula:

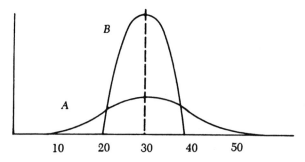

Fig. 11.8   Illustration of two normal data distributions that have the same mean but differ in variability.

$$s = \sqrt{\frac{\Sigma(x_i - \bar{X})^2}{N-1}} \text{ or } \sqrt{\left[\frac{\Sigma x_i^2 - \frac{(\Sigma x_i)^2}{N}}{N-1}\right]}$$

Procedure:

1 Compute the sample mean, $(\bar{X})$
2 Calculate the deviation from the mean for each measurement, $(x_i - \bar{X})$
3 Square each of the deviations, $(x_i - \bar{X})^2$
4 Sum all the squared deviations, $\Sigma(x_i - \bar{X})^2$
5 Divide the sum of the squared deviations by the sample size minus one,

$$\frac{\Sigma(x_i - \bar{X})^2}{N-1}$$

6 Take the square root of the number computed in Step 5.

The standard deviation can be used to reflect the distribution of the data. If the sample data are normally distributed (Figure 11.9), the range included in the mean ± 1 standard deviation includes approximately 68% of the data, the mean ± 2 standard deviations includes about 96% of the data, and the mean ± 3 standard deviations includes 99.7% of the data (Figure 11.9).

11.7.7   Sample mean confidence interval

The sample mean usually only approximates the true population mean, since it is generally based only on a sample from the entire population. We can, however, calculate a range around the sample mean in which we feel confident the population mean lies.

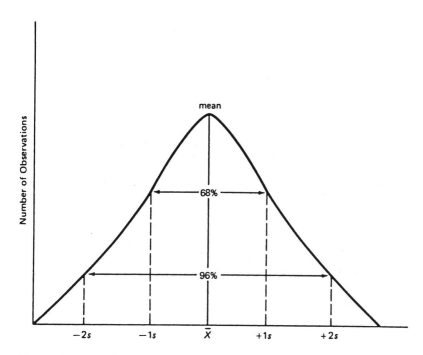

Fig. 11.9 Frequency distribution of a *normally distributed* sample of measurements (observations) and the percentage of those measurements encompassed by the mean ± 1s and 2s.

The *sample mean confidence interval* (C) is computed by dividing the standard deviation of the sample mean (s) by the square root of the number of measurements (N) and multiplying by a factor (t) based on the confidence level (probability level) desired and the number of measurements.

$$C = +t\left(\frac{s}{\sqrt{N}}\right)$$

The value of $s/\sqrt{N}$ is also referred to as the standard error of the mean ($SE_{\bar{X}}$):

$$SE_{\bar{X}} = \frac{s}{\sqrt{N}}$$

Therefore the confidence interval (C) can also be calculated by multiplying the standard error of the mean by t:

$$C = \pm t(SE_{\bar{X}})$$

The value for t is obtained from Table A1 in Appendix A. The confidence level is determined first (e.g. 90% confidence level=0.10 in Table A5) and the degrees of freedom (df)$=N-1$.

Table 11.4 *Hypothetical measurements of the durations of 10 singing bouts by a male bird*

Bout No.	Duration (s)
1	4.6
2	5.3
3	4.4
4	3.1
5	6.4
6	5.3
7	4.7
8	4.8
9	5.0
10	4.4
	Total = 48.0
	$\overline{X}$ = 4.8

Table 11.5 *Calculation of the standard devition of the 10 singing bouts from Table 11.4*

Bout No.	$(x_i - \overline{X})$	$(x_i - \overline{X})^2$
1	-0.2	0.04
2	0.5	0.25
3	-0.4	0.16
4	-1.7	2.89
5	1.6	2.56
6	0.5	0.25
7	-0.1	0.01
8	0.0	0.00
9	0.2	0.04
10	-0.4	0.16
	Total = $\Sigma(x_i - \overline{X})^2$ = 6.36	

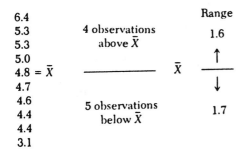

Fig. 11.10   Ordering of the data from Table 11.5 to obtain an initial indication of normality (i.e. number and range of observations on either side of the mean).

EXAMPLE: We measure the duration of 10 singing bouts in a male bird as in Table 11.4. We calculate the standard deviation as in Table 11.5.

$$\frac{\Sigma(x_i - \bar{X})^2}{N-1} = \frac{6.36}{9} = 0.71$$

The square root of this number provides the standard deviation.

$$s = \sqrt{0.71}$$

$$= 0.84$$

We can look for normality in the data by ranking the observations as in Figure 11.10.

The measurements appear to be normally distributed with four observations above the mean and five below, the ranges above and below the mean are almost equal, being 1.6 and 1.7, respectively. However, we do not know if the data are really normally distributed without also knowing the actual frequency distribution on each side of the mean. Nevertheless, we can observe how our data are being distributed with regard to the standard deviation as in Figure 11.11.

To define the confidence limits for the mean we begin by calculating the standard error of the mean:

$$SE_{\bar{X}} = \frac{s}{\sqrt{N}} = \frac{0.84}{3.16} = 0.27$$

The confidence interval is then calculated by multiplying the $SE_{\bar{X}}$ by t. We set our confidence level at 95% (0.05), and × our degrees of freedom are $N-1=9$. The tabular value for t is 2.262 (Table A5).

$$C = \pm t(SE_{\bar{X}}) = 2.262(0.27) = +0.61$$

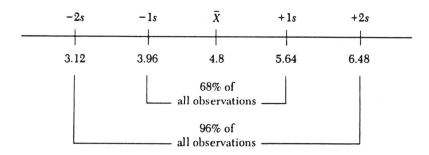

Fig. 11.11   Use of the standard deviation to illustrate the variability in the data from the Table 11.4.

Then we are confident at the 95% level that the population mean lies between:

4.19 and 5.41   i.e. $4.19 \xleftarrow{-0.61} 4.8 \xrightarrow{+0.61} 5.41$

11.7.8 Coefficient of variation

We may wish to compare the amount of variation about the mean for two or more samples of data. For example, is the variation in song-bout duration for male B different from that for male A, measured previously? We now must calculate the standard deviation for Male B (see Table 11.6).

Male A: $X = 4.8$ $s = 0.84$

By comparing the ten bouts for each individual we can see that the durations for male B are more variable, and our sample statistic ($s = 1.72$ versus 0.84) bears this out. However, the mean for male B (5.4 seconds) is also larger than that for male A and may contribute to the larger variation. That is, it is possible to have greater variation around larger means than around smaller means. This is called the *floor effect*; since zero is the bottom limit on durations, a smaller mean is closer to this limit. The converse is the *ceiling effect*, where there is an upper limit to the data. We therefore generate the sample statistic *coefficient of variation* (CV) that expresses the standard deviation as a percentage of the mean. The greater the CV, the greater the variability in the data.

$$CV = \frac{s}{X} \times 100$$

Male A: $CV = \frac{0.84}{4.80} \times 100 = 0.175 \times 100 = 17.5\%$

Male B: $CV = \frac{1.72}{5.40} \times 100 = 0.318 \times 100 = 31.8\%$

Table 11.6 *Hypothetical measurements of the durations of 10 singing bouts by a male bird 'B' for comparison with the measurements for male bird 'A' (Table 11.5)*

Bout No.	Duration (s)
1	5.7
2	3.2
3	7.5
4	6.9
5	3.4
6	4.9
7	7.7
8	6.9
9	3.8
10	4.5
	Total 54.5
	$\overline{X} = 5.4$ $s = 1.72$

Even after adjusting for the differences in means, the CV's demonstrate that male B's song duration is much more variable than male A's song duration.

Significant differences between CVs can be determined using the test statistic C (Dawkins and Dawkins, 1973):

$$C = \frac{(CV_1 - CV_2)}{\sqrt{(Scv_1^2 + Scv_2^2)}}$$

where $Scv = \dfrac{CV}{\sqrt{2N}}$

The probability associated with the test statistic C is obtained from the table for the distribution for t (Table A5). Using this method we can test for a significant difference between the CVs for song duration between Male A and Male B, above.

$$C = \frac{(0.175 - 0.318)}{\sqrt{(Scv_1^2 + Scv_2^2)}}$$

$N = 10$

$$Scv_1^2 = \frac{0.175^2}{2N} = \frac{0.175^2}{20} = (0.039)^2 = 0.001$$

$$S_{cv_2}^2 = \frac{0.318^2}{2N} = \frac{0.318^2}{20} = (0.071)^2 = 0.005$$

$$C = \frac{(0.175 - 0.318)}{0.006} = \frac{0.143}{0.077} = 1.86$$

Since 1.86 is less than the tabular value of 2.26 (9 df, 0.05 level), we conclude that there is no significant difference in variability in the duration of songs between Males A and B.

The coefficient of variation has also been used to determine the extent to which a behavior is 'fixed' or 'stereotyped' (e.g. Wiley, 1973). Some behavior patterns show very little variation ($<2\%$), often at the limit of the equipment used to obtain the measurements (Slater, 1981).

Barlow (1977) has proposed a measure related to the coefficient of variation which he calls *stereotypy* (ST):

$$ST = \frac{\bar{X}}{s} + 0.01\bar{x}$$

The maximum values of ST that are allowable in order to refer to a behavior pattern as 'stereotyped' are undecided and relatively arbitrary. Since the communicative value of many displays varies with context, guidelines for the use of ST measures are difficult to formulate (Bekoff, 1977b).

12 Selection of a statistical test

Kolmogorov–Smirnov test: Assay for the purity of vodka.
Mann–Whitney test: Determination whether a cotton gin were
transported across state lines for immoral purposes.
Rank correlation: Stinkingly low.   *[Norman and Streiner 1986]*

Statistical tests are used to test hypotheses about one or more samples of data. The results of these tests will also add to our current knowledge about the scientific questions you are investigating, whether they result in rejection of the null hypothesis ('significant results'), or not (Rosenthal and Rubin, 1985).

The statistical test you select for analysis of your data will be dictated primarily by your experimental design and scale of measurement; therefore, the type of statistical analysis should be considered when designing the data-collection format. Neither the statistical test nor the experimental design should entirely dictate the other, but they should be coordinated.

Table 12.1 will assist in selecting appropriate statistical tests for: 1. completely randomized designs, and 2. randomized block, matched pair or repeated measures designs. When using this table, it might help to refer back to the experimental designs section in Chapter 6. Many statistical analysis computer programs lead the researcher through a step-by-step decision-making process of selecting the appropriate test. Also, several statistics textbooks provide tables and charts that assist in selecting statistical tests, including Glantz (1992), Krauth (1988), Meyer (1976), Robson (1973), Siegel (1956), Siegel and Castellan (1988), and Sokal and Rohlf (1981b). Martin and Bateson (1986,1993) provide a series of questions which will help direct you to the pertinent tests.

12.1 PARAMETRIC VERSUS NONPARAMETRIC STATISTICAL TESTS

12.1.1 Parametric Tests

There are generally four assumptions inherent in the parametric statistical model that cannot always be met in ethological research, hence necessitating the use of

Table 12.1. *Experimental designs, scales of measurement and corresponding statistical tests*

| | Scales of measurement | | |
| | Parametric | Nonparametric | |
Experimental designs	Ratio or interval	Ordinal	Nomimal
I. Completely randomized design (each measurement x_{ij} is from a different individual)			
A. One variable			
1. *One sample*			
$\dfrac{A_1}{\begin{array}{l} x_1 \\ x_2 \\ \vdots \end{array}}$		One sample runs test Kolmogorov–Smirnov one sample test	One sample Chi-square test Binomial test
2. *Two independent samples*			
$\dfrac{\begin{array}{cc} A_1 & A_2 \end{array}}{\begin{array}{cc} x_{11} & x_{21} \\ x_{12} & x_{22} \\ \vdots & \vdots \end{array}}$	Student's *t*-test	Mann–Whitney *U*-test Kolmogorov–Smirnov two sample test Wald–Wolfowitz runs test	Chi-square goodness-of-fit test Test of two percentages
3. *Three or more independent*			
$\dfrac{\begin{array}{cccc} A_1 & A_2 & A_3 \; \cdots & A_n \end{array}}{\begin{array}{cccc} x_{11} & x_{21} & x_{31} & x_{n1} \\ x_{12} & x_{22} & x_{32} & x_{n2} \\ \vdots & \vdots & \vdots & \vdots \end{array}}$	One-way analysis of variance (ANOVA)	Kruskal–Wallis one-way ANOVA Dunn's multiple comparison test	Chi-square goodness-of-fit test

Table 12.1. (cont.)

| | Scales of measurement | | |
| | Parametric | Nonparametric | |
Experimental designs	Ratio or interval	Ordinal	Nominal
B. Two variables (independent and/or dependent)			
1. One sample			
A_1 versus B_1	Student's t-test	Mann–Whitney U-test Kolmogorov–Smirnov two sample test Wald–Wolfowitz runs test	Chi-square goodness-of-fit test Test of two percentages
2. Two independent samples	Two-way ANOVA	Chi-square test of independence 2×2 Test of two percentages	Chi-square test of independence 2×2 Fisher's exact test
3. Three or more independent samples	Two-way ANOVA	Kendall's coefficient of concordance	Chi-square test of independence ($r \times k$) Contingency coefficient

2. Two independent samples

	A_1	A_2
B_1	x_{11}	x_{21}
B_2	x_{12}	x_{22}

3. Three or more independent samples

	A_1	A_2	A_3	$\cdots$	A_n
B_1	x_{11}	x_{21}	x_{31}		x_{n1}
B_2	x_{12}	x_{22}	x_{32}		x_{n2}
$\vdots$	$\vdots$	$\vdots$	$\vdots$		$\vdots$
B_n	x_{1n}	x_{2n}	x_{3n}		x_{nn}

Table 12.1. (cont.)

	Scales of measurement		
	Parametric	Nonparametric	
Experimental designs	Ratio or interval	Ordinal	Nomimal

II. *Randomized block, matched pairs or repeated measures design (each measurement x_{ij} may be from different individuals, or from the same individual for each sample)*

One variable

1. *Two matched samples*

	Parametric	Nonparametric	Nominal
	Paired t-test	Sign test	McNemar's test
	Walsh test	Wilcoxon matched-pairs signed-ranks test	
	Correlation measures:		
	Pearson's product moment correlation coefficient	Spearman's rho	
		Kendall's tau	

$$
\begin{array}{c|cc}
 & A_1 & A_2 \\
\hline
B_1 & x_{11} & x_{21} \\
B_2 & x_{12} & x_{22} \\
\vdots & \vdots & \vdots \\
B_n & \cdot & x_{nn}
\end{array}
$$

2. *Three or more matched samples*

	Parametric	Nonparametric	Nominal
	One-way repeated measure ANOVA	Friedman's two-way ANOVA	Cochran's Q test
		Dunnett's multiple comparison test	

$$
\begin{array}{c|ccccc}
 & A_1 & A_2 & A_3 & \cdots & A_n \\
\hline
B_1 & x_{11} & x_{21} & x_{31} & \cdots & x_{nl} \\
B_2 & x_{12} & x_{22} & x_{32} & \cdots & x_{n2} \\
\vdots & \vdots & \vdots & \vdots & & \vdots \\
B_n & x_{1n} & x_{2n} & x_{3n} & \cdots & x_{nn}
\end{array}
$$

nonparametric statistical tests. Even slight deviations from the assumptions listed below may make the use of some parametric statistical tests not valid (Siegel and Castellan, 1988).

1   *The measures to be analyzed are continuous measures with equal intervals* (i.e. interval or ratio scales of measurement; see Chapter 6). Behind this assumption is the need, in parametric statistics, to apply arithmetic operations of adding, subtracting, multiplying and dividing. This necessitates the use of nonparametric statistical tests when the scale of measurement is nominal or ordinal (Siegel, 1956; Siegel and Castellan, 1988; Kerlinger, 1973; Drew and Hardman, 1985; Norman and Streiner, 1986; Gibbons, 1993). However, some statisticians have argued that parametric tests could be applied to ordinal data except that they are usually not normally distributed (e.g. Anderson, 1961; Gaito, 1959; Zar, 1984).

2   *The measurements must be independent.* No measurement should be influenced by any other measurement.

3   *Samples are drawn from populations that are normally distributed* (bell-shaped curve, Figure 11.9). The test for normality (see below) can be used to test for a normal distribution. Data which are not normally distributed can sometimes be transformed to make them normally distributed (see section 12.1.2).

4   *There is homogeneity of variance between the groups.* That is, the variances of groups must be the same, statistically. The F-max test (below) and Bartlett's test (Zar, 1984) can be used to determine whether there is homogeneity of variance between sample populations. Data can sometimes be transformed to make their variance homogeneous (see section 12.1.2).

12.1.1a *Test for normality*

Sokal and Rohlf (1981a) and Zar (1984) provide several tests of normality. The Kolmogorov–Smirnov test (described in section 14.1.1b) can also be used to test for normality (Sokal and Rohlf, 1981a; its use in an ethological study is illustrated by Tokarz and Beck, 1987).

The calculations below are used to test for a significant difference from normality in: 1. *skewness* (curve tailing off toward lower or higher values; section 11.7), and 2. *kurtosis.* A *leptokurtic* distribution has more measurements clustered around the mean than normal; a *platykurtic* distribution has fewer observations around the mean than normal.

The skewness and kurtosis tests involve the following calculations of moments (M_i):

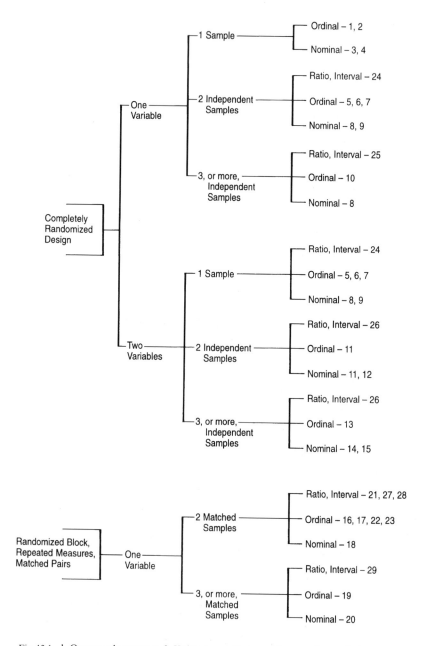

Fig. 12.1   1, One sample runs test, 2, Kolmogorov–Smirnov One sample test; 3, One–sample
chi-square test; 4, Binomial test; 5, Mann–Whitney U-test; 6,
Kolmogorov–Smirnov two sample test; 7, Wald–Wolfowitz runs test; 8, Chi-square
goodness-of-fit test; 9, Test of two proportions; 10, Kruskal–Wallis one-way

$$M_i = \frac{\Sigma(x - \bar{x})^i}{N}$$

therefore:

$$M_2 = \frac{\Sigma(x - \bar{x})^2}{N}$$

$$M_3 = \frac{\Sigma(x - \bar{x})^3}{N}$$

$$M_4 = \frac{\Sigma(x - \bar{x})^4}{N}$$

Skewness (SK) is calculated as:

$$SK = \frac{M_3}{M_2 \sqrt{M_2}}$$

Kurtosis (KUR) is calculated as:

$$KUR = \frac{M_4}{M_2^2}$$

To test for significant skewness (i.e. SK differs significantly from zero) we calculate the variance of SK, then calculate the Z value.

$$\text{Variance of } SK = VSK = \frac{6N(N-1)}{(N-2)(N+1)(N+3)}$$

$$Z = \frac{SK}{VSK}$$

If $Z \geq \pm 1.96$ then skewness is significantly different from zero at the 0.05 level.

To test for significant kurtosis (i.e. KUR differs significantly from zero) we calculate the variance of KUR, then calculate the Z value.

$$\text{Variance of } KUR = VKUR = \frac{4(N^2-1)VSK}{(N-3)(N+5)}$$

$$Z = \frac{KUR}{\sqrt{VKUR}}$$

ANOVA; 11, Chi-square test of independence (2×2); 12, Fisher's exact test; 13, Kendall's coefficient of concordance; 14, Chi-square test of independence (r×k); 15, Contingency coefficient; 16, Sign test; 17, Wilcoxon matched-pairs signed-ranks test; 18, McNemar's test; 19, Friedman's two-way ANOVA; 20, Cochran's Q-test; 21, Pearson's product moment correlation coefficient; 22, Spearman's rho; 23, Kendall's tau; 24, Student's t-test; 25, One-way ANOVA; 26, Two-way ANOVA; 27, Paired t-test; 28, Walsh test; 29, One-way repeated measures ANOVA.

Table 12.2. *Hypothetical data used to illustrate the measures of skewness and kurtosis*

Song duration (s) x	$(x-\bar{x})^2$	$(x-\bar{x})^3$	$(x-\bar{x})^4$
4.7000	0.6400	0.5120	0.4096
5.1000	1.4400	1.7280	2.0736
3.2000	0.4900	-0.3430	0.2401
4.2000	0.0900	0.0270	0.0081
3.7000	0.0400	-0.0080	0.0016
4.1000	0.0400	0.0080	0.0016
4.5000	0.3600	0.2160	0.1296
3.6000	0.0900	-0.0270	0.0081
2.9000	1.0000	-1.0000	1.0000
3.0000	0.8100	-0.7290	0.6561
39.0000	5.0000	0.3840	4.5284

If $Z \geq \pm 1.96$ then kurtosis is significantly different from zero at the 0.05 level.

As an example we will use the data on song duration in two populations of birds (p. 382) and test whether the measurements in Population B are normally distributed (Table 12.2).

Test for significant skewness:

$$SK = \frac{M_3}{M_2\sqrt{M_2}} = \frac{0.0384}{0.50(0.7071)} = 0.1086$$

$$VSK = \frac{6N(N-1)}{(N-2)(N+1)(N+3)} = \frac{6(10)(9)}{8(11)(13)} = \frac{540}{1144} = 0.4720$$

$$Z = \frac{SK}{\sqrt{VSK}} = \frac{0.1086}{\sqrt{0.4720}} = \frac{0.1086}{0.6870} = 0.158$$

Since the calculated Z of 0.158 is not $\geq \pm 1.96$, the skewness is not significantly different from zero at the 0.05 level; that is, this distribution of song durations is not significantly skewed.

Test for significant kurtosis:

$$KUR = \frac{M_4}{M_2{}^2} - 3 = \frac{0.4528}{0.5^2} - 3 = -1.1888$$

$$\text{VKUR} = \frac{4(N^2-1)\text{VSK}}{(N-3)(N+5)} = \frac{4(99)(0.4720)}{7(15)} = \frac{186.912}{105} = 1.7801$$

$$Z = \frac{\text{KUR}}{\sqrt{\text{VKUR}}} = \frac{-1.1888}{\sqrt{1.7801}} = \frac{-1.1888}{1.3342} = -0.891$$

Since the calculated Z of -0.891 is not $\geq \pm 1.96$, the kurtosis is not significantly different from zero at the 0.05 level; that is, this distribution of song durations is neither significantly leptokurtic nor platykurtic.

12.1.1b F-max test for homogeneity of variance

This test determines whether the variances in the measurements from two samples are significantly different. From the example in Chapter 13 on p. 382 we can determine whether there is a significant difference in song-duration variability between Population A and Population B. The F-max test assumes that the data are normally distributed, but it is robust and valid even when this assumption is violated slightly.

Procedure:

1 Determine the variance for each sample population by squaring the standard deviation:

variance of Pop. $A = S_A^2 = (0.82^2) = 0.68$
variance of Pop. $B = S_B^2 = (0.74^2) = 0.55$

2 Calculate F:

$$F = \frac{\text{largest variance}}{\text{smallest variance}} = \frac{0.68}{0.55} = 1.23$$

3 Obtain the tabular value of F from Table A4. Two different degrees of freedom are needed. Those across the top of the table refer to the sample which had the larger variance; those on the side are for the sample with the smaller variance. In our case both degrees of freedom are 9 ($\text{df} = N - 1$). The tabular F value for the 95% confidence level is 4.03.

Since our calculated value of 1.23 is not larger than the tabular value (4.03), we conclude that there is no statistically significant difference in the variances of the samples; that is, the song durations do not vary significantly more in one population than the other.

12.1.2 Data transformations

Data are sometimes transformed in order to meet the assumptions of parametric statistical tests (discussed above). There are three major reasons for using data transformations in the analysis of variance (Kirk, 1968):

1 To achieve homogeneity of error variance.
2 To achieve normality of measurement distributions within samples.
3 To obtain additivity of treatment effects.

Kirk (1968) also states that a transformation that achieves one of the goals above will usually accomplish the other two.

Some researchers have used transformations on nominal or ordinal data in order to achieve a normal distribution and then proceeded to use parametric statistics (e.g. Mendl, 1988). However, nominal and ordinal data should not be used in parametric tests, regardless of the transformations they receive. *No transformation can create interval or ratio data out of nominal or ordinal data.* Some of the more common transformations are given below.

12.1.2a *Square root transformation*

The square root transformation is used to create more *homogeneity of variance*. The transformed measurement x_{ij}' is calculated by the following:

If *all* measurements ≥ 10 use: $x_{ij}' = \sqrt{x_{ij}}$

If *any* measurement < 10 use: $x_{ij}' = \sqrt{(x_{ij} + 0.5)}$

This transformation has also been used to create a normal distribution (e.g. Mendl, 1988).

12.1.2b *Logarithmic transformation*

The logarithmic transformation is generally used to create a *normal distribution* in the measurements (e.g. Lawrence 1985); it is commonly used when the measurements are skewed to the right. The transformed measurement is given by:

$$x_{ij}' = \log_{10}(x_{ij})$$

If some of the measurements are zero, or very small, use:

$$x_{ij}' = \log_{10}(x_{ij} + 1)$$

12.1.2c *Arcsine transformation*

The arcsine transformation is used with percentages and proportions to create a *normal distribution* (e.g. Sherry *et al.*, 1981; Mendl, 1988), especially when the distribution of measurements is binomial. The transformed measurement is given by:

$$x_{ij}' = 2 \text{ arcsine } \sqrt{x_{ij}}$$

12.1.3 Nonparametric tests

Nonparametric statistical tests are distribution-free tests which do not demand that the assumptions of parametric tests be met.

> A nonparametric statistical test is a test whose model does not specify conditions about the parameters of the population from which the sample was drawn.   *[Siegel, 1956:31]*

Since there are fewer constraints on nonparametric tests, they are usually less powerful when used to analyze data where parametric tests are applicable (however see Blair and Higgins, 1985). Therefore, researchers often proceed with parametric tests without having necessarily satisfied the four criteria listed above (p. 373).

This is supportable, in part, by the fact that some parametric tests are robust; that is, they can be used with reasonable validity even when some of the parametric model assumptions are violated. For example, Student's *t*-test can be used even when there is considerable deviation from normality and/or homogeneity of variance, except in an independent-samples design with unequal numbers of scores; however, analysis of variance is highly sensitive to the kurtosis of a population (Govindarajulu, 1976).

Overall, there are several factors which should be considered when selecting between parametric and nonparametric tests. Gibbons (1993) has compiled a list which serves as the basis for a safe (yet sometimes conservative) guideline. According to Gibbons (1993) *Use a nonparametric statistical test when any of the following are true:*

1   The data are counts or frequencies of different types of outcomes (i.e. categories; nominal scale of measurement).
2   The data are measured on an ordinal scale.
3   The assumptions required for the validity of the corresponding parametric procedure are not met or cannot be verified.
4   The shapes of the distributions from which the samples are drawn are unknown.
5   The sample size is small.
6   The measurements are imprecise.
7   There are outliers and/or extreme values in the data, making the median more representative than the mean.

If the data meet the assumptions for parametric tests then parametric tests will be more powerful; however, the more the data violate the assumptions for parametric tests, the more powerful nonparametric tests become relative to parametric tests (Zar, 1984).

Power-efficiency is a measure of the amount of increase in sample size necessary to make test *B* as powerful as test *A* (Siegel, 1956). It can be used to compare any two tests which can be validly applied to the data, such as comparable parametric and nonparametric tests (Welkowitz *et al.*, 1976). Given that the data meet the criteria for use of parametric test, then for a given difference between population means, a given alpha level and a specified power, power-efficiency is a ratio expressed as a percent as follows:

$$\text{Power-efficiency of nonparametric test} = \frac{N_p}{N_{np}} \times 100\%$$

Where: N_p = sample size for the parametric test.

N_{np} = sample size necessary for the nonparametric test to make it as powerful as the parametric test.

For example, if the parametric test requires a sample size of 80 and the nonparametric test requires a sample size of 100 to make it as powerful as the parametric test, then the power-efficiency is:

$$\text{Power-efficiency of nonparametric test} = \frac{80}{100} \times 100\% = 80\%$$

In this case, the nonparametric test is 80% as powerful as the parametric test. The power-efficiency is provided for several of the nonparametric tests discussed in the following chapters.

13 Parametric statistical tests

Since much of the data gathered in ethology do not meet the assumptions necessary to use parametric statistical tests, only a few of the more commonly applied parametric tests are described below. Also, nonparametric statistical tests can be applied to data which meet the assumptions for parametric tests; however, in those cases the parametric tests will be more powerful.

13.1 COMPLETELY RANDOMIZED DESIGN

13.1.1 Two independent samples

13.1.1a *Standard error of the difference between means*

We can compare means from two samples and determine if they are significantly different, that is, whether they came from significantly different populations or whether there was a significant treatment effect. The standard error of the difference of the means is computed according to the following formula:

$$SE_{\bar{x}_1 \bar{x}_2} = \sqrt{\left(\frac{s_1^2}{N_1} + \frac{s_2^2}{N_2}\right)}$$

The symbols s_1^2, s_2^2 and N_1, N_2 represent the variances and sample sizes of samples 1 and 2, respectively. If the difference between the two means is larger than two times the standard error of the difference, they are significantly different.

For example, assume we want to test the research hypothesis that the mean duration of song bouts in Population A of a bird species is significantly different than it is in Population B. We randomly sample 10 males from each population and record the duration of one randomly selected song bout from each male (see below; we would normally take a much larger sample than this).

1 Calculate the total and mean song bout duration (in seconds) for each population.

Sample Population A	Sample Population B
5.2	4.7
4.8	5.1
6.4	3.2
5.3	4.2
3.1	3.7
5.0	4.1
4.4	4.5
5.2	3.6
4.9	2.9
4.7	3.0
Total=49.0	39.0
Mean=$\bar{X}$= 4.9	3.9

2 Calculate the *standard deviation* for both samples.

a Sample: Population A

$(x_i-\bar{X})$	$(x_i-\bar{X})^2$
0.3	0.09
−0.1	0.01
1.5	2.25
0.4	0.16
−1.8	3.24
0.1	0.01
−0.5	0.25
0.3	0.09
0.0	0.00
−0.2	0.04
	Total=6.14=$\Sigma(x_i-\bar{X}^2)$

$$\frac{\Sigma(x-\bar{X})^2}{N-1}=\frac{6.14}{9}=0.68$$

$$S_A=\sqrt{\left[\frac{(x_i-\bar{X})^2}{N-1}\right]}=\sqrt{0.68}=0.82$$

b Sample: Population B:

$(x_i-\bar{X})^2$	$(x_i-\bar{X})$
0.8	0.64
1.2	1.44
−0.7	0.49
0.3	0.09
−0.2	0.04
0.2	0.04
0.6	0.36
−0.3	0.09
−1.0	1.00
−0.9	0.81

$$\text{Total}=\overline{5.00}=\Sigma(x_i-\bar{X})^2$$

$$\frac{\Sigma(x_i-\bar{X})^2}{N-1}=\frac{5.00}{9}=0.55$$

$$S_B=\sqrt{\left[\frac{(x_i-\bar{X})^2}{N-1}\right]}=\sqrt{0.56}=0.74$$

We can then calculate the standard error of the difference of the means:

$$\text{SE}_{\bar{x}_A\bar{x}_B}=\sqrt{\frac{S_A^2}{N_A}+\frac{S_B^2}{N_B}}=\sqrt{\left(\frac{0.68}{10}+\frac{0.56}{10}\right)}=\sqrt{0.124}=0.124=0.35$$

The difference between the means $(4.9-3.9=1.0)$ is larger than twice the $\text{SE}_{\bar{x}_A\bar{x}_B}=0.35\times2=0.70$; therefore, the difference between the means is statistically significant.

13.1.1b Student's t-test

Student's t-test is also used to test for significant differences between two sets of data and is based on a comparison of means. We will use the same data on song duration from the two populations that we used in the previous examples. In those examples we tested for the assumptions of most parametric tests and found: 1. there is homogeneity of variance between Populations A and B, and 2. the data from Population B are neither significantly skewed nor leptokurtic or platykurtic. We should check the normality of the data in Population A before we proceed, but we also know that the t-test is sufficiently robust so that these assumptions can be violated to a reasonable extent without affecting the validity of the test. Also, all the factors in the formula have already been calculated (above).

$$t = \frac{(X_A - X_B)\sqrt{\left(\dfrac{N_A N_B}{N_A + N_B}\right)}}{\sqrt{\dfrac{(N_A - 1)(S_A{}^2) + (N_B - 1)(S_B{}^2)}{N_A + N_B - 2}}}$$

$$= \frac{4.9 - 3.9\sqrt{\left(\dfrac{10 \times 10}{10 + 10}\right)}}{\sqrt{\dfrac{(10 - 1)(0.71) + (10 - 1)(0.56)}{10 + 10 - 2}}}$$

$$= \frac{1.0\sqrt{\left(\dfrac{100}{20}\right)}}{\sqrt{\left[\dfrac{(9)(0.68) + (9)(0.56)}{18}\right]}}$$

$$t = \frac{(1.0)(2.24)}{\sqrt{\left(\dfrac{6.12 + 5.4}{18}\right)}} = \frac{2.24}{\sqrt{0.62}} = \frac{2.24}{0.79} = 2.84$$

We then obtain the tabular value for t from Table A5 for 18 degrees of freedom (df) and a significance level of $P = 0.05$. Since our calculated t of 2.84 is larger than the tabular t value of 2.101, we conclude that the data are from two distinctly different populations. That is, song duration in Population A is statistically greater than it is in Population B; this agrees with our comparison using the standard error of the difference between means (above).

13.1.2 Three or more independent samples

13.1.2a One-way analysis of variance

One-way analysis of variance (ANOVA) tests for significant differences between three or more independent samples of measurements. Variations of this test can be applied to a wide variety of experimental designs; see Meddis (1973) for a clear and concise overview. The one-way ANOVA described in the example below is for the completely randomized design. We will once again use the hypothetical data on song duration that we used in the examples above; however, we will expand our hypothesis and samples from Populations A and B to include two more populations C and D (Table 13.1).

1 Complete the analysis of variance (Table 13.2) below by making the calculations in Steps 2 – 13. Results are found in Table 13.3.

2 Calculate the totals for each sample (column): $t_1 \ldots t_4$

3 Calculate the totals for each row: $r_1 \ldots r_{10}$

4 Calculate the grand total (GT): $GT = 174.7$

Table 13.1. *Song bout duration (s)*

| (r) | Population samples | | | | Row totals |
	A	B	C	D	
	5.2	4.7	3.9	5.1	18.9
	4.8	5.1	4.2	5.9	20.0
	6.4	3.2	3.9	4.8	18.3
	5.3	4.2	3.1	5.2	17.8
	3.1	3.7	3.6	4.9	15.3
	5.0	4.1	4.1	5.3	18.5
	4.4	4.5	3.2	5.4	17.5
	5.2	3.6	3.0	4.8	16.6
	4.9	2.9	2.7	5.2	15.7
	4.7	3.0	2.9	5.5	16.1
Column totals(t)=49.0		39.0	34.6	52.1	174.7=GT

Table 13.2

Source of variation	df	Sum of squares	Mean square	F
Between samples (columns)		BSSS	BSMS	
Within samples (rows)		WSSS	WSMS	
Total		TSS		

5 Calculate the Correction Term: $CT = \dfrac{GT^2}{N} = \dfrac{174.7^2}{40} = 763$

where: N=total number of measurements=40

6 Calculate the sum of squares of the measurements (Σx_{ij}^2); that is, square each of the individual measurements and sum them.

$\Sigma x_{ij}^2 = 5.2^2 + 4.8^2 + 6.4^2 \ldots 5.5^2$
$= 27.04 + 23.04 + 40.96 \ldots 30.25$
$= 798.21$

7 Calculate the total sum of squares (TSS):

$TSS = \Sigma x_{ij}^2 - CT = 798.21 - 763 = 35.21$

8 Calculate the between sample sum of squares (BSSS):

$$BSSS = \frac{t_1^2 + t_2^2 + t_3^2 + t_4^2}{n_c} - CT$$

where: n_c = number of scores in each column (sample)

$$= \frac{2401 + 1521 + 1197.16 + 2714.41}{10} - 763 = 20.36$$

9 Calculate the within samples sum of squares (WSSS):

$$WSSS = TSS - BSSS$$
$$= 35.21 - 20.36 = 14.85$$

10 Calculate the degrees of freedom (df):

Between-samples df = Number of samples (columns) − 1 = 3
Within-samples df = (Number of rows − 1)(Number of columns) = (10−1)(4) = 36
Total df = Number of measurements (N) − 1 = 39

11 Calculate each mean square (MS) by dividing the sum of squares by the corresponding df

Between samples mean square (BSMS) = $\dfrac{20.36}{3}$ = 6.78

Within samples mean square (WSMS) = $\dfrac{14.85}{36}$ = 0.4125

12 To test the hypothesis that there is a significant difference in song bout duration between the populations calculate the between-samples F value:

Between-samples $F = \dfrac{\text{Between-samples MS}}{\text{Within-samples MS}} = 16.445$

13 Compare the calculated F value to the tabular value (Table A6) using the between-samples df (3), the within-samples df (36) and the appropriate alpha level. The tabular value for P=0.05 is 2.87 (extrapolated from 2.92 and 2.84). Since our calculated F of 16.445 is larger than the tabular value (2.87), we reject the null hypothesis of no significant difference between the samples.

13.2 RANDOMIZED BLOCK, MATCHED PAIRS OR REPEATED MEASURES DESIGN

13.2.1 Two related or matched samples

13.2.1a *Paired t-test*

The paired *t*-test is used to test for significant differences between two related or matched samples. These could be measurements on the same individuals under both conditions (treatments) or they could be measurements paired by some characteristic (e.g. litter, time, location).

1 Calculate:

$$t=\frac{\bar{D}}{\sqrt{\frac{\Sigma D^2 - N(\bar{D}^2)}{N-1}}}$$

Where:

D=difference between each pair of measurements

$\bar{D}$=mean of the difference between each pair of measurements

N=number of pairs of measurements

2 Compare the calculated t to the tabular value (Table A5), where df=$N-1$. If the calculated value of $t\geq$tabular value, then the null hypothesis of no significant difference between the samples is rejected.

Table 13.3

Source of variation	df	Sum of squares	Mean square	F
Between-samples (columns)	3	20.36	6.78	16.44
Within-samples (rows)	36	14.85	0.41	
Total	39	35.21		

As an example, we will provide hypothetical data on songbird species x (Table 13.4), similar to that recorded by Reid (1987) for Ipswich sparrows (*Passerculus sandwichensis princeps*). Our research question is whether time spent singing is greater than time spent foraging in a habitat with an abundant food supply. We randomly selected 10 individual males from a population of 18 and took focal-animal/all-occurrences samples, measuring the time spent singing and foraging during the hours 0600 to 0900. Total observation time for each male was 10 hours.

Table 13.4. *Hypothetical data on time spent singing and foraging by ten male songbirds*

| | Total time (min.) | | | |
| | Singing | Foraging | | |
Individual			D	D^2
A	105	152	47	2 209
B	97	202	105	11 025
C	115	117	2	4
D	95	233	138	19 044
E	120	105	15	225
F	87	275	188	35 344
G	103	176	73	5 329
H	89	260	171	29 241
I	112	131	19	361
J	109	139	30	900
Totals:	1 032	1 790	788(ΣD)	103 682(ΣD^2)

$$\bar{D}=\frac{\Sigma D}{N}=\frac{788}{10}=78.8$$

$$t=\frac{78.8}{\sqrt{\dfrac{\left[\dfrac{103\,682-10(78.8^2)}{10-1}\right]}{10}}}$$

$$=\frac{78.8}{\dfrac{\sqrt{\left(\dfrac{41\,588}{9}\right)}}{3.162}}$$

$$=\frac{78.8}{\dfrac{67.97}{3.162}}=\frac{78.8}{21.5}=3.66$$

The tabular t value (Table A5) for df=9 and $P=0.05$ is 2.262. Since our calculated t (3.66) is larger than the tabular value, we reject the H_0, and conclude that there is a significant difference between time spent foraging and singing.

As another example, Oberhauser (1988) studied mating strategies in male monarch butterflies and found that a lower percentage of males mated in 1986 than in 1985 (Table 13.5; paired $t=3.3$, df=8, $P=0.011$), a result she attributed to an unusually cool summer. Note that the measurements are paired by the number of days since the last mating.

Table 13.5. *Copulations by males with different mating histories to virgin female monarch butterflies*

Time since last mating (days)	1985		1986	
	Number tested	% Mated	Number tested	% Mated
0*	396	33	489	30
1	49	57	36	31
2	88	65	36	56
3	52	56	27	30
4	40	63	42	40
5	29	48	32	44
6	19	37	38	29
7	21	62	27	30
8	8	38	24	42

Note: * Virgins.
Source: Copyrighted by Bailliere Tindall.

13.2.1b *Measure of association*

Pearson product moment correlation coefficient

The Pearson correlation coefficient (r) is used to determine if there is a relationship between two sets of paired data. The data must be either interval or ratio.

$$r = \frac{N\Sigma XY - (\Sigma X)(\Sigma Y)}{\sqrt{[N\Sigma X^2 - (\Sigma X)^2][N\Sigma Y^2 - (\Sigma Y)^2]}}$$

Where:

N = Number of pairs of scores

XY = sum of the products of the paired scores

X = sum of the scores of one variable (X)

Y = sum of the scores of the other variable (Y)

X^2 = sum of the squared scores of the X variable

Y^2 = sum of the squared scores of the Y variable

The range of r is -1.00 to $+1.00$, and the sign of r denotes whether the correlation is positive or negative. The larger the r value, the more highly correlated are the two sets of data. The significance of r (from $r=0$) can be determined by comparing it to the value of r_c in Table A7. This is a two-tailed test for a significant correlation, regardless of the sign of r. If r is larger than r_c then you reject the H_0 and conclude that the two variables (X, Y) are significantly correlated.

As an example, we will use the same hypothetical data on a songbird species that was used for the paired t-test (above). However, since this is only a hypothetical example, we will divide each of the measurements by 10 in order to keep the calculations more manageable. All the variables necessary to calculate Pearson's coefficient are provided in Table 13.6. We will arbitrarily assign singing as the X variable and foraging as the Y variable.

$$r = \frac{10(1787.60) - (103.2)(179.0)}{\sqrt{\{[10(1076.08) - 103.2^2][10(3535.94) - 179.0^2]\}}}$$

$$= \frac{17\,876.00 - 18\,472.80}{\sqrt{[(10\,760.80 - 10\,650.24)(35\,359.40 - 32\,041.00)]}}$$

$$= \frac{-596.80}{\sqrt{[(110.56)(3318.40)]}} = \frac{-596.80}{605.71} = -0.98$$

Table 13.6

Male	Total time (min.) Singing (X)	Foraging (Y)	XY	X^2	Y^2
A	10.5	15.2	159.60	110.25	231.04
B	9.7	20.2	195.94	94.09	408.04
C	11.5	11.7	134.55	132.25	136.89
D	9.5	23.3	221.35	90.25	542.89
E	12.0	10.5	126.00	144.00	110.25
F	8.7	27.5	239.25	75.69	756.25
G	10.3	17.6	181.28	106.09	309.76
H	8.9	26.0	231.40	79.21	676.00
I	11.2	13.1	146.72	125.44	171.61
J	10.9	13.9	151.51	118.81	193.21
Sums:	103.2 (ΣX)	179.0 (ΣY)	1\,787.60 (ΣXY)	1\,076.08 (ΣX^2)	3\,535.94 (ΣY^2)

The calculated coefficient of -0.98 indicates a very strong negative correlation. Since the calculated r of -0.98 is larger than the tabular r_c of 0.632, we reject the H_0 of no correlation and conclude that time spent foraging and singing are significantly, negatively correlated.

14 Nonparametric statistical tests

Nonparametric statistical tests are the most commonly used statistical tests in ethological research. They are simple tests that can easily be calculated by hand or with a hand-held calculator. This means they can be used reliably while you are in the field without access to a computer.

Since nonparametric tests are so easily conducted, it is tempting to apply them to all data. However, parametric statistical tests should be used when the data meet the criteria (see section 12.1.1), since they are more powerful.

14.1 COMPLETELY RANDOMIZED DESIGN

14.1.1 One variable

14.1.1a One sample

$\dfrac{S_A}{x_1}$ S_A = Sample of variable A (or Treatment)

x_2 x_1 = Measurement on Individual No. 1

x_3

$\vdots$

x_n

One sample runs test

This test determines whether a sequence of two different items (in time or space) is non-random.

Sample sequence: A A B B B A B B B B A A B A

As examples, the sequence above could be the:

1 Sequence of occurrence of two behaviors (A,B).
2 Sequential order of male (A) and female (B) starlings perched along a telephone wire.
3 Sequential order of sheep (A) and cows (B) lined up at a feed bunk.

In any case we want to determine whether the two items are clustered or separated in the sequence more than would be expected by chance. The *two non-random extremes*, for the example above, would be:

Non-randomly *clustered*: $\underline{A\ A\ A\ A\ A\ A}\ \underline{B\ B\ B\ B\ B\ B\ B\ B}$

Non-randomly *separated*: $\underline{B}\ \underline{A}\ \underline{B}\ \underline{A}\ \underline{B}\underline{A}\ \underline{B}\underline{A}\ \underline{B}\underline{A}\ \underline{B}\underline{A}\ \underline{B}\ \underline{B}$

Example:

H_0: The sheep (A) and cows (B) are randomly distributed along the feed bunk.

  1 Determine the number of runs (r).

$$\underset{1}{\underline{A\ A}}\ \underset{2}{\underline{B\ B\ B}}\ \underset{3}{\underline{A}}\ \underset{4}{\underline{B\ B\ B\ B}}\ \underset{5}{\underline{A\ A}}\ \underset{6}{\underline{B}}\ \underset{7}{\underline{A}}\ r = 7$$

  2 Determine:

N_A = number of A items = 6
N_B = number of B items = 8

  3 Compare r with the tabular values (Tables $A8_1$, $A8_2$) for N_A and N_B. The number of runs (r) is significant at the 0.05 level, if: r is less than or equal to the value in Table $A8_1$ *or* r is greater than or equal to the value in Table $A8_2$.

     In our example, 7 is larger than 3 (Table $A8_1$) and smaller than 12 (Table $A8_2$); therefore, we cannot reject the H_0 that the sheep and cows are randomly distributed.

One sample chi-square test

This test is used to analyze data in which individuals can make one of only two responses (e.g. accept or reject, turn left or right, fly or don't fly). It should be used only when the expected values are ≥ 5. If they are less than 5, use the *binomial test* (described below). Additional precautions when using the chi-squared statistic in ethology are given by Kramer and Schmidhammer (1992).

$$\chi^2 = \Sigma \frac{[(\text{observed–expected})–0.5]^2}{\text{expected}}$$

For example, Butlin *et al.* (1985) measured whether female grasshoppers (*Chorthippus brunnneus*) that did not give response chirps to males subsequently mated (Table 14.1).

H_0: Females that did not give response chirps to males were equally likely to mate, as not mate.

  1 Complete the (Table 14.2).

    Observed (O) = is the data from Table 14.1.

Table 14.1. *Numbers of female grasshoppers that had not given response chirps to males that subsequently mated or did not mate*

	Trials which ended with:	
	Mating	No mating in 2 h
Female did not give response stridulation	11	23

Source: From Butlin *et al.* (1985).

Table 14.2. *Calculation of chi-square from data in Table 14.1*

Response	O	E	$(O-E)$	$(O-E)^2$	$\dfrac{(O-E)^2}{E}$
Mating	11	17	−6	36	2.118
No mating	23	17	+6	36	2.118
Totals	34	34			

Expected (E)=total number of responses equally distributed between each category of response:

$11+23=34 \qquad 34/2=17$

We would expect 17 to mate and 17 not to mate if they had an equal (50:50) chance of reponding either way.

2 Calculate chi-square, $\chi^2 = \Sigma = 4.236$

3 Compare the calculated χ^2 to the the value in Table A9 with a degree of freedom of 1 (df=no. categories−1). If the calculated χ^2 is greater than or equal to the tabular value, then the H_0 that there is no significant difference in the females' responses is rejected.

The calculated χ^2 of 4.236 is larger than the tabular value of 3.841 (0.05 level of significance); therefore, we reject the H_0 and conclude that the females mated less often than would be expected by chance.

Binomial test

Like the chi-square test above, the binomial test is a one-sample goodness-of-fit test in which there is a predicted distribution between two discrete categories. It is a good replacement for the chi-square test when any of the expected frequencies are <5. The binomial test determines the probability of obtaining x events (the smallest observed value), or fewer, in one category and $N-x$ events in the other category, out of a total of N events. The researcher must specify the expected probabilities.

1. Procedure for determining the probability of obtaining exactly x events in one category and $N-x$ events in the other category out of a total of N events.

Calculate the probability, $p(x)$, using the following formula:

$$p(x)=\frac{N!}{x!\,(N-x)!}P^x\,Q^{N-1}$$

where:

$N!$ means the factorial of N

$P=$ expected proportion of x

$Q=1-P$

$$\frac{N!}{x!\,(N-x)!}=\binom{N}{x}=\text{binomial coefficient}$$

This part of the formula can be calculated using factorials (Table A1) or by determining the binomial coefficient using Table A3.

As an example, Hews (1988) tested the alarm reaction of toad (*Bufo boreas*) larvae to chemical cues released from predation by a waterbug on conspecific and heterospecific larvae. The data in Table 14.3 below are only for the tests with predation on a conspecific larva in the experimental half of the test tank. The 20 larvae could choose the half of the tank where the predation was occurring (experimental half) or the other half (control half).

We will use an expected probability of 0.50 in each half of the tank. That is, $P=0.5$ and $Q=1-P=0.5$

$N=20$

$x=$ smallest number of larvae choosing half of tank $=3$

$$p(x)=\binom{N}{x}P x Q^{N-x}$$

$$=(1140)(0.5^3)(0.5^{17})$$

$$=(1140)(0.125)(7.63^{-6})=0.001$$

Therefore, 0.001 is the probability of having exactly three larvae choose the experimental half of the tank, if the expected number is 10 (based on an equal distribution).

Table 14.3. *Alarm reaction of toad larvae to chemical cues*

No. of larvae spending most of time (>600 s) towards		
Experimental half	Control half	*N*
3	17	20

Source: From Hews (1988).

2. A more appropriate probability to calculate for most ethological experiments is the probability of obtaining x, or fewer, events in one category and $N-x$ events in the other category out of a total of N events.

To calculate this probability, sum the probability of the observed events x and all the more extreme distributions. This is accomplished by successively reducing x by 1 and calculating the probability for each value including zero. This total probability is the probability of obtaining the observed distribution of events, or more extreme values. For our example, the probability of having 3, 2, 1 or 0 larvae choose the experimental half of the tank is the sum of each of the individual probabilities.

$$p(x) = \binom{N}{i} P^i Q^{N-i}$$

$p(3) = 0.001$ (calculated above)

$$p(2) = \binom{20}{2}(0.5^2)(0.5^{18}) = 190(0.25)(3.81^{-6}) = 1.81^{-4}$$

$$p(1) = \binom{20}{1}(0.5^1)(0.5^{19}) = 20(0.5)(1.91^{-6}) = 1.90^{-5}$$

$$p(0) = \binom{20}{0}(0.5^0)(0.5^{20}) = 1(1)(9.53^{-7}) = 9.53^{-7}$$

$$p(\leq 3) = 0.001 + 1.81^{-4} + 1.90^{-5} + 9.53^{-7} = 0.0012$$

Therefore, the probability of having three, or fewer, larvae choose the experimental half of the tank is 0.0012.

14.1.1b *Two independent samples*

$$A_1 \quad A_2 \quad A_1 = \text{Sample or Treatment No.1}$$

$$
\begin{array}{ll}
x_{11} & x_{21} \\
x_{12} & x_{22} \\
x_{13} & x_{23} \\
\vdots & \vdots \\
x_{1n} & x_{2n}
\end{array}
$$

$A_{11} = \text{Measurement on Individual No.1 in Sample No.1}$

Mann–Whitney U *test*

The Mann–Whitney U test is the nonparametric counterpart of Student's t-test for independent samples. Whereas the t-test determines significant differences between means, the Mann–Whitney U test uses the medians to test for a significant difference in the location of the sample data. It is 95% as powerful as the Student's t-test (Mood, 1954).

This test can be used when the data are, at least, ordinal. For large samples, the Mann–Whitney U Test is more powerful than the Kolmogorov–Smirnov Test; for very small samples, the Kolmogorov–Smirnov Test is more powerful (Siegel and Castellan, 1988). If the samples are correlated (paired or matched), use the Wilcoxon matched-pairs signed-rank test. The use of both the Mann–Whitney U test and the Wilcoxon matched-pairs signed-rank test on independent and paired data, respectively, is illustrated by Breitwisch's (1988) study of parental defense in mockingbirds.

As an example, we will use samples of song durations from two populations in which the variances are obviously different; that is, an F-max test would be expected to show a significant difference in the variation of bird-song duration from these two populations. Therefore, we will use a nonparametric statistical test.

Procedure:

1 Rank the data (Table 14.4) using both samples. The smallest measurement gets rank no.1.

2 Determine the sum of the ranks in the smaller sample (T; in our example it is Population B). If both samples are the same size, sum the ranks of the first sample.

$$T = 68$$

3 Calculate the U_S and U_L statistics:

$$U_S = N_S N_L + \frac{N_S(N_S+1)}{2} - T$$

Table 14.4. *Hypothetical data on song durations from two populations of birds and rankings used in the Mann–Whitney* U *test (see text)*

Song duration (s)		Ranks	
Population A sample	Population B sample	Population A sample	Population B sample
4.7	8.1	10	15
5.3	4.2	12	7
3.6	6.7	3	13
5.1	9.5	11	16
4.0	2.7	5	2
4.1	1.8	6	1
3.8	7.8	4	14
4.3		8	
4.4		9	
		Sum of the ranks $= T = 68$	

where:

N_S = number of measures in the smaller sample

N_L = number of measures in the larger sample

$$U_S = (7)(9) + \frac{7(7+1)}{2} - 68 = 63 + 28 - 68 = 23$$

$$U_L = N_S N_L - U_S = (7)(9) - 23 = 40$$

4 Obtain the tabular value for $N_S = 7$, $N_L = 9$ in Table A10.

tabular value = 12

$$U_S = 23$$
$$U_L = 40$$

There is a significant difference if either of the observed values (U_S or U_L) is *equal to or less than* the tabular value. Hence, in our example the song durations are not significantly different between the two populations. This you would suspect, since there is enough variability in Sample B to overlap the values in Sample A; we can see this better in graphic form (Figure 14.1).

Kolmogorov–Smirnov two sample test

This test is a substitute for the Mann–Whitney U test, especially when the sample size is small. It determines whether two samples differ significantly in either form or location. That is, it tests the H_0 that the two samples are not significantly different.

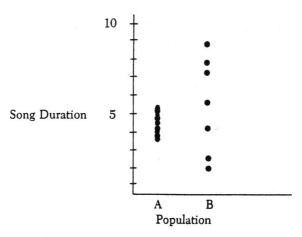

Fig. 14.1 Graph of hypothetical data on song duration from two populations of songbirds.

The scale of measurement must be at least ordinal. Even if the data are interval or ratio, they are analyzed as an ordinal scale, which means that resolution is lost. Therefore, a parametric statistical test (Student's *t*-test) would normally be used if the criteria (section 12.1.1) are met; however, when the sample size is small the Kolmogorov–Smirnov test is 96% as powerful as Student's *t*-test (Dixon, 1954).

Small samples   The procedure below is used when the number of measurements is ≤25 in both samples. As a hypothetical example, we will determine whether the duration of feeding bouts are significantly different between two small herds of deer from different habitats but with the same sex and age composition (Table 14.5).

> Procedure:
> 1 Convert ratio or interval data to an ordinal scale. Since we are working with ratio data, the scale of measurements must be divided into intervals. An interval should be selected such that no single interval contains more than two to three measurements. Each sample is then arranged in order of magnitude, and the cumulative frequencies of observations for each sample up to each interval are determined (Table 14.6).
> 2 Calculate the ratios of the cumulative frequencies to the total number of measurements in each sample.
>
> where: m=no. measurements in Sample 1 (Herd A)
> n=no. measurements in Sample 2 (Herd B)
> S_m=ratio of cumulative frequency to m
> S_n=ratio of cumulative frequency to n

Table 14.5. *Hypothetical data on the duration of feeding bouts in tow herds of deer*

Feeding bout duration (min.)	
Herd A	Herd B
43	24
62	31
81	19
69	47
73	35
11	29
64	18
18	21
89	43
67	17
59	65
12	28

3 Find D, the largest difference between S_m and S_n.
For a one-tailed test, D is calculated as the maximum difference in the *predicted* direction. For example, if our research hypothesis was that the feeding durations in Herd A were significantly longer than in Herd B, the formula would be:

$D=$ maximum $[S_n - S_m]$

Test Statistic $= D \times m \times n$

The largest difference in the *predicted* direction is at the intervals of 45–49 and 50–54 minutes where $S_n = 0.916$ and $S_m m = 0.333$.

$D = 0.916 - 0.333 = 0.583$

Test statistic $= 0.583 \times 12 \times 12 = 83.95$

Compare the calculated test statistic to the tabular value (Table A11) for the appropriate values of m, n and level of significance. The tabular value for $m=12$, $n=12$ at $P=0.05$ is 72. Since our calculated value of 83.95 is larger than the tabular value of 72, we reject the H_0 that the feeding durations in Herd A are not longer than they are in Herd B.

Table 14.6. *Calculations for the Kolmogorov–Smirnov two sample test for small samples using the data from Table 14.5*

Interval (no. minutes)	Rearranged measurements		Cumulative frequencies		Ratios (S_m)	(S_n)
	Herd A	Herd B	Herd A	Herd B	Herd A	Herd B
0–4			0	0	0.000	0.000
5–9			0	0	0.000	0.000
10–14	11,12		2	0	0.166	0.000
15–19	18	17,18,19	3	3	0.250	0.250
20–24		21,24	3	5	0.250	0.417
25–29		28,29	3	7	0.250	0.583
30–34		31	3	8	0.250	0.667
35–39		35	3	9	0.250	0.750
40–44	43	43	4	10	0.333	0.833
45–49		47	4	11	0.333	0.916
50–54			4	11	0.333	0.916
55–59	59		5	11	0.417	0.916
60–64	62,64		7	11	0.583	0.916
65–69	67,69	65	9	12	0.750	1.000
70–74	73		10	12	0.833	1.000
75–79			10	12	0.833	1.000
80–84	81		11	12	0.916	1.000
85–89	89		12	12	1.000	1.000
Totals:			$m=12$	$n=12$		

For a two-tailed test, D is calculated as the maximum *absolute* difference between S_m and S_n.

$$D = \text{maximum} \mid S_m - S_n \mid$$

Test statistic $= D \times m \times n$

As we stated at the beginning of this example our question was whether there is a significant duration in the feeding bouts of the deer in these two herds; therefore, our H_0 is that there is no significant difference between the feeding durations in the two herds. The largest *absolute* difference is at the intervals of 45–49 and 50–54 minutes (the same as it was for our one-tailed test).

Table 14.7. *The number of days it took reed warblers to reject model cuckoo eggs by either ejection or desertion*

| Rejected during | Number of nests with rejection | |
	By ejection	By desertion
Day 1	10	4
Day 2	4	7
Day 3	21	14
Day 4	21	18
Day 5	25	20
Day 6	29	23
Day 7	31	24
Total rejected:	31	24

Source: From Davies and Brooke (1988).

$D = 0.916 - 0.333 = 0.583$

Test statistic $= 0.583 \times 12 \times 12 = 83.95$

Compare the calculated test statistic to the tabular value (Table A12) for the appropriate values of m, n and level of significance. The tabular value for $m = 12$, $n = 12$ at $P = 0.05$ is 72. Since our calculated value of 83.95 is larger than the tabular value of 72 we reject our H_0 of no significant difference in duration of feeding bouts between Herd A and Herd B.

Large samples   If the number of measurements in either sample is ≥ 25, a different table and procedures are necessary depending on whether you are conducting a one-tailed or two-tailed test.

As an example, Davies and Brooke (1988) studied nest parasitism by cuckoos (*Cuculus canarus*) on reed warblers (*Acrocephalus scirpaceus*). As part of the research, they measured the number of days it took reed warblers to reject model cuckoo eggs by two methods, ejection of the egg from the nest and desertion of the nest (Table 14.7).

The research question was whether the distribution of times to reject the model eggs by desertion was significantly different than by ejection.

Procedure:
1   Determine the cumulative frequencies and ratios for each period, as in Table 14.8.

Table 14.8. *Calculations for the Kolmogorov–Smirnov two sample test for large samples using the data from Table 14.7*

Rejected within	Cumulative no. nests with rejection by:		Ratios of cumulative frequencies to total	
	Ejection	Desertion	Ejection (S_m)	Desertion (S_n)
1 day	10	4	0.322	0.166
2 days	14	11	0.451	0.458
3 days	21	14	0.677	0.583
4 days	21	18	0.677	0.750
5 days	25	20	0.806	0.833
6 days	29	23	0.935	0.958
7 days	31	24	1.000	1.000
Total rejected:	$31=m$	$24=n$		

2 Find D, the largest difference between S_m and S_n. For a one-tailed test, D is calculated as the maximum difference in the *predicted* direction, just as in the small sample case above. For example, if our hypothesis is that the nests are rejected by desertion sooner than by ejection, the formula for D would be:

$$D = \text{maximum } (S_n - S_m)$$

The largest difference in the predicted direction is at 4 days where $S_n = 0.750$ and $S_m = 0.677$.

$$D = 0.750 - 0.677 = 0.073$$

3 For a large sample, one-tailed test, a chi-square value is now calculated as follows:

$$\chi^2 = 4D^2 \frac{m \times n}{m+n}$$

$$= 4(0.073)^2 \frac{31 \times 24}{31+24}$$

$$= 4(0.005)\,(13.53) = 0.271$$

4 Compare the calculated χ^2 to the tabular value (Table A9) for df$=2$. The tabular value at P$=0.05$ is 5.99. Since our calculated χ^2 of 0.271 is smaller

than the tabular value (5.99), we fail to reject our hypothesis that the nests are rejected by desertion sooner than by ejection.

For a two-tailed test, D is calculated as the maximum *absolute* difference between S_m and S_n. Since we are testing whether there is a significant difference in either direction, we will use the maximum absolute difference.

$$D = \text{maximum} \mid S_m - S_n \mid$$

The largest *absolute* difference is at Day 1 where $S_m = 0.322$ and $S_n = 0.166$.

$$D = 0.322 - 0.166 = 0.166$$

This calculated D is compared to the D value obtained by entering the observed values of m and n in the expression given in Table A13 at the appropriate P value. For our example, we will use the expression in Table A13 for $P = 0.05$ as follows:

$$\text{Critical value of } D = 1.36 \sqrt{\left(\frac{m+n}{m \times n}\right)}$$

$$= 1.36 \sqrt{\left(\frac{31+24}{31 \times 24}\right)}$$

$$= (1.36)\sqrt{0.0739}$$

$$= 0.370$$

5 Since our calculated D of 0.166 is smaller than 0.370 we fail to reject the H_0 that there is no significant difference between the distribution of times to reject the model eggs by desertion and by ejection.

Wald–Wolfowitz two sample runs test

Like the Mann–Whitney U test and the Kolmogorov–Smirnov test this test is used to test the H_0 that that there is no significant difference between two independent samples. It will reject H_0 if the two samples differ significantly in either form or location. It is approximately 75% as powerful as Student's t-test for sample sizes of approximately 20 (Smith, 1953). Since it is less powerful than the Mann–Whitney U test and the Kolmogorov–Smirnov test, the primary advantage of this test is its simplicity.

As an example, we will determine whether the hypothetical frequency of agonistic behavior of songbirds differs significantly between birds at feeder Type A and at feeder Type B. We will assume that we collected the hypothetical data during nine one-hour sampling periods at each feeder type (Table 14.9).

Table 14.9. *Hypothetical data on the frequency of agonistic behavior of songbirds at two types of feeders*

Feeder Type A sample	Feeder Type B sample
15	6
16	11
9	6
18	3
17	12
13	8
10	7
15	5
14	5

Procedure:

1 Rank all the measurements in order of increasing size and cast them into a single order. Then identify the population from which each score comes and determine the number of runs accordingly.

Measurements listed in order and their corresponding populations:

```
3 5 5 6 6 7 8   9 10   11 12   13 14 15 16 17 18
B B B B B B B   A A    B  B    A  A  A  A  A  A
```

Runs: 1   2   3   4

Number of runs $(r) = 4$

2 Obtain the tabular value from Table A8$_1$. If the observed value is *equal to or less than* the tabular value, then the H_0 is rejected at the 0.05 level.

n_1 = number of measurements in the first sample = 9
n_2 = number of measurements in the second sample = 9

In our example the tabular value = 5. Since our calculated $r(4)$ is smaller than the tabular value of 5, we reject the H_0 that there is no significant difference between the frequency of agonistic behaviors between songbirds at the two feeder types.

Chi-square goodness-of-fit test, two samples

The chi-square goodness-of-fit test can be used to determine whether two, or more, independent samples are significantly different. The importance of having independent measurements when using chi-square has been emphasized by many authors, including Kramer and Schmidhammer (1992). Whether the chi-square is a goodness-of-fit test or a test of independence, the measurements that are summed to provide the observed cell frequencies must be independent in order to have a valid test. The assumption of independence may be violated if an individual contributes more than once to a data set.

The application of chi-square with two samples is described below; its use with three, or more, samples will be discussed later in this chapter. It is used with nominal data and compares observed frequencies with frequencies that would be expected in a uniform or random distribution.

As an example, Vives (1988) studied parent choice by larval cichlids (*Cichlasoma nigrofasciatum*) that were reared under two treatments: 1. in the presence of predators of fry; and 2. not in the the presence of predators of fry. Later, the free-swimming fry were placed in an aquarium where they could choose to stay in close proximity to their mother, father or neither (see the analysis of Vives' 2×2 matrix of data in the discussion of Fischer's exact test later in this chapter). Vives combined the data from the two treatments and tested whether the larval cichlids chose to stay in proximity to either their maternal or paternal parent significantly more often:

- Parental choice:
- Maternal, 35
- Paternal, 12
- Total, 47

Procedure:

1 Determine the expected by either assuming: 1. a random expected distribution and randomly assigning each of the 47 measurements to one of the two categories (maternal or paternal); or 2. a uniform expected distribution and assigning 50% of the 47 measurements to each category, as has been done in Table 14.10.

2 Calculate the χ^2 for each category (as shown in the Table 14.11):

$$\frac{(O-E)^2}{E}$$

It is often recommended that *Yates' correction for continuity* be used in chi-square tests where the degree of freedom is 1 (e.g. Parker, 1979). This consists of reducing the numerator by 0.5 before it is squared, as follows:

$$\frac{[(Observed-Expected)-0.5]^2}{Expected}$$

Table 14.10. *Calculation of the Chi-square goodness-of-fit test on data for parental choice by larval cichlids*

Choice	Observed (O)	Expected (E)	$\dfrac{(O-E)^2}{E}$
Maternal	35	23.5	5.63
Paternal	12	23.5	5.63
Total	47	47	11.26

Source: From Vives (1988).

The use of Yates' correction is considered to be overly conservative and unnecessary by many statisticians (see additional discussion on the chi-square 2×2 test of independence in section 14.1.2b).

3 Obtain the Chi-square value by summing the figures calculated in Step 2 (as shown in the table):

$$\Sigma \frac{(O-E)^2}{E} = 5.63 + 5.63 = 11.26$$

4 Compare the calculated χ^2 to the tabular value (Table A9) for $p=0.05$ and df=no. categories$-1=2-1=1$. The tabular value is 3.84. Since our calculated χ^2 of 11.26 is larger than the tabular value (3.84) we reject the H_0 of a uniform distribution of choices between the maternal and paternal parent, and we conclude that the larval cichlids chose the maternal parent significantly more often than would be expected by chance.

Test of two percentages

This test is used to determine whether there is a significant difference in the percentages (proportions) of a response in two samples (or treatments).

Calculate:

$$Z = \frac{P_1 - P_2}{\sqrt{\left[\dfrac{p(1-p)}{N_1} + \dfrac{p(1-p)}{N_2}\right]}}$$

where:

P_1 = Percentage of response in Sample No. 1
N_1 = Number of measurements in Sample No. 1

$$p = \frac{N_1 P_1 + N_2 P_2}{N_1 + N_2}$$

Table 14.11. *Response of experimental and control toads to droneflies (see text)*

Response	Experimental toads	Control toads	Total
Ate dronefly	9	19	28
	(41%)	(86%)	
Rejected dronefly	13	3	16
	(59%)	(14%)	
Totals	22	22	44

Source: From Brower and Brower (1962).

If $Z \geq +1.96$ *or* ≤ -1.96, then the proportions are significantly different at the 0.05 level.

For example, Brower and Brower (1962) determined whether experimental toads (fed intact, live honeybees) and control toads (fed dead honeybees, with stinging apparatus removed) ate droneflies (honeybee mimic) in different proportions (Table 4.11).

$$N_1 = 22 \; P_1 = 0.41$$

$$N_2 = 22 \; P_2 = 0.86$$

$$p = \frac{N_1 P_1 + N_2 P_2}{N_1 + N_2} = \frac{22(.41) + 22(.86)}{22 + 22} = 0.635$$

$$Z = \frac{p_1 - p_2}{\sqrt{\left[\dfrac{p(1-p)}{N_1} + \dfrac{p(1-p)}{N_2} \right]}}$$

$$= \frac{0.41 - 0.86}{\sqrt{\left[\dfrac{0.635(1-0.635)}{22} + \dfrac{0.635(1-0.635)}{22} \right]}}$$

$$= \frac{-0.45}{0.145} = -3.103$$

Since -3.103 falls outside the limits of $+1.96$ to -1.96, we conclude that the two proportions are significantly different.

V. DeGhett (pers. commun.) offers the following caution when using this test: take the smaller of the two values p or $1-p$ and multiply it by the smaller N; if the product is ≥ 5 then the ratio can be interpreted as a Z value; if not, the statistic

Table 14.12. *Hypothetical data on birdsong duration (in seconds) from three habitats*

Habitat A sample	Habitat B sample	Habitat C sample
4.4	6.9	9.2
3.4	7.1	8.1
6.7	5.2	8.3
3.8	4.3	7.2
4.1	8.2	9.1
5.0		8.9

cannot be interpreted. In the example above, $p=0.635$ and $1-p=0.365$. When 0.365 is multiplied by the smaller N (22) the product is 8.03. Since 8.03 is ≥ 5 we can properly interpret -3.103 as a Z value.

14.1.1c Three or more independent samples

Kruskal–Wallis one-way analysis of variance
The Kruskal–Wallis one-way analysis of variance is a test for determining whether three, or more, independent samples are significantly different. It is the nonparametric counterpart of the parametric one-way analysis of variance (discussed in Chapter 13) and is 95% as powerful (Andrews, 1954). It tests for differences in location and requires at least ordinal measurement.

As an example, we will use hypothetical song durations from different populations, as we did with the parametric one-way analysis of variance. The hypothetical data in Table 14.12 are mean song durations from one-hour samples of the same species of songbird from three different habitats. Since our samples are small we will not calculate the normality or homogeneity of variance to determine whether we can use a parametric statistical test; instead we will proceed with the nonparametric test, even though we will be losing resolution in the data when we reduce this ratio data to an ordinal scale of measurement.

Procedure:
1 Rank all the measurements from the three populations as one group beginning with rank 1 for the smallest measurement. If ties occur between two or more measurements, assign each measurement the mean of the ranks for which it is tied. Then calculate the sum of the ranks for each column (R_j) (Table 14.13).

Table 14.13. *Ranks of data from Table 14.12 for calculation of the Kruskal–Wallis one-way analysis of variance*

Habitat A	Habitat B	Habitat C
5	9	17
1	10	12
8	7	14
2	4	11
3	13	16
6		15
$R_A = 25$	$R_B = 43$	$R_C = 85$

2 Divide the square of each of the R_js by the number of measurements in that sample.

$$\frac{R_j^2}{N_j}$$

$$\frac{(25)^2}{6} = 104.17 \qquad \frac{(43)^2}{5} = 369.80 \qquad \frac{(85)^2}{6} = 1204.17$$

3 Sum the figures just calculated.

$$\Sigma \frac{R_j^2}{N_j} = 104.17 + 369.80 + 1204.17 = 1678.14$$

4 Calculate the test statistic H.

$$H = \left(\frac{12}{N(N+1)} \times \frac{R_j^2}{N_j} \right) - 3(N+1)$$

where: N = total number of measurements

$$H = \left[\frac{12}{17(17+1)} \times (1678.14) \right] - 3(17+1) = 11.78$$

5 When there are three samples and the number of measurements in each of the samples is ≤ 5, compare the calculated H to the value in Table A11. When there are three or more samples and >5 measurements in each sample (as in our example), compare the calculated H to the chi-square value in Table A9; df = no. of samples $- 1$.

$df = 3 - 1 = 2$

Tabular $\chi^2_{0.05,2} = 5.99$

Since our calculated H (11.78) is larger than the tabular chi-square value (5.99), we reject our H_0 and conclude that there is a statistically significant difference in song duration between the three samples (habitats). In order to determine which pairs of habitats (A versus B, A versus C, B versus C) are significantly different use Dunn's test below.

Dunn's multiple comparison test

This test can be used to determine which pairs of samples differ significantly when the null hypothesis is rejected by the Kruskal–Wallis one-way analysis of variance test.

Procedure:

1 Calculate the test statistic Q:

$$Q = \frac{\bar{R}_B - \bar{R}_A}{SE}$$

Where:

$\bar{R}_J =$ mean rank for the Jth sample

$$\bar{R}_A = \frac{R_A}{N_A}$$

$$\bar{R}_B = \frac{R_B}{N_B}$$

where:

$N_A =$ number of measurements in Sample A

$N_B =$ number of measurements in Sample B

$$SE = \sqrt{\left[\frac{N(N+1)}{12} \left(\frac{1}{N_A} + \frac{1}{N_B} \right) \right]}$$

where:

$N =$ total number of measurements in all the samples

2 Compare the calculated Q to the value in Table A15 for the desired level of significance and where k is the total number of samples. If the calculated $Q \geq$ tabular Q, then the H_0 of no significant difference between the two samples is rejected.

For example, we will use the example from the Kruskal–Wallis test (above) and test for a significant difference in song duration between the samples from Habitat A and Habitat B.

$$\bar{R}_A = \frac{R_A}{N_A} = 25/6 = 4.2$$

$$\bar{R}_B = \frac{R_B}{N_B} = 43/5 = 8.6$$

$$N = 17 \qquad N_A = 6 \qquad N_B = 5 \qquad k = 3$$

$$SE = \sqrt{\left(\frac{17(18)}{12} \frac{1}{6} + \frac{1}{5} \right)} = 3.005$$

$$Q = \frac{\bar{R}_B - \bar{R}_A}{SE} = \frac{8.6 - 4.2}{3.055} = 1.44$$

The calculated Q of 1.44 is smaller than the tabular Q ($2.394_{0.05,3}$). Therefore, we cannot reject the H_0 of no significant difference between the samples from Habitats A and B. This implies that the difference that contributed to the significant Kruskal–Wallis test probably came from Habitat A versus Habitat C and perhaps from Habitat B versus Habitat C. You can again use Dunn's test to find out.

Chi-square goodness-of-fit test, three or more samples

This test is used to determine whether three, or more, independent samples are significantly different. As a hypothetical example, we observe male songbirds of species X singing from three different species of trees. This test is merely an extension of the two-sample test described earlier in section 14.1.1b. It is used with nominal data and compares observed frequencies with frequencies that would be expected in a uniform or random distribution. That is, we can uniformly (equally) distribute the total number of observations between the three tree species, randomly assign each observation to a tree species, or generate random expected values via calculation, such as the negative exponential distribution used by Krebs (1974).

We have determined that the three species of trees are equally distributed throughout each male's territory, so that if there were no preference for any tree species each male would sing equally often from each species of tree. We collect the data shown in Table 14.14 from 60 males and then assume a uniform expected distribution by assigning equal numbers (20) to each tree species.

The number of degrees of freedom in a goodness-of-fit test equals the number of categories minus one. In our case we have three tree species (A, B and C); therefore df $= 3 - 1 = 2$. Looking at Table A9, with 2 df, we see that our calculated χ^2 of 24.4

Table 14.14. *Hypothetical data on the frequency of birds singing from three different species of trees*

Tree Species	No. males observed singing observed frequency (O)	Expected frequency (E) based on equal distribution	$\dfrac{(O-E)^2}{E}$
A	38	20	16.2
B	10	20	5.0
C	12	20	3.2
Tables	60	60	$\chi^2 = 24.4$

exceeds the tabular values even at alpha level 0.001. Therefore, we conclude that the male songbirds do not sing equally from all three tree species.

14.1.2 TWO VARIABLES

14.1.2a One sample

One sample of each of two independent variables (A_1, B_1) corresponds in design to two samples of one independent variable (A_1, A_2; discussed earlier in this chapter). These two designs are compared Table 14.15.

Since both designs are testing two independent samples, the same statistical tests are used. Therefore, for one sample of two independent variables, the appropriate nonparametric tests are:

- For ordinal data: Mann–Whitney U test
  Kolmogorov–Smirnov two sample test
  Wald–Wolfowitz runs test

- For nominal data: chi-square goodness-of-fit test
  test of two percentages

All of these tests were described earlier in this chapter.

14.1.2b Two independent samples

Chi-square test of independence (2×2)

This test is used to determine the degree of association among measures in two independent samples of one variable and two or more samples of another variable. It is

Table 14.15. *Comparison of two experimental designs: one sample of two independent variables, and two samples of one independent variable*

One sample of two independent variables		Two samples of one independent variable	
A_1	B_1	A_1	A_2
x_{11}	x_{21}	x_{11}	x_{21}
x_{12}	x_{22}	x_{12}	x_{22}
$\vdots$	$\vdots$	$\vdots$	$\vdots$
x_{1n}	x_{2n}	x_{1n}	x_{2n}

Table 14.16. *The priority ratios for obtaining food relative to the hatching order within sibling cattle egret dyads*

		Priority ratio		
		Elder sib	Younger sib	Totals
Hatching order		525	362	887
	1–2	(539)	(348)	
		99	40	139
	3–4	(85)	(54)	
Totals		624	402	1026

Note:
The Observed (O) measures are the upper figure in each cell. The Expected (E) values are provided for each cell (in parentheses)
Source: From Fujioka (1985)

based on the degree of difference between the observed measures and what would be expected by chance. For our example we will use research which involved two samples each of two variables (2×2).

Large samples   As an example, Fujioka (1985) studied sibling competition in the cattle egret (*Bubulcus ibis*). The priority ratios for obtaining food relative to the hatching order within sibling dyads were measured (Table 14.16). The priority ratio was the percentage of time one sib obtained food from the parent prior to the other sib when both siblings begged simultaneously.

1 Expected values for each cell are calculated as follows:

$$\text{Expected} = \frac{\text{Row total} \times \text{Column total}}{\text{Grand total}}$$

For example, for the upper left cell (elder sib×hatching order 1–2):

$$\text{Expected} = \frac{887 \times 624}{1026} = 539$$

2 Calculate chi-square (x^2):

$$x^2 = \Sigma \frac{(\text{observed} - \text{expected})^2}{\text{expected}}$$

Cell	$\dfrac{(O-E)_2}{E}$
Elder×1–2	0.363
Elder×3–4	2.305
Younger×1–2	0.563
Younger×3–4	3.629

$$x^2 = 6.86$$

3 Compare the calculated x^2 value to the tabular value (Table A9).

Degrees of freedom = (No. rows − 1)(No. columns − 1) = 1

Since the calculated x^2 of 6.86 is larger than the tabular value ($3.84_{0.05}$) the H_0 of no association is rejected.

Small samples   The expected values in the example above are large; however, it is commonly recommended that when expected values are smaller than 5, Fisher's exact test (described below) should be used in place of chi-square. Several authors have investigated the validity of this recommendation (see Everitt, 1977) and suggested that expected values as small as 0.5 may be acceptable (Van Hoof, 1982). The format for two samples of two variables ($A_{1,2}$ and $B_{1,2}$) is:

	A_1	A_2	Row totals
B_1	A	B	$A+B$
B_2	C	D	$C+D$
Column totals	$A+C$	$B+D$	N

The following is a simple formula for chi-square with small sample sizes:

$$x^2 = \frac{(AD - BC)^2 N}{(A+B)(C+D)(A+C)(B+D)}$$

Another common recommendation for small sample sizes, small expected values, and all 2×2 matrices (Denenberg 1976) is to use *Yates' Correction for Continuity* when calculating the chi-square value for each cell. The correction involves reducing the numerator by 0.5 before squaring, as follows:

$$\frac{[(\text{observed–expected})-0.5]^2}{\text{expected}}$$

Once again, there is controversy over the necessitity of this correction, including recommendations that it not be used when the sample size is larger than 20 (e.g. Sokal and Rohlf, 1981) or not used at all (e.g. Howell, 1992). If the researcher wishes to be conservative, use Yates' correction or Fisher's exact test.

Fisher's exact test

Fisher's exact test, like the chi-square test of independence (above), is used to analyze contingency tables for significant associations. However, in contrast to the chi-square test it can be used when one, or more, expected values are less than 5. We use the same format as above:

	A_1	A_2	Row totals
B_1	A	B	RT_1
B_2	C	D	RT_2
Column totals	CT_1	CT_2	Grand total (GT)

One method of calculating the probability (P) of a set of observed values (A,B,C,D) is by using the factorials (!) of the observed values and the marginal totals as follows:

$$P=\frac{(RT_1!)\ (RT_2!)\ (CT_1!)\ (CT_2!)}{(A!)\ (B!)\ (C!)\ (D!)\ (GT!)}$$

The probability can be calculated using a hand-held calculator or the factorial table (Table A1). The probability should be <0.05 to reject the null hypothesis of a random distribution.

A second method is based on the logs of the factorials (Table A2) and an antilog (Sokal and Rohlf, 1981). In this method:

$$P=\text{antilog}\ (T-C)$$

where:

$T=(\log RT_1!+\log RT_2!+\log CT_1!+\log CT_2!)-\log GT!$
$C=\log A!+\log B!+\log C!-\log D!$

Table 14.17. *The effect of rearing condition of parental choice by larval cichlids*

| Rearing condition | Parental choice | | Totals |
	Maternal	Paternal	
Predators Present	23	10	33
Predators Absent	12	2	14
Totals	35	12	47

Source: From Vives (1988).

As an example, Vives (1988) studied parent choice by larval cichlids that were reared under two treatments: 1. in the presence of predators of fry, and 2. in the absence of predators of fry. Later, the free-swimming fry were placed in an aquarium where they could choose to stay in close proximity to their mother, father or neither. The data in Table 14.17 are for the young that chose one of the parents.

We can see that if we attempted to use the chi-square test, the expected frequency in cell D (predators absent×paternal choice) would be less than 5; $(12×14)/47=3.57$. Therefore, we will use Fisher's exact test.

$$T=(36.938+10.940+40.014+8.680)-59.413=37.159$$
$$C=22.412+6.559+8.680+0.301=37.952$$
$$P=\text{antilog }(37.159-37.952)=\text{antilog }-0.793=0.161$$

Since the calculated probability of 0.161 is larger than 0.05 we cannot reject the hypothesis of chance distribution. That is, we conclude, as did Vives (1988), that being reared with, or without, predators of fry had no effect on which parent the larval cichlids chose.

For a test of significance of a one-tailed test, we must obtain the total probability for the observed values plus the probability for all the more extreme values (as in the binomial test). This can be done as follows:

1 Keeping the marginal totals fixed, reduce the smallest value by 1, adjusting the other values in the cells accordingly, and calculate the new probability. Note that the numerator stays the same.

2 Repeat step 1 reducing the value by one until you have calculated the probability when the smallest cell is 0.

3 Add these probabilites together to obtain the total probability for the one-tailed test. Multiply this probability by 2 to obtain the probability for a two-tailed test (Siegel, 1956; Sokal and Rohlf, 1981).

4 Determine if this probability is sufficiently small (e.g. <0.05 or <0.01) to reject the null hypothesis of a random distribution.

A second, and perhaps simpler, method for obtaining the more extreme probabilities (Feldman and Kluger, 1963; Zar, 1984) is as follows:

1 Designate the smallest observed value as a and the observed value in the diagonal cell as d.

2 Designate the observed value in the remaining cell in row 1 as b and the observed value in the remaining cell in row 2 as c.

3 P is the probability of a given table of observed values, and P' is the probability of the next more extreme table. In the next more extreme table, b becomes b' and c becomes c'.

4 Calculate P' for all the more extreme tables and add them to the P for the observed values.

$$P' = \frac{ad}{b'c'}(P)$$

3 *Three or more independent samples*

	A_1	A_2	A_3	$\cdots$	A_n
B_1	x_{11}	x_{21}	x_{31}	$\cdots$	x_{n1}
B_2	x_{12}	x_{22}	x_{32}	$\cdots$	x_{n2}
B_3	x_{13}	x_{23}	x_{33}	$\cdots$	x_{n3}
$\vdots$	$\vdots$	$\vdots$	$\vdots$	$\vdots$	
B_n	x_{1n}	x_{2n}	x_{3n}	$\cdots$	x_{nn}

Kendall's coefficient of concordance

Kendall's coefficient of concordance is a measure of association. It determines the extent of correlation among several sets of rankings (ordinal scale). For example, we might be interested in how well three measures of dominance compare in ranking individual wolves in a pack's dominance heirarchy.

As another example, Whitfield (1986) ranked the quality of 19 turnstone (*Arenaria interpres*) territories using five measures; he then determined the correlation among the ranks for the different measures (Table 14.18).

To illustrate the use of Kendall's coefficient of concordance we'll reconstruct the table as if only seven territories ($Y1$–7) had been ranked using only three measures: number and density of chironomids, and *Larus* density (Table 14.19).

K = number sets of rankings = 3

N = number of items (territories) ranked = 7

Table 14.18. *Ranking of territories using several measures of territory quality*

Chironomid no.	Chironomid density	*Larus* density	*Sterna* density	Larid density
N1	Y2	Y4	Y1	Y4
Y5	Y3	Y2	Y4	Y1
Y1	Y1	Y3	Y3	Y3
Y4	Y5	Y5	Y6	Y2
Y3	Y4	Y6	Y5	Y5
Y6	Y6	O6	O1	Y6
Y2	O2	O3	T2	O1
O2	N1	O4	T1	T2
O5	O5	O7	Y2	O4
T1	N2	Y1	O4	T1
N2	O6	N1	N1	O6
O6	T1	O1	N3	O3
O3	Y7	N3	O2	N1
T2	O7	O2	O5	O7
O4	O3	O5	N2	N3
N3	O1	T1	O6	O2
O7	T2	N2	O3	O5
O1	O4	T2	O7	N2
Y7	N3	Y7	Y7	Y7

Note:
The rankings are essentially the same by Kendall's coefficient of concordance:
$W = 0.698$, $\chi^2 = 62.82$, df = 18, $P < 0.001$.
Source: From Whitfield (1986). Copyrighted by Bailliere Tindall.

Table 14.19. *Use of three measures to rank the quality of seven turnstone territories*

	Territory						
	Y1	Y2	Y3	Y4	Y5	Y6	Y7
Chironomid number	2	6	4	3	1	5	7
Chironomid density	3	1	2	5	4	6	7
Larus density	6	2	3	1	4	5	7
$R_j =$	11	9	9	9	9	16	21

Source: Portion of Table I in Whitfield (1986).

1 Calculate Kendall's coefficient of concordance (W):

$$W = \frac{S}{MP}$$

where:

  S = sum of squares of observed deviations from mean of R_j

$$= \Sigma(R_j - R_j/N)^2$$
$$[R_j/N = 84/7 = 12]$$
$$= (11-12)^2 + (9-12)^2 + \ldots (21-12)^2 = 134$$

MP = maximum possible sum of the squared deviations; that is, if the rankings were in perfect agreement.

$$MP = \frac{1}{12}K^2(N^3 - N) = \frac{1}{12}9(343-7) = 252$$

$$W = \frac{S}{MP} = \frac{134}{252} = 0.53$$

The value of 0.53 reflects the degree of agreement between the three measures of territory quality.

2 For small samples ($N \leq 7$) the significance of W can be determined by comparing the value of S to the tabular value (Table A16) at the appropriate level of significance. Since the calculated S of 134 is smaller than the tabular value of $157_{0.05}$, we fail to reject the H_0 that the rankings by the three measures of territory quality are independent. That is, in order to show a significant association the H_0 must be rejected (i.e. the calculated S must be larger than the tabular S).

When $N > 7$, significance can be determined by calculating the chi-square value, as follows:

$$\chi^2 = \frac{S}{\frac{1}{12}K(N)(N+1)} = K(N-1)W$$

df $= N - 1$

The calculated chi-square value is then compared to the tabular value (Table A9).

Chi-square test of independence (r×k)

This test is used to determine whether there is a significant association between measures in two variables, with r samples of one variable and k samples of the other. It is calculated in the same way as was described previously for the 2×2 Chi-square test of independence, except that there is a larger number of cells.

Table 14.20. *The Number of song types shared between replacement males and previous owners; replacements and other males in the preceding year; and replacements and the previous male's neighbors*

	Number of songs shared			
	0	1	2	3
Previous owner	15	5	5	1
Other males	276	177	62	11
Neighbors	54	37	9	0

Notes:

$\chi^2_{6 \, df} = 6.57$, $P \approx 0.4$.

Figures given in the table are the numbers of comparisons between replacement males ($N=26$), previous owners of the territory ($N=26$), other males breeding in the same year as the previous owner (the number varies from year to year, range $= 17$–28), and neighbors of the previous males (varies from 2 to 7, $\bar{X} \pm {\rm SE} = 4.1 \pm 0.31$

Source: From McGregor and Krebs (1984). Copyrighted by Bailliere Tindall.

For example, McGregor and Krebs (1984) studied song learning and deceptive mimicry in great tits (*Parus major*). As part of their research they measured the number of songs shared by new territory owners with: 1. previous owners of the same territory; 2. all other males of the previous owner's breeding year; and 3. all neighbors of the previous owner. Their data form a 3×4 matrix of nominal data (Table 14.20).

Procedure review:

1 Calculate the row totals (RT), column totals (CT) and grand total (GT).

2 Calculate the expected values for each cell:

$$\text{Expected} = \frac{\text{RT} \times \text{CT}}{\text{GT}}$$

3 Calculate the χ^2 value for each cell from the cell's observed (O) and expected (E) values:

$$\text{Cell } \chi^2 = \frac{(O-E)^2}{E}$$

4 Obtain the total χ^2 by summing the individual cell χ^2s.

5 Compare the total χ^2 with the tabular value in Table A9.

14.2 RANDOMIZED BLOCK, MATCHED PAIRS AND REPEATED MEASURES DESIGNS

14.2.1 One variable

14.2.1a *Two matched samples*

$$\text{Samples}$$

Blocks	A_1	A_2
B_1	x_{11}	x_{21}
B_2	x_{12}	x_{22}
$\vdots$	$\vdots$	$\vdots$
B_n	x_{1n}	x_{2n}

In the *randomized block design*:

1 Each x_{ij} is a measurement from a different individual.
2 Individuals are blocked across (e.g. block B_1) according to some characteristic such as sex, age, litter, place or time.

In the *repeated measures design*:

1 Each block (e.g. B_1) is an individual.
2 Each x_{ij} is a measurement made on individual B_j in each sample (such as different treatments) A_i.

The *matched-pairs design* is a special case of either the *randomized block or repeated measures* designs in which there are two samples; hence, the matched pairs.

Sign test

The sign test is used when the measurements in the two samples are matched (i.e. blocked or paired). It tests for significant differences in form or location between the two samples' measurements. The sign test can be considered a first-approximation test which is less powerful than the Wilcoxon matched-pairs signed-rank test since it does not take into account the magnitude of the difference in the paired measurements. It is 95% as powerful as the paired *t*-test when the sample size is six, but power-efficiency decreases to an asymptote of 63% as sample size increases (Siegel, 1956). An illustration of the use of the sign test is found in Riechert and Hedrick's (1993) study of fitness-linked behavior traits in the spider, *Agelenopsis aperta*.

As another example, we will use data similar to those we compared with the standard error of the difference between the means, earlier in this chapter (Table 14.21).

Table 14.21. *Hypothetical data on song durations from two populations of birds and signs used in the sign test (see text)*

	Song duration (s)		
Bout No.	Population A sample	Population B sample	Sign
1	4.6	4.7	−
2	5.3	5.1	+
3	4.4	3.2	+
4	3.1	4.2	−
5	6.4	3.7	+
6	5.3	4.1	+
7	4.7	4.5	+
8	4.8	3.6	+
9	5.0	2.9	+
10	4.4	3.0	+

Procedure:

1 Score each pair of measurements as a plus ($+$) if the measurement in the left-hand column exceeds that in the right-hand column; score a minus ($-$) if the reverse is true, or score a zero (0) if there is no difference (see Table 14.21).

L=no. of times least frequent sign occurs=2
T=total no. of pluses and minuses=10

2 Determine from Table A17 the probability of obtaining L or fewer of the less frequent sign out of a total of T signs.

p=0.110 for L=2 and T=10

Since this probability is greater than the significance level decided upon (e.g. 95%=0.05), we conclude that the difference in song duration between the two populations is not significant.

Visual inspection of the data suggests that song duration for population A is really greater. Another way to test for significant differences between song durations from the two populations is to utilize a test for significant difference between the means that also takes into account the variability in the data. One such measure is

Table 14.22. *Hypothetical data on song duration for 16 adult male songbirds matched according to age and the ranks assigned for the Wilcoxon matched-pairs signed-ranks test (see text)*

Age (y)	Song duration (s) maintained in:		Differences	Rank
	Sound-enriched	Sound-impoverished		
1	8.2	6.2	−2.0	4
	7.1	4.3	−2.8	6
2	7.5	5.4	−2.1	5
	6.8	1.3	−5.5	8
3	7.8	7.9	0.1	1
	8.1	7.1	−1.0	3
4	6.9	2.0	−4.9	7
	7.4	8.0	0.6	2

the standard error of the difference between the means that we used on similar data, earlier in this chapter.

Wilcoxon matched-pairs signed-ranks test

This test, like the sign test (above), is used when the measurements in the two samples are matched (i.e. blocked or paired). It tests for significant differences in locations (medians) of the two samples' measurements. It is the nonparametric counterpart of the paired samples t-test (above). Mood (1954) concluded that this test is 95% as powerful as the paired t-test, but Blair and Higgins (1985) demonstrated that the power advantages of the paired t test over the Wilcoxon test are small even for normal distributions, and overall the Wilcoxon test is more often the more powerful test. For an example of its use in the ethology literature see Slotow *et al.* (1993).

As a hypothetical example, we will use 16 adult male songbirds which were matched according to age, four to each age group, and randomly assigned to one of two treatments: 1. maintained in a sound-enriched environment; and 2. maintained in a sound-impoverished environment. After they had been in the treatment conditions for two weeks we measured their song durations when placed individually in an observation room (Table 14.22).

Procedure:

1 Determine the differences between each pair of scores (right column score−left column score; shown in Table 14.22).

2 Rank the scores (ignoring the signs). The lowest score gets rank no. 1 (shown in Table 14.22).
3 Calculate T:

T = sum of ranks for differences with less frequent sign.
$T = 1 + 2 = 3$

4 Compare the calculated T to the tabular value (Table A18).
    where: N = number of paired scores = 8

Tabular $T_{0.05} = 4$

If the calculated T is less than or equal to the tabular value, then there is a significant difference between the two samples. Since our calculated T (3) is smaller than the tabular T (4), we reject the H_0 of no significant difference between the two samples and conclude that the type of sound environment had a significant effect on the birds' song duration.

McNemar's test

The McNemar test is a variation of the chi-square test used to determine the direction and extent of change in pairs of repeated measures (e.g. the same individuals are measured before and after treatment). For example, Tokarz (1985) used McNemar's test to analyze data on whether male brown anoles (*Anolis sagrei*) perched higher or lower in their cage, before and after encounters with larger and smaller males.

Procedure:
1 Cast the data into a 2×2 matrix as shown below.
    This creates two cells (B and C) which reflect change:

Cell B (+ to −) and Cell C (− to +)

		After	
		+	−
Before	+	A	B
	−	C	D

As an example, we will generate hypothetical data for Tokarz's (1985) experiment on lizards (see Table I in Tokarz). The matrix below is for the smaller males of the pairs. It shows the number of males on (+) or off (−) the perch, before the test and 15 min. after removal of the partition between the two males.

After 15 min.

		On Perch	Off Perch
		+	−
Pre-test	On perch +	$A=19$	$B=17$
	Off perch −	$C=5$	$D=1$

2 Calculate the χ^2:

$$\chi^2 = \frac{[(B-C)-1]^2}{B+C}$$

$$= \frac{[(17-5)-1]^2}{17+5}$$

$$= \frac{121}{22} = 5.5$$

3 Compare the calculated χ^2 to the value in Table A9, where df=(no. of rows−1)×(no. columns−1)=1. Since our calculated χ^2 (5.5) is larger than the tabular value at $P=0.05$ (3.84) we reject the H_0 of no significant difference and concur with Tokarz's (1985:749) conclusion that 'A significant change occurred in the perching location of the smaller males from before the test to 15 min. after removal of the partition . . .'

McNemar's test is 95% as powerful as the paired t-test when the sum of the A and D cells is six, but the power-efficiency decreases to an asymptote of 63% as the sum of A and D increases (Siegel, 1956).

Measures of association

These analyses are used to test for the association between paired measurements in two samples. The measurements might be made on individuals of the same age or at the same places and/or times (*randomized block*: each measurement on a different individual) or one measurement in each sample can be made on the same individual (*repeated measures*).

We can also examine the correlation between paired measurements in two samples in which Sample 1 is the independent variable and Sample 2 is the dependent variable. These are *matched pairs* in which a measure in Sample 2 (dependent variable) is obtained for each measure in Sample 1 (independent variable). Researchers should first prepare a scattergram of their data for visual inspection. Scattergrams provide a good indication of: 1. whether a correlation exists; and 2. whether it is positive or negative.

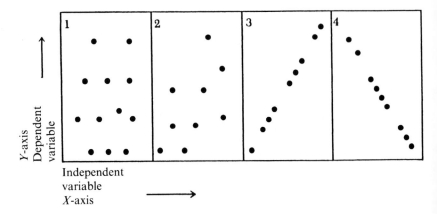

Fig. 14.2 Scattergrams of hypothetical data illustrating: (1) no correlation; (2) low positive correlation; (3) perfect positive correlation; and (4) perfect negative correlation.

Figure 14.2 illustrates how the data are plotted with the dependent variable on the Y-axis and the independent variable on the X-axis. The examples illustrate: 1. no correlation; 2. low positive correlation; 3. perfect positive correlation; and 4. perfect negative correlation.

The two correlations described below, Kendall's tau and Spearman's rho, are both used with ordinal, interval or ratio data. Both correlations generally have the same power to detect an association (Siegel, 1956). However, if there are no tied ranks Spearman's rho is the preferred statistic since it uses both the direction and the difference in magnitude of the ranks. If there are several tied ranks use Kendall's tau (Norman and Streiner, 1986). Both Kendall's tau and Spearman's rho are approximately 91% as powerful as Pearson's product moment correlation coefficient (Hotelling and Pabst, 1936).

Spearman's rho   Spearman's rho (ρ_s) is a nonparametric correlation coefficient which measures the covariation between two rank-ordered variables. Only an ordinal scale of measurement is necessary, as it is with Kendall's tau.

As an example, we will use the hypothetical data on activity level and frequency of song in males of a species of territorial songbird in the Table 14.23. This data set consists of matched pairs of measurements of two dependent variables: 1. activity units (flights/min.); and 2. song frequency (songs/ min.). We will assume that the data are medians from several one-hour samples for each bird. The data are similar to the example we used for Pearson's product moment correlation (section 13.2.2) except that data were in the ratio scale of measurement.

We believe that these two dependent variables are correlated so we begin by constructing a scattergram (Figure 14.3). From the scattergram alone, there would

Table 14.23. *Hypothetical data on activity level and song frequency in males of a territorial songbird species*

Individual	Activity units (flights/5 min.)	Song frequency (songs/min.)
A	4	6
B	7	8
C	12	14
D	3	5
E	18	16
F	15	11
G	9	13

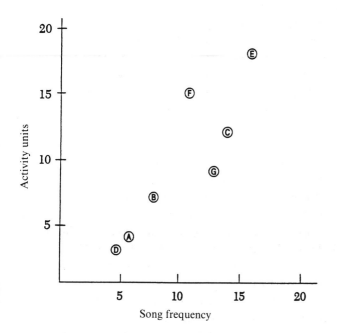

Fig. 14.3 Scattergram of hypothetical data on songbird activity and song frequency. (See text for explanation.)

Table 14.24. *Ranks of measurements from the data in Table 14.23 for calculation of Spearman's rho (see text)*

| Individual | X variable | | Y variable | |
	Activity units	Rank	Song frequency	Rank
A	4	2	6	2
B	7	3	8	3
C	12	5	14	6
D	3	1	5	1
E	18	7	16	7
F	15	6	11	4
G	9	4	13	5

appear to be a positive correlation; that is, as activity increases, song frequency increases. However, we can proceed with the calculation of Spearman's rho and determine whether it is a statistically significant correlation.

Procedure:

1 Rank the measurements for each variable, arbitrarily assigned X and Y (Table 14.24).

   If two or more measurements within a variable are equal, then either assign to each of them the average of the ranks that would have been assigned had they not been equal, or use Kendall's tau (below), especially if there are several ties.

2 Determine the difference between the ranks for each pair of measurements (d) and calculate the square of that difference (d^2) (Table 14.25).

3 Determine the total for the d^2s (as in Table 14.25).

4 Calculate Spearman's rho:

$$p_s = 1 - \frac{6(\Sigma d^2)}{N^3 - N}$$

where: N = no. of paired measurements = 7

$$p_s = 1 - \frac{6(6)}{343 - 7} = 1 - \frac{36}{336} = 1 - 0.107 = 0.89$$

5 When $N \le 50$:

Compare the calculated rho (0.89) with the tabular value (Table A19) for N = no. of paired measurements = 7. Note that this is a one-tailed test.

Table 14.25. *Differences between the ranks of the measurements in Table 14.24 for calculation of Spearman's rho (see text)*

Individual	X variable rank	Y variable rank	d	d^2
A	2	2	0	0
B	3	3	0	0
C	5	6	1	1
D	1	1	0	0
E	7	7	0	0
F	6	4	2	4
G	4	5	1	1
				Total = 6 = Σd^2

Since our calculated rho of 0.89 is larger than the tabular value ($0.714_{0.05}$), we reject the H_0 of no correlation and conclude that there is a statistically significant correlation between activity and song frequency.

Alternatives when N is >20:

a Convert rho to the 't' statistic:

$$t = \text{rho} \sqrt{\left(\frac{N-2}{1-\text{rho}^2}\right)}$$

Compare the calculated 't' with the tabular value (Table A5), where: df$=N-2$. If the calculated t is larger than the tabular t, at the appropriate P value, then the H_0 of no correlation is rejected. Although some statisticians have recommended always converting rho to 't', Siegel and Castellan (1988) recommend using Table A19 whenever N is less than 50.

b Convert rho to the 'z' statistic:

$$z = \text{rho} \sqrt{N-1}$$

For a two-tailed test, rho is significant at the 0.05 level if z is >1.96. For a one-tailed test, rho is significant at the 0.05 level if z is >1.64.

Kendall's tau Like Spearman's rho, Kendall's tau determines the tendency of two rank orders of data to be similar. Kendall's tau is preferred when there are several tied ranks.

As an example, we will use the same hypothetical data that we used with

Table 14.26. *Arrangement of the ranks from Table 14.24 for calculation of Kendall's tau (see text)*

Individual	Column 1 Activity units (flights/5 min.)	Column 2 Song frequency (songs/min.)
D	1	1
A	2	2
B	3	3
G	4	5
C	5	6
F	6	4
E	7	7

Spearman's rho in order to compare the two measures of association. Although they are considered to be equally powerful, Kendall's tau will yield a lower coefficient value than Spearman's rho, on the same data, because they have different underlying scales.

Procedure:

1 Arrange the ranks in order along one of the variables (i.e. put them in rank order in Column 1). In our example (Table 14.26) we will rank order them by the activity units, but either could be used.

2 For each individual (A through G) count the number of individuals below it in the table with both larger activity unit ranks and song frequency ranks (Table 14.27). Do not count ties. For example, below individual A there are five individuals (B, G, C, F, and E) that have both larger activity unit ranks and song frequency ranks.

   Another method is to use the scattergram (Figure 14.3) and for each individual count the number of individuals which have larger X and Y variable measurements (i.e. those individuals to the right and above in the scattergram) (Table 14.27).

3 Calculate the total (S^+): from Table 14.27 = 19

4 Calculate S:

$$S = 2S^+ - \frac{N(N-1)}{2}$$

where: N = no. of pairs of scores = 7

Table 14.27. *Calculations for Kendall's tau using data from the scattergram (Figure 14.3; see text)*

Individual	Number of individuals with higher ranks in both variables
A	5
B	4
C	1
D	6
E	0
F	1
G	2

$$S = 2(19) - \frac{7(7-1)}{2} = 38 - \frac{42}{2} = 17$$

If there are no tied ranks in either column continue with Step 5; if there are tied ranks go to Step 5A

5 Calculate tau:

$$tau = \frac{S}{N(N-1)/2} = \frac{17}{7(7-1)/2} = \frac{17}{21} = 0.81$$

As predicted the tau of 0.81 is smaller than the rho of 0.89 calculated above on the same data.

6 Compare the calculated tau to the tabular value (Table A20). If the calculated value is more than the tabular value then the H_0 of no significant correlation is rejected. Since our calculated tau (0.81) is larger than the tabular tau (0.63), we reject the H_0 of no significant correlation and conclude that there is a significant positive correlation between activity and frequency of song. The size of the calculated tau is, of course, a measure of the strength of the correlation, with a tau of 1.00 indicating a perfect correlation. When tau is positive the relationship is direct (i.e. the variables increase together). An inverse relationship is indicated by a negative tau.

Calculations when there are tied ranks in either column:

5A For Column 1 determine the number of measurements that are tied, and categorize each tie as being a set of two, three, etc. For each set of ties calculate:

$$X(X-1)$$

where: $X =$ the number of measurements tied (i.e. set of two, three, etc.)

6A   Add the $[X(X-1)]$ for each set of ties together and divide by 2 to obtain R_1:

$$R_1 = \frac{\Sigma[X(X-1)]}{2}$$

7A   Calculate T_1:

$$T_1 = \frac{N(N-1)}{2} - R_2$$

8A   Repeat steps 5A and 6A for Column 2, and calculate T_2.

$$T_2 = \frac{N(N-1)}{2} - R_2$$

9A   Calculate W:

$$W = \sqrt{V} \qquad \text{where: } V = T_1 \times T_2$$

10A   Obtain the value for S from Step 4 above.

11A   Calculate tau:

$$\text{tau} = \frac{S}{W}$$

12A   Compare the calculated tau with the tabular tau (Table A20), as in Step 5 above.

Other indices of association   There are several other indices of association between paired measurements in two samples. Four examples of coefficients of association are given below (from V. DeGhett, pers. commun.). Each coefficient will be calculated from the following hypothetical data:

Frog A vocalizes?	Frog B vocalizes?
Yes	Yes
No	Yes
Yes	Yes
Yes	Yes
No	No
No	No
Yes	No
No	No
Yes	Yes

		Frog B	
		Yes	No
		a	*b*
Frog A	Yes	4	1
		c	*d*
	No	1	3

$$M=a+d=4+3=7$$
$$U=b+c=1+1=2$$
$$N=M+U=7+2=9$$

Coefficients of association:

Dice's:

$$\frac{2a}{2a+U}=\frac{2(4)}{2(4)+2}=0.80$$

Yule's:

$$\frac{ad-bc}{ad+bc}=\frac{4(3)-1(1)}{4(3)+1(1)}=\frac{11}{13}=0.85$$

Jaccard's:

$$\frac{a}{a+U}=\frac{4}{4+2}=0.67$$

Sokal and Michener's:

$$\frac{M}{N}=\frac{7}{9}=0.78$$

The different values these coefficients yield from the same data reflect each researcher's concept of how an association should be measured.

Interpretation of an association   A significantly large tau or rho indicates a high degree of association whether positive or negative. There are basically three interpretations for a high degree of association (i.e. high correlation):

1 Change in variable A causes a change in variable B.
2 Change in variable B causes a change in variable A.
3 Neither interpretations 1 or 2 are correct, but rather a change in some other variable (C) causes a change in both A and B.

The correlation coefficients will not tell us which of these interpretations is correct. We can only judge (with varying degrees of validity) which is correct according to our prior knowledge of the variables and their relationships.

14.2.1b Three or more matched samples

	Samples		
	A_1	A_2	A_3 ... A_n
B_1	x_{11}	x_{21}	x_{31} ... x_{n1}
B_2	x_{12}	x_{22}	x_{32} ... x_{n2}
⋮	⋮	⋮	⋮ ⋮⋮⋮ ⋮
B_n	x_{1n}	x_{2n}	x_{3n} ... x_{nn}

Friedman's two-way analysis of variance

The Friedman two-way analysis of variance is a nonparametric test used to determine if several (three or more) samples with blocked measurements or repeated measures on the same individuals are significantly different. It requires ordinal data (at least) and tests for differences in location. Compared to the parametric ANOVA, it is 72% as powerful for three samples and 87% as powerful for ten samples (van Elteren and Noether, 1959).

As an example, we will again sample song durations, from four habitats; but this time we will block our samples into three time periods: sunrise, noon, and sunset (Table 14.28).

Procedure:
1 Rank the measurements for each row (blocks) separately (Table 14.29). This minimizes the between-row differences and provides a more powerful test of the between-column (sample; habitat) differences.
2 Calculate each of the rank sums (i.e. column totals; R_J, Table 14.29).
3 Calculate each of the R_J^2's.

$$R_A^2 = 25 \qquad R_B^2 = 81 \qquad R_C^2 = 144 \qquad R_D^2 = 16$$

4 Sum the R_J^2s:

$$\Sigma R_J^2 = 25 + 81 + 144 + 16 = 266$$

5 Calculate χ_r^2 according to the formula:

$$\chi_r^2 = \left[\frac{12}{kN(N+1)} \Sigma R_J^2 \right] - 3k(N+1)$$

where:
   k = number of rows, or the number of times the rank order system was used
   N = number of colums, or number of ranks

Table 14.28. *Hypothetical data on song duration of brids from four habitats blocked into three time periods*

	Mean song duration (s)			
	Habitat A	Habitat B	Habitat C	Habitat D
Sunrise	6.7	8.6	10.2	6.5
Noon	3.4	4.3	6.5	2.5
Sunset	6.3	8.3	9.9	7.1

Table 14.29. *Ranks of each row of measurements in Table 14.28 for calculation of Friedman's two-way analysis of variance (see text)*

	Habitat A	Habitat B	Habitat C	Habitat D
Sunrise	2	3	4	1
Noon	2	3	4	1
Sunset	1	3	4	2
	$R_A = 5$	$R_B = 9$	$R_C = 12$	$R_D = 4$

$$\chi_r^2 = \left(\frac{12}{12(5)} \times 266\right) - 9(5) = 53.20 - 45 = 8.20$$

6 Compare the calculated χ_r^2 to the tabular value (Table A9) for $df = N - 1 = 2$

tabular $\chi_{0.05}^2 = 5.99$

Since our calculated χ_r^2 (8.20) is larger than the tabular value (5.99), we reject the H_0 of no difference in song durations in birds from the different habitats and conclude that there is a significant difference in song duration in birds from the different habitats.

Table 14.30 illustrates Friedman's two-way analysis of variance with replications:

Table 14.31 is the result of ranking the measurements from Table 14.30.

$$\Sigma R_j^2 = 196 + 676 + 1225 + 225 = 2322$$

$$\chi_r^2 = \left[\frac{12}{kN(N+1)}\Sigma R_j^2\right] - 3k(N+1) = \frac{12}{36(5)}2322 - 27(5)$$

Table 14.30. *Hypothetical data on song duration (in seconds) of birds from four habitats blocked into three time periods illustrating Friedman's two-way analysis of variance with replications*

	Mean song duration (s)			
	Habitat A	Habitat B	Habitat C	Habitat D
Sunrise	6.7	8.6	10.2	6.5
	6.4	6.9	8.4	5.8
	5.5	7.0	9.8	5.9
Noon	3.4	4.8	4.6	2.5
	2.7	4.7	7.5	0.5
	4.1	3.9	7.4	2.9
Sunset	6.3	8.3	9.9	7.1
	5.8	7.0	11.1	6.1
	4.9	6.6	8.1	6.7

Table 14.31. *Ranks of each row of measurements in Table 14.30 for calculation of Friedman's two-way analysis of variance with replications (see text)*

	Habitat A	Habitat B	Habitat C	Habitat D
Sunrise	2	3	4	1
	2	3	4	1
	1	3	4	2
Noon	2	4	3	1
	1	3	4	2
	3	2	4	1
Sunset	1	3	4	2
	1	3	4	2
	1	2	4	3
	$R_A = 14$	$R_B = 26$	$R_C = 35$	$R_D = 15$
	$R_A^2 = 196$	$R_B^2 = 676$	$R_C^2 = 1225$	$R_D^2 = 225$

$$= 154.88 - 135.00 = 19.88$$

tabular $\chi_{0.05}^2 = 15.51$

Since our calculated χ_r^2 (19.88) is larger than the tabular value (15.51), we reject the H_0 that there is no significant difference in song duration between birds in the four habitats.

Dunnett's multiple comparison test   This test can be used to determine which pairs of samples differ significantly when the null hypothesis is rejected in Friedman's two-way analysis of variance test (Glantz, 1992). When the two samples have unequal sample sizes use Dunn's Test (section 14.1.1).

Procedure:

1 Calculate q:

$$q = \frac{|R_1 - R_2|}{\sqrt{\left[\dfrac{pn(p+1)}{6}\right]}}$$

Where:

$|R_1 - R_2|$=the absolute difference between the rank sums for the two samples being compared.

p=the number of samples spanned by the comparison after ranking the rank sums in order (ascending or descending)

n=the number of blocks in each sample, or the number of individuals in a repeated measures design

Using the same example we used for the Friedman's two-way analysis of variance (above), we will compare the samples from Habitats A and C.

$| R_1 - R_2 | = R_C - R_A = 12 - 5 = 7$ (From Table 14.29)

Rank of rank sums:

(1) $R_C = 12$ (2) $R_B = 9$ (3) $R_A = 5$ (4) $R_4 = 4$

$p = 3$ $df = \infty$

$n = 3$

$$q = \frac{7}{\sqrt{\left[\dfrac{3 \times 3(4)}{6}\right]}} = \frac{7}{\sqrt{6}} = 2.86$$

Table 14.32. *Hypothetical data on the number of trials in which each of five juvenile coyotes howled in response to each of three treatments (see text for explanation)*

Individuals	Treatments (samples)			Row Totals (RT)
	A_1	A_2	A_3	
1	4	3	0	7
2	4	2	1	7
3	5	2	0	7
4	5	3	0	8
5	4	2	1	7
Column totals (CT)	22	12	2	36=Grand total (GT)

2 Compare the calculated q to the tabular value (Table A21) for p comparisons. Since our calculated q (2.86) is larger than the tabular q at $P=0.05$ (2.21), we reject the H_0 of no significant difference between the two samples and conclude that there is a significant difference in mean song durations between songbirds in Habitats A and C.

Cochran's Q-test

Cochran's Q-test is an extension of McNemar's test, described earlier in this chapter. It tests whether three or more samples of repeated measures differ among themselves. It is used with nominal data in the form of frequencies or proportions in a randomized block or repeated measures design.

As a hypothetical example, we will assume we have measured whether juvenile coyotes (*Canis latrans*) (B_{1-5}) howled in response to playbacks of howls of: littermates (A_1), adults of their pack (A_2), and adults of a distant pack (A_3). We gave each juvenile coyote 5 trials of each treatment (A_{1-3}) when it was alone and measured whether it howled in response, or not (Table 14.32).

Procedure:

1 Calculate Q:

$$Q=\frac{(k-1)[(k)(SS_{CT})-GT^2]}{(k)(GT)-SS_{RT}}$$

where:

k=number of columns (i.e. Treatments $A_1 \ldots A_n$)

df=k−1

SS_{CT} = sum of squares of column totals
$$= \Sigma(CT_1{}^2 + CT_2{}^2 \ldots CT_n{}^2)$$

SS_{RT} = sum of squares of row totals
$$= \Sigma(RT_1{}^2 + RT_2{}^2 \ldots RT_n{}^2)$$

For our example:

$k = 3$

$df = 2$

$SS_{CT} = 22^2 + 12^2 + 2^2$
$$= 484 + 144 + 4 = 632$$

$SS_{RT} = 7^2 + 7^2 + 7^2 + 8^2 + 7^2$
$$= 49 + 49 + 49 + 64 + 49 = 260$$

$$Q = \frac{(k-1)[(k)(SS_{CT}) - GT^2]}{(k)(GT) - SS_{RT}}$$

$$= \frac{2[3(632) - 1296]}{(3)(36) - 260} = \frac{1200}{152} = 7.894$$

2 Compare the calculated Q to the tabular chi-square value (Table A9).
   Since our calculated Q (7.894) is larger than the tabular value at $P = 0.05$
   (5.99) we reject the H_0 of no significant difference between the treatments
   and conclude that the juvenile coyotes' howling responses to the different
   playbacks were significantly different.

15 Rates of behavior and analysis of sequences

15.1 RATES OF BEHAVIOR

Temporal patterns of behavioral events are generally described in terms of bouts (Chapter 6), time between successive events (period), or number of events per specified interval of time (rate), with rate being inversely related to period (Gaioni and Evans, 1984).

Rates of behavior are usually calculated from data collected by all-occurrences sampling (Chapter 8); however, Altmann and Wagner (1970) described a method, based on the Poisson distribution, for estimating rates from one–zero samples when the sample period duration is small (also see Chapter 8).

The intuitive method for determining the rate of a behavior from continuous samples is to divide the number of occurrences of the behavior by the length of the sample period. This provides a valid measure of rate for behaviors that occur at a steady rate (i.e. the periods between successive events are the same or very similar). For most behaviors, however, the events will not occur at a steady rate and better measures are needed.

Analysis of repetition rate of avian vocalizations provides a good example, which will be applicable to other behavioral acts, as well. Gaioni and Evans (1984) noted that two methods of calculating repetition rate from period have been traditionally used. First, localized estimates of rate are calculated from individual periods, and these localized rates are averaged to provide an overall estimate of repetition rate. Second, an average period length is calculated and this is used to calculate overall rate. It is this second method that was recommended by Scoville and Gottlieb (1978), as follows:

$$\text{Repetition rate } (R; \text{notes per second}) = \frac{1}{d+g}$$

where: d = average note duration
g = average inter-note interval

Gaioni and Evans (1984) argued that temporally patterned acts (e.g. stimuli used for communication) should be described in terms of period, rather than repetition rate; however, Miller and Blaich (1984) argued that for acts segregated into discrete bouts, repetition rate is the better measure.

More complex problems associated with the analysis of *rates of behavior* were addressed by Altmann and Altmann (1977). They identified the following questions that commonly arise when dealing with rates of behavior:

1 Can the frequency distribution of the observations be accounted for by the population composition (age and sex class distribution)?
2 What are the expected values of these frequencies if the mean rate of behavior is independent of class?
3 How can reasonable estimates of class-specific behavior rates be obtained from a set of data and tested in a new sample?
4 What are the expected frequencies, for dyadic interactions, with pairwise independence?

a. The expected frequency of behavior per class when the rates are unknown but are assumed to be constant or independent of class, and the population composition is stable is calculated according to the following formula:

$$E_a = \frac{Nm_x}{M}$$

where: E_a = expected frequency of behavior a for members of class x
N = total number of occurrences of behavior a for all individuals of all classes in the sample
m_x = number of individuals in class x
M = number of individuals in the sampled population

In order to illustrate the use of the formula we will use the hypothetical data on the number of threats in a herd of 50 deer in Table 15.1. See Altmann and Altmann (1977) for an additional example.

We can determine the expected frequency of threat behavior in adult males as follows:

$$N = 100$$

$$M = 50$$

$$m_x = 15$$

$$E_a = \frac{Nm_x}{M} = \frac{100 \times 15}{50} = \frac{1500}{50} = 30$$

Therefore, the expected frequency of threats in adult males is 30 compared to the observed frequency of 80. This can be done for all age classes, and then a chi-square test can be conducted to determine whether the differences between the observed and expected frequencies are statistically significant.

Table 15.1. *Hypothetical data on the number of threats in a herd of 50 deer*

	Adult males	Adult females	Immatures	Total
Number of individuals	15	20	15	50
Observed threats	80	15	5	100

b. The expected frequency of behavior per class when the the conditions are the same as in (a) (above), except that the population composition changes, is calculated according to the following formula:

$$E_a = N \frac{\Sigma_j t_j m_{xj}}{\Sigma_i \Sigma_j t_j m_{ij}}$$

where:

E_a = expected frequency of behavior a for members of class x

N = total number of occurrences of behavior a for all individuals of all classes in the sample

$\Sigma_j t_j m_{xj}$ = total sample time for all individuals of class x in the entire study

= (time in sample period one × number of individuals in class x) + (time in sample period two × number of individuals in class x) + ... (time in final sample period × number of individuals in class x)

$\Sigma_i \Sigma_j t_j m_{ij}$ = total sample time for all individuals of all classes for the entire study

= (time in sample period one × number of individuals in all classes) + (time in sample period two × number of individuals in all classes) + ... (time in final sample period × number of individuals in all classes)

As an illustration, we will use the hypothetical data in Table 15.2 (see Altmann and Altmann, 1977, for another example). We can determine the expected frequency of threats in adult males in this population of changing composition as follows:

N = 260

$\Sigma_j t_j m_{xj}$ = [(5×10)+(11×15)+(4×12)]
= 50+165+48=263

$\Sigma_i \Sigma_j t_j m_{ij} = 263 + [(5×18)+(11×20)+(4×16)]$

$$+[(5\times 13)+(11\times 15)+(4\times 12)]$$
$$=263+374+278=915$$

$$E_a \quad =N\frac{\Sigma_j t_j m_{xj}}{\Sigma_i \Sigma_j t_j m_{ij}}=260(263/915)=75$$

Therefore, the expected frequency of threats in adult males in this changing population is 75 compared with the observed number of 210. Once again we could make the same calculations for the adult females and immatures, and then use a chi-square test to determine if the differences are significant.

Uniform, class-specific rates of behavior are expressed as a mean number of occurrences per individual per unit time, according to the formula below:

$$E_a \quad =\Sigma_x \Sigma_j t_j m_{xj}$$

E_a = the hypothetical expected frequency of behavior a for members of class x

Lambda$_x$ = hypothetical mean participation rate per member of class x

$\Sigma_j \Sigma_j m_{xj}$ = total sample time for all individuals of class x in the entire study

As an illustration we can obtain the hypothetical mean participation rate (lambda) for adult males by referring to Table 15.1.

$$\text{Lambda}_x = \frac{80}{100}=0.80$$
$$\Sigma_j t_j m_{xj} \quad =(5\times 10)+(11\times 15)+(4\times 12)=263$$
$$E_a \quad =0.80\times 263=210$$

Therefore, the hypothetical expected frequency of threats for adult males during the entire sample period (20 hours, Table 15.2) is 210. The total number of males observed was 37, so that the mean rate per individual was $210/37=5.7$ threats for the 20-hour sample. This hypothetical mean rate can then be compared to the observed mean rate in this sample or from observations gathered later from the same or a different population.

Altmann and Altmann (1977) proceed to the consideration of interactions between individuals. They provide procedures for calculating expected rates of behavior for symmetric and asymmetric interactions at constant rates and interactions with hypothetical class-specific rates of behavior. Michener (1980) noted that the assumption in Altmann and Altmann's (1977) formulae that for any given period of time any individual has the potential to interact with any other individual, while true for gragarious species, is not true for many species where certain individu-

Table 15.2. *Hypothetical data on the number of threats in a herd of mule deer*

	Adult males	Adult females	Immatures	Totals
Sample period 1 (5 h)				
Number of individuals	10	18	13	41
Observed number of threats	60	13	2	75
Sample period 2 (11 h)				
Number of individuals	15	20	15	50
Observed number of threats	80	15	5	100
Sample period 3 (4 h)				
Number of individuals	12	16	12	40
Observed number of threats	70	14	1	85
Total number of observed threats (Sample periods 1, 2 and 3)	210	42	8	260

als never occupy the same space. Michener (1980) goes on to point out the importance of choosing the correct denominator for the formulae (i.e. the number of individuals, or number of dyads).

> Rates expressed as the number of social acts observed per animal per unit time are appropriate when information is required on the proportion of time that individuals spend behaving socially. . . . When information is required on interaction rates, numbers of interactions must be expressed relative to the number of pairs of animals available to interact.
> *[Michener, 1980: 382]*

The procedures for determining interaction rates (Altmann and Altmann, 1977; Michener, 1980) are somewhat complex extensions of the formulae for individual rates of behavior, discussed above, and will not be described here.

15.2 ANALYSIS OF SEQUENCES

The temporal relationships between two, or more, behaviors (from the same or different individuals) are often complex and difficult to analyze. For example, in some bird species the male and female of a pair sing separate portions of a duet, either simultaneously (polyphonically) or successively (antiphonally). However, the temporal patterning of each individual's contribution to the duet is quite constant.

The following discussion of behavior sequence analysis is cursory and meant only as an overview. More detailed discussions of the general topic of sequence analysis, including in-depth descriptions of specific methods not discussed in this book, can be found in: Bakeman and Gottman (1986), Castellan (1979), Fagen and Young (1978), Gottman and Roy (1990), Haccou and Meelis (1992), and Van Hoof (1982).

Golani (1976) developed a vocabulary to describe temporal patterns for two limb segments in his analysis of social interactions in golden jackals and Tasmanian devils. This same terminology can be applied to behavioral units performed by one or more individuals. This terminology has not received wide acceptance and use, but is included here as a descriptive format which is available for use or modification. Golani considered all of these relationships to be variations of 'simultaneous' events, even when one immediately preceded (prevened) or succeeded (supervened) another.

We will use Golani's definitions; but let p and q stand for two behavioral units:

1. If p and q are temporally contiguous, i.e. follow each other immediately, the relationships will be designated by the suffix '-vene'.
2. If during every 'instant' of the occurrence of p, q occurs, i.e. p either starts together with or later than the start of q, and p ends together with or earlier than the end of q, the relationship will be designated by the suffix '-dure'.
3. If only during the later part of the occurrence of p, q or part of it also occurs, i.e. p starts before q and ends after the start of q but before or together with the end of q, the relationship will be designated by the suffix '-vade'.
4. If p starts after or together with the start of q, but before the end of q, and ends after the end of q, the relationship will be designated by '-cede'.
5. If q occurs only during the middle part of p, i.e. p starts before the start of q and ends after the end of q, the relationship will be designated by the suffix '-case'.

Variants within the five groups are distinguished by the prefixes pre- (before) and super- (after); in- (going in) and ex- (going out); pri- (prior), and post- (later); con- (together); end- (within); and en- (around). (Golani, 1976)

These prefixes and suffixes can then be combined to describe the various temporal relationships shown in Table 15.3.

Fentress and Stillwell (1973) used a grammar to describe the hierarchial organization of sequences of behavioral acts used in face grooming sequences by mice. They explained the principle, as follows:

Table 15.3. *Vocabulary of terms used to describe the temporal patterning of two behavioral units*

	p ends just before *q* starts (contiguous)	*p* ends after *q* started but before *q* ended	*p* ends together with *q*	*p* ends after *q* ended
p starts before *q* started	prevene	invade	convade	encase
p starts together with *q*		pridure	condure	concede
p starts after *q* started but before it ended		entdure	postdure	excede
p starts immediately after *q* ended (contiguous)				supervene

Source: Adapted from Golani (1976).

> We suggest that an analogy to human grammar, in which individual
> letters form different combinations in different words which in turn are
> sequentially arranged into phrases, may provide a more realistic model
> of control than attempts to explain coding simply on an element by
> element level.
> *[Fentress and Stillwell, 1973:53]*

Likewise, Rodgers and Rosebrugh (1979) described a procedure for finding grammars (i.e. theories of sequence generation) for sequences of discrete, mutually exclusive behavioral acts. They describe the use of two computer programs which assist in finding an appropriate grammar for specific sequences.

Most researchers have found it sufficient to examine sequences of behavior in a more general way than those proposed by Golani (1976), Fentress and Stillwell (1973), and Rodgers and Rosebrugh (1979). For example, Bakeman (1978) has found it useful to make two basic distinctions in his data. First, are the behavior patterns sequential (mutually exclusive) or concurrent? Second, is the behavior recorded in an event or time base; that is, are they occurrences or occurrences plus durations, respectively? These two dichotomies combine to form four types of data:

1 Type I data (event base, sequential). The observer records the order of mutually exclusive behavior units, ignoring their duration (Sackett, 1978; Van Der Kloot and Morse, 1975).

2 Type II data (event base, concurrent). The observer records the order of behavior units, ignoring their duration; however, they can occur simultaneously (S.A. Altmann, 1965).

3 Type III data (time base, sequential). The observer records the order and duration of mutually exclusive behavior units (Dane and Van Der Kloot, 1964).

4 Type IV data (time base, concurrent). The observer records the order and duration of behavior units; however, they can occur simultaneously (Verberne and Leyhausen, 1976).

The ethologist may suspect that behaviors occur in rather stereotyped sequences which may be the result of: 1. a common causal factor; 2. one stimulating the other; 3. differing thresholds; or 4. one priming the other. Predictable sequences are to be expected since animals are certainly not random behavior generators. However, it is the extent of predictability that A will be quickly followed by B that is important to our analysis.

15.2.1 Intra-individual sequences

A simple form of analysis is to measure the frequencies or conditional probabilities associated with two or more behaviors. If the conditional probability that A is followed by B is 100%, then it is a deterministic sequence. This is rare in behavior. More often A is followed by B at some level of probability less than 100%. This is called a probabilistic sequence (or stochastic sequence).

$$
\text{Deterministic sequence (rare)} \qquad A \xrightarrow{\quad 100\% \quad} B
$$

$$
\begin{array}{l}
\text{Probabilistic sequence} \\
\text{(stochastic sequence; common)}
\end{array} \qquad A \xrightarrow{\quad <100\% \quad} B
$$

Kinematic graphs (Chapter 18) can be generated which show the conditional probabilities of several different behaviors. This is a useful procedure when sequential effects are strong.

There are several explanations for why one behavior is often found to follow another (see above). Another factor is the relative frequency with which particular behaviors occur in the animal's repertoire. The more frequent the two behaviors, the more likely they are to occur together. An extreme example of this is a hypothetical animal which is capable of only two behaviors, A and B. Only the four sequences given below are possible.

1 A⟶A
2 A⟶B
3 B⟶B
4 B⟶A

The frequency of occurrence of these sequences is determined not only by the size of the repertoire, but also by impossible combinations and the observer's criterion for determining the beginning and end of a sequence. It is also often difficult to determine whether a behavior was repeated (A⟶A) or whether it was simply a single occurrence (A).

15.2.1a *Markov chains*

The sequences discussed above considered only single transitions from one behavior to another. They are referred to as dyads. Higher-level sequences are shown below:

Dyad   A⟶B
Triad   A⟶B⟶C
Tetrad   A⟶B⟶C⟶D

Markov chains are sequences of behaviors in which it can be shown that the transitions between two or more of the behaviors are dependent on one another at some level of probability greater than chance; also, the level or probability is assumed to be stationary (see below). Markov chain analysis should take into account the expected frequency of occurrence of sequences based on the frequency of occurrence in the repertoire.

Ashby (1963:165–166) has provided the following illustrative example of the application of Markov chains.

> Suppose an insect lives in and about a shallow pool – sometimes in the water (W), sometimes under pebbles (P), and sometimes on the bank (B). Suppose that, over each unit interval of time, there is a constant probability that, being under a pebble, it will go up on the bank; and similarly for the other possible transitions. (We can assume if we please, that its actual behaviour at any instant is determined by minor details and events in its environment). Thus a protocol of its positions might read:
>
> W B W B W P W B W B W B W P W B B W B W P W B W P W
> B W B W B B W B W B W B W P P W P W B W B B B W
>
> Suppose, for definiteness, that the transition probabilities are as shown below.

↓	B	W	P
B	1/4	3/4	1/8
W	3/4	0	3/4
P	0	1/4	1/8

> These probabilities would be found . . . by observing the animal's behavior over long stretches of time, by finding the frequency of, say B——→W, and then finding the relative frequencies, which are the probabilities. Such a table would be, in essence, a summary of actual past behaviour, extracted from the protocol.

Note that Markov chains are referred to in 'orders' rather than dyads, triads, etc. (Table 15.4).

What constitutes a sufficient *sample size* for analyzing Markov chains? Fagen and Young (1978) provided the following 'rule-of-thumb' (based on simulations run by Fagen) for analyses of first-order Markov chains (Table 15.4):

> For: R = number of individual behaviors for which sequences will be measured
>
> Then: $2R^2$: insufficient sample size
> $5R^2$: borderline sample size
> $10R^2$: sufficient sample size

Table 15.4. *Orders of Markov chains*

Order of Markov chain	Definition
'zeroth-order' (A,B,C)	The behavioral events are independent
first-order ($B \rightarrow C$)	The probability of occurence of a particular behavior is dependent on only the immediately preceding behavior
second-order ($A \rightarrow B \rightarrow C$)	The probability of occurence of a particular behavior is dependent on the two immediately preceding behaviors
nth-order ($\ldots X \rightarrow Y \rightarrow Z$)	The probability of occurence of a particular behavior is dependent on the 'n' immediately preceding behaviors

15.2.1b *Transition matrices*

Markov chains are generally analyzed through the use of a transition matrix and comparison to a random model. The random model is one that generates expected frequencies of occurrence based on observed number of occurrences. It should be noted that the transition matrix is not a true contingency table since the events included are not independent of each other. Below is a hypothetical transition matrix for three behaviors.

Transition matrix
Observed occurrences

Following behavior

		A	B	C	Row totals
	A	6	20	9	35
Preceding behavior	B	9	4	17	30
	C	19	5	8	32
Column totals		34	29	34	97

Have we collected enough data? According to Fagen and Young's (1978) rule-of-thumb (previous section), if the number of measurements (sequences recorded) equals, or exceeds, $10R^2$ (R = number of different behaviors) then the sample size

Table 15.5. *Transition frequencies among behavior pattern categories in contests involving supplemented and unsupplemented owners of poor-quality territories (N=25 contests for each; major between-sample group differences in bold-face*

	Locate	Signal	Threat	Contact	Retreat
Supplemented owners					
Locate	**130**	**80**	0	0	0
Signal	**73**	**435**	15	10	30
Threat	0	8	30	3	10
Contact	0	0	3	0	10
Retreat	3	8	0	0	5
Unsupplemented owners					
Locate	**25**	**25**	3	0	5
Signal	**13**	**85**	20	0	28
Threat	0	8	18	0	20
Contact	0	0	0	0	0
Retreat	3	8	0	0	5

Source: From Riechert (1984). Copyrighted by Bailliere Tindall.

should be sufficient. In our example, $R=3$ and 10 $R^2=90$; therefore, since our sample size of 97 is larger than $10R^2$, it should be sufficient.

The matrix of observed frequencies can initially be inspected for cells which show large frequencies. For example, Lemon and Chatfield (1971) contructed a transition matrix of preceding and following song types for cardinals (*Richmondena cardinalis*); their initial inspection of the matrix showed that switches between certain song types occurred very frequently. Two matrices, each from a different value of the independent variable, or treatment, can be initially examined for cells that appear to differ greatly between the two samples. For example, Reichert (1984) measured transition frequencies of agonistic behaviors in spiders that owned poor-quality territories which were either food supplemented, or not supplemented. Table 15.5 shows the transition matrix for each treatment; a chi-square test showed that the boldfaced cells were significantly different between the treatments.

Chi-square test

After examining the observed frequency transition matrix (p. 450) for large differences between cells, the expected frequency matrix is constructed by calculating the expected frequency for each cell according to the formula:

$$\text{expected frequency} = \frac{\text{row total} \times \text{column total}}{\text{grand total}}$$

Transition matrix

Expected occurrences

Following behavior

		A	B	C	Row totals
Preceding behavior	A	12.3	10.4	12.3	35
	B	10.5	9.0	10.5	30
	C	11.2	9.6	11.2	32
Column totals		34	29	34	97

The individual cells in both tables can now be searched for those where the observed frequently is much larger or smaller than the expected frequency. For example, the observed frequency of the A———▶B transition in our example (20), is much larger than the expected frequency (10.4).

This provides a first approximation look at the data. The researcher can then proceed with a chi-square test for the entire matrix (e.g. Lemon and Chatfield 1971), segments of the matrix (e.g. Dawkins and Dawkins, 1976), or reduce it to the most important cells (e.g. 2×2 matrix) and conduct a chi-square test (e.g. Stokes, 1962). Some transition matrices will contain sequences of only two behaviors to begin with (e.g. Fincke, 1985; Table 15.6).

When conducting a chi-square test, if any of the calculated expected frequencies are very small (≤ 5), the sample size may be too small for a valid test. In our example above, we decided that the sample size of 80 approached a sufficient sample, but we also notice that the expected frequency for the BB sequence is only 9.0. Bakeman and Gottman (1986) developed a formula for calculating sample size based on expected frequencies; their formula was derived from a calculation suggested by Siegel (1956). The minimum number of sequences that need to be recorded (N_S) is determined by:

$$N_S = \frac{9}{P(1-P)}$$

where:

P = calculated expected probability of the *least frequent* sequence

For our example:

$$P = \frac{9.0}{97} = 0.092$$

Table 15.6. *Pattern of successful use of searching and waiting tactics*

Initial tactic used	Subsequent tactic used	
	Searching	Waiting
Searching	53	39
	(53.5)	(38.4)
Waiting	36	25
	(35.5)	(25.5)

Expected values are in parentheses. There was no significant effect of the initial tactic used successfully on the subsequent tactic used successfully ($x^2=0.03$, df=1, NS).

Source: From Fincke (1985).

$$N_S=\frac{9}{0.092(0.908)}=108$$

Since our number of sequences observed (97) is close to the number suggested by our calculations (108), and our sample is sufficient according to Fagen and Young's 'rule-of-thumb', we will conclude that we probably have a large enough sample. Therefore, we will proceed with the chi-square analysis (see the discussion of the Chi-square $r \times k$ test of independence described in section 14.1.2).

Cell	$\left[\dfrac{(\text{Observed}-\text{expected})^2}{\text{Expected}}\right]$
AA	3.22
AB	8.86
AC	0.88
BA	0.21
BB	2.77
BC	4.02
CA	5.43
CB	2.20
CC	0.91

$$x^2=\Sigma\frac{(O-E)^2}{E}=28.50$$

The tabular value for χ^2 for 4 df at $P=0.05$ is 9.49. Since our calculated value of $28.50 \geq 9.49$, we infer that the frequencies of the sequences observed are significantly different from random chance. In fact, you can see from Table A9, that our calculated χ^2 is larger than the tabular value at $P=0.001$.

Cramer's phi

Another convenient measure of association between cells in a matrix of any size is Cramer's phi. It converts the χ^2 to a coefficient that varies from 0 to 1 and is calculated as follows (Zar, 1984):

$$\Phi_C = \sqrt{\frac{\chi^2}{N}}$$

where: N=sample size

For the example above: $N=97$

$$\chi^2 = 28.50$$

$$\Phi_C = \sqrt{\frac{28.50}{97}} = 0.54$$

Although the chi-square test, above, was significant, the phi value does not reflect a very high degree of association. Cramer's phi can also be calculated as follows (Howell, 1992; Castellan, 1979):

$$\Phi_C = \sqrt{\left[\frac{\chi^2}{N(k-1)}\right]}$$

where: k=the smaller of the number of rows or the number of columns

For our example, $k=3$, therefore:

$$\Phi_C = \sqrt{\left[\frac{28.50}{97(3-1)}\right]} = 0.38$$

This method produces an even smaller phi coefficient, so it is necessary to indicate which method of calculation was used.

Stationarity (the probability of one behavior following another not changing over time) is assumed in the analysis of transition matrices; however, we know that stationarity rarely exists in the behavior of an animal over time. For example, daily cycles (circadian rhythms) are likely to cause trends in the data; that is, the frequency of occurrence of most behaviors is not likely to be stable throughout the day. Staddon (1972) concluded that Markov chain analysis was not applicable in his study of behavioral sequences in *Columba livia* because of a lack of stationarity. Lemon and Chatfield (1971) tested their data for stationarity before proceeding with Markov chain analysis. They tested for significant differences in the probabilities of occurrence of different song types between the first and second halves of their sample periods. To overcome the stationarity problem, Oden (1977) developed a method for analyzing behavior sequences based on fitting ascending-order non-stationary Markov processes to the data.

For a more complete discussion of the analysis of sequences using Markov chains and transition matrices see Gottman and Notarius (1978), Gottman and Roy (1990), and Haccou and Meelis (1992).

15.2.1c Lag sequential analysis

Sackett (1974, 1978, 1979) developed a method called *lag sequential analysis* for measuring the frequency with which selected behaviors precede or follow themselves and each other at various lag steps. Lags are the number of event, or time-unit, steps between sequential behaviors. An overview of the method was given by Bakeman (1978) as follows:

> The analysis begins by designating one behavior the 'criterion behavior' (this procedure can be repeated as many times as there are behaviors, so that each behavior can serve as the criterion). Then a set of 'probability profiles' is constructed, one for each of the other behaviors. Each profile graphs the conditional probabilities for that particular behavior immediately following the criterion (lag 1), following an intervening behavior(lag 2), following two intervening behaviors (lag 3), and so forth ... Peaks in the profile indicate sequential positions following the criterion at which a given behavior is more likely to occur, whereas valleys indicate positions at which it is less likely to occur.
>
> *[Bakeman, 1978:71]*

For example, in the following sequence of four behaviors (A,B,C,D) we have designated behavior A as the *criterion behavior* and examine the behaviors that occur at Lag 1, Lag 2 and so on:

Lag: 1   2   3
Behavior sequence: A   B   A   C   B   D   A   B   D   A
                     ↑   ↑   ↑
             Positions of criterion behavior A

Autolag analysis involves determining the conditional probability of a behavior following itself at each lag. Behavior A occurs for four out of the ten behavioral acts observed in the example above; therefore, the unconditional probability for behavior A is 4/10=0.40. We can use three positions of A in the sequence (indicated by arrows) to determine the conditional probabilities for behavior A following itself at lags 1,2 and 3; the three lags in the example are for A in the first position.

Using all three positions of A in the example, it can be seen that A never followed itself at lag 1, so the conditional probability of A following itself at lag 1 is zero. Behavior A followed itself at lag 2 one out of the three possible times, so the conditional probability for A following A at lag 2 is $\frac{1}{3}=0.33$. Behavior A followed itself at lag 3 one out the 3 possible times, so that conditional probability is $\frac{1}{3}=0.33$. From this very simple example, we would suggest that A has a greater probability of following itself at lags 2 or 3 than at lag 1. Gottman and Notarius (1978) illustrate how the standard deviation of the expected probability (based on a large sample) can be used to test the null hypothesis that the conditional probability at each lag is not significantly larger than the unconditional probability.

The same procedure used for the autolag analysis above, can be used for *cross lag* analysis of each behavior following each of the other behaviors at the various lags.

Lag sequential analysis can be used with intra-individual (above) or inter-individual sequences. For example, Sackett (1974) applied lag-sequential analysis to sequential data from a crab-eating macaque mother and her infant.

More complete discussions of Lag sequential analysis can be found in Bakeman and Gottman (1986), Gottman and Bakeman (1979), Gottman and Notarius (1978), Gottman and Roy (1990), and Sackett (1979, 1980). Computer programs for conducting lag sequential analyses are provided by Sackett *et al.* (1979) and Bakeman (ELAG program; 1983).

15.2.1d *Multivariate analyses*

Another method of analyzing sequences is the use of multivariate analysis. The applications of factor analysis and multidimensional scaling to sequences of behavior are discussed in Chapter 16.

15.2.2 Inter-individual sequences

In the analysis of sequences discussed above we have been assuming that an individual's behavior is its primary source of stimulation for subsequent behavior. That is, the behavior of other individuals is disregarded as an important variable. That decision has traditionally been a subjective one left to the experience and discretion of the researcher. The other side of the coin is where the researcher suspects that the most important stimuli are originating from other individuals, a situation we can

Table 15.7. *An example of a sociometric matrix showing the results of encounters between male dark-eyed juncos in an aviary*

		Loser					
		A1	A2	A3	A4	A5	A6
	A1	—	43	40	61	158	58
	A2	1	—	34	10	14	9
Winner	A3	1	2	—	16	10	8
	A4	0	6	6	—	21	7
	A5	1	2	4	8	—	21
	A6	0	1	1	0	4	—

Source: Abridged from Yasukawa and Bick (1983). Copyrighted by Bailliere Tindall.

loosely describe as communication. Under those conditions we cast the data into a contingency table called a sociometric matrix.

15.2.2a *Sociometric matrices*

A sociometric matrix is a special type of transition matrix which reflects interactions (Table 15.7), including sequential behavior, between two individuals. A sociometric matrix is often the format in which data on inter-individual sequences are collected during sociometric matrix sampling (see section 9.2.4). An example of a hypothetical sociometric matrix implying communication is shown below.

Sociometric matrix

Observed frequencies

Receiver's behavior

		A	B	C	Row totals
	A	3	18	27	48
Transmitter's behavior	B	24	7	35	66
	C	29	19	6	54
Column totals		56	44	68	168

Once again we can use a chi-square test to analyze for significantly large correlations between the transmitter's behavior and the receiver's behavior. In inter-individual matrices there is no problem involved with measurements found along the diagonal; however, stationarity can again be a problem, as it is in all contingency tables. Bekoff (1977b) listed the following five conditions in which social-behavior data are often collected and which, generally, do not satisfy the assumption of stationarity:

1 Two or more individuals are interacting (sociometric matrix).
2 Lumping data for different individuals (see Chatfield, 1973)
3 Developmental studies; individuals forming social relations and perceptual motor skills being acquired.
4 Motivational studies (see Slater, 1973).
5 Studies of signals having a cumulative 'tonic' effect (Schleidt, 1973).

As with the intra-individual analyses discussed earlier, associations found between a transmitter's behavior and a receiver's behavior are only that; associations. No causal relationship is explicitly demonstrated, although it is implied when we use the terms transmitter and receiver.

15.2.2b *Information analysis*

Another way to analyze sequences, both intra- and inter-individual sequences, is through the application of information theory. Information analysis has been used with a wide diversity of sequential data. As examples, information analysis has been applied to aggressive communication in mantis shrimps (*Gonodactylus bredini*) (Dingle, 1969, 1972), grooming behavior in flies (Dawkins and Dawkins, 1976), odor trail laying by fire ants (Wilson, 1962), leadership-rank in fallow deer (Gilbert and Hailman, 1966), to movements of mice (Fentress and Stillwell, 1973) and rat fetuses (Robinson & Smotherman, 1987).

This approach consists of calculating the *uncertainty* (U) in predicting what behavior will occur next; it is measured in *bits of information* required to make the prediction. Like Markov chain analysis, it assumes stationarity.

Uncertainty
Uncertainty can be measured for the following conditions:

1 When the behaviors are equiprobable or non-equiprobable.
2 When the antecedent (preceding) behavior is not known.
3 When the antecedent behavior is known.
4 When the two, or more, antecedent behaviors are known.

Given the following behavior sequence:

$$w \longrightarrow x \longrightarrow y$$

x = a behavioral act = monad

$$x \longrightarrow y = \text{dyad}$$
$$w \longrightarrow x \longrightarrow y = \text{triad}$$

P_x = probability of each behavior (x; monad)
$P_{x,y}$ = probability of each dyad ($x \longrightarrow y$)
$P_{w,x,y}$ = probability of each triad ($w \longrightarrow x \longrightarrow y$)

a. For *equiprobable* monads, dyads, or triads, etc:

U = uncertainty = $\log_2 N$

where: N = the number of different monads, dyads or triads

$\log_2 N$ can be found in Table A22. If $N \geq 100$ use the formula for non-equiprobable behaviors, below.

b. For *non-equiprobable* monads, dyads, or triads, etc:
1. Uncertainty, when the antecedent behavior is *not known*:
   (i) For monads:

$U_{(x)}$ = average uncertainty in predicting the next monad (x)

$$= -\Sigma P_x \log_2 P_x$$

Determine the $P_x \log_2 P_x$ for each of the individual behaviors (x) with their respective probabilities (P_x) and then sum them to calculate $U_{(x)}$. The $P_x \log_2 P_x$ can be found for each P_x in Table A22.

As an example, 101 'vocalizations' were elicited from a 'Bugs Bunny' toy. Each different 'vocalization' type was coded (A, B . . .) as it occurred producing a continuous sequence (A,B,C,D,E,F,B,G,C,H,B,G . . .) of 101 vocalizations of 10 different vocalization types (A–J; Table 15.8). For the information analysis, only the first 100 vocalizations were used so that the frequency of occurrence (n) equalled the probability (P_x) for each vocalization type (x). From Table 15.8, it can be seen that the information analysis for non-equiprobable monads produced an uncertainty of 3.1726 bits.

(ii) For dyads:

$U_{(x,y)}$ = average uncertainty in predicting the next dyad ($x \longrightarrow y$)

$$= -\Sigma P_{x,y} \log_2 P_{x,y}$$

Table 15.8. *Information analysis for 10 different 'vocalization' types given by a 'Bugs Bunny' toy. Since vocalizations were elicited 100 times, the number of occurences (*n*) also equals the probability of occurrence (*P*$_x$) for each vocalization type (see text for further explanation)*

Vocalization types (monads)	$n=P_x$	$-P_x \log_2 P_x$
A	21	0.4728
B	11	0.3503
C	13	0.3826
D	11	0.3503
E	11	0.3503
F	6	0.2435
G	4	0.1858
H	9	0.3127
I	9	0.3127
J	5	0.2161
	100	

Uncertainty $= -\Sigma P_x \log_2 P_x = 3.1726$ bits $= U_{(x)}$

Determine the $P_{x,y} \log_2 P_{x,y}$ for each of the individual dyads (x,y) with their respective probabilities $(P_{x,y})$ and then sum them to calculate $U_{(x,y)}$. The $P_{(x,y)} \log_2 P_{(x,y)}$ can be found in Table A22.

For example, the sequence of 101 'vocalizations' elicited from 'Bugs Bunny' (above) were divided into the 50 dyads of the 10 'vocalization' types. The results are shown in Table 15.9. It can be seen that the information analysis for non-equiprobable dyads produced an uncertainty of 4.8198 bits.

(iii) For triads:

$U_{(w,x,y)}$ = average uncertainty of predicting the next Triad $(w \longrightarrow x \longrightarrow y)$

$$= -\Sigma P_{w,x,y} \log_2 P_{w,x,y}$$

Determine $U_{(w,x,y)}$ using the same procedure (using Table A22) as described for the monads and dyads, above.

Table 15.9. *Information analysis for the different dyads from a sequence of 101 'vocalizations' given by a 'Bugs Bunny' toy. Only the 36 different dyads that occurred in the sample of 100 dyads are included in the table. Since 100 dyads were sampled, the number of occurences (n) for each different dyad (e.g. AB, AD) equals its probability (*$P_{x,y}$*) (see text for further explanation)*

Vocalization dyads	$n = P_{x,y}$	$-P_{x,y} \log_2 P_{x,y}$
AA	1	0.0664
AB	2	0.1129
AD	7	0.2686
AE	7	0.2686
AF	1	0.0664
AJ	3	0.1518
BC	5	0.2161
BH	1	0.0664
BI	5	0.2161
CA	9	0.3127
CC	1	0.0664
CE	1	0.0664
CF	1	0.0664
CJ	1	0.0664
DB	1	0.0664
DC	5	0.2161
DG	2	0.1129
DH	3	0.1518
EB	5	0.2161
EE	1	0.0664
EH	5	0.2161
FA	1	0.0664
FB	1	0.0664
FD	3	0.1518
FJ	1	0.0664
GA	3	0.1518
GI	1	0.0664
HA	4	0.1858

Table 15.9. (*cont.*)

Vocalization dyads	$n=P_{x,y}$	$-P_{x,y} \log_2 P_{x,y}$
HC	2	0.1129
HI	3	0.1518
IA	2	0.1129
IE	3	0.1518
IF	4	0.1858
JB	2	0.1129
JG	2	0.1129
JH	1	0.0664
	100	

Uncertainty$=-\Sigma P_{x,y} \log_2 P_{x,y}=4.8198$ bits$=U_{(x,y)}$

2 Uncertainty, when the antecedent behavior is *known*:
(i) For dyads:

$U_x(y)=$average uncertainty of predicting behavior (x) when the antecedent behavior (y) is known.

$=U_{(x,y)}-U_{(x)}$; these two variables were calculated above.

From the 'Bugs Bunny' example above, the average uncertainty in predicting vocalization (x) when the antecedent vocalization (y) is known is:

$U_x(y)=4.8198-3.1726=1.6472$ bits

(ii) For triads:

$U_{w,x}(y)=$average uncertainty of predicting a behavior (y) when the two antecedent behaviors (w,x) are known.

$=U_{(w,x,y)}-U_{(x,y)}$

The objective of information analysis is to determine the number of behavioral units which must be included in a sequence (e.g. dyads, triads, etc.) in order to reduce the uncertainty to an acceptably low level, perhaps <0.10. Thus, the difference between the measures of uncertainty yielded by successive models will give a measure of the decrease in uncertainty (or conversely the extra amount of information) yielded by that model. However, soon the law of diminishing returns comes into play and the uncertainty decreases only slightly for successive models.

Reduction in uncertainty (T)

If a behavior (y) is contingent, in part, upon the immediately preceding behavior (x) then:

$$U_x(y) < U_{(y)}$$

a. Reduction from monads to dyads:

    $T_{(x;y)}$ = decrease in uncertainty about behavior y that is derived by observing behavior x.

$$= U_{(x)} - U_x(y)$$

From the 'Bugs Bunny' example above, the decrease in uncertainty about vocalization y that is derived by hearing vocalization x is:

$$T_{(x;y)} = 3.1726 - 1.6472 = 1.5254$$

b. Reduction from dyads to triads:

    $T_{(w,x;y)}$ = decrease in uncertainty about behavior y that is derived by observing behaviors w and x.

$$= U_x(y) - U_{w,x}(y)$$

c. Reduction from triads to tetrads:

    $T_{(v,w,x;y)}$ = decrease in uncertainty about behavior y that is derived by observing behaviors v, w and x.

$$= U_{w,x}(y) - U_{v,w,x}(y)$$

As an example, S.A. Altmann (1965) terminated his analysis of sequences of behavior in the rhesus monkey at tetrads (sequence of four). Altmann used Shannon's measure of redundancy to calculate an 'index of stereotypy' (see below) for each order of approximation. This varied from zero for the zero-order of approximation (all behaviors equiprobable) to over 0.9 for his fourth order of approximation (tetrad). Since a behavior of the rhesus monkey is almost completely determined by the three preceding behavioral events, Altmann chose to go no further.

Stereotypy index (S)

The stereotypy index is used to compare the conditional uncertainties for monads, dyads and triads (calculated above) with their maximum possible values. The stereotypy index will increase from monads to dyads to triads.

For monads:

$$S_x = 1 - \frac{U_{(x)}}{\max U_{(x)}}$$

where: $\max U_{(x)} =$ equiprobable $U_{(x)} = \log_2 N$

For the 'Bugs Bunny' example above, the stereotypy index for the monads (Table 15.8) is:

$$\max U_{(x)} = \log_2 10 = 3.322$$

$$S_x = 1 - \frac{3.1726}{3.322} = 1 - 0.955 = 0.045$$

For dyads:

$$S_x(y) = 1 - \frac{U_{(x)}(y)}{\max U_x(y)}$$

where: $\max U_x(y) = \max U_{(x,y)} - \max U_{(x)}$

From the 'Bugs Bunny' example above, the stereotypy index for dyads (Table 15.9) is as follows:

The 10 different vocalization types (A–J) can be given in 100 possible dyadic combinations (e.g. AA,AB,AC . . . JJ). Therefore:

$$\max U_{(x,y)} = \log_2 100 = 6.644$$

$$\max U_x(y) = 6.644 - 3.322 = 3.322$$

$$S_x(y) = 1 - \frac{1.6472}{3.322} = 1 - 0.496 = 0.504$$

For triads:

$$S_{w,x}(y) = 1 - \frac{U_{w,x}(y)}{\max U_{w,x}(y)}$$

where: $\max U_{w,x}(y) = \max U_{(w,x,y)} - \max U_{(x,y)}$

Additional discussions of the use of information analysis in ethology can be found in Drickamer and Vessey (1992), Grier and Burk (1992), Hailman (1977), Losey (1978), Steinberg (1977), Van Hoof (1982), and Wilson (1975). Note, however, that the utility and applicability of information analysis to ethological data has been questioned by some ethologists, including Bekoff (1976) and Fagen (1977).

16 Multivariate analyses

Multivariate analyses treat several variables and compare two or more groups. They can be used for: 1. initial data exploration and hypothesis development; 2. classification (grouping according to similarities); and 3. limited hypothesis testing.

These techniques are useful in helping to clarify results through illustrative visual representations, such as dendrograms and three-dimensional diagrams. However, they should be used to express results, not to impress readers. In addition, the researcher should fit the method of analysis to the animal and the problem under investigation, not the animal to the method (Aspey and Blankenship, 1977; Bekoff, 1977b; Tinbergen, 1951).

This section is a brief overview which explores selected multivariate analysis techniques by: 1. explaining what they will do for you in terms of how they treat your data; 2. illustrating how they have been used; and 3. describing or referencing the methodologies for this use (Table 16.1). For more extensive coverage see Aspey and Blankenship (1977; 1978), Manly (1986), Maxwell (1977), Morgan *et al.* (1976), Sneath and Sokal (1973), and Sparling and Williams (1978). Keep in mind that, overall, 'multivariate analyses are powerful diagnostic tools: (1) for uncovering homogeneous subgroups from naturally-selected heterogeneous samples; and (2) for identifying relationships among multiple variables when the underlying source, or biological basis, or individual variation is unknown' (Aspey and Blankenship, 1977:77).

16.1 MATRICES

Many ethological data are gathered, or can be organized, into matrices in which several individuals or behaviors are being studied (Figure 16.1). These data may be widely variable, be scaled in arbitrary units, and frequently include interacting variables (Aspey and Blankenship, 1977). Initial visual inspection of these matrices is often confusing; that is where multivariate analyses come into play.

The use of six multivariate techniques (*R*-factor analysis, *Q*-factor analysis, principal component analysis, cluster analysis, discriminant function analysis, and multidimensional scaling) in the analysis of data found in matrices like those in Figure 16.1 will be discussed. However, the application of these analyses is not restricted to those discussed, but may be used to analyze any correlation matrix. Nevertheless,

Table 16.1. *Selected references on the use of multivariate analyses*

Matrix type[1]	Multivariate analysis	Example references	Procedure references
A	Principal components analysis	Huntingford (1976) three-spined stickleback reproductive behavior	Seal (1964)
		Halliday (1975) newt sexual behavior	Cooley and Lohnes (1971) Fortran program
		Aspey and Blankenship (1976) *Aplysia* burrowing behavior	
		Bekoff (1978a) ontogeny of Adélie penguin	Overall and Klett (1972) computer program
		Sparling and Williams (1978) components of avian vocalizations	Frey and Pimental (1978)
		Dudzinski and Norris (1970) rabbit behavior	Pimentel and Frey (1978)
		DeGhett (1980) maternal behavior of gerbils	
		Gadagkar and Joshi (1983) caste differentiation in wasps	
A	Discriminant analysis	Falls and Brooks (1975) avian song recognition	Cooley and Lohnes (1971) Fortran program, BMD program
		Aspey (1977c) dominance in spiders	
		Bekoff *et al.* (1975) behavioral taxonomy in canids	Nie *et al.* (1975) SPSS program
		Bekoff (1978a) ontogeny of Adélie penguin behavior	Pimentel and Frey (1978)
		Sparling and Williams (1978) components of avian vocalizations	
A	Cluster analysis	Sparling and Williams (1978) avian vocalizations	DeGhett (1978)
		DeGhett (1980) maternal behavior of gerbils	
		Gadagkar and Joshi (1983) caste differentiation in wasps	
A	Factor analysis	Svendsen and Armitage (1973) marmot response to mirror image	Fruchter (1954)
		Aspey and Blankenship (1976) *Aplysia* burrowing behavior	Comrey (1973)
			Schmitt (1977) BASIC program
			Overall and Klett (1972) computer program

Table 16.1. (*cont.*)

Matrix type[1]	Multivariate analysis	Example references	Procedure references
A	Multidimensional scaling	Aspey and Blankenship (1976) *Aplysia* burrowing behavior	Overall and Free (1972)
B	Cluster analysis	Morgan *et al.* (1976) chimpanzee social associations Ralston (1977) horse social associations	Morgan *et al.* (1976) SLCA Jardine and Sibson (1968) Sibson (1973) Fortran program
B	Multidimensional scaling	Morgan *et al.* (1976) chimpanzee social associations	Kruskal (1964) Shephard *et al.* (1972) computer program Nie *et al* (1975) SPSS program
C	Factor analysis	Baerends and Van der Cingel (1962) heron snap display Baerends *et al.* (1970) herring-gull incubation behavior Wiepkema (1961) bitterling reproductive behavior Van Hooff (1970) chimpanzee social behavior	Nie *et al* (1975) SPSS program
C	Cluster analysis	Dawkins and Dawkins (1976) grooming in flies Maurus and Pruscha (1973) squirrel monkey communication	DeGhett (1978)
C	Multidimensional scaling	Golani (1973) jackal precopulatory behavior Guttman *et al.* (1969) mouse behavioral sequences	Lingoes (1966) computer program Spence (1978)
C	Principal components analysis	Giles and Huntingford (1984) anti-predator behavior in sticklebacks	Huntingford (1976, 1982)

Note:
[1] See Figure 16.1.

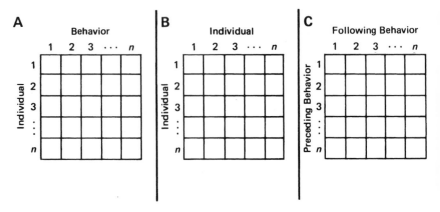

Fig. 16.1   Three typical matrices in which ethological data can be organized: A. Different behaviors performed by different individuals; B. Interactions between individuals; C. Inter- or intra-individual behavior sequences.

the characteristics of the particular analysis should be matched to the objectives of the study (see Gottman, 1978).

Since one of the advantages of these multivariate methods is that they allow visualization of the variables through graphical displays (Rohlf, 1968), selected examples will be provided.

16.2   GROUPING BEHAVIORS

R-factor analysis organizes a large number of variables (e.g. behaviors in Matrix **A**, Figure 16.1) into a smaller number of 'factors' based on their underlying similarities. When applied to Matrix **A**, this analysis extracts the behavior-related factors which account for a large percentage of the total variance.

For example, Aspey (1977c) recorded the occurrence of 20 different behaviors in 40 individual spiders. He used R-factor analysis to extract four behavior-related factors (groups from the 20 different behaviors) which accounted for 74.3% of the total variance. Aspey then descriptively labeled the factors after examining the behaviors comprising them. Consequently, Factor I was labeled 'approach/signal,' Factor II 'vigorous pursuit,' Factor III 'run/retreat.' Factor IV at first seemed biologically uninterpretable, but was then labeled 'non-linking', since none of the composite behaviors was significantly linked with any other behavior during inter-individual interactions. Although it is commonly stated that use of factor analysis assumes the existence of a common 'motivational state' underlying each extracted factor (Slater, 1973), note that Aspey has provided fairly descriptive, rather than functional, labels. In constrast, Wiepkema

(1961) labeled his three major extracted factors as 'aggressive, flight, and sexual tendencies'.

Principal component analysis (also called principal-axis factor analysis) utilizes an orthogonal rotation of the data (e.g. Huntingford, 1982). Factors analyzed using rotation account for maximum possible variance among the observed behaviors. Cooley and Lohnes (1971) provide a FORTRAN program listing for Varimax rotation. Giles and Huntingford (1984) use principal components analysis to describe the anti-predator responses of sticklebacks to a model heron or a live pike.

16.3 GROUPING INDIVIDUALS

If it is suspected that groups of individuals are behaving in a similar way, then Q-factor analysis can be applied to the data in Matrix **A**. The analysis extracts individual-related factors based on their observed behaviors.

In contrast to principal component analysis, powered-vector factor analysis extracts factors without rotation. It places maximum emphasis on biological relevancy, whereas the principal-axis method emphasizes parsimony (i.e. maximum variance accounted for by a few factors) (Aspey and Blankenship, 1977).

Aspey and Blankenship (1976) studied burrowing behavior in the marine mollusc *Aplysia brasiliana*. They recorded the occurrence of 10 burrowing parameters in 32 individuals (Matrix Type **A**). Q-factor analysis extracted three factors (groups of individuals) which accounted for 80.2% of the burrowers. Burrowing characteristics, interpreted relative to efficiency, were examined and the three groups were labeled 'inefficient burrowers', 'efficient burrowers' and 'intermediate burrowers'. The three-dimensional representation of these 32 individuals relative to the three extracted factors illustrates their location into three distinct groups (Figure 16.2).

Hierarchial cluster analysis groups variables (e.g. individuals or behaviors) on the basis of similarities (or differences) among common characteristics. Simple distance-function cluster analysis is sensitive to variability in the data and tends to 'split' the variables into more groups (Aspey, pers. commun.). Cluster analysis often makes fewer assumptions than other methods and therefore is easier to understand (Morgan *et al.*, 1976; Sparling and Williams, 1978).

Hierarchial cluster analyses begin with a matrix of similarities (or dissimilarities), and variables are sequentially joined on the basis of their relative similarities into dendograms (Figure 16.3), which are simple, visually interpretable representations of the results of the groupings.

Morgan *et al.* (1976) used hierarchial cluster analysis on data from a matrix of Type B and provided a straightforward description of the mechanics of conducting a single-link cluster analysis (SLCA), as well as discussing the positive and negative

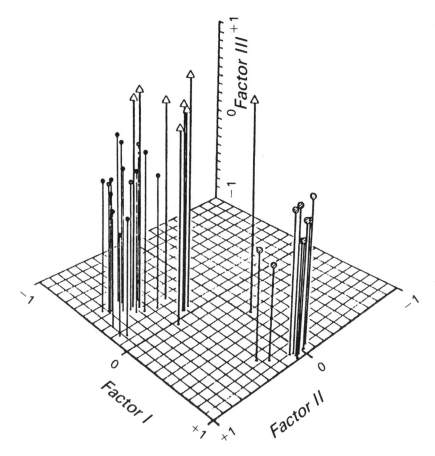

Fig. 16.2   Factor loadings for *Aplysia* burrowing behavior projected onto coordinate axes corresponding to the three factors extracted by *Q*-factor analysis. The origin falls in the center of the three-dimensional space. Factor I (large circles) represents 'inefficient burrowers'; Factor II (small stippled circles), 'efficient burrowers'; and Factor III (triangles), 'intermediate burrowers' (from Aspey and Blankenship, 1976).

attributes of the method. Their paper formed the basis for the discussion which follows.

In conducting a single-link cluster analysis, let us assume that we want to determine the relative association among individuals in a herd of eight adult doe deer in a wildlife preserve. We observe them for a total of 500 hours and record the amount of time that individuals are closer than four meters through scan sampling every minute. Each occurrence of an association is assumed to represent one minute of

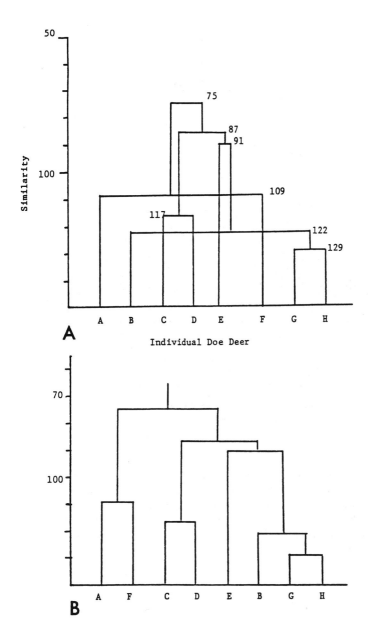

Fig. 16.3   A. Initial dendrogram for associations in a hypothetical herd of eight adult doe
deer; B. Final dendrogram for the associations in A.

Table 16.2. *Hypothetical association data for a herd of eight adult doe deer. Data are the number of hours observed in association*

	A	B	C	D	E	F	G	H
A								
B	21							
C	18	31						
D	30	21	48					
E	16	10	22	28				
F	38	22	11	17	11			
G	37	67	48	41	47	31		
H	31	30	25	23	34	27	77	
Total hours observed	191	202	203	208	168	157	348	247

association (see section 8.3 on sampling for a discussion of the hazards of this assumption). The procedure is as follows:

1 The data are first organized into the association shown in Table 16.2. This table, a matrix of Type B, can be read both vertically and horizontally in order to determine associations. However, it is still difficult to see much pattern of associations from the data in the table. Hierarchial cluster analysis will provide a better visual presentation of the associations.

2 The next step is to generate a triangular table of similarities (associations) among deer. Since each deer was seen for different total periods of time, we must first normalize the data to adjust for those differences.

Morgan *et al.* (1976) describe a method for normalizing data for time observed, which they derived from Dice's coefficient of association (1945). Note that this is essentially the same as Cole's coefficient of association (1949) (section 17.4.1b).

$$\text{Similarity} = \frac{XY}{X+Y}$$

where: XY = total time when individuals X and Y were observed together
X = total time X was observed
Y = total time Y was observed

Therefore, the similarity for individuals A and B in Table 16.2 is:

$$\text{Similarity } A+B = \frac{21}{191+202} = 0.053$$

Table 16.3. *Similarities for the associations in the hypothetical herd of eight adult doe der in Table 16.2*

	A	B	C	D	E	F	G	H
A								
B	53							
C	45	76						
D	75	51	117					
E	44	27	59	74				
F	109	61	30	46	34			
G	69	122	87	74	91	61		
H	71	69	56	50	82	67	129	

The similarity is then multiplied by 1000 for convenience, so that we can deal with whole numbers.

similarity $= 0.053 \times 1000 = 53$

We then complete the table of similarities (Table 16.3).

3 Next, we search the table for the range of similarities. The highest is 129 for H+G, and the lowest is 27 for E+B. The vertical axis of our dendrogram should include this range, so for convenience we will set up a vertical scale of from 0 to 150; note that the scale increases from top to bottom (Figure 16.3A).

4 We then begin linking individuals on the basis of similarities, starting with the largest similarity and working to the smallest. Individuals H+G provide our first association. The next similarity 122 is between G and B. This means we link B up with both G and H at the 122 level. This is an example of 'chaining', which is an undesirable characteristic of the method, since it links through intermediates (Jardine and Sibson, 1968) and is difficult to interpret visually. That is, the apparent association between B and H in Figure 16.3A is really due to B's association with G. Morgan *et al.* (1976) discuss the problems of 'chaining' in more detail.

The next association is D+C at the 117 level of similarity. We then continue the similarities (Table 16.3) until all the individuals have been linked together. This happens when we make the association of similarity 75 between D and A. If there are ties in similarities between two or more pairs with a common individual, then they are all linked together at the same level. The association between E and G is difficult to graph, because

E must be linked to the association between B and G+H (Figure 16.3A). This type of association becomes clearer when the individuals are rearranged on the abscissa.

5 It can be seen that we did not use over half of the ordinate in our dendrogram and that the ordering of individuals on the abscissa makes interpretation of the dendrogram difficult. Therefore, we correct these shortcomings by rearranging as in Figure 16.3B.

6 The dendrogram can now be used in a visual inspection of the associations. We can easily see the relative 'strength' of the various individual associations and the associations among three or more individuals.

Because of the chaining which occurs in this two-dimensional representation, distortion occurs and increases toward the lower levels of similarity. This distortion can be measured, but not tested statistically, by Sokal and Rohlf's (1962) cophenetic correlation coefficient, and Jardine and Sibson's (1971) distortion measure.

Maternal behavior of Mongolian gerbils was studied by DeGhett (1980) using hierarchial cluster analysis and principal components analysis. Both analyses revealed the existence of a multi-cluster or multi-factor set of systems in maternal behavior. Gadagkar and Joshi (1983) used principal components analysis and hierarchial cluster analysis to examine the behavior of 20 wasps (*Ropalidia marginata*) over time. They found that three behavioral castes existed (Figure 16.4), although there was no morphological caste differentiation.

DeGhett (1978) presents an excellent discussion of the use of hierarchial cluster analysis in ethology.

16.4 DESCRIBING DIFFERENCES AMONG INDIVIDUALS

Once the individuals in Matrix Type **A** (Figure 16.1) have been grouped by Q-factor analysis, we might want to know more about the differences among individuals and about the parameters (components) of their behavior that are most important in distinguishing differences.

Principal components analysis separates individuals in a sample in terms of a few composite components. The first principle component is the one which accounts for the maximum individual difference, and the second principal component is that combination of variables (e.g. behaviors), uncorrelated with the first principal component, that accounts for the largest proportion of the remaining individual differences. The analysis can be extended to additional components, if necessary, to explain the greatest proportion of individual differences.

Individuals may be represented on a multidimensional figure to provide a visual image of the results of analysis. Subgroups can also be delineated for a clearer presentation (Figures 16.4a, 16.5).

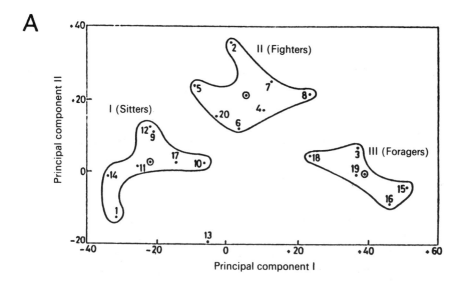

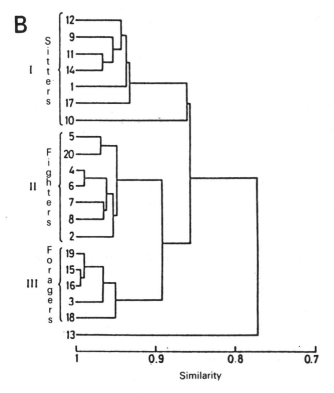

Fig. 16.4 Behavioral castes of 20 wasps shown by principal components analysis (A) and hierarchial cluster analysis (B) (from Gadagkar and Joshi, 1983). Copyrighted by Bailliere Tindall.

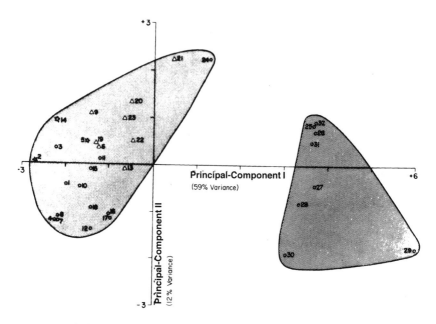

Fig. 16.5   Principal components analysis of 32 burrowing *Aplysia brasiliana* showing two
extracted subgroups. The first principal component (shaded area on right)
accounted for 59% of the variance and generally corresponded to individuals
which had higher Factor I (approach/signal) and Factor II (vigorous pursuit)
values. The second principal component accounted for 12% of the variance and
corresponded to individuals with low Factor II values and high Factor III
(run/retreat) values (from Aspey and Blankenship, 1978).

Good references on the use of principal component analysis in ethology are
Dudzinski and Norris (1970) and Frey and Pimentel (1978).

16.5   DISCRIMINATING AMONG GROUPS OF INDIVIDUALS OR BEHAVIORS

Discriminant analysis: 1. determines relationships among several identified groups
(e.g. predefined or resulting from Q-factor analysis or principal component analysis); 2. assesses the discriminability between the groups; and 3. places individuals or
behaviors in the appropriate groups (Sparling and Williams, 1978).

Aspey (1977c) used multiple stepwise discriminant analysis to discriminate
further between three groups of 40 spiders (*Schizocosa crassipes*) which Q-factor
analysis had extracted and Aspey had labeled as 'Dominant', 'Intermediate', and
'Subordinate'. In his analysis 'a sequence of discriminant equations was computed
in a stepwise manner so that one variable was added to the equation at each step,

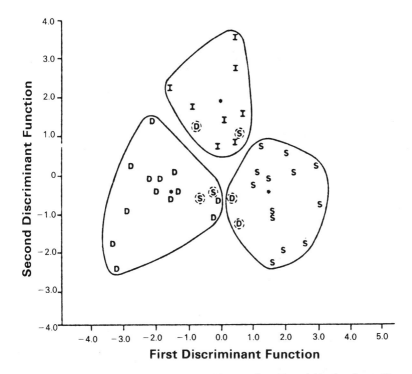

Fig. 16.6   The locations of 16 dominant (D), 8 intermediate (I), and 16 subordinate (S) adult male *Schizocosa crassipes* plotted in a geometric space of minimum dimensionality by multiple stepwise discriminant analysis on the basis of the frequency of 20 behaviors observed during agonistic encounters. The first discriminant function (abscissa) is plotted against the second discriminant function (ordinate) and * denotes group means. The spiders were initially grouped as dominant, intermediate, or subordinate by a dominance index (DI). Dotted circles represent spiders 'misclassed' by the DI relative to the discriminant analysis (from Aspey, 1977c).

and a one-way analysis of variance *F*-statistic was then used to determine which variable (i.e., behavior) should join the function next. The variable added is the one making the greatest reduction in the error sum of squares' (Aspey and Blankenship 1977:90). The results were then plotted (Figure 16.6). The first discriminant function (abscissa) separated the 'dominant' and 'subordinate' spiders, but it took the second discriminant function (ordinate) to separate out the 'intermediate' group from the other two groups. The three groups were encircled in the figure for added clarity.

In contrast to Aspey's (1977c) use of multiple stepwise discriminant analysis, Bekoff *et al.* (1975) used linear discriminant analysis to assess the taxonomic rela-

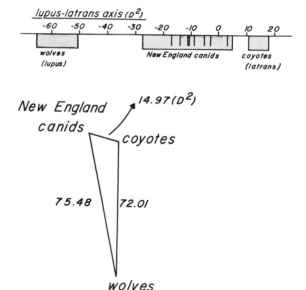

Fig. 16.7 The results of a behavioral taxonomy study on infant coyotes (*Canis latrans*)
wolves (*C. lupus*) and New England canids (eastern coyotes). The relative
frequencies of occurrence of social play and agonistic behavior were used as
behavioral characters (Bekott, *et al.*, 1975). Top: Linear discriminant values of
known *C. lupus* and *C. latrans* litters cast on a *lupus–latrans* discriminant axis.
New England canids are projected onto this–they fall between *lupus* and *latrans*,
but closer to *latrans*. Bottom: Distances (D^2) in discriminant function units based
on pairwise analyses of *lupus, latrans* and New England canids. Note the close
relationship between coyotes and New England canids and that both fall
approximately the same distances from wolves. This is an example of the
application of multivariate analysis of behavior to taxonomy, an area where these
analyses are commonly used on morphological data (from Jardine and Sibson,
1971; Sneath and Sokal, 1973).

tionships among infant wolves, coyotes, and 'New England Canids' (Silver and
Silver, 1969). Linear discriminant analysis compares the means of variables (e.g.
behaviors) from two populations (e.g. individuals from a canid taxon) and produces
a discriminant function which is a relative measure of the difference between the
populations. Bekoff *et al.* (1975) used two behaviors (social play and agonistic
behavior) in their analysis of differences among the three canid 'types'. Their
results, when plotted on a linear discriminant function axis (Figure 16.7), show that
the wolves and coyotes were clearly separated, and that the 'New England Canids'
were intermediate, but fell closest to the coyotes. Interestingly, their results are in full

agreement with the same type of analysis done on various anatomical measurements.

In contrast to the other multivariate analyses discussed in this section, discriminant analysis can be used to test hypotheses (Sparling and Williams, 1978).

16.6 ANALYZING SEQUENCES OF BEHAVIOR

Multivariate analyses are only one method for analyzing sequences of behavior (see Chapter 15).

16.1.1 Factor analysis

Whereas Markov chain analysis tends to emphasize 'sequential effects' (Slater, 1973), factor analysis assumes that the measured variables (e.g. behaviors) do not depend causally on each other (Blalock, 1961), but rather that there are underlying 'motivational processes' common to the behaviors grouped around the extracted factors (Andrew, 1972; Hutt and Hutt, 1974). Slater (1973) points out that both motivational changes and sequence effects are probably present in all behaviors so that neither analysis is perfect.

Factor analysis was used to analyze behavioral sequences (Matrix type C, Figure 16.1) by Wiepkema (1961) for bitterling (*Rhodeus amarus* Bloch) reproductive behavior, and Baerends and Van der Cingel (1962) for common-heron (*Ardea cinerea* L.) snap displays. Both started with matrices of type C, and then generated transition frequencies on the basis of observed/expected frequencies. Expected frequencies are calculated the same as for a chi-square test:

$$\text{expected frequencies} = \frac{\text{row total} \times \text{column total}}{\text{grand total}}$$

Calculating transition frequencies on this basis provided a ratio which indicated relative frequency with which each of the behaviors preceded or followed the other, but it also included transitions between a given behavior and itself which may be impossible or difficult to interpret. Slater and Ollason (1972) discuss these difficulties and suggest another method provided by Goodman (1968). Wiepkema (1961) and Baerends and Van der Cingel (1962) then ranked the transition frequencies and generated correlation coefficients using Spearman's rank correlation, in contrast to the Pearson product-moment correlations used by Aspey (1977c) and Van Hooff (1970). Factor analysis was then used to extract the minimum number of common causal factors necessary to explain the sequences observed. Since three factors explained the majority of the variability in the data, a three-dimensional vector model was used to illustrate the clustering of the behaviors around the three factors (Figure 16.8).

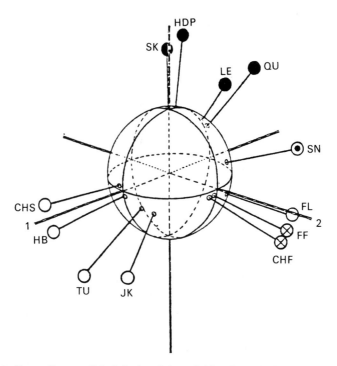

Fig. 16.8 Vector diagram of the behavior of the male bitterling (*Rhodeus amarus*).
CHF=chafing; CHS=chasing; FF=fin flickering; FL=fleeing; HB=head
butting; HDP=head-down posture; JK=jerking; LE=leading; QU=quivering;
SK=skimming; SN=snapping; TU=turning beat (from Wiepkema, 1961).

The angle between any two vectors represents the extent of their correlation. For
vectors of unit length, the correlation is represented by an angle determined according to the following formula:

Correlation Coefficient=COS φ_{AB}

Therefore, the vector angle equals that angle whose cosine equals the correlation
coefficient (Table A23). If the correlation is perfect ($r=1.00$) the two vectors will be
the same (1 in Figure 16.9). If there is no correlation ($r=0.00$) the two vectors will
diverge by $90°$ (2 in Figure 16.9). If there is a positive correlation the angle will be
between 0 and $90°$ (3 in Figure 16.9), and if it is negative it will be between $90°$ and
$180°$ (4 in Figure 16.9).

In the vector diagram (Figure 16.8) it can be seen that the behaviors cluster
around three axes. These three axes are the causal factors: 1 is the positive side of the
aggressive factor; 2 is the positive side of the nonreproductive factor; 3 is the posi-

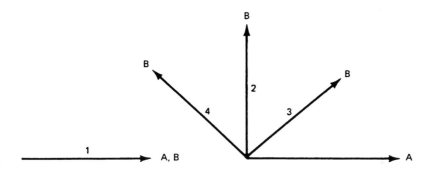

Fig. 16.9   Examples of vector angles in factor analysis.

tive side of the sexual factor. The length of each vector represents the amount of the common variance of the behavior accounted for by the factors.

The use of factor analysis in ethological studies in general has been questioned by several investigators (Andrew, 1972; Morgan *et al.*, 1976; and Maurus and Pruscha, 1973). Factor analysis often makes more assumptions than some other multivariate analyses (e.g. cluster analysis) (Morgan *et al.*, 1976; Sparling and Williams, 1978). Once causal factors have been extracted, some researchers question what they really mean. Slater (1973:145) concludes that 'It is doubtful whether the extraction of factors which are themselves of complex causation advanced understanding'. However, it does allow us to visualize our data more clearly and generate hypotheses and experiments that truly will advance our understanding. For example, Aspey and Blankenship's (1976) experimental study of *Aplysia* grew out of factor analysis of groups and advanced our understanding of the function of burrowing and its relationship to reproduction behavior.

Short and Horn (1984) make the following four points which will serve as cautions to ethologists using factor analysis:

1 Usually it is unwise to use factor analysis when the sample size is small.
2 One should ensure that factored variables are reliable.
3 Derived variables can hopelessly confound any basis for interpretation of factor analysis results.
4 One should be wary about using product moment correlations in factor matching.

16.6.2 Multidimensional scaling

Multidimensional scaling orders individuals or behaviors from a matrix and discriminates between them in terms of distances along coordinate axes. Golani (1973) used multidimensional scaling in his analysis of precopulatory behavioral sequences in two pairs of golden jackals (*Canis aureus*). He utilized the Guttman–Lingoes multidimensional scalogram analysis (L. Guttman, 1966), of which a computer program is available (Lingoes, 1966). Guttman *et al.* (1969) had previously used the analysis to measure sequential behavior in mice.

Morgan *et al.* (1976) reanalyzed Wiepkema's (1961) data, using multidimensional scaling, and discussed their different results. They suggested that Wiepkema's grouping of behaviors around three causal factors may tend to oversimplify the situation.

Multidimensional scaling is too complex a method to discuss in detail here; consult Shepard (1980) and the references above.

16.7 SUMMARY

Multivariate analyses provide a diversity of methods to treat ethological data in matrices. They are fundamentally descriptive techniques that should be used primarily: 1. to aid in interpreting relationships in the data through grouping and visual representation; and 2. to generate hypotheses for further testing. In order to interpret the results of multivariate analyses, researchers must have their objective clearly in mind and match the analyses to the objectives and type of data.

Finally, Aspey and Blankenship (1977:78) 'caution that multivariate analyses are simply one research tool available to the animal behaviorist for determining inherent data structure . . . misusing multivariate analyses in ethology could most certainly lead to a bewildering array of extraordinarily sterile papers.'

Regardless of the mechanics, analysis must be accurate. Moving up the scale of sophistication from paper and pencil to calculator to large computer will speed up the analysis and may improve its accuracy, but it will not change the data. Computers are tremendous tools for the storage and analysis of data, and their use should be encouraged. However, poorly collected and invalid data cannot be laundered in a computer, although some researchers will try to sell it by wrapping it in complex analyses, tying it with a computer and presenting it with complicated diagrams.

Notterman cautions us to be on the lookout for 'computeritis disease', which has the following symptoms:

> (1) The diarrhea symptom – this is manifested in the researcher's assumptions that the more data, the better the experiment. (2) The

displacement symptom – the more time spent in manipulating the computer, the more incisive the psychological research. (3) The Lorenzian-territorial symptom – the more expensive the computer, and the larger the laboratory in which it is housed, the greater the personal authority exerted either in the academic or in the institutional setting.

[Notterman, 1973]

In short, use the best methods and equipment available both to generate and to test hypotheses; however, avoid methodological overkill. As Fagen and Young (1978:114) suggested 'The future may well belong to those who can use simple methods effectively to test a theoretical hypothesis, or who are prepared to construct original models of behavior should no simple technique be available'.

16.8 SOFTWARE PACKAGES FOR STATISTICAL ANALYSES

Data analysis is generally performed using commercial statistics packages, although there are some statistics used by ethologists that are generally not available, such as circular statistics (see Chapter 17).

Statistics software usually has several modules: database manager, graphics and statistical tests. The database manager (often in the form of a spreadsheet) is usually relatively rudimentary and is meant to handle data entered directly from the keyboard, and to provide a method of performing manipulations on the data, such as transformations (e.g. arc-sine, log). It is recommended that any reasonably large set of data be handled through a true database manager or spreadsheet, as described above, and these files be simply read into the statistical program.

Most statistical packages come with some form of graphic presentation ability. In the minimum form, simple low-resolution scatter plots and histograms can be generated for a quick survey of the data before testing. Larger packages, particularly those custom written for microcomputers, may have much better graphic abilities. Generally, presentation-quality graphics must be prepared in specialized graphics software (see Chapter 18).

Finally, all of these statistical packages contain a set of statistical tests and manipulations. The important criterion is that the package perform the tests you need (Table 12.1). A complete package, like those listed below, contain at least the following categories of statistics: descriptive, nonparametric, simple regression, *t*-tests, analysis of variance, and multivariate analyses. Multivariate analyses may include: discriminant function, factor analysis, principle components analysis, general linear modeling, time series analysis, multiple and stepwise regression, and multiple factor analysis of variance. The largest packages (usually older packages converted from mainframe computers) will contain additional statistics. Besides confirming that a package performs the tests you require, check on the range of transformations that are available, the ability of the software to handle missing data (many packages don't), the ease of importing 'foreign' data files, and the ease of performing last-minute data manipulations or 'quick-and-dirty' graphical analyses.

Statistics software comes in several types (Appendix B). The best known packages are those which have been modified from mainframe computer versions. Software in this category is generally well-known and tested, and often uses commands which are familiar to users of the software in mainframe versions. These programs include *SAS-PC* (SAS Institute, Inc.), *SPSS/PC+* (SPSS Inc.), *BMDP-PC* (BMDP Statistical Software, Inc.). Programs written expressly for microcomputers include the most popular statistics package, *Systat* (Systat, Inc.).

Reviews of these programs are often published by computer magazines and weeklies (e.g. *PC Week*). Generally, the mainframe-based programs perform a larger variety of statistical tests, are well tested and familiar, but are memory-hungry and non-intuitive to use. PC-based programs are more efficiently designed for microcomputer hardware and for non-statistically-minded users. They are often less expensive, but are less complete in their repertoire of statistical tests, and are less well recognized by journal reviewers. A number of low-cost (but incomplete) packages are available for microcomputer users (e.g. *Mystat* from Systat Inc.), and free (or 'public- domain') software is available and surprisingly good, if limited (e.g. *A-STAT*). The user must decide what statistical tests are required, and locate a package which provides the best combination of required tests, price, compatibility, and ease-of-use. *Systat* (Systat, Inc.), while not inexpensive, is complete, easy to use, and works well on microcomputers. Check for academic site license versions of many of these programs at your home institution.

17 Spatial orientation and time: circular statistics and spatial patterns

Circular statistics are used to analyze data points that are distributed on a circle, such as direction and time (Figure 17.1).

As examples, these data points may represent angular degrees from magnetic north taken by homing pigeons when they disappear from sight, or the periods of activity for gerbils during a 24-hour period. The calculations described below will only be applied to examples of directional data, but the reader will recognize how they can also be easily applied to points in time using the same methods (Table A23). Readers who wish to pursue circular statistics beyond this cursory introduction should consult Batschelet (1965, 1972, 1981).

The use of basic circular statistics proceeds in two steps:

1 *Determine whether the angular directions (or times) measured are randomly distributed.* You can continue analyses only if the sample directions are *not* randomly distributed. If the sample directions are in a unimodal distribution, then the tests will also determine whether the sample directions are significantly clustered about the mean direction.

2 If the sample directions are not randomly distributed, you can determine whether:

   a *the sample mean direction (or time) differs significantly from a specified direction (or time),*

   and/or

   b *there is a significant difference in sample mean directions (or times) between two, or more, groups of animals.*

These steps are summarized in the flow chart (Figure 17.2).

17.1 TESTS FOR RANDOMNESS OF DIRECTIONS (OR TIMES)

The chi-square test can be used for nominal data such as the number of observations in segments of a circle (e.g 0–120°, 121°–240°,and 241°–360°; or between

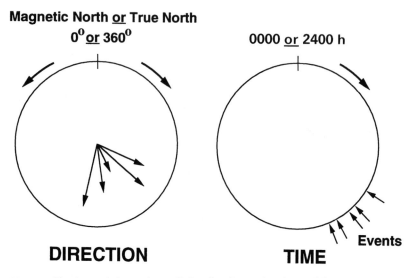

Fig. 17.1 Circular statistics can be applied to direction or time (e.g. activity patterns; drawing by Lori Miyasato).

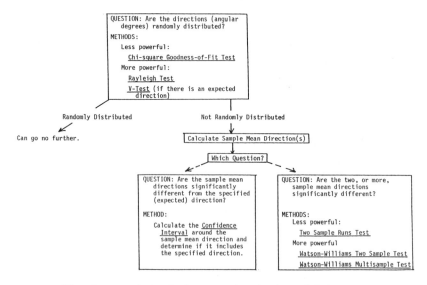

Fig. 17.2 Flow chart to assist in selecting appropriate circular statistical tests.

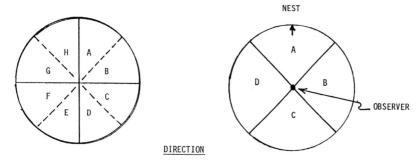

NEST

DIRECTION

OBSERVER

Fig. 17.3   Directions can be divided into quadrants and equal divisions of quadrants for
analysis with Chi-square. Orientation of the quadrants is dependent upon the
research question, such as whether distraction displays are given primarily in the
quadrant including the nest or in quadrants away from the nest (see text for
explanation).

North and East, East and South, South and West, and West and North). If you have
interval data, such as angular degrees (or hours and minutes) then the Rayleigh and
V tests are more powerful. When there is an expected direction such as would occur
when homing pigeons are clock-shifted (i.e. photoperiod altered), then the V test is
more powerful than the Rayleigh test.

17.1.1   Chi-square goodness-of-fit test

This test (described in Chapter 14) can easily be applied to circular data. It tests
whether the sample of directions (angular degrees, times) differs significantly from
randomness. It is not as powerful as the Rayleigh test or V test if you have interval
data (e.g. degrees from magnetic north), since that data is reduced to nominal data
in this test.

Procedure:
1   Divide the compass (or clock) into equal sectors (e.g. Figure 17.3). The
number of sectors and their orientation relative to direction is not
restricted, except as stated in No. 2, below.
2   Assign the expected number of observations to each sector according to
the hypothesis of randomness. That is, the total number of your observations should be randomly assigned to the sectors or equally divided
between the sectors (uniform distribution).

$$\text{expected number/sector} = \frac{\text{total no. observations}}{\text{no. of sectors}}$$

Table 17.1. *Layout for chi-square goodness-of-fit test of data from a circular distribution divided into four or eight sectors (see Figure 17.3)*

Sector	Expected		Observed		$(O-E)^2/E$	
A	25%	12.5%	—	—	—	—
B	25	12.5	—	—	—	—
C	25	12.5	—	—	—	—
D	25	12.5	—	—	—	—
E		12.5		—		—
F		12.5		—		—
G		12.5		—		—
H		12.5		—		—
			$\chi^2 = \Sigma(O-E)^2/E =$		— or —	

The expected number/sector must ≥ 4.

3   Calculate the Chi-square value (see Table 17.1):

$$\chi^2 = \Sigma \frac{(O-E)^2}{E}$$

4   Compare the calculated chi-square value to the tabular value (Table A9). If the calculated value is $\geq$ the tabular value then the sample directions are not randomly distributed but are significantly clustered.

5   If your data are angular degrees from magnetic north (or time) then also conduct a V test (below) for confirmation of significant clustering.

As an example, Sordahl (1986) tested whether the distraction displays of 14 avocets and 5 stilts were directional, especially whether they would lead an observer away from the nest. He stood 10–20 m from the nests and recorded the position of colored-banded, displaying birds in a circle with him at the center. For analysis, he divided the circle into four equal quadrants with one of them subtended by an angle of 45° on either side of a line from him to the nest (Figure 17.3). He observed 81 distraction displays from the 19 birds. The observed and expected (based on a uniform distribution) display distributions is given in Table 17.2.

The calculated chi-square value of 5.94 is smaller than the tabular value (3 df, alpha level=0.05) of 7.81. This means that although most of the distraction displays were in the quadrant away from the nest, the direction of all distraction displays was not significantly different from random.

If you use only the number of displays toward (quadrant A) and away from the nest (quadrant C), the chi-square test shows that those displays were distributed significantly different than random (see Table 17.3). However, disregarding the dis-

Table 17.2. *The chi-square goodness-of-fit table for date on the distribution of avocet and stilt distraction displays relative to the direction towrds their nests (see text)*

Sector	Expected		Observed	$(O-E)^2/E$
A	(25%)	20.25	14	1.92
B	(25%)	20.25	20	0.00
C	(25%)	20.25	29	3.78
D	(25%)	20.25	18	0.24
			81	$\chi^2=5.94$

Source: Data from Sordahl (1986).

Table 17.3. *The chi-square goodness-of-fit table using only the data from Sectors A (toward the nest) and C (away from the nest) from Table 17.2*

Sector	Expected		Observed	$(O-E)^2/E$
A	(50%)	21.5	14	2.61
C	(50%)	21.5	29	2.61
			43	$\chi^2=5.22$

plays in the other two quadrants would be considered misleading, if not invalid, by most researchers.

The calculated chi-square of 5.22 is larger than the tabular value of 3.84 (1 df, alpha level=0.05); therefore, the number of displays in these two quadrants was significantly different.

17.1.2 Rayleigh test

This test is used to determine whether the sample of directions (or time) differs significantly from random. The data must be in angular degrees from north (0°).

Procedure:

1 Calculate the sum of the *sines* and *cosines* of the sample directions. Table A23 can be used for either angular degrees or time.

$$S=\Sigma \text{sine}_{ai} \qquad C=\Sigma \text{cosines}_{ai}$$

Table 17.4. *Critical values of* Z_a

	Alpha level		
N	0.05	0.01	0.001
30	2.97	4.50	6.62
50	2.98	4.54	6.74
100	2.99	4.57	6.82
200	2.99	4.59	6.87
500 (or larger)	2.99	4.60	6.89

2 Calculate R.

$$R = \sqrt{(S^2 + C^2)}$$

3 Calculate the test statistic Z.

$$Z = R^2/N \qquad \text{where: } N = \text{number of sample directions}$$

4 Compare the calculated Z to the tabular Z_a in Table 17.4.

If the calculated Z is greater than the tabular Z_a, then the directions are non-random. If the distribution of the sample directions appear to be unimodal, then significance in the Rayleigh test also proves that the sample directions are significantly clustered about the mean direction. When the sample directions are nonrandom, you can go on to calculate the sample mean direction and determine whether it is significantly different from the predicted direction (e.g. home).

As an example, I collected data on the directional orientation abilities of 13 students in my ethological methods class when they were deprived of normal visual cues (landmarks and sun postion). They were blindfolded and transported in a van to a secluded location approximately 10 miles northeast of the Colorado State University campus. The exercise was conducted to provide data for the students to analyze and to allow them to critique the methods. It was not a valid test of their ability to use other cues, such as the earth's magnetic field (see Baker, 1980, 1987; Gould and Able, 1981); for example, the van windows were not covered, and some of the students sitting next to the windows reported using the heat from the afternoon sun as a cue to the westerly direction. The students were individually led from the van while still blindfolded and asked the direction to the campus from their present position. They were then asked to point towards campus, first with the blindfold still on and then with the blindfold removed. The directions the students pointed while blindfolded, and the sines and cosines of those angular degrees are given in Table 17.5.

Table 17.5. *Directions 13 blindfolded students pointed to indicate the direction to the Colorado State University campus after having been driven approximately 10 miles northeast of campus while blindfolded*

Directions pointed while blindfolded (angular degrees from magnetic north)	Sine	Cosine
86	+0.9976	+0.0698
164	+0.2756	−0.9613
88	+0.9994	+0.0349
144	+0.5875	−0.8090
328	−0.5299	+0.8480
290	−0.9397	+0.3420
114	+0.9135	−0.4067
180	0.0000	−1.0000
128	+0.7880	−0.6157
152	+0.4695	−0.8829
108	+0.9511	−0.3090
178	+0.0349	−0.9994
208	−0.4695	−0.8829
	Totals: $S=+4.078$	$C=-5.572$

$$R=\sqrt{(S^2+C^2)}=\sqrt{(16.630+31.047)}=\sqrt{47.677}=6.904$$

$$Z=R^2/N=47.677/13=3.667$$

Since the calculated Z of 3.667 is larger than $Z_{(a)}$ (2.97) we reject the H_0 that the directions pointed by the students are random.

17.1.3 *V* test

When there is an expected direction, the V test is preferable to the Rayleigh test since it is more powerful. For example, when homing pigeons are clock-shifted 6 hours late, their expected disappearance direction upon release is 90° clockwise from the home direction. The V test will lead to significance only if there is sufficient clustering around the predicted direction. However, the V test should only be used to test for randomness. It does not test whether the sample mean direction deviates significantly from the predicted direction; the confidence interval should be used for that purpose (Batschelet 1981; see section 17.2.2).

Procedure:

1 Calculate V^1.

$$V^1 = R \cos(\bar{\Phi} - \theta_0)$$

where: θ_0 = expected direction (angular degrees)

$\bar{\Phi}$ = observed sample mean direction (angular degrees); see below for calculation of sample mean direction.

R was calculated previously (see Rayleigh Test)

2 Calculate the test statistic u.

$$u = \sqrt{\left(\frac{2}{N}\right)\left(V^1\right)}$$

where: N = number of sample directions

3 Compare the calculated u to the tabular value u_a (Table A24). If the calculated $u \geq u_a$ then the sample directions are not randomly distributed and are clustered significantly about the expected direction.

For example, since there was a predicted direction (Colorado State University campus) for the data on the students used in the Rayleigh test example above, we could have used the V test as a more powerful test. Even though the H_0 of randomness was rejected using the Rayleigh test, we'll apply the V test to the same data to determine whether there is sufficient clustering around the predicted direction to reject the H_0 of randomness.

1 First we calculate the sample mean direction ($\bar{\Phi}$):

First method:

$$\bar{\Phi} = \text{arc Tan } (\overline{\sin}/\overline{\cos})$$

= angle whose Tangent equals $(\overline{\sin}/\overline{\cos})$ (see Table A23)

where:

$$\overline{\sin} = S/N = +4.078/13 = +0.314$$
$$\overline{\cos} = C/N = -5.572/13 = -0.429$$
$$\overline{\sin}/\overline{\cos} = -0.7319$$

Using Table A23 we find that $143°$ has a tangent of 0.7536 and $144°$ has a tangent of -0.7265. Since our calculated tangent of -0.7319 is approximately midway between the two tangents in the table, we approximate the sample mean angle at $143.5°$.

Second method, as a check on our use of the first method:

We determine the sine and cosine of the sample mean angle and then use Table A23 to find the sample mean angle.

$r = R/N$ R was calculated in our example using the Rayleigh test.

$= 6.904/13 = 0.5311$

Sine of the mean angle $= \overline{\sin}/r = +0.314/5.311 = +0.5912$
Cosine of the mean angle $= \overline{\cos}/r = -0.429/0.5311 = -0.8077$
Using Table A23 we find that the angle whose sine approximates $+0.5912$ and whose cosine approximates -0.8077 is $143.5°$. This concurs with our previous calculation.

2 Calculate test statistic u.

$\bar{\Phi} = $ sample mean angle $= 143.5°$

$\theta_0 = $ predicted direction (campus) $= 196°$

$V^1 = R \cos(\bar{\Phi} - \theta_0)$

$\qquad = 6.904 \cos (143.5° - 196°)$

$\qquad\qquad 143.5° - 196° = -52.5° = 307.5°$

$\qquad = 6.904 (\cos 307.5°)$

$\qquad = 6.904 (0.6090) = 4.204$

$u = \sqrt{\left(\dfrac{2}{N}\right)\left(V^1\right)} = \sqrt{\dfrac{2}{13}} (4.204) = 0.3922 (4.204) = 1.648$

3 We compare the calulated u with the tabular u_a (Table A24). Since the calculated u (1.648) is larger than the tabular u_a of 1.647 we reject the H_0 that the sample directions are random. This concurs with the results from the Rayleigh test.

17.2 DIFFERENCE BETWEEN SAMPLE MEAN DIRECTION (OR TIME PERIOD) AND SPECIFIED DIRECTION (OR SPECIFIED TIME PERIOD)

The confidence interval is used to determine whether the sample mean direction from a non-randomly distribute sample (determined above) is significantly different from a predicted or specified direction, such as homeward direction. The first step is to determine the sample mean direction (angular degrees).

17.2.1 Sample mean direction

The mean of the sample directions cannot be determined by taking the arithmetic mean of the angles (especially if the sample is larger than three), even if you assign minus degrees to angles between 180° and 360° (see Batschelet, 1981, for examples). Use one of the methods described below.

First method:
1 Calculate:

   Mean sine $=\overline{sine}$ $= S/N$

   Mean cosine $=\overline{cosine} = C/N$

   S and C were calculated in the Rayleigh test, above.

2 Calculate the sample mean direction (mean angle; $\bar{\Phi}$)

   $\bar{\Phi} = \arctan(\overline{sine}/\overline{cosine})$

   Simply stated, the sample mean direction is the angle in Table A23 whose tangent equals the mean sine divided by the mean cosine.

Second method:
This method is a good check on the first method.
1 Calculate r (length of the mean vector).

   $r = R/N$

   R was calculated previously (see Rayleigh test).
   $N =$ number of sample directions.

2 Calculate:

   sine of the mean angle $= \overline{sine}/r$

   cosine of the mean angle $= \overline{cosine}/r$

3 The sample mean angle is the angle whose sine and cosine equal those calculated in Step 2 above. Use Table A23.

17.2.2 Confidence interval

Confidence intervals about the sample mean direction (angular degrees) are used to test whether the sample mean direction is significantly different from the predicted direction.

Procedure:

1 Calculate the length of the mean vector (r). This was calculated above using the second method to determine the sample mean angle.

$r = R/N$

2 From Figures A1 or A2 (for 95 and 99% confidence, respectively), use r and N to determine δ, the interval in degrees about the mean.

Confidence interval (95 or 99%) $= \bar{\Phi} \pm \delta$

3 If the predicted direction (e.g. home direction) lies outside the confidence interval then the sample mean direction is significantly different than the specified direction.

As an example, we will test whether the sample mean angle for the 13 blindfolded students (see example in Rayleigh and V tests above) is significantly different from the predicted direction (Colorado State University campus).

1 We already calculated r as 0.5311, and $N = 13$.
2 Using r and N we extrapolate on Figure A1 and find that δ is approximately 44° for the 95% confidence limit.
3 The 95% confidence interval $= \bar{\Phi} \pm \delta$
$$= 143.5° \pm 44°$$
$$= 99.5° \text{ to } 187.5°$$
Since the predicted direction (196°) lies outside the confidence interval for the sample mean angle, we reject the H_0 that the two directions (angular degrees) are not significantly different. That is, the mean angle of the directions pointed by the blindfolded students differs significantly from the direction to the campus.

17.3 DIFFERENCE IN SAMPLE MEAN DIRECTIONS BETWEEN TWO, OR MORE, GROUPS OF ANIMALS

These tests are used to determine whether there is a significant difference between the sample mean directions (or times) of two, or more, groups of animals, such as a control group and one, or more, treatment group(s).

17.3.1 Two sample runs test

This is a quick and simple two sample test to conduct. It is merely an application of two sample runs test, described earlier in Chapter 14. This test reduces interval data to ordinal data and should not be used with large samples since its power is low (Batschelet, 1981).

Procedure:

1 Plot all the sample directions, for both groups, on a circle *or* list them all in order, indicating which group each direction belongs to.

2 Determine the total number of runs (h) in the two samples. If the same direction (angular degrees) occurs in both samples (ties), their positions in the order must be randomly allocated (e.g. coin toss).

3 Find the probability for h in Table A25, where m and n are the sample sizes of each group; if there are different sample sizes, m is used for the smaller sample. You may have to extrapolate the probability for your calculated h.

As an example, we will use the data from the 13 blindfolded students that we used in the Rayleigh and V tests. We will test whether the directions the students pointed with the blindfolds on differed significantly from the directions pointed after the blindfolds had been removed. Table 17.6 shows the angular degrees from magnetic north listed in order for each sample.

The total number of runs (h) is 13. Using Table A25, we can see that when both sample sizes (m and n) are 13, the number of runs would have to be 9, or less, in order to have a probability less than 0.05. Therefore, we cannot reject the H_0 that the two samples are the same (i.e. drawn from the same circular population).

17.3.2 Watson–Williams two sample test

This is also a test for whether two samples of directions (angular degrees) are significantly different. However, this test uses the interval scale of measurement (angular degrees) resulting in a more powerful test than the two sample runs test described above.

Procedure:

1 Calculate for each sample:

S_1 = sum of the sines for Sample 1

S_2 = sum of the sines for Sample 2

C_1 = sum of the cosines for Sample 1

C_2 = sum of the cosines for Sample 2

$R_1 = \sqrt{(S_1^2 + C_1^2)}$

$R_2 = \sqrt{(S_2^2 + C_2^2)}$

Table 17.6. *Directions 13 students pointed, still blindfolded and after the blindfolds were removed, to indicate the direction to the Colorado State University after having been driven approximately 10 miles northeast of campus while blindfolded. The measurements are ordered to determine the total number of runs for a Two Sample Runs Test of significant difference in directions pointed under the two conditions (see text)*

Blindfolded	Not blindfolded	Runs
86		1
88		
	106	2
108		3
	110	4
	112	
114		5
128		
	132	
	140	6
	142	
144		
152		7
164		
	170	8
	178	
178		9
180		
	182	
	188	10
	194	
	198	
208		11
290		
	302	12
328		13

2 Calculate for the combined samples:

$$N_c = N_1 + N_2$$

$$S_c = S_1 + S_2$$

$$C_c = C_1 + C_2$$

$$R_c = \sqrt{(S_c^2 + C_c^2)}$$

3 Obtain the value g.

a. Calculate the mean vector length ($\bar{r}$).

$$\bar{r} = \frac{(R_1 + R_2)}{N_c}$$

b. Calculate the mean sample size.

$$\bar{N} = \frac{N_c}{2}$$

c. From Table A26 obtain an *estimated* value of k by using the mean vector length and mean sample size (calculated above) instead of r and n, respectively.

d. Calculate g:

$$g = 1 + \frac{3}{8k}$$

4 Calculate the test statistic F:

$$F = g(N_c - 2)\frac{R_1 + R_2 - R_c}{N_c - (R_1 + R_2)}$$

5 Compare the calculated F with the tabular value (Table A6) where: df$=1$ (for the numerator) and $N_c - 2$ (for the denominator). Note that all the tabular values for this test are in the first column of the table. If the calculated F is larger than the tabular F then the two sample mean directions are significantly different.

As an example, we will again test whether the 13 students pointed in significantly different directions when blindfolded and with the blindfolds removed.

1 We will designate the blindfolded group as sample 1; therefore, referring back to the Rayleigh test example, we have already calculated S_1 and C_1.

$$S_1 = +4.078 \qquad C_1 = -5.572$$

2 We now calculate the sum of the sines and cosines for the sample without blindfolds (sample 2) (Table 17.7).

Table 17.7. *The sine and cosine for the angular degrees (directions) the 13 students pointed after their blindfolds were removed (see Table 17.5). These data are used in the Watson–Williams two sample test (see text)*

Angular degrees	Sine	Cosine
110	+0.9397	−0.3420
194	−0.2419	−0.9703
112	+0.9272	−0.3746
140	+0.6428	−0.7660
302	−0.8480	+0.5299
142	+0.6157	−0.7880
106	+0.9613	−0.2756
198	−0.3093	−0.9511
132	+0.7431	−0.6691
178	+0.0349	−0.9994
182	−0.0349	−0.9994
170	+0.1736	−0.9848
188	−0.1392	−0.9903
Sums: $S_2 = +3.465$		$C_2 = -8.581$

$$R_1 = \sqrt{(S_1^2 + C_1^2)}$$

$$= \sqrt{(+4.078^2 + -5.572^2)}$$

$$= \sqrt{(16.630 + 31.047)} = \sqrt{47.677} = 6.904$$

$$R_2 = \sqrt{(S_2^2 + C_2^2)}$$

$$= \sqrt{(+3.465^2 + -8.581^2)}$$

$$= \sqrt{(12.006 + 73.633)} = \sqrt{85.639} = 9.254$$

3 We now calculate for the combined samples:

$$N_c = N_1 + N_2 = 13 + 13 = 26$$

$$S_c = S_1 + S_2$$

$$= +4.078 + (+3.465) = 7.543$$

$$C_c = C_1 + C_2$$

$$=(-5.572)+(-8.581)=-14.153$$

$$R_c=\sqrt{(S_c{}^2+C_c{}^2)}$$

$$=\sqrt{[(+7.543)^2+(-14.153)^2]}$$

$$=\sqrt{(56.896+200.307)}=\sqrt{257.203}=16.037$$

4 We then obtain the value of g.

$$\bar{r}=\frac{(R_1+R_2)}{N_c}=\frac{6.904+9.254}{26}$$

$$=\frac{16.158}{26}=0.621$$

$$\bar{N}=\frac{N_c}{2}=\frac{26}{2}=13$$

From Table A26, using $\bar{r}$ and $\bar{N}$ instead of r and n, respectively, we obtain an estimated value of k of 1.50.

5 We then calculate g.

$$g=1+\frac{3}{8k}=1+\frac{3}{12}=1+0.25=1.25$$

6 We then calculate the test statistic F.

$$F=g\,(N_c-2)\frac{(R_1+R_2)-R_c}{N_c-(R_1+R_2)}$$

$$=1.25\,(26-2)\frac{(6.904+9.254)-16.037}{26-(6.904+9.254)}$$

$$=1.25\,(24)\frac{0.128}{9.742}=(30)\,(0.013)=0.39$$

7 We then compare the calculated F (0.39) with the tabular value (Table A6) where the df=1 and $N_c-2=26-2=24$. Since our calculated F of 0.39 is smaller than the tabular value (4.26) we fail to reject the H_0 of no significant difference between the directions pointed when blindfolded and not blindfolded.

17.3.3 Watson–Williams multisample test

This test is used to determine whether the sample mean directions from three, or more, groups are significantly different. Proceed the same way as for the Watson–Williams two sample test above, except for k samples.

1 Calculate S, C and R for each group (sample). Use the same calculations as were used in the Rayleigh test.

$S_1, S_2 \ldots S_k$

$C_1, C_2 \ldots C_k$

$R_1, R_2 \ldots R_k$

2 Calculate for the combined groups (samples):

$N_c = N_1 + N_2 \ldots + N_k$

$S_c = S_1 + S_2 \ldots + S_k$

$C_c = C_1 + C_2 \ldots + C_k$

$R_c = \sqrt{S_c^2 + C_c^2}$

3 Obtain the value of g.

a. Calculate the mean vector length ($\bar{r}$).

$$\bar{r} = \frac{(R_1 + R_2 \ldots + R_k)}{N_c}$$

b. Calculate the mean sample size ($\bar{N}$).

$$\bar{N} = N_c/k$$

where: k = number of groups (samples)

c. From Table A26 obtain the *estimated* value of k by using $\bar{r}$ and $\bar{N}$ (calculated above) as estimates of r and n in the table.

d. Calculate g:

$$g = 1 + \frac{3}{8k}$$

4 Calculate the test statistic F:

$$F = g \frac{(N_c - k)(\Sigma R_i - R_c)}{(k-1)(N_c - \Sigma R_i)}$$

5 Compare the calculated F with the tabular value (Table A6), where: df = 1 and $N_c - 2$. If the calculated F is larger than the tabular F then the sample mean directions are not all equal. This is a test for homogeneity amongst all the sample mean directions (Batschelet, 1981) and does not allow for discrimination among individual sample mean directions.

17.4 SPATIAL PATTERNS: ANIMAL–ANIMAL AND ANIMAL–ENVIRONMENT

Ethologists generally measure the location of animals relative to: 1. other animals (animal–animal; e.g. mother–infant); and 2. the environment (animal–environment; e.g. home range). Although animal–environment spatial relationships are measured relative to the physical environment, they often reflect the effects of other animals (intra- or interspecific). For example, Lockie (1966) found that when the density of weasels (*Mustela nivalis*) decreased dramatically, they rarely encountered each other, and their territorial system disintegrated; then, rather than each weasel's home range being restricted to its territory, they wandered throughout the study area and all their home ranges overlapped.

The following discussion is a brief overview of the analysis of two types of spatial patterns with selected references to the literature. An excellent synopsis of spatial data analysis is provided by Upton and Fingleton (1985). Circular statistics for analyzing animal orientation data are discussed earlier in this chapter, and the use of film, videotape, digitizing and computers in spatial analysis are discussed in Chapter 9.

17.4.1 Animal–animal spatial relationships

Animal–animal spatial relationships can involve intra- or interspecific associations between individuals or groups. Additionally, the research question can require the measurement of distances or associations between known individuals over time, or spatial relationships among all members of the group being sampled simultaneously.

The observability of the animals and your ability to accurately determine their positions, either in the environment or relative to each other, will influence the type of sampling method employed. The methods basically involve two different procedures: 1. determine associations by recording when (frequency and/or duration) two or more individuals are together (operationally defined by the researcher); or 2. locate the postion of all animals periodically and determine the actual (or relative) distances between individuals. The first procedure is often used in field studies and generally involves all-occurrences or focal-animal sampling. The second procedure is more common in laboratory or enclosure studies and uses scan sampling.

When the first procedure is used, the criterion distance between individuals (e.g. Mitani *et al.*, 1991), or the distance between individuals used to define an association, depends on the species under study and the researcher's objectives and experience with that species. In some studies, less specific distances between individuals and more generalized locations within the groups will suffice. For example, Collins

(1984) was interested in the association of individual yellow baboons with different groups of individuals (defined spatially) within the moving troop. He used instantaneous samples of focal animals to record the amount of time they spent in each part of the moving troop (front, side, rear, middle or clusters; see Collins, 1984, for a diagram).

When associations between individuals are based on distances, the researcher must decide on a criterion distance between individuals in which the probability of their interacting greatly increases and beyond which they commonly approach one another in order to re-establish the association; this decision is based on the researcher's experience. For example, Grant (1973) used as his 'measure of association' between individual grey kangaroos (*Macropus giganteus*) the number of times each animal occurred within 120 cm of another at set 15–minute intervals (instantaneous samples). With this procedure, the accuracy of observers in determining the distances between individual animals can be a problem, especially in field studies involving distant observations. For example, Morton (1993) measured the accuracy of observers in determining the locations of individual elk in small herds. The distance discrepancy between observed animal locations (from observer diagrams) and actual animal locations (from aerial photographs taken simultaneously with ground observations) averaged 5.6 body lengths. Sullivan and Morton (1994) measured the ability of ground observers to judge inter-animal distances in groups of life-size artifical deer. They took photographs from different observer viewing angles of different herd sizes with the animals oriented in different directions (facing away or perpendicular to the line of sight); then observers diagrammed the animals' locations from the photographs. Larger herds, lower viewing angles, and perpendicular orientation of the animals produced greater discrepancies between observer perceived and actual animal locations.

Enclosures or small field study sites are often gridded to improve the observer's accuracy in determining animal locations. For example, Vastrade (1987) studied the spacing behavior of nine free-ranging domestic rabbits in a meadow gridded into 1.25 m×1.25 m squares; he scan sampled the animals' locations every 15 min. for three 24 h periods and recorded their locations within individual quadrats. To assist in measuring the spacing pattern of butterflyfish (*Chaetodon trifasciatus*) at coral reefs, Sutton (1985) surveyed an area of approximately 8000 square feet at each of his study sites into a grid system of 10 m×10 m quadrats; the grid line intersections were marked with coded subsurface buoys.

The resolution of association data taken from animals in a gridded area is proportional to the size of the quadrats relative to the size of the animals, since an animal is recorded only as being present in a particular grid, regardless of its actual position in that grid. Figure 17.4 illustrates how a large grid system may not clearly reflect the actual distances between animals. Individuals A and B are in adjoining

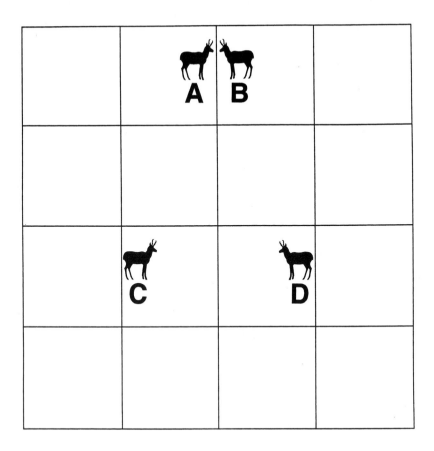

Fig. 17.4 Hypothetical positions of eight individual pronghorns (A–H) in two gridded
enclosures (see text for explanation; drawing by Lori Miyasato).

quadrats as are individuals C and D. Although data analysis would treat these pairs
of individuals as being at equal distances from each other, in fact A and B are much
closer together than C and D. The opposite problem is illustrated by pairs of individuals E and F, and G and H, which are actually the same distance apart but separated by one and two quadrats, respectively. If the grid on the left was divided into
an 8×8 format, then grid columns 2 and 3 would be halved, separating individuals
C and D by two quadrats, but leaving A and B in adjoining quadrats; this would
more accurately reflect their true distances. However, smaller quadrats don't always
improve resolution and accuracy since it may be difficult to determine which
quadrat an animal is in.

The accuracy of animal-location data from grids is limited, in part, by the size of
the study area, the density of animals and the size of the quadrats relative to the size

of the animal. To control for this potential problem, Burgess (1979) compared the spatial behavior of rhesus monkeys, neon tetras, communal and solitary spiders, cockroaches and gnats by observing them in gridded open-field arenas, scaled to the animal's size. He scan sampled the animals' locations by photographing them at intervals.

Accuracy is also affected by the criteria for deciding when an animal is in a quadrat and the observer's ability to make that determination (affected by viewing angle, distance and orientation of the animal). Researchers have often designed special equipment and techniques to give them more accurate animal location data. For example, Pitcher (1973) used a gridded flow tank, mirror and camera (Figure 17.5) to measure animal-animal spatial relationships in his research on minnow schooling.

Animals can be accurately located in the laboratory, enclosures, or small study areas, by filming or videotaping from directly above (or a high viewing angle) and then digitizing each animal's location from a screen or video monitor. For example,

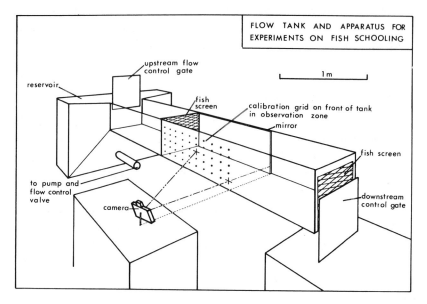

Fig. 17.5   Diagram of the flow tank and apparatus used in experiments on minnow
schooling. The X axis was taken as running along the length of the tank; the Y
axis across it; and the Z axis was taken as vertical. Therefore the photographs
showed fish and their reflections in the XZ plane only (from Pitcher, 1973).

Mankovich and Banks (1982) used time-lapse photography and computerized film
digitization to analyze the position and social orientation of five female domestic
fowl in a flock. (See Chapter 9 for additional discussions of the use of films and
videotapes for spatial analyses.)

Analysis of location data from animals in a gridded area can be based on distances (quadrats) between known individuals or overall spatial patterns between all
members of the group. Aspey (1977a,b) has written computer programs in BASIC
for computing inter-individual distances within a gridded rectangular area
(RECDIS) and a gridded circular area (CIRDIS). Stricklin *et al.* (1977) described a
FORTRAN program which analyzes relative distances between individual animals
within a gridded enclosure (square, circular, or rectangular) and also considers each
animal's angle of orientation on a 360° scale. This allows the calculation of angular
relationships and a determination of the angles any two individuals would have to
turn in order to be facing each other. Ludwig and Reynolds (1988) describe computer programs in BASIC for several types of spatial pattern analysis, such as
paired-quadrat variance.

Regardless of how it is obtained (e.g. grids, biotelemetry, digitized from videotape), animal-location data can be used to analyze: 1. the spatial relationships of all

members of the group or aggregation simultaneously (e.g. nearest neighbor analysis); or 2. the degree of spatial association between individuals, groups or species (e.g. association indices or coefficients of association).

17.4.1a Nearest neighbor analysis

There are several formulas available for calculating indices of 'aggregation', 'cohesion', 'crowding' and 'dispersion' which reflect the overall spatial relationships of animals in a group or population (Southwood, 1966). One method is *nearest neighbor analysis* (Clark and Evans, 1954); for example, Campbell (1990) used this analysis in describing the spatial relationships of singing crickets.

There are two basic approaches to nearest neighbor analysis: 1. select an individual at random and measure the distance between it and its nearest neighbor (true nearest neighbor technique); and 2. select a point and measure the distance to the nearest or nth nearest individuals (closest individual techniques). The simplest formula is from Clark and Evans (1954).

$$\text{density per unit area} = m = \frac{1}{4\bar{r}^2}$$

where: $\bar{r}$ = mean distance between nearest neighbors

When second, third, . . . nth nearest neighbors are measured, the following statistic (Thompson, 1956) can be calculated, which is distributed as a χ^2 with $2N$ degrees of freedom:

$$2\,\pi\,m\frac{\Sigma r_n^2}{N}$$

Southwood (1966) can be consulted for more detail on basic nearest neighbor analysis. Ripley (1979) described the complication of edge effects caused by nearest neighbor analyses being applied in small areas. Donnelly (1978) provided formulae to correct for edge effects, and DeGhett (pers. commun.) described methods for dealing with both the effects of edges and small numbers. Stapanian *et al.* (1982) suggested using sampling grids (Figure 17.4) where edge effects are a problem.

17.4.1b Association indices

Several *association indices* used to measure the frequency of association between individuals were reviewed by Cairns and Schwager (1987) and Ginsberg and Young (1992). Four commonly used indices are presented below, following the terminology of Cairns and Schwager (1987).

Half-weight association index:

$$\frac{x}{(n_a+n_b)/2}$$

where: x=number of observation periods during which individuals A and B are observed together

n_a=total number of observation periods during which A is observed

n_b=total number of observation periods during which B is observed

This association index is also known as Cole's, Dice's, Sorenson's and the coherence association index. It is the association index most commonly used by ethologists (Cairns and Schwager, 1987). As an example, Penzhorn (1984) used this index to measure individual associations in Cape Mountain zebras.

Twice-weight association index:

$$\frac{T_t}{T_t+T_a+T_b}$$

where: $T_t=x$=number of observation periods during which individuals A and B are observed together

T_a=number of observation periods during which A is observed in the absence of B

T_b=number of observation periods during which B is observed in the absence of A

As an example, Myers (1983) used this index to measure associations between individual sanderlings.

Simple Ratio Association Index:

$$\frac{x}{x+y}$$

where: y=number of observation periods during which A and/or B are observed in separate groups

As an example, Clutton-Brock *et al.* (1982) used this index to measure the associations between individual red deer. Poole (1989) used the following modification of the simple ratio index to measure associations between pairs of sexually active male elephants (*Loxodonta africana*).

$$\text{Association index}=\frac{T}{(D_I+D_J)-T}$$

where:

T=number of days the pair was seen together in a group of females

D_I=number of days individual I was seen in female groups

D_J=number of days individual J was seen in female groups

Square root association index:

$$\frac{x}{\sqrt{[(x+y_a+y_{ab})(x+y_b+y_{ab})]}}$$

where:

y_a=number of observation periods during which only A is observed

y_b=number of observation periods during which only B is observed

y_{ab}=number of observation periods during which A and B are both observed in separate groups

Lott and Minta (1983) derived this index to measure associations between individual American bison.

The Half-weight, simple ratio and square root indices are all the same when both individuals (A and B) are seen during every observation period; the twice-weight and simple ratio indices are identical if A and B are never observed in separate groups (y_{ab}=0; Cairns and Schwager, 1987). The biases inherent in these indices are discussed by Cairns and Schwager (1987) and Ginsberg and Young (1992).

Several researchers have calculated the probability of associations due to chance and used those as expected values to compare to their observed values. For example, Festa-Bianchet (1991) measured the association of kin in her study of bighorn sheep sociality. She recorded group composition (which could include ewes and their sons and daughters) during censuses of the winter range and searches of the summer range. She calculated the probability that a ewe would be in the same group as her daughter if they were distributed at random (T) as:

$$T=(n-1)/N-1)$$

where: N=number of ewes two years of age and older in the population

n=number in the group

The probability that a ewe would occur by chance in the same group as her yearling son or daughter (T) was calculated as:

$$T=n/N$$

The expected number of times that a mother and her daughter would have been seen together (E) if they behaved independently was calculated as:

$$E=\sum_{i=1}^{t}(n_i-1)/(N-1)$$

Festa-Bianchet also established criteria for determining valid sightings of associations. The number of valid sightings when mother and offspring were seen together was compared to the expected number (E) with Wilcoxon matched-pairs signed-ranks tests.

Mitani *et al.* (1991) calculated expected rates, durations and proportions of total time of associations between individual orang-utans by adapting models developed by Waser (1982). The average duration of associations expected by chance was calculated as:

$$T_{exp} = 2.467 \, r(v_i^2 + v_j^2)^{-1/2}$$

where:

r = the distance criterion used to define associations (30 m in this study)

v_i = the mean travel velocity of individual i

v_j = the velocity of all other conspecifics or age–sex class of individuals

An estimate of the expected proportion of the time each individual spent in association with conspecifics was calculated as:

$$P_{exp} = 4.934 r^2 p_j$$

where: p_j = the density of groups of species j

The observed versus expected durations of associations and proportions of time in associations were tested using the Wilcoxon test.

Cairns and Schwager (1987) describe how assumptions about biases in the data collection can be used to develop formulae for expected values which can then be used to derive a maximum likelihood estimator for the individual associations. Ginsburg and Young (1992) suggest that several maximum likelihood estimators may be needed for each study. Further, they recommend that if researchers choose not to use maximum likelihood estimators, they should use the simple ratio index, the most accurate of the indices described above.

A final caution on the analysis of animal–animal spatial relationships is provided by Bekoff and Corcoran (1975). They stress the importance of knowing not only the spatial relationships between individuals, but also the behavior of each. For example, copulation, fighting, and nursing all require close spatial relationships; but their interpretation is quite different.

17.4.2 Animal–environment spatial relationships

The procedures described above for measuring animal–animal spatial patterns can also be used in studies of animal–environment spatial relationships in the laboratory, enclosures or field. For example, Weeden (1965) used a grid system in the field to measure spatial patterns of tree sparrows (see below). Likewise, Kleerekoper

(1969), in his laboratory study of fish–environment spatial relationships, used a grid system different than that in Figure 17.5. He used photosensors in a 50×50 X–Y grid to detect the position of fish relative to olfactory gradients in a large tank. An on-line computer then analyzed the X and Y coordinate positions of the fish, the time of occurrence of a change in position, velocity of movement, distance covered, angle of turns, left- and right-handedness, and the statistical significance of changes in these parameters. (Similar systems using films or videotapes, digitizing, and computer storage and analysis are described in Chapter 9.)

Long-term animal–environment spatial patterns are generally reflected as *home ranges* and/or *territories*. The home range of an individual is that area covered during normal daily activities (Blair, 1953); researchers have sometimes chosen to refer to these distributions of an animal's time relative to space as 'activity fields' (e.g. Smith and Dobson, 1994). An animal's home range may overlap the home ranges of other individuals (or groups, Figure 17.6) and may, or may not, contain a territory, an area defended against members of the same species and occasionally other species. The designation of territories is often the result of finding contiguous and non-overlapping home ranges, suggesting that the entire home ranges are mutually exclusive and are thus assumed to be territories (e.g. weasels, Lockie, 1966).

Home ranges and territories are calculated from successive locations of individuals, or groups, obtained through: 1. *direct observations* (continuous or time samples); or 2. *indirect methods*, including (a) location of natural signs (e.g. tracks), (b) capture-recapture, (c) radioactive material, (d) dyes for urine and feces, (e) photographic devices, and (f) radiotelemetry. Some of these techniques are discussed in Chapter 9, along with recommended references.

Weeden (1965) used a gridded area to measure the territories of tree sparrows (*Spizella arborea*) (Figure 17.7). An observer spent an entire four-hour observation period with a single pair of sparrows (focal pair) plotting every location visited on a map. Weeden then used the observation-area curve method to determine the number of observations necessary to calculate a reasonably accurate territorial area.

The observation-area curve method (Figure 17.8) was developed by Odum and Kuenzler (1955) to assist in determining the number of observations (locations) necessary to determine the territory sizes of several bird species. It is based on the same principle as the curves used to construct an ethogram and assess the repertoire size (Chapter 4); that is, as the number of observations increases, the rate of increase in area visited by the bird decreases, so that an asymptotic curve results (Figure 17.8). Odum and Kuenzler (1955) arbitrarily selected the one-percent level on the curve as the point at which the territory size was determined. The 1% level is that point on the curve where each additional observation produces less than a 1%

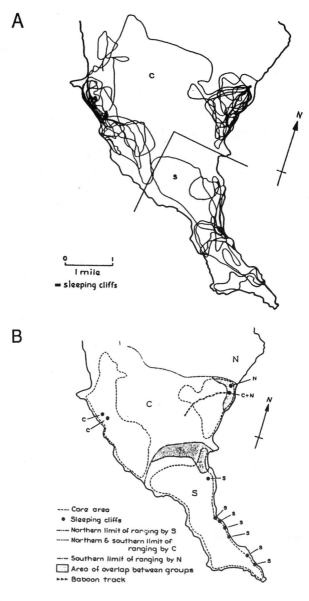

Fig. 17.6 Baboon group home ranges and core areas. A. Ten-day ranges for group S and
21–day ranges for group C, Cape Reserve. The sharp line indicates the
approximate limit of each group's home range. B. Areas occupied by C and S
groups and southern limit of N group's range, indicating amount of overlap in
home ranges and location of core areas (from DeVore and Hall, 1965).

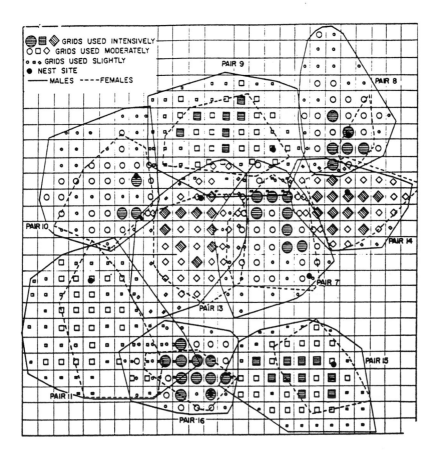

Fig. 17.7 Location and utilization of total activity spaces of males, location of total
activity spaces of females, and location of nest sites of tree sparrows (from
Weeden, 1965).

increase in the area calculated as the territory. The number of observations necessary to reach the 1% level varied with the species and the stage of the nesting cycle.
Sanderson (1966) suggested that live-trapping mammals would probably provide
insufficient data to apply the observation-area curve method, but that radiotelemetry probably would.

The validity of the calculated home range, or territory, will depend on the technique employed. When direct observations are made, the animal's location can be
considered along an essentially continuous distribution. That is, it may be continuously observed throughout, and found at any point within, its true home range. This
sampling method can be considered instantanous sampling of a focal animal using
very small sample intervals. Weeden (1965; Figure 17.7) sampled using focal-pair,

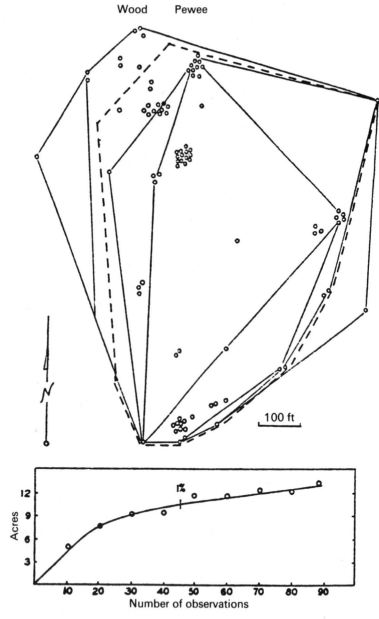

Fig. 17.8 Observed positions of a male wood pewee (*Contopus virens*) at five-minute
intervals (small circles; above), with maximum observed area enclosed in solid
lines after successive tens of observations. The broken line in the upper diagram
encloses the calculated maximum territory size (10.8 acres) at the 1% level (see
text for explanation) as shown on the observation area curve below (from Odum
and Kuenzler, 1955).

all-occurrences of visits to different quadrats. Some indirect measures, such as tracking in snow, can also provide continuous data, and biotelemetry can provide nearly continuous data by rapidly scan sampling individual locations (see Chapter 9).

Other methods, such as the commonly used capture–recapture method for small mammals, provides a discontinuous spatial distribution of locations restricted to trap sites. Therefore, the validity of this method is affected by trap spacing (Stickel, 1954), as well as trapping interval (time of trap set to trap check), sample size (number of trapping intervals) and the responses of individual animals to the traps (Balph, 1968; see below). This sampling method can be considered a one–zero sample (Chapter 8) for each trap site. That is, for each trap site each individual is either captured, or not captured, during the trapping interval.

Most small mammal trapping designs used to determine home ranges and territories are grids similar to that in Figure 17.9, but with more traps. However, Lockie (1966) placed his traps near features in the environment likely to be frequented by weasels (*Mustela nivalis*) and stoats (*M. erminea*).

With capture–recapture methods, the animal is individually marked (Chapter 8) on the first capture, and its location recorded for that and subsequent captures. The following assumptions are inherent in the use of capture–recapture data to calculate home ranges:

1 The animal will be trapped over all of the ecologically significant area of its home range; that is, it will be trapped wherever it goes given the following conditions:

   (a) The grid being as large as (or larger than) its home range.

   (b) On encountering a trap the probability of capture being high.

   (c) The probability of capture on encountering a trap being the same throughout its home range.

2 The frequency of capture at a particular trap site reflects the frequency of visits by the animal.

3 Each animal has an equal chance of being captured upon encountering a trap.

Unfortunately, these assumptions are probably never completely met and perhaps seldom approximated.

The following are the more common, simple methods of describing home ranges based on data from trap grids (Stickel 1954):

1 Minimum area method. The outermost capture sites are connected by straight lines (e.g. Figure 17.9A; also called the minimum convex polygon method).

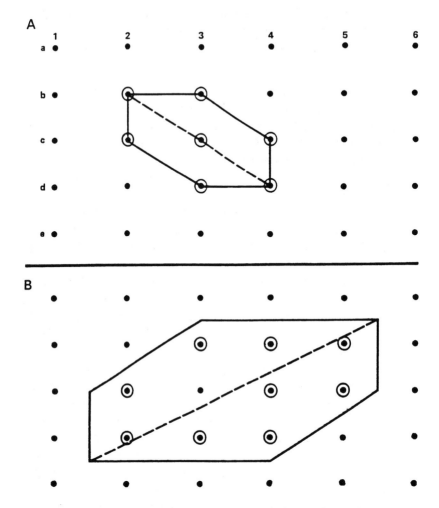

Fig. 17.9 Hypothetical captures of an individual small mammal in a grid of live traps. Circled dots denote sites of captures. Home range are indicated as: A. minimum area and observed range length; B. boundary strip and adjusted range length.

2 Boundary strip method. Points halfway between the outermost capture sites and the next closest trap sites are connected by straight lines (e.g. Figure 17.9B). The rationale behind this method is that on the average the animal will have traveled halfway to the next trap site during its movements (see Stickel, 1954, for the inclusive and exclusive variations on this method).

3 Observed range length. The distance between the two most widely separated capture sites is measured (e.g. Figure 17.9A).

4 Adjusted range length. The furthest distance across the home range calculated by the boundary strip method is measured (e.g. Figure 17.9B).

Another method which does not incorporate unoccupied grid cells, or traps where the animal was not caught, has been recommended by Waser and Wiley (1979), Getty (1981) and Lair (1987). Stickel (1954) concluded that the boundary strip method and the adjusted range length provide closer estimates of the true home range than the other methods. Stickel also found that trap spacing altered the apparent size of the home range even when trap visitations are random and biological factors are excluded.

These two-dimensional methods can be extended to a third dimension, and home range volume can then be calculated for arboreal species (e.g. cricetid rodents, Meserve, 1977). Koeppl *et al.* (1977) describe a three-dimensional home range model.

An individual will occasionally be caught at a great distance from the cluster of other captures. These 'outliers' may reflect occasional excursions out of the animal's true home range and are often disregarded in delineating the home range using the descriptive methods discussed above, or statistical procedures (Samuel and Garton 1985; examples listed below).

An individual animal's behavioral response to a trap reflects its own unique predispositions and responses to experience (Chapter 2). Balph (1968) observed the behavioral responses of uinta ground squirrels (*Citellus armatus*) to live traps. He found that the trap was initially an attractant which could be enhanced by baiting; however, because of the configuration of the trap there was an equal probability of capture on the first encounter whether the trap was baited or not. Capture appeared to be punishing while the bait served as a reward (see model; Chapter 2); this produced a conflict between the tendencies to approach and avoid the trap on subsequent encounters, and recaptures were influenced by the relative strengths of these tendencies in different individuals (Figure 17.10).

Getz (1972) used multiple captures (more than one individual in the same trap) to infer associations between sex and age groups in a population of *Microtus pennsylvanicus*. He compared the number of multiple captures of the different sex and age categories to that expected from their relative frequency in the population. He used a chi-square test to determine whether they were found together more frequently or less frequently than expected and inferred attraction and avoidance, respectively. Slade (1976) elaborated further on Getz's procedure.

Individuals are often captured more than once in the same trap, perhaps reflecting a disproportionate use of those portions of its home range. Hayne (1949) calcu-

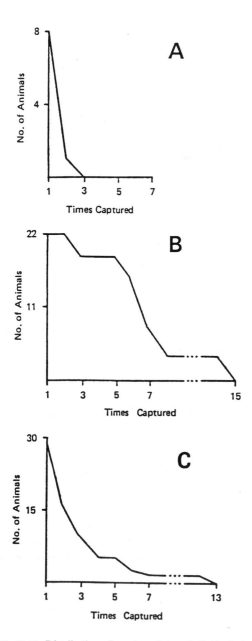

Fig. 17.10 Distribution of number of times individual uinta ground squirrels were 'captured' in an unbaited and functional trap (A); a baited, but nonfunctional trap (B); and a baited and functional trap (C). Shows the effect of reward (bait) and punishment (capture in a functional trap) on different individual squirrels' behaviors (from Balph, 1968).

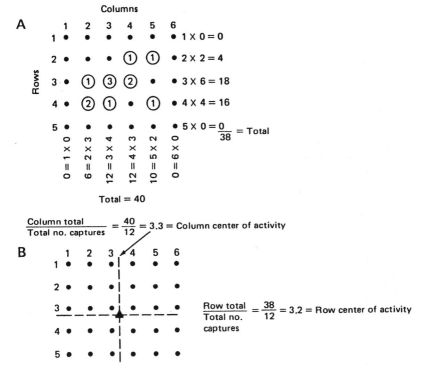

Fig. 17.11 Calculation of *center of activity* (triangle, grid B) based on hypothetical
recaptures (circled numbers, grid A). The numbers in the circles denote the
number of captures at each trap site (see text for explanation).

lated the *center of activity*, the mathematical center of the distribution of total captures within the grid, taking into account the number of captures at each trap site
(Figure 17.11). This can be obtained by weighting the rows and columns of the trap
grid and multiplying the number of captures in each row and column by its respective weight. The totals for all rows is then divided by the total number of captures in
the grid to determine the center of activity for the rows; the same procedure is
repeated for the columns. The point at which the centers of activity (means) for the
rows and columns intersect is the center of activity for the animal's home range
(Figure 17.11). Rather than using the means of captures along the coordinates of
the grid, Mohr and Stumpf (1966) suggested using the medians of captures in the x
and y coordinates of the grid to determine the median center of activity. In contrast
to the rather simple calculations necessary to determine the mean or median center
of activity, Dixon and Chapman (1980) proposed a more complex method for calculating the center of activity based on the harmonic mean. Lair (1987) compared

the three methods using data from red squirrels and found that the harmonic mean was the only method that generated a center of activity which coincided with the behavioral focal center, whenever the latter could be identified by direct observation.

The center of activity does not necessarily reflect the location of anything specific (e.g. the animal's nest or burrow), but the distance between centers of activity for residents of adjacent territories, or home ranges, might be used to infer their relative avoidance throughout the year (see Clark and Evans, 1954).

Koeppl *et al.* (1975) suggested that the center of activity might better be called the 'center of familiarity', based on Ruff's (1969) correlation of uinta ground squirrels' (*Spermophilus armatus*) heart rates with their locations in their home ranges. Koeppl and his colleagues also demonstrated how confidence ellipses can be calculated around the center of activity in elliptical home ranges and then used to determine the probability of finding the resident at any given location. Weeden (1965) used the relative number of visitations to the different quadrats in the tree sparrow's territory to determine 'central cores of more concentrated use' (Figure 17.7).

Disproportionate use of the home range is also revealed by continual following, or tracking, of individual animals (or groups) for several days by direct observation or radiotelemetry (e.g. Sargent 1972; Chapter 9). Primatologists often observe groups of primates for ten-day periods in order to construct ten-day ranges, within which are generally found areas of heavy usage designated 'core areas' (e.g. DeVore and Hall, 1965; Figure 17.6). White and Garrott (1990) provide statistical methods for comparing utilization distributions.

Beyond the graphic methods discussed above (e.g. boundary strip method), several mathematical/statistical models and methods have been developed for calculating home range and center of activity from large data sets of animal movements and locations. These methods include the use of: Fourier estimator (Anderson, 1982), harmonic mean (Dixon and Chapman, 1980), Jennrich–Turner estimator (Jennrich and Turner, 1969), Dunn estimator (Dunn and Gipson, 1977), bivariate (Koeppl *et al.*, 1975), weighted bivariate normal estimator (Samuel and Garton, 1985), and minimum convex polygon (Southwood, 1966). White and Garrott (1990) describe the application of several methods to radiotelemetry data.

Software packages are available for home range calculations on microcomputers. *McPaal* is a menu-driven software package for analyzing animal location data on IBM-PCs and compatibles; it calculates home ranges using the convex polygon, concave polygon, 95% ellipse, fourier and harmonic mean methods. McPaal is available from Michael Stuwe, Conservation and Research Center, National Zoological Park, Front Royal, VA 22630. *Wildtrak* is a package of nonparametric home range analyses for the Apple Macintosh. Analyses include animation, autocorrelation, drift, dynamic interaction, grid cell analysis, movements, polygon analysis and static interaction. *Wildtrak 2* will have additional analyses including habitat preference and harmonic mean. Wildtrak is available from Dr Ian Todd, 6, Sollereshott House, Linkside Ave., Oxford, OX2 8JA, United Kingdom.

Once home ranges have been calculated, overlap of home ranges is sometimes determined (e.g. Geffen and Macdonald, 1992; Swihart, 1992). As an example, Lazo (1994) calculated the home ranges for feral cattle using the minimum convex polygon method, then calculated the home range overlap for all possible dyads of individuals using the following formula:

$$\text{Home range overlap} = V = \frac{2P_{\text{I}}}{(P_{\text{A}} + P_{\text{B}})}$$

where:

P_{I} = area of the polygon delimited by the intersection of the home ranges
P_{A} = area of the home range of individual A
P_{B} = area of the home range of individual B

The methods commonly used to measure home range overlap, such as Lazo's (above) does not take into account the frequency of use of that area by each individual. Smith and Dobson (1994) describe a method for calculating asymmetrical weighted overlap values between neighboring individuals, including a computer program written for Statistical Analysis Software (SAS Institute, Cary, NC) which will calculate those values.

The method(s) selected for sampling animal locations and describing or calculating the home range (or territory) should be based on a knowledge of the species' behavior. As examples: 1. How easily can they be observed (nocturnal? dense vegetation?), and what is the relative size of their suspected home range (e.g. m^2, hectares, km^2)? Will these conditions require the use of radiotelemetry?, 2. How rapidly and continuously do you suspect they move throughout their home range? How does this affect your choice of sampling intervals (Swihart and Slade, 1985)?

Sampling methods will be limited by constraints on your time, equipment and abilities, and determination of home range will be limited by the quality and quantity of the animal location data. For example, based on simulations, Bekoff and Mech (1984) suggest that fieldworkers should ascertain 100 to 200 animal locations in order to estimate reliably home range area (also see White and Garrott, 1990).

Researchers should constantly assess the validity of both their sampling and home range determination methods. For example, Jones and Sherman (1983) compared the home ranges of meadow voles using: 1. grid trapping (capture–recapture) and radiotelemetry to obtain the location data; and 2. several methods to calculate the home ranges. Only the convex polygon method gave approximately the same estimate for an individual's home range whether grid trapping or radiotelemetry data were used.

18 Interpretation and presentation of results

Discovery consists in seeing what everybody else has seen and thinking
what nobody else has thought. [Albert Szent-Gyorgy]
'Science' interprets. That means that a number of minds agree that in a
given phenomenon there is something that occurs with regularity, can be
reproduced, and can be traced back to recognizable causes, something,
indeed, that can be 'interpreted'. [Eigen and Winkler, 1981:21]

18.1 WHAT DO YOUR RESULTS MEAN?

The end point (temporary pause in the ethological approach cycle) in your research
is not the results of your data analyses, but their interpretation.

Have you been able to reject your null hypothesis (H_0)? If so, you can accept the
alternate hypothesis (H_1) and take temporary pleasure in your accomplishment. If
you were not able to reject your H_0, then you have negative results which are difficult
to interpret (Kerlinger, 1964; Chapter 11). Most importantly, note that failure to
reject the H_0 does not mean that you automatically accept it. Rather, it means that
several factors could have contributed to your results, only one of which was that H_0
was true. Also consider your techniques – were they reliable and valid? Did you
overlook an important parameter in your original design? Are you now convinced
that the H_0 is true or should you design another experiment to test the H_0 in what
you consider to be a more valid approach?

If you were merely making reconnaissance observations or using analyses (e.g.
cluster analysis) in hypothesis seeking, can you now generate a meaningful and
testable hypothesis?

Once you have your results it is often useful to prepare a visual representation for
further inspection.

18.2 VISUAL REPRESENTATIONS

Most researchers can learn more about the results of a research project by studying
visual presentations (other than tables) of results. Figures and graphs are not only

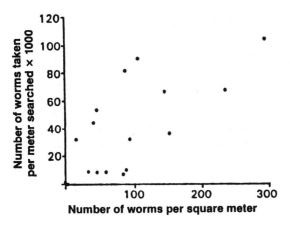

Fig. 18.1 Redshank (*Tringa totanus*) foraging behavior. The numbers of worms above 30
mg dry weight taken per meter searched in relation to their density in the mud
(from Goss-Custard, 1977).

useful in presenting your results to other people, but also help you interpret the
results and see relationships that were not apparent in the tabular data (Wolff and
Parsons, 1983). Increased insight into interpretation and new hypotheses are often
the result of careful contemplation of visual representations; Cleveland (1993)
shows the relationship between visualization and classical methods of data analysis.
Different visual presentations of the same data sometimes allow you to recognize
subtleties in relationships which were previously hidden; Tufte (1983) illustrates
numerous ways to display quantitative information visually.

The dendrogram, described previously for cluster analysis (Chapter 18), is obviously a valuable tool. Others which are useful are described below.

18.2.1 Graphs and figures

There are several types of graphic formats which have proven valuable in interpreting results. The simple scatter diagram (scattergram) is generally used to graph correlation data. The interpretation of the graph is dependent on the location and
distribution of the points relative to the axes (Chapter 14).

For example, Goss-Custard (1977) examined the hypothesis that the redshank
(*Tringa totanus*) selects the sizes of polychaete worms (prey) that maximize the
biomass ingested per unit time. He plotted the number of large worms taken relative
to their density in the mud (Figure 18.1). The graph revealed an apparent positive
correlation that was then shown to be statistically significant.

Frequency distributions are plotted in two common formats – frequency polygons and histograms.

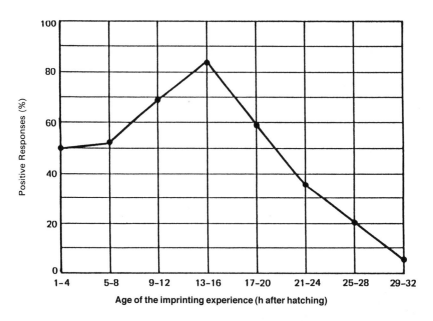

Fig. 18.2   Mean scores made in testing, 24 hours after the imprinting experience, by ducklings which had been given the imprinting experience at different ages (from Hess, 1962).

Frequency polygons are usually produced by connecting points along a continuous frequency distribution. Points should generally not be connected if the distribution is not continuous or if the sample points are distantly separated. Connection implies that the line between the points is a reasonable representation of the missing data.

Hess (1962, 1972), in his studies of imprinting, exposed ducklings to an adult model (decoy) for a limited period early in life and later tested them for the imprinting response (Figure 18.2). The frequency polygon shows that the highest percentage of positive responses was given by ducklings exposed to the model at 13–16 hours after hatching, the 'critical period' or 'sensitive period' for this species.

Frequency polygons are sometimes used with discontinuous data to show relative changes from one condition to another, while the identity of several individuals or levels of another variable are maintained. Figure 18.3 shows the mean number of courtship displays for 11 male guppies (*Poecilia reticulata*) at different population densities. As the population density increased beyond two pairs, the mean number of displays for most males increased or decreased in an unpredictable manner (Farr and Herrnkind, 1974).

Several frequency polygons can be combined into a single graph to show changes in several dependent variables relative to an independent variable. Figure 18.4 illus-

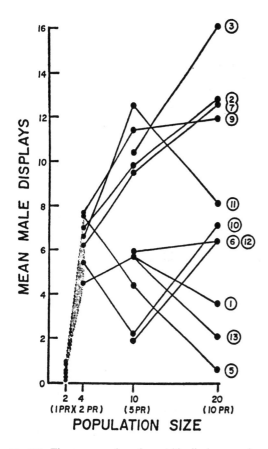

Fig. 18.3   The mean number of courtship displays per observation period for specific male
guppies. Each line represents a single male (from Farr and Herrnkind, 1974).

trates the change in relative occurrence of five behaviors in female woodchucks
(*Marmota monax*) as their infants grew older (Barash, 1974). See Figure 18.7 for
another approach to illustrating similar data.

Some ethologists have found it illustrative to present results in a three-dimensional frequency polygon. For example, Figure 18.5 demonstrates that the probability of a stickleback eating a food item is due, in part, to a complex relationship
between the cumulative number of eats and rejects which precede the encounter
with the food item (Thomas, 1977). It can be seen that, in general, the probability of
an eat occurring is greatest when the number of prior rejects is low, irrespective of
the number of prior eats.

Frequency distributions within categories of discontinuous (i.e. discrete) independent variable are often illustrated with histograms. When the means of a single

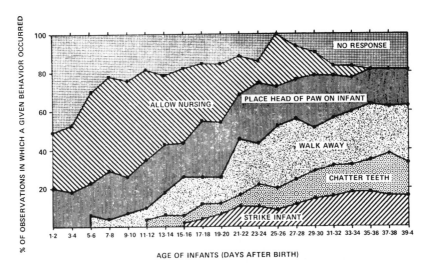

Fig. 18.4 Behavior of mother woodchucks immediately following onset of squeaking by infants (cumulative data for three litters, based on a total of 387 incidents, with a minimum of 14 per two-day interval) (from Barash, 1974).

variable are illustrated, the standard error of the mean (Figure 18.6A; generally preferred to the standard deviation) is informative relative to the statistical significance of the difference between the means. For example, Nyby *et al.* (1977) showed that male mice made significantly more ultrasounds in response to facial chemicals from females than from either males or controls (i.e. clean surgical cotton swabs), and their responses to male and control facial chemicals were not significantly different (Figure 18.6A).

It is often illustrative to incorporate two or more groups within an independent variable (e.g. dominant or subordinate within sex) into the same histogram (Figure 18.6B).

A type of histogram can be employed to show the emergence and disappearance of behavior over time. For example, Figure 18.7 illustrates the timing of emergence of postural, locomotor, and related skills in the laboratory rat (Altman and Sudarshan, 1975). Note that the presentation of results in this figure are similar to those presented in the combined frequency polygon (Figure 18.4).

The basics of graphing data are provided by Cleveland (1985). There is a plethora of computer software programs available for generating two- and three-dimensional color graphs of all types; consult other ethologists and computer store personnel for software packages that are compatible with your database and produce the variety of graphs you want.

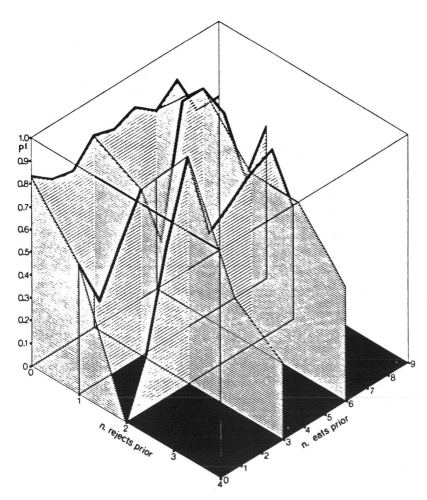

Fig. 18.5 Stickleback feeding behavior. Probabilities of an eat occurring (P^E) in any
encounter correlated against the accumulative numbers of foregoing eats on one
plane and foregoing rejects on the other (from Thomas, 1977).

18.2.2 Vector diagrams

Vector diagrams are used to illustrate the distribution of data relative to two or
more coordinate axes. Recall that three-dimensional vector diagrams were previously discussed as a useful method for presenting and interpreting the results of
factor analysis (Chapter 16).

Two-dimensional vector diagrams are often used to illustrate the directional
responses of individual animals in orientation studies. For example, Figure 18.8

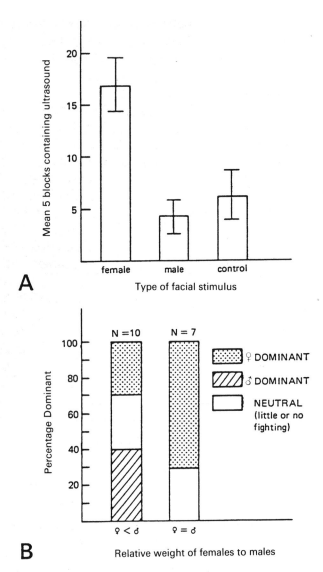

Fig. 18.6 A. Mean number of 5–second blocks containing ultrasound from DBA/J2 inbred male house mice in response to facial chemicals on a cotton swab (measurements=mean ± se) (from Nyby et al., 1977). B. Male–female dominance record in hamsters as a function of relative body weight. A female was considered lighter if its weight was 20 g or more below that of the male. (from Marques and Valenstein, 1977).

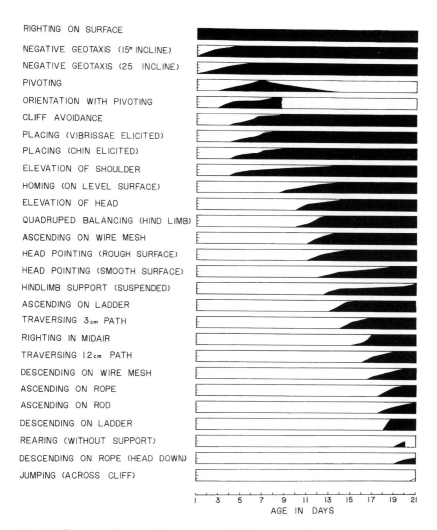

RIGHTING ON SURFACE
NEGATIVE GEOTAXIS (15° INCLINE)
NEGATIVE GEOTAXIS (25 INCLINE)
PIVOTING
ORIENTATION WITH PIVOTING
CLIFF AVOIDANCE
PLACING (VIBRISSAE ELICITED)
PLACING (CHIN ELICITED)
ELEVATION OF SHOULDER
HOMING (ON LEVEL SURFACE)
ELEVATION OF HEAD
QUADRUPED BALANCING (HIND LIMB)
ASCENDING ON WIRE MESH
HEAD POINTING (ROUGH SURFACE)
HEAD POINTING (SMOOTH SURFACE)
HINDLIMB SUPPORT (SUSPENDED)
ASCENDING ON LADDER
TRAVERSING 3 cm PATH
RIGHTING IN MIDAIR
TRAVERSING 1.2 cm PATH
DESCENDING ON WIRE MESH
ASCENDING ON ROPE
ASCENDING ON ROD
DESCENDING ON LADDER
REARING (WITHOUT SUPPORT)
DESCENDING ON ROPE (HEAD DOWN)
JUMPING (ACROSS CLIFF)

AGE IN DAYS

Fig. 18.7 Summary diagram of the emergence of different postural, locomotor and related
skills in the laboratory rat. in the majority of instances performance level
(vertical axis of each graph; 0, 25, 50, 75, and 100%) refers to the percentage of
animals successful in the full display of the response. In a few instance the
reference is to level of performance with respect to asymptotic response
frequency (from Altman and Sudarshan, 1975).

shows the compass directions in which individual hen mallards vanished from sight
after being released (Matthews and Cook, 1977). The effect of overcast conditions
on the mean direction can easily be seen.

McKinney (1975) used vector diagrams to illustrate the orientations, distances

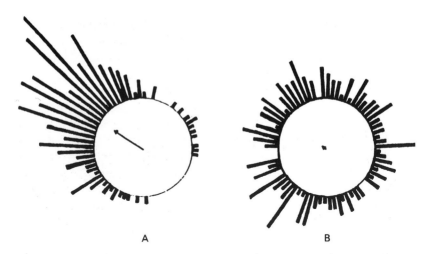

Fig. 18.8 The superimposed vanishing bearings of Borough Fen mallards released under (a) sunny and (b) overcast conditions. The centrifugal arrow indicates the mean vector (m) whose length (r) increases the tighter the bearings cluster about the mean. (from Matthews and Cook 1977).

and positions of the male green-winged teal relative to the female during selected courtship displays (Figure 18.9).

18.2.3 Kinematic graphs

Kinematic graphs (often called flow diagrams) are useful to illustrate transitions between behaviors (see Figure 8.3). Sustare (1978) discussed, in detail, the use of various systems diagrams including information networks (e.g. sociogram Figure 10.14), association diagrams (Figure 18.12), state-space diagrams and kinematic graphs. Halliday (1975) used two types of kinematic graph to show the sexual behavior sequence in the smooth newt (*Triturus vulgaris*). Figure 18.10 includes drawings of the male and female, which increases the ability of the reader to 'visualize' the sequence through the orientation of the two sexes.

Figure 18.11 provides increased information on the probability of particular transitions occurring (width of arrows). Altering the size of the arrows can be supplemented or replaced with the actual number (Massey, 1988) or probabilities of transitions in percentages. Malafant and Tweedie (1982) describe computer production of kinetograms in which each state is depicted as a circle and the number of lines between circles indicating the magnitude of the transition probability between the states.

GRUNT-WHISTLE   DOWN-UP

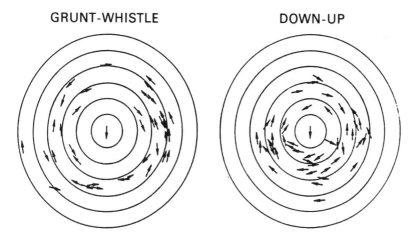

Fig. 18.9 Orientations, distances and positions of male green-winged teal in relation to the
female (center arrow) during performances of gruntwhistle, down-up, bridling,
and turn-back-of-head displays (see Figure 8.3). Note the precise lateral body
orientation of males when performing the grunt-whistle, and shorter distance
from the female in the case of the down-up. Grunt-whistle can occur when only
one male is present, but down-up is performed only when a second male is
present. The distance between concentric circles is one foot; a swimming teal
measures slightly less from bill-tip to tail-tip (from McKinney, 1975).

18.2.4 Conceptual models

Conceptual models are a means of maintaining perspective about the entire context
in which the behavior(s) of interest occur(s) (see the model discussed in Chapter 2).
They allow you to fit together pieces of information about a behavioral system (e.g.
reproductive behavior) in an attempt to understand better their causes and functions and illustrate the interrelationships between behaviors. Models are generally
hypothetical and temporary, being changed as new results come forth. For example,
Baerends (1976) proposed a model (Figure 18.12) to explain the occurrence of interruptive behavior during the incubation in herring gulls. Baerends broke the model
into 'systems', 'subsystems', and 'acts' which he relates to Tinbergen's (1950) earlier
conceptual model of the hierarchical organization of behavior (see also Dawkins,
1976a and Gass, 1985). Huntingford (1984) illustrates and discusses several additional conceptual models of motivation. McFarland (1971), McFarland and
Houston (1981) and Toates (1980) contain additional examples of systems models.
    Sometimes other concepts and mechanisms can be used as analogies or

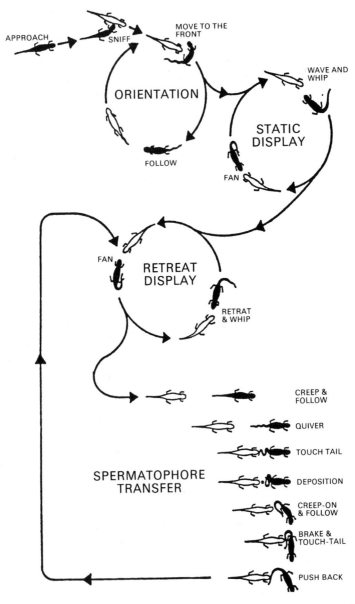

Fig. 18.10 Kinematic graph of the sexual behavior sequence of the smooth newt. The male
is in black (from Halliday, 1975).

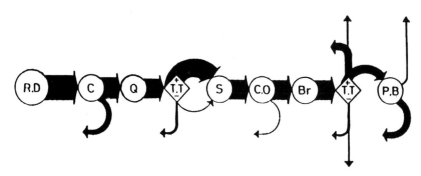

Fig. 18.11   Kinematic graph of the spermatophore-transfer phase of the smooth newt's sexual behavior sequence (Figure 18.10). Width of arrows is proportional to the frequency of transition. Arrows pointing to the left are returning to retreat display; arrows pointing outwards are leaving sexual behavior, for example, to breathe. Br=brake; C=creep; C.O.=creep-on; P.B.=push-back; Q=quiver; R.D.=retreat display; S=spermatophore deposition; T.T.=touch-tail (from Halliday, 1975).

metaphors to help visualize, and often better understand, behavioral processes. For example, Lorenz's original (1950), and revised (1981), psycho-hydraulic model of motivation has appeared to some to be analogous to a flush toilet (e.g. Goodenough *et al.*, 1993); however, it served as the basis for much early theorizing about innate animal behavior. For example, discussing Lorenz's early models, Thorpe (1979) stated:

> Some of his models were obviously analogous only – but the very essence of 'analogy' is its imperfections which challenges rethinking. One did not suppose them to be 'true' but they were valuable in being highly suggestive.
> *[Thorpe, 1979:103]*

Likewise, game theory models, such as Prisoner's Dilemma (e.g. Axelrod, 1984), have served as useful metaphors (Sigmund, 1993) for envisioning animal conflict from an evolutionary perspective (e.g. Maynard Smith, 1982).

Further perspective is provided by generalized conceptual models which help the ethologist visualize the complex of variables which impinge on behavior (Chapter 2). Some models aid researchers in recognizing how their research fits within the 'big picture' and assists in identifying important variables to investigate in future studies. A very general model of this type was described and discussed in detail by Crook *et al.* (1976). The broad use allowed by their model is illustrated in Figure 18.13. Colgan (1978) provides a good overall discussion of the role of modeling in ethological research.

Conceptual models often lead to *predictive models*, which are expressed in mathe-

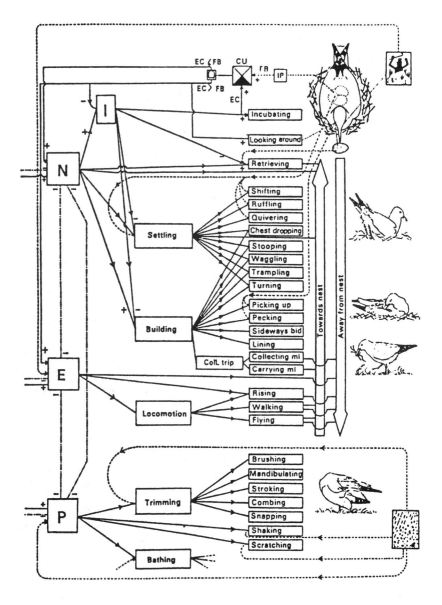

Fig. 18.12 Model for the explanation of the occurrence of interruptive behavior during the incubation of a herring gull. The fixed action patterns are in the right column and superimposed control systems of first and second order are represented left of them (N=incubation system, E=escape system, P=preening system). The large vertical arrows represent orientation components with regard to the nest. Incubating is the consummatory act. Feedback stimulation from the clutch, after being processed in IP, flows to a unit (CU) where it is compared with expectancy,

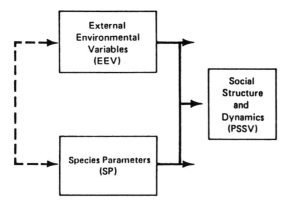

Fig. 18.13   A conceptual model showing how external environmental variables (EEV) are
expected to interact with species parameters (SP, for example, morphological and
physiological characteristics) to determine social structure (measured as the
principal social system variables (PSSV) and social dynamics (changes in PSSV
over time)). The dotted arrow takes note of the fact that EEVs also affect SP, but
on a slower (evolutionary) time scale than the effects on PSSVs, which may
change within the lifespan of an individual through learning (from Crook *et al.*,
1976).

matical terms to enable tests of their validity; that is, they should result in falsifiable
hypotheses (e.g. Drickamer and Vessey, 1982). Predictive models are built from data
sets. Generally, the larger and more accurate the data set, the more accurate the
model; however, Gauch (1993) has argued that a model can be more accurate than the
data used to build it since the model amplifies hidden patterns and discards noise.

Predictive models can be rather general, such as Regelmann's (1984) model for
how competing individuals should distribute themselves between food resource
patches, or they can be more specific such as Altmann's (1980) mathematical model
expressing the relationship of a baboon mother's feeding time requirement to her
infant's age. These models are beyond the scope of this book, but good discussions
can be found in Colgan (1978), Hazlett and Bach (1977) and Mangel and Clark
(1988).

---

an efference copy or corollary on the input for incubation. This input is fed
through a unit (*I*), necessary to explain the inhibition of settling and building
when feedback matches expectancy. The effect of feedback discrepancy on *N*
(and *I*), *E*, and *P*, can be read from the arrows. The main systems mutually
suppress one another; *P* is thought to occur as interruptive behaviour through
disinhibition of *N* and *E*. *P* can be activated directly by external stimuli like dust,
rain, or parasites; *E* can also be stimulated by disturbances other than deficient
feedback from the clutch (from Baerends, 1976).

18.2.5 Other illustrations

The type of visual representation employed and its value in interpreting results are limited only by the ingenuity of the researcher. Simplicity in illustrations is generally a virtue worth pursuing. For example, Bercovitch (1988) used a pie-chart to illustrate the percentages of the different types of consort change-overs (e.g. feed, fight) in adult male baboons. Patterson (1977) used a simple diagram which clearly demonstrates the rank-order changes of male shelducks (*Tadorna tadorna*) over a two-year observation period (Figure 18.14). The positional and relative extent of the changes in rank order are obvious and conducive to further interpretation.

Hutt and Hutt (1970) followed up on a suggestion by Altmann (1965) and described the application of a phase structure grammar model to the analysis of behavioral sequences. The model was first developed by Chomsky (1957) for the study of psycholinguistics. The model consists of the sequential partitioning of a sentence into its constituent parts based on its explicit meaning. The result is a tree diagram of sequentially smaller clusters of words that together carry the meaning of the sentence. This hierarchical model, discussed by R. Dawkins (1976a) and Westman (1977), was used by Marshall (1965) in his study of syntax in the reproductive behavior of the pigeon (Figure 18.15).

The 'Catch-22' of this method for the ethologist is that to apply the model to gain understanding of the message in communication, we must first understand the message.

This difficulty is illustrated by applying the analysis to the sentence 'We fed her dog bones', which can have two meanings; hence it can be diagrammed in two ways (Figure 18.16).

As Dale (1976) states, the ambiguity does not arise from a difference in words or in their order, but rather from a difference in their constituent structure. Do we have this level of resolution in analyzing sequences of animal behavior? Altmann (1965a) suggests that with sufficient experience it can be done.

> If one's goal is to draw up an exclusive and exhaustive classification of the animals' repertoire of socially significant behaviour patterns, then these units of behaviour are not arbitrarily chosen. On the contrary,they can be empirically determined. One divides up the continuum of action wherever the animals do. If the resulting recombination units are themselves communicative, that is, if they affect the behaviour of other members of the social group, then they are social messages. Thus, the splitting and lumping that one does is, ideally, a reflection of the splitting and lumping that the animals do.   [*Altmann, 1965a:492*]

To summarize, all of these techniques of visual representation (and others not discussed) can aid in the interpretation of results. They should be examined not only

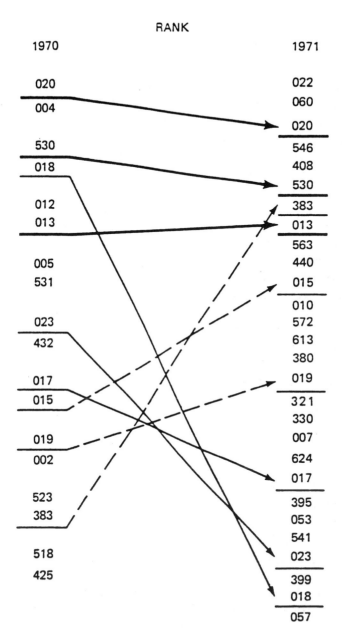

Fig. 18.14 Changes in rank order of shelducks between years. The figures are the serial
numbers of individual marked males arranged in rank order. Birds which were
ranked in both years are joined by arrows. The higher ranking birds in 1970
tended to remain high in 1971 (heavier solid arrows) but the middle birds in 1970
(thinner solid arrows) tended to lose rank relative to those low in 1970 (dashed
arrows) (from Patterson, 1977).

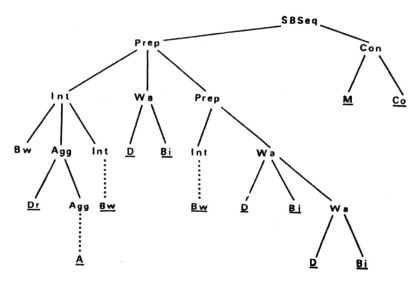

Fig. 18.15   Tree diagram showing application of generative grammar in its recursive form to reproductive behavior of the male pigeon. SBSeq=sexual behavior sequence; Prep=preparatory behavior; Con=consummatory behavior; Int=introduce; Wa=warm up; Agg=aggressive behavior; Bw=bowing; Dr=driving; A=attacking; D=displacement preening; Bi=billing; M=mounting; Co=copulation. The underlining represents the final behavior that results from the previous steps. The dots indicate where the pigeon can backtrack in the sequence (from Hutt and Hutt, 1970).

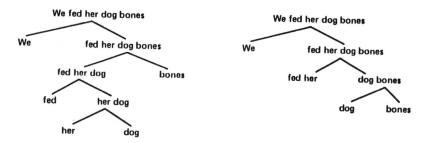

Fig. 18.16   Two-phase-structure grammar models of a single sentence to illustrate the different meanings (adapted from Dale, 1976).

to understand better the particular behavior studied, but also to put it in perspective relative to the various levels of behavior (Chapter 1). Are the results similar to those seen for other behaviors, for other species, and under other environmental conditions? Are the results consistent with a conceptual model or valuable for use in a predictive model? Does your interpretation help you to develop new models and generate new hypotheses? At this point it is again important to consider what other research has shown.

18.3 COMPARISONS WITH PREVIOUS RESULTS

How do your results compare with those of other researchers? Discuss your results with the same researchers you consulted before beginning your study (Chapter 4). Even though you reviewed the literature before beginning your study, it is wise to again search for relevant material in the light of your results. You may want to know more about similar behavior in other species or different behavior in the same species. You may discover that your results have a bearing on a general concept or current theoretical issues. The importance of results are often unforeseen when a study begins, but become apparent as the study proceeds and finally come to light as the results are carefully interpreted.

18.4 RE-EVALUATION

You've now reached the point where you can re-evaluate the entire study. You know where you began, and you think you know what your results mean. Even though your results were seemingly conclusive, your study could have been better. Re-evaluate the economics, efficiency and validity of your methods. Did you select the proper species, study area, behavioral units, data-collection method, analytical tests, etc.? Re-evaluate your study at each phase of the ethological approach (Chapter 1). You should improve your methods with each study, but this can only come through a critical re-evaluation of each study as it is completed.

18.5 REVISING AND RESTATING HYPOTHESES

You may want to revise or restate your hypotheses whether your results were positive or negative. Testing revised hypotheses can help reinforce positive results and isolate the source of negative results. You might choose to isolate additional variables or test the external validity of your results on other species.

Whether you revise, restate, or generate new hypotheses, you are now back at the beginning of the ethological approach cycle (Chapter 1), ready to begin again. This time you are more experienced, and, hopefully, wiser.

Speaking as an evolutionary biologist, E.O. Wilson offered the following insight:

> Love the animals for themselves first, then strain for general
> explanations, and, with good fortune, discoveries will follow. If they
> don't, the love and the pleasure will have been enough. *[Wilson, 1994:191]*

For ethologists, having had the pleasure of observing animals and having learned *what* they do is generally exceedingly rewarding without having yet fully understood *why*.

Statistical figures and tables

Table A1. *Factorials. Values of* n*!*

n	$n!$
0	1
1	1
2	2
3	6
4	24
5	120
6	720
7	5 040
8	40 320
9	362 880
10	3 628 800
11	39 916 800
12	479 001 600
13	6 227 020 800
14	87 178 291 200
15	1 307 674 368 000
16	20 922 789 888 000
17	355 687 428 096 000
18	6 402 373 705 728 000
19	121 645 100 408 832 000
20	2 432 902 008 176 640 000

Table A2. *Logarithms of factorials for n=1 to 199*

n	log n!	n	log n!	n	log n!	n	log n!	n	log n!	n	log n!	n	log n!	n	log n!
		25	25.19065	50	64.48307	75	109.39461	100	157.97000	125	209.27478	150	262.75689	175	318.05094
1	0.00000	26	26.60562	51	66.19064	76	111.27543	101	159.97432	126	211.37515	151	264.93587	176	320.29645
2	0.30103	27	28.03698	52	67.90665	77	113.16192	102	161.98293	127	213.47895	152	267.11771	177	322.54443
3	0.77815	28	29.48414	53	69.63092	78	115.05401	103	163.99576	128	215.58616	153	269.30241	178	324.79485
4	1.38021	29	30.94654	54	71.36332	79	116.95164	104	166.01280	129	217.69675	154	271.48993	179	327.04770
5	2.07918	30	32.42366	55	73.10368	80	118.85473	105	168.03399	130	219.81069	155	273.68026	180	329.30297
6	2.85733	31	33.91502	56	74.85187	81	120.76321	106	170.05929	131	221.92796	156	275.87338	181	331.56065
7	3.70243	32	35.42017	57	76.60774	82	122.67703	107	172.08867	132	224.04854	157	278.06928	182	333.82072
8	4.60552	33	36.93869	58	78.37117	83	124.59610	108	174.12210	133	226.17239	158	280.26794	183	336.08317
9	5.55976	34	38.47016	59	80.14202	84	126.52038	109	176.15952	134	228.29949	159	282.46934	184	338.34799
10	6.55976	35	40.01423	60	81.92017	85	128.44980	110	178.20092	135	230.42983	160	284.67346	185	340.61516
11	7.60116	36	41.57054	61	83.70550	86	130.38430	111	180.24624	136	232.56337	161	286.88028	186	342.88468
12	8.68034	37	43.13874	62	85.49790	87	132.32382	112	182.29546	137	234.70009	162	289.08980	187	345.15652
13	9.79428	38	44.71852	63	87.29724	88	134.26830	113	184.34854	138	236.83997	163	291.30198	188	347.43067
14	10.94041	39	46.30959	64	89.10342	89	136.21769	114	186.40544	139	238.98298	164	293.51683	189	349.70714
15	12.11650	40	47.91165	65	90.91633	90	138.17194	115	188.46614	140	241.12911	165	295.73431	190	351.98589
16	13.32062	41	49.52443	66	92.73587	91	140.13098	116	190.53060	141	243.27833	166	297.95442	191	354.26692
17	14.55107	42	51.14768	67	94.56195	92	142.09476	117	192.59878	142	245.43062	167	300.17714	192	356.55022
18	15.80634	43	52.78115	68	96.39446	93	144.06325	118	194.67067	143	247.58595	168	302.40245	193	358.83578
19	17.08509	44	54.42460	69	98.23331	94	146.03638	119	196.74621	144	249.74432	169	304.63033	194	361.12358
20	18.38612	45	56.07781	70	100.07840	95	148.01410	120	198.82539	145	251.90568	170	306.86078	195	363.41362
21	19.70834	46	57.74057	71	101.92966	96	149.99637	121	200.90818	146	254.07004	171	309.09378	196	365.70587
22	21.05077	47	59.41267	72	103.78700	97	151.98314	122	202.99454	147	256.23735	172	311.32931	197	368.00034
23	22.41249	48	61.09391	73	105.65032	98	153.97437	123	205.08444	148	258.40762	173	313.56735	198	370.29701
24	23.79271	49	62.78410	74	107.51955	99	155.97000	124	207.17787	149	260.58080	174	315.80790	199	372.59586

Table A3. *Binomial coefficients*

n	$\binom{n}{0}$	$\binom{n}{1}$	$\binom{n}{2}$	$\binom{n}{3}$	$\binom{n}{4}$	$\binom{n}{5}$	$\binom{n}{6}$	$\binom{n}{7}$	$\binom{n}{8}$	$\binom{n}{9}$	$\binom{n}{10}$
0	1										
1	1	1									
2	1	2	1								
3	1	3	3	1							
4	1	4	6	4	1						
5	1	5	10	10	5	1					
6	1	6	15	20	15	6	1				
7	1	7	21	35	35	21	7	1			
8	1	8	28	56	70	56	28	8	1		
9	1	9	36	84	126	126	84	36	9	1	
10	1	10	45	120	210	252	210	120	45	10	1
11	1	11	55	165	330	462	462	330	165	55	11
12	1	12	66	220	495	792	924	792	495	220	66
13	1	13	78	286	715	1287	1716	1716	1287	715	286
14	1	14	91	364	1001	2002	3003	3432	3003	2002	1001
15	1	15	105	455	1365	3003	5005	6435	6435	5005	3003
16	1	16	120	560	1820	4368	8008	11440	12870	11440	8008
17	1	17	136	680	2380	6188	12376	19448	24310	24310	19448
18	1	18	153	816	3060	8568	18564	31824	43758	48620	43758
19	1	19	171	969	3876	11628	27132	50388	75582	92378	92378
20	1	20	190	1140	4845	15504	38760	77520	125970	167960	184756

Source: Reprinted from Burington; reproduced with permission of McGraw-Hill, Inc.

Table A5. (*cont.*)

df	| P													
	0.90	0.80	0.70	0.60	0.50	0.40	0.30	0.20	0.10	0.05	0.02	0.01	0.001
20	0.127	0.257	0.391	0.533	0.687	0.860	1.064	1.325	1.725	2.086	2.528	2.845	3.850
21	0.127	0.257	0.391	0.532	0.686	0.859	1.063	1.323	1.721	2.080	2.518	2.831	3.819
22	0.127	0.256	0.390	0.532	0.686	0.858	1.061	1.321	1.717	2.074	2.508	2.819	3.792
23	0.127	0.256	0.390	0.532	0.685	0.858	1.060	1.319	1.714	2.069	2.500	2.807	3.767
24	0.127	0.256	0.390	0.531	0.685	0.857	1.059	1.318	1.711	2.064	2.492	2.797	3.745
25	0.127	0.256	0.390	0.531	0.684	0.856	1.058	1.316	1.708	2.060	2.485	2.787	3.725
26	0.127	0.256	0.390	0.531	0.684	0.856	1.058	1.315	1.706	2.056	2.479	2.779	3.707
27	0.127	0.256	0.389	0.531	0.684	0.855	1.057	1.314	1.703	2.052	2.473	2.771	3.690
28	0.127	0.256	0.389	0.530	0.683	0.855	1.056	1.313	1.701	2.048	2.467	2.763	3.674
29	0.127	0.256	0.389	0.530	0.683	0.854	1.055	1.311	1.699	2.045	2.462	2.756	3.659
30	0.127	0.256	0.389	0.530	0.683	0.854	1.055	1.310	1.697	2.042	2.457	2.750	3.646
40	0.126	0.255	0.388	0.529	0.681	0.851	1.050	1.303	1.684	2.021	2.423	2.704	3.551
60	0.126	0.254	0.387	0.527	0.679	0.848	1.046	1.296	1.671	2.000	2.390	2.660	3.460
120	0.126	0.254	0.386	0.526	0.677	0.845	1.041	1.289	1.658	1.980	2.358	2.617	3.373
∞	0.126	0.253	0.385	0.524	0.674	0.842	1.036	1.282	1.645	1.960	2.326	2.576	3.291

Source: Adapted from Campbell (1974).

Table A3. *Binomial coefficients*

n	$\binom{n}{0}$	$\binom{n}{1}$	$\binom{n}{2}$	$\binom{n}{3}$	$\binom{n}{4}$	$\binom{n}{5}$	$\binom{n}{6}$	$\binom{n}{7}$	$\binom{n}{8}$	$\binom{n}{9}$	$\binom{n}{10}$
0	1										
1	1	1									
2	1	2	1								
3	1	3	3	1							
4	1	4	6	4	1						
5	1	5	10	10	5	1					
6	1	6	15	20	15	6	1				
7	1	7	21	35	35	21	7	1			
8	1	8	28	56	70	56	28	8	1		
9	1	9	36	84	126	126	84	36	9	1	
10	1	10	45	120	210	252	210	120	45	10	1
11	1	11	55	165	330	462	462	330	165	55	11
12	1	12	66	220	495	792	924	792	495	220	66
13	1	13	78	286	715	1287	1716	1716	1287	715	286
14	1	14	91	364	1001	2002	3003	3432	3003	2002	1001
15	1	15	105	455	1365	3003	5005	6435	6435	5005	3003
16	1	16	120	560	1820	4368	8008	11440	12870	11440	8008
17	1	17	136	680	2380	6188	12376	19448	24310	24310	19448
18	1	18	153	816	3060	8568	18564	31824	43758	48620	43758
19	1	19	171	969	3876	11628	27132	50388	75582	92378	92378
20	1	20	190	1140	4845	15504	38760	77520	125970	167960	184756

Source: Reprinted from Burington; reproduced with permission of McGraw-Hill, Inc.

Table A4. Critical values for the F-Max test for homogeneity of variance. Alpha = 0.05 for two-tailed test. N_1 are the degrees of freedom for the larger variance. N_2 are the degrees of freedom for the smaller variance

N_2	N_1																		
	1	2	3	4	5	6	7	8	9	10	12	15	20	24	30	40	60	120	∞
1	648	800	864	900	922	937	948	957	963	969	977	985	993	997	1001	1006	1010	1014	1018
2	38.51	39.00	39.16	39.25	39.30	39.33	39.36	39.37	39.39	39.40	39.42	39.43	39.45	39.46	39.46	39.47	39.48	39.49	39.50
3	17.44	16.04	15.44	15.10	14.88	14.74	14.62	14.54	14.47	14.42	14.34	14.25	14.17	14.12	14.08	14.04	13.99	13.95	13.90
4	12.22	10.65	9.98	9.60	9.36	9.20	9.07	8.98	8.90	8.84	8.75	8.66	8.56	8.51	8.46	8.41	8.36	8.31	8.26
5	10.01	8.43	7.76	7.39	7.15	6.98	6.85	6.76	6.68	6.62	6.52	6.43	6.33	6.28	6.23	6.18	6.12	6.07	6.02
6	8.81	7.26	6.60	6.23	5.99	5.82	5.70	5.60	5.52	5.46	5.37	5.27	5.17	5.12	5.07	5.01	4.96	4.90	4.85
7	8.07	6.54	5.89	5.52	5.29	5.12	4.99	4.90	4.82	4.76	4.67	4.57	4.47	4.42	4.36	4.31	4.25	4.20	4.14
8	7.57	6.06	5.42	5.05	4.82	4.65	4.53	4.43	4.36	4.30	4.20	4.10	4.00	3.95	3.89	3.84	3.78	3.73	3.67
9	7.21	5.71	5.08	4.72	4.48	4.32	4.20	4.10	4.03	3.96	3.87	3.77	3.67	3.61	3.56	3.51	3.45	3.39	3.33
10	6.94	5.46	4.83	4.47	4.24	4.07	3.95	3.85	3.78	3.72	3.62	3.52	3.42	3.37	3.31	3.26	3.20	3.14	3.08
12	6.55	5.10	4.47	4.12	3.89	3.73	3.61	3.51	3.44	3.37	3.28	3.18	3.07	3.02	2.96	2.91	2.85	2.79	2.72
15	6.20	4.76	4.15	3.80	3.58	3.41	3.29	3.20	3.12	3.06	2.96	2.86	2.76	2.70	2.64	2.58	2.52	2.46	2.40
20	5.87	4.46	3.86	3.51	3.29	3.13	3.01	2.91	2.84	2.77	2.68	2.57	2.46	2.41	2.35	2.29	2.22	2.16	2.09
24	5.72	4.32	3.72	3.38	3.15	2.99	2.87	2.78	2.70	2.64	2.54	2.44	2.33	2.27	2.21	2.15	2.08	2.01	1.94
30	5.57	4.18	3.59	3.25	3.03	2.87	2.75	2.65	2.57	2.51	2.41	2.31	2.20	2.14	2.07	2.01	1.94	1.87	1.79
40	5.42	4.05	3.46	3.13	2.90	2.74	2.62	2.53	2.45	2.39	2.29	2.18	2.07	2.01	1.94	1.88	1.80	1.72	1.64
60	5.29	3.93	3.34	3.01	2.79	2.63	2.51	2.41	2.33	2.27	2.17	2.06	1.94	1.88	1.82	1.74	1.67	1.58	1.48
120	5.15	3.80	3.23	2.89	2.67	2.52	2.39	2.30	2.22	2.16	2.05	1.94	1.82	1.76	1.69	1.61	1.53	1.43	1.31
∞	5.02	3.69	3.12	2.79	2.57	2.41	2.29	2.19	2.11	2.05	1.94	1.83	1.71	1.64	1.57	1.48	1.39	1.27	1.00

Table A5. *Critical values of the t-statistic for a two-tailed test.* P=*alpha level*

df	P													
	0.90	0.80	0.70	0.60	0.50	0.40	0.30	0.20	0.10	0.05	0.02	0.01	0.001
1	0.158	0.325	0.510	0.727	1.000	1.376	1.936	3.078	6.314	12.706	31.821	63.657	636.619
2	0.142	0.289	0.445	0.617	0.816	1.061	1.386	1.886	2.920	4.303	6.965	9.925	31.598
3	0.137	0.277	0.424	0.584	0.765	0.978	1.250	1.638	2.353	3.182	4.541	5.841	12.924
4	0.134	0.271	0.414	0.569	0.741	0.941	1.190	1.533	2.132	2.776	3.747	4.604	8.610
5	0.132	0.267	0.408	0.559	0.727	0.920	1.156	1.476	2.015	2.571	3.365	4.032	6.869
6	0.131	0.265	0.404	0.553	0.718	0.906	1.134	1.440	1.943	2.447	3.143	3.707	5.959
7	0.130	0.263	0.402	0.549	0.711	0.896	1.119	1.415	1.895	2.365	2.998	3.499	5.408
8	0.130	0.262	0.399	0.546	0.706	0.889	1.108	1.397	1.860	2.306	2.896	3.355	5.041
9	0.129	0.261	0.398	0.543	0.703	0.883	1.100	1.383	1.833	2.262	2.821	3.250	4.781
10	0.129	0.260	0.397	0.542	0.700	0.879	1.093	1.372	1.812	2.228	2.764	3.169	4.587
11	0.129	0.260	0.396	0.540	0.697	0.876	1.088	1.363	1.796	2.201	2.718	3.106	4.437
12	0.128	0.259	0.395	0.539	0.695	0.873	1.083	1.356	1.782	2.179	2.681	3.055	4.318
13	0.128	0.259	0.394	0.538	0.694	0.870	1.079	1.350	1.771	2.160	2.650	3.012	4.221
14	0.128	0.258	0.393	0.537	0.692	0.868	1.076	1.345	1.761	2.145	2.624	2.977	4.140
15	0.128	0.258	0.393	0.536	0.691	0.866	1.074	1.341	1.753	2.131	2.602	2.947	4.073
16	0.128	0.258	0.392	0.535	0.690	0.865	1.071	1.337	1.746	2.120	2.583	2.921	4.015
17	0.128	0.257	0.392	0.534	0.689	0.863	1.069	1.333	1.740	2.110	2.567	2.898	3.965
18	0.127	0.257	0.392	0.534	0.688	0.862	1.067	1.330	1.734	2.101	2.552	2.878	3.922
19	0.127	0.257	0.391	0.533	0.688	0.861	1.066	1.328	1.729	2.093	2.539	2.861	3.883

Table A5. (*cont.*)

df	\(P\)												
	0.90	0.80	0.70	0.60	0.50	0.40	0.30	0.20	0.10	0.05	0.02	0.01	0.001
20	0.127	0.257	0.391	0.533	0.687	0.860	1.064	1.325	1.725	2.086	2.528	2.845	3.850
21	0.127	0.257	0.391	0.532	0.686	0.859	1.063	1.323	1.721	2.080	2.518	2.831	3.819
22	0.127	0.256	0.390	0.532	0.686	0.858	1.061	1.321	1.717	2.074	2.508	2.819	3.792
23	0.127	0.256	0.390	0.532	0.685	0.858	1.060	1.319	1.714	2.069	2.500	2.807	3.767
24	0.127	0.256	0.390	0.531	0.685	0.857	1.059	1.318	1.711	2.064	2.492	2.797	3.745
25	0.127	0.256	0.390	0.531	0.684	0.856	1.058	1.316	1.708	2.060	2.485	2.787	3.725
26	0.127	0.256	0.390	0.531	0.684	0.856	1.058	1.315	1.706	2.056	2.479	2.779	3.707
27	0.127	0.256	0.389	0.531	0.684	0.855	1.057	1.314	1.703	2.052	2.473	2.771	3.690
28	0.127	0.256	0.389	0.530	0.683	0.855	1.056	1.313	1.701	2.048	2.467	2.763	3.674
29	0.127	0.256	0.389	0.530	0.683	0.854	1.055	1.311	1.699	2.045	2.462	2.756	3.659
30	0.127	0.256	0.389	0.530	0.683	0.854	1.055	1.310	1.697	2.042	2.457	2.750	3.646
40	0.126	0.255	0.388	0.529	0.681	0.851	1.050	1.303	1.684	2.021	2.423	2.704	3.551
60	0.126	0.254	0.387	0.527	0.679	0.848	1.046	1.296	1.671	2.000	2.390	2.660	3.460
120	0.126	0.254	0.386	0.526	0.677	0.845	1.041	1.289	1.658	1.980	2.358	2.617	3.373
∞	0.126	0.253	0.385	0.524	0.674	0.842	1.036	1.282	1.645	1.960	2.326	2.576	3.291

Source: Adapted from Campbell (1974).

Table A6₁. Critical values of **F** for the analysis of variance. Alpha=0.05

df for denominator	df for numerator															
	1	2	3	4	5	6	7	8	9	10	15	20	25	30	40	50
1	161.40	199.50	215.80	224.80	230.00	233.80	236.50	238.60	240.10	242.10	245.20	248.40	248.90	250.50	250.80	252.60
2	18.51	19.00	19.16	19.25	19.30	19.33	19.35	19.37	19.38	19.40	19.43	19.44	19.46	19.47	19.48	19.48
3	10.13	9.55	9.28	9.12	9.01	8.94	8.89	8.85	8.81	8.79	8.70	8.66	8.63	8.62	8.59	8.58
4	7.71	6.94	6.59	6.39	6.26	6.16	6.09	6.04	6.00	5.96	5.86	5.80	5.77	5.75	5.72	5.70
5	6.61	5.79	5.41	5.19	5.05	4.95	4.88	4.82	4.77	4.74	4.62	4.56	4.52	4.50	4.46	4.44
6	5.99	5.14	4.76	4.53	4.39	4.28	4.21	4.15	4.10	4.06	3.94	3.87	3.83	3.81	3.77	3.75
7	5.59	4.74	4.35	4.12	3.97	3.87	3.79	3.73	3.68	3.64	3.51	3.44	3.40	3.38	3.34	3.32
8	5.32	4.46	4.07	3.84	3.69	3.58	3.50	3.44	3.39	3.35	3.22	3.15	3.11	3.08	3.04	3.02
9	5.12	4.26	3.86	3.63	3.48	3.37	3.29	3.23	3.18	3.14	3.01	2.94	2.89	2.86	2.83	2.80
10	4.96	4.10	3.71	3.48	3.33	3.22	3.14	3.07	3.02	2.98	2.85	2.77	2.73	2.70	2.66	2.64
11	4.84	3.98	3.59	3.36	3.20	3.09	3.01	2.95	2.90	2.85	2.72	2.65	2.60	2.57	2.53	2.51
12	4.75	3.89	3.49	3.26	3.11	3.00	2.91	2.85	2.80	2.75	2.62	2.54	2.50	2.47	2.43	2.40
13	4.67	3.81	3.41	3.18	3.03	2.92	2.83	2.77	2.71	2.67	2.53	2.46	2.41	2.38	2.34	2.31
14	4.60	3.74	3.34	3.11	2.96	2.85	2.76	2.70	2.65	2.60	2.46	2.39	2.34	2.31	2.27	2.24
15	4.54	3.68	3.29	3.06	2.90	2.79	2.71	2.64	2.59	2.54	2.40	2.33	2.28	2.25	2.20	2.18
16	4.49	3.63	3.24	3.01	2.85	2.74	2.66	2.59	2.54	2.49	2.35	2.28	2.23	2.19	2.15	2.12
17	4.45	3.59	3.20	2.96	2.81	2.70	2.61	2.55	2.49	2.45	2.31	2.23	2.18	2.15	2.10	2.08
18	4.41	3.55	3.16	2.93	2.77	2.66	2.58	2.51	2.46	2.41	2.27	2.19	2.14	2.11	2.06	2.04
19	4.38	3.52	3.13	2.90	2.74	2.63	2.54	2.48	2.42	2.38	2.23	2.16	2.11	2.07	2.03	2.00

Table A6$_1$. (cont.)

df for denominator	df for numerator															
nator	1	2	3	4	5	6	7	8	9	10	15	20	25	30	40	50
20	4.35	3.49	3.10	2.87	2.71	2.60	2.51	2.45	2.39	2.35	2.20	2.12	2.07	2.04	1.99	1.97
22	4.30	3.44	3.05	2.82	2.66	2.55	2.46	2.40	2.34	2.30	2.15	2.07	2.02	1.98	1.94	1.91
24	4.26	3.40	3.01	2.78	2.62	2.51	2.42	2.36	2.30	2.25	2.11	2.03	1.97	1.94	1.89	1.86
26	4.23	3.37	2.98	2.74	2.59	2.47	2.39	2.32	2.27	2.22	2.07	1.99	1.94	1.90	1.85	1.82
28	4.20	3.34	2.95	2.71	2.56	2.45	2.36	2.29	2.24	2.19	2.04	1.96	1.91	1.87	1.82	1.79
30	4.17	3.32	2.92	2.69	2.53	2.42	2.33	2.27	2.21	2.16	2.01	1.93	1.88	1.84	1.79	1.76
40	4.08	3.23	2.84	2.61	2.45	2.34	2.25	2.18	2.12	2.08	1.92	1.84	1.78	1.74	1.69	1.66
50	4.03	3.18	2.79	2.56	2.40	2.29	2.20	2.13	2.07	2.03	1.87	1.78	1.73	1.69	1.63	1.60
60	4.00	3.15	2.76	2.53	2.37	2.25	2.17	2.10	2.04	1.99	1.84	1.75	1.69	1.65	1.59	1.56
120	3.92	3.07	2.68	2.45	2.29	2.18	2.09	2.02	1.96	1.91	1.75	1.66	1.60	1.55	1.50	1.46
200	3.89	3.04	2.65	2.42	2.26	2.14	2.06	1.98	1.93	1.88	1.72	1.62	1.56	1.52	1.46	1.41
500	3.86	3.01	2.62	2.39	2.23	2.12	2.03	1.96	1.90	1.85	1.69	1.59	1.53	1.48	1.42	1.38
1000	3.85	3.01	2.61	2.38	2.22	2.11	2.02	1.95	1.89	1.84	1.68	1.58	1.52	1.47	1.41	1.36

Source: From Howell (1992).

Table A6$_2$. *Critical values of F for the analysis of variance. Alpha=0.01*

df for denominator	df for numerator															
nator	1	2	3	4	5	6	7	8	9	10	15	20	25	30	40	50
1	4048	4993	5377	5577	5668	5924	5992	6096	6132	6168	6157	6209	6235	6261	6287	6300
2	98.50	99.01	99.15	99.23	99.30	99.33	99.35	99.39	99.40	99.43	99.43	99.45	99.46	99.47	99.47	99.50
3	34.12	30.82	29.46	28.71	28.24	27.91	27.67	27.49	27.34	27.23	26.87	26.69	26.58	26.51	26.41	26.36
4	21.20	18.00	16.69	15.98	15.52	15.21	14.98	14.80	14.66	14.55	14.20	14.02	13.91	13.84	13.75	13.69
5	16.26	13.27	12.06	11.39	10.97	10.67	10.46	10.29	10.16	10.05	9.72	9.55	9.45	9.38	9.29	9.24
6	13.75	10.92	9.78	9.15	8.75	8.47	8.26	8.10	7.98	7.87	7.56	7.40	7.30	7.23	7.14	7.09
7	12.25	9.55	8.45	7.85	7.46	7.19	6.99	6.84	6.72	6.62	6.31	6.16	6.06	5.99	5.91	5.86
8	11.26	8.65	7.59	7.01	6.63	6.37	6.18	6.03	5.91	5.81	5.52	5.36	5.26	5.20	5.12	5.07
9	10.56	8.02	6.99	6.42	6.06	5.80	5.61	5.47	5.35	5.26	4.96	4.81	4.71	4.65	4.57	4.52
10	10.04	7.56	6.55	5.99	5.64	5.39	5.20	5.06	4.94	4.85	4.56	4.41	4.31	4.25	4.17	4.12
11	9.65	7.21	6.22	5.67	5.32	5.07	4.89	4.74	4.63	4.54	4.25	4.10	4.01	3.94	3.86	3.81
12	9.33	6.93	5.95	5.41	5.06	4.82	4.64	4.50	4.39	4.30	4.01	3.86	3.76	3.70	3.62	3.57
13	9.07	6.70	5.74	5.21	4.86	4.62	4.44	4.30	4.19	4.10	3.82	3.66	3.57	3.51	3.43	3.38
14	8.86	6.51	5.56	5.04	4.69	4.46	4.28	4.14	4.03	3.94	3.66	3.51	3.41	3.35	3.27	3.22
15	8.68	6.36	5.42	4.89	4.56	4.32	4.14	4.00	3.89	3.80	3.52	3.37	3.28	3.21	3.13	3.08
16	8.53	6.23	5.29	4.77	4.44	4.20	4.03	3.89	3.78	3.69	3.41	3.26	3.16	3.10	3.02	2.97
17	8.40	6.11	5.18	4.67	4.34	4.10	3.93	3.79	3.68	3.59	3.31	3.16	3.07	3.00	2.92	2.87
18	8.29	6.01	5.09	4.58	4.25	4.01	3.84	3.71	3.60	3.51	3.23	3.08	2.98	2.92	2.84	2.78
19	8.18	5.93	5.01	4.50	4.17	3.94	3.77	3.63	3.52	3.43	3.15	3.00	2.91	2.84	2.76	2.71

Table A6$_2$, (cont.)

df for denominator	df for numerator															
nator	1	2	3	4	5	6	7	8	9	10	15	20	25	30	40	50
20	8.10	5.85	4.94	4.43	4.10	3.87	3.70	3.56	3.46	3.37	3.09	2.94	2.84	2.78	2.69	2.64
22	7.95	5.72	4.82	4.31	3.99	3.76	3.59	3.45	3.35	3.26	2.98	2.83	2.73	2.67	2.58	2.53
24	7.82	5.61	4.72	4.22	3.90	3.67	3.50	3.36	3.26	3.17	2.89	2.74	2.64	2.58	2.49	2.44
26	7.72	5.53	4.64	4.14	3.82	3.59	3.42	3.29	3.18	3.09	2.81	2.66	2.57	2.50	2.42	2.36
28	7.64	5.45	4.57	4.07	3.75	3.53	3.36	3.23	3.12	3.03	2.75	2.60	2.51	2.44	2.35	2.30
30	7.56	5.39	4.51	4.02	3.70	3.47	3.30	3.17	3.07	2.98	2.70	2.55	2.45	2.39	2.30	2.25
40	7.31	5.18	4.31	3.83	3.51	3.29	3.12	2.99	2.89	2.80	2.52	2.37	2.27	2.20	2.11	2.06
50	7.17	5.06	4.20	3.72	3.41	3.19	3.02	2.89	2.78	2.70	2.42	2.27	2.17	2.10	2.01	1.95
60	7.08	4.98	4.13	3.65	3.34	3.12	2.95	2.82	2.72	2.63	2.35	2.20	2.10	2.03	1.94	1.88
120	6.85	4.79	3.95	3.48	3.17	2.96	2.79	2.66	2.56	2.47	2.19	2.03	1.93	1.86	1.76	1.70
200	6.76	4.71	3.88	3.41	3.11	2.89	2.73	2.60	2.50	2.41	2.13	1.97	1.87	1.79	1.69	1.63
500	6.69	4.65	3.82	3.36	3.05	2.84	2.68	2.55	2.44	2.36	2.07	1.92	1.81	1.74	1.63	1.57
1000	6.67	4.63	3.80	3.34	3.04	2.82	2.66	2.53	2.43	2.34	2.06	1.90	1.79	1.72	1.61	1.54

Source: Adapted from Howell (1992). Copyrighted by the Wadsworth Publishing Company.

Table A7. *Critical values for significance of the Pearson product moment correlation coefficient,* r_c

n	Alpha 0.05	Alpha 0.01	n	Alpha 0.05	Alpha 0.01	n	Alpha 0.05	Alpha 0.01	n	Alpha 0.05	Alpha 0.01	n	Alpha 0.05	Alpha 0.01
1	0.997	1.000	21	0.413	0.526	41	0.301	0.389	72	0.229	0.298	130	0.171	0.223
2	0.950	0.990	22	0.404	0.515	42	0.297	0.384	74	0.226	0.294	135	0.168	0.219
3	0.878	0.959	23	0.396	0.505	43	0.294	0.380	76	0.223	0.290	140	0.165	0.215
4	0.811	0.917	24	0.388	0.496	44	0.291	0.376	78	0.220	0.286	145	0.162	0.212
5	0.755	0.875	25	0.381	0.487	45	0.288	0.372	80	0.217	0.283	150	0.159	0.208
6	0.707	0.834	26	0.374	0.479	46	0.285	0.368	82	0.215	0.280	160	0.154	0.202
7	0.666	0.798	27	0.367	0.471	47	0.282	0.365	84	0.212	0.276	170	0.150	0.196
8	0.632	0.765	28	0.361	0.463	48	0.279	0.361	86	0.210	0.273	180	0.145	0.190
9	0.602	0.735	29	0.355	0.456	49	0.276	0.358	88	0.207	0.270	190	0.142	0.185
10	0.576	0.708	30	0.349	0.449	50	0.273	0.354	90	0.205	0.267	200	0.138	0.181
11	0.553	0.684	31	0.344	0.442	52	0.268	0.348	92	0.203	0.264			
12	0.532	0.661	32	0.339	0.436	54	0.263	0.341	94	0.201	0.262			
13	0.514	0.641	33	0.334	0.430	56	0.259	0.336	96	0.199	0.259			
14	0.497	0.623	34	0.329	0.424	58	0.254	0.330	98	0.197	0.256			
15	0.482	0.606	35	0.325	0.418	60	0.250	0.325	100	0.195	0.254			
16	0.468	0.590	36	0.320	0.413	62	0.246	0.320	105	0.190	0.248			
17	0.456	0.575	37	0.316	0.408	64	0.242	0.315	110	0.186	0.242			
18	0.444	0.561	38	0.312	0.403	66	0.239	0.310	115	0.182	0.237			
19	0.433	0.549	39	0.308	0.398	68	0.235	0.306	120	0.178	0.232			
20	0.423	0.537	40	0.304	0.393	70	0.232	0.302	125	0.174	0.228			

Source: Adapted from Zar (1984).

Table A8$_1$. *Critical values of* r *for the one-sample and two-sample runs tests. Alpha=0.05. For the one-sample runs test, any value of* r *which is equal to or smaller than the value in Table A8$_1$ or equal to or larger than the value in Table A8$_2$ is significant. For the Wald–Wolfowitz two-sample runs test, any value of* r *which is equal to or smaller than the value in Table A8$_1$ is significant*

n_2

n_1	2	3	4	5	6	7	8	9	10	11	12	13	14	15	16	17	18	19	20
2											2	2	2	2	2	2	2	2	2
3					2	2	2	2	2	2	2	2	3	3	3	3	3	3	3
4				2	2	2	3	3	3	3	3	3	3	3	4	4	4	4	4
5			2	2	3	3	3	3	3	4	4	4	4	4	4	4	5	5	5
6		2	2	3	3	3	3	4	4	4	4	5	5	5	5	5	5	6	6
7		2	2	3	3	3	4	4	5	5	5	5	5	6	6	6	6	6	6
8		2	3	3	3	4	4	5	5	5	6	6	6	6	6	7	7	7	7
9		2	3	3	4	4	5	5	5	6	6	6	7	7	7	7	8	8	8
10		2	3	3	4	5	5	5	6	6	7	7	7	7	8	8	8	8	9
11		2	3	4	4	5	5	6	6	7	7	7	8	8	8	9	9	9	9
12	2	2	3	4	4	5	6	6	7	7	7	8	8	8	9	9	9	10	10
13	2	2	3	4	5	5	6	6	7	7	8	8	9	9	9	10	10	10	10
14	2	3	3	4	5	5	6	7	7	8	8	9	9	9	10	10	10	11	11
15	2	3	3	4	5	6	6	7	7	8	8	9	9	10	10	11	11	11	12
16	2	3	4	4	5	6	6	7	8	8	9	9	10	10	11	11	11	12	12
17	2	3	4	4	5	6	7	7	8	9	9	10	10	11	11	11	12	12	13
18	2	3	4	5	5	6	7	8	8	9	9	10	10	11	11	12	12	13	13
19	2	3	4	5	6	6	7	8	8	9	10	10	11	11	12	12	13	13	13
20	2	3	4	5	6	6	7	8	9	9	10	10	11	12	12	13	13	13	14

Source: From Siegel (1956); reproduced with permission of McGraw-Hill, Inc.; adapted from Swed and Eisenhart (1943). Reproduced courtesy of the Institute of Mathematical Statistics

Table A8$_2$. *Critical values of r for the one-sample runs test (see Table A8$_1$)*

n_1 \ n_2	2	3	4	5	6	7	8	9	10	11	12	13	14	15	16	17	18	19	20
2																			
3																			
4				9	9														
5			9	10	10	11	11												
6			9	10	11	12	12	13	13	13	13								
7				11	12	13	13	13	14	14	14	15	15	15					
8				11	12	13	14	14	15	15	16	16	16	16	17	17	17	17	17
9					13	14	14	15	16	16	16	17	17	18	18	18	18	18	18
10					13	14	15	16	16	17	17	18	18	18	19	19	19	20	20
11					13	14	15	16	17	17	18	19	19	19	20	20	20	21	21
12					13	14	16	16	17	18	19	19	20	20	21	21	21	22	22
13						15	16	17	18	19	19	20	20	21	21	22	22	23	23
14						15	16	17	18	19	20	20	21	22	22	23	23	23	24
15						15	16	18	18	19	20	21	22	22	23	23	24	24	25
16							17	18	19	20	21	21	22	23	23	24	25	25	25
17							17	18	19	20	21	22	23	23	24	25	25	26	26
18							17	18	19	20	21	22	23	24	25	25	26	26	27
19							17	18	20	21	22	23	23	24	25	26	26	27	27
20							17	18	20	21	22	23	24	25	25	26	27	27	28

Source: From Siegel (1956); reproduced with permission of McGraw-Hill, Inc. Reproduced courtesy of the Institute of Mathematical Statistics

Table A9. *Critical values of the chi-square distribution for one-tailed tests, the primary use of the distribution*

df	0.99	0.95	0.10	0.05	0.01	0.001
1	0.0^3157	0.00393	2.71	3.84	6.63	10.83
2	0.0201	0.103	4.61	5.99	9.21	13.81
3	0.115	0.352	6.25	7.81	11.34	16.27
4	0.297	0.711	7.78	9.49	13.28	18.47
5	0.554	1.15	9.24	11.07	15.09	20.52
6	0.872	1.64	10.64	12.59	16.81	22.46
7	1.24	2.17	12.02	14.07	18.48	24.32
8	1.65	2.73	13.36	15.51	20.09	26.12
9	2.09	3.33	14.68	16.92	21.67	27.88
10	2.56	3.94	15.99	18.31	23.21	29.59
11	3.05	4.57	17.28	19.68	24.73	31.26
12	3.57	5.23	18.55	21.03	26.22	32.91
13	4.11	5.89	19.81	22.36	27.69	34.53
14	4.66	6.57	21.06	23.68	29.14	36.12
15	5.23	7.26	22.31	25.00	30.58	37.70
16	5.81	7.96	23.54	26.30	32.00	39.25
17	6.41	8.67	24.77	27.59	33.41	40.79
18	7.01	9.39	25.99	28.87	34.81	42.31
19	7.63	10.12	27.20	30.14	36.19	43.82
20	8.26	10.85	28.41	31.41	37.57	45.31

df	0.99	0.95	0.10	0.05	0.01	0.001
21	8.90	11.59	29.62	32.67	38.93	46.80
22	9.54	12.34	30.81	33.92	40.29	48.27
23	10.20	13.09	32.01	35.17	41.64	49.73
24	10.86	13.85	33.20	36.42	42.98	51.18
25	11.52	14.61	34.38	37.65	44.31	52.62
26	12.20	15.38	35.56	38.89	45.64	54.05
27	12.88	16.15	36.74	40.11	46.96	55.48
28	13.56	16.93	37.92	41.34	48.28	56.89
29	14.26	17.71	39.09	42.56	49.59	58.30
30	14.95	18.49	40.26	43.77	50.89	59.70
40	22.16	26.51	51.81	55.76	63.69	73.40
50	29.71	34.76	63.17	67.50	76.15	86.66
60	37.48	43.19	74.40	79.08	88.38	99.61
70	45.44	51.74	85.53	90.53	100.4	112.3
80	53.54	60.39	96.58	101.9	112.3	124.8
90	61.75	69.13	107.6	113.1	124.1	137.2
100	70.06	77.93	118.5	124.3	135.8	149.4

Source: Adapted from Campbell (1974).

Table A10. *Critical values of U for the Mann–Whitney test*

N_S	N_L																
	4	5	6	7	8	9	10	11	12	13	14	15	16	17	18	19	20
2	—	—	—	—	0	0	0	0	1	1	1	1	1	2	2	2	2
3	—	0	1	1	2	2	3	3	4	4	5	5	6	6	7	7	8
4	0	1	2	3	4	4	5	6	7	8	9	10	11	11	12	13	13
5	—	2	3	5	6	7	8	9	11	12	13	14	15	17	18	19	20
6	—	—	5	6	8	10	11	13	14	16	17	19	21	22	24	25	27
7	—	—	—	8	10	12	14	16	18	20	22	24	26	28	30	32	34
8	—	—	—	—	13	15	17	19	22	24	26	29	31	34	36	38	41
9	—	—	—	—	—	17	20	23	26	28	31	34	37	39	42	45	48
10	—	—	—	—	—	20	23	26	29	33	36	39	42	45	48	52	55
11	—	—	—	—	—	23	26	30	33	37	40	44	47	51	55	58	62
12	—	—	—	—	—	26	29	33	37	41	45	49	53	57	61	65	69
13	—	—	—	—	—	28	33	37	41	45	50	54	59	63	67	72	76
14	—	—	—	—	—	31	36	40	45	50	55	59	64	67	74	78	83
15	—	—	—	—	—	34	39	44	49	54	59	64	70	75	80	85	90
16	—	—	—	—	—	37	42	47	53	59	64	70	75	81	86	92	98
17	—	—	—	—	—	39	45	51	57	63	67	75	81	87	93	99	105
18	—	—	—	—	—	42	48	55	61	67	74	80	86	93	99	106	112
19	—	—	—	—	—	45	52	58	65	72	78	85	92	99	106	113	119
20	—	—	—	—	—	48	55	62	69	76	83	90	98	105	112	119	127

Source: Adapted from Mann and Whitney (1947) and Robson (1973)

Table A11. *Critical values for the Kolmogorov–Smirnov two-sample, one-tailed test for small samples. The upper, middle, and lower values are for alpha=0.10, 0.05 and 0.01, respectively*

													m											
n	3	4	5	6	7	8	9	10	11	12	13	14	15	16	17	18	19	20	21	22	23	24	25
3	9	10	11	15	15	16	21	19	22	24	25	26	30	30	32	36	36	37	42	40	43	45	46
	9	10	13	15	16	19	21	22	25	27	28	31	33	34	35	39	40	41	45	46	47	51	52
					19	*22*	*27*	*28*	*31*	*33*	*34*	*37*	*42*	*43*	*43*	*48*	*49*	*52*	*54*	*55*	*58*	*63*	*64*
4	10	16	13	16	18	24	21	24	26	32	29	32	34	40	37	40	41	48	45	48	49	56	53
	10	16	16	18	21	24	25	28	29	36	33	38	38	44	44	46	49	52	52	56	57	60	61
			17	*22*	*25*	*32*	*29*	*34*	*37*	*40*	*41*	*46*	*46*	*52*	*53*	*56*	*57*	*64*	*64*	*66*	*69*	*76*	*73*
5	11	13	20	19	21	23	26	30	30	32	35	37	45	41	44	46	47	55	51	54	56	58	65
	13	16	20	21	24	26	28	35	35	36	40	42	50	46	49	51	56	60	60	62	65	67	75
	13		*17*	*26*	*29*	*33*	*36*	*40*	*41*	*46*	*48*	*51*	*60*	*56*	*61*	*63*	*67*	*75*	*75*	*76*	*81*	*82*	*90*
6	15	16	19	24	24	26	30	32	33	42	37	42	45	48	49	54	54	56	60	62	63	72	67
	15	18	21	30	25	30	33	36	38	48	43	48	51	54	56	66	61	66	69	70	73	78	78
		17	*26*	*36*	*31*	*38*	*42*	*44*	*49*	*54*	*54*	*60*	*63*	*66*	*68*	*78*	*77*	*80*	*84*	*88*	*91*	*96*	*96*
7	15	18	21	24	35	28	32	34	38	40	44	49	48	51	54	56	59	61	70	68	70	72	74
	16	21	24	25	35	34	36	40	43	45	50	56	56	58	61	64	68	72	77	77	79	83	85
	19	*25*	*29*	*31*	*42*	*42*	*46*	*50*	*53*	*57*	*59*	*70*	*70*	*71*	*75*	*81*	*85*	*87*	*98*	*97*	*99*	*103*	*106*
8	16	24	23	26	28	40	33	40	41	48	47	50	52	64	57	62	64	72	71	74	76	88	81
	19	24	26	30	34	40	40	44	48	52	53	58	60	72	65	72	73	80	81	84	89	96	95
	22	*32*	*33*	*38*	*42*	*48*	*49*	*56*	*59*	*64*	*66*	*72*	*75*	*88*	*81*	*88*	*91*	*100*	*100*	*106*	*107*	*120*	*118*

9	88	87	82	79	78	73	70	72	65	61	60	54	51	51	45	43	45	33	32	30	26	21	21
	101	99	94	91	90	83	80	81	74	68	69	63	57	57	51	46	54	40	36	33	28	25	21
	124	*123*	*117*	*111*	*111*	*103*	*99*	*99*	*92*	*86*	*84*	*77*	*73*	*73*	*62*	*61*	*63*	*49*	*46*	*42*	*36*	*29*	*27*
10	100	92	88	86	80	90	74	72	69	66	65	60	55	52	48	50	43	40	34	32	30	24	19
	110	106	101	98	91	100	85	82	77	76	75	68	62	60	57	60	46	44	40	36	35	28	22
	140	*130*	*125*	*120*	*118*	*120*	*104*	*104*	*97*	*94*	*90*	*84*	*78*	*74*	*69*	*70*	*61*	*56*	*50*	*44*	*40*	*34*	*28*
11	100	98	95	99	85	84	79	76	72	69	66	63	59	54	66	48	45	41	38	33	30	26	22
	116	111	108	110	101	95	92	87	83	80	76	72	67	64	66	57	51	48	43	38	35	29	25
	143	*138*	*132*	*143*	*124*	*117*	*114*	*108*	*104*	*100*	*95*	*88*	*85*	*77*	*88*	*69*	*62*	*59*	*53*	*49*	*41*	*37*	*31*
12	106	108	100	98	93	92	85	84	77	76	72	68	61	72	54	52	51	48	40	42	32	32	24
	120	132	113	110	108	104	98	96	89	88	84	78	71	72	64	60	57	52	45	48	36	36	27
	153	*156*	*138*	*138*	*132*	*128*	*121*	*120*	*111*	*108*	*102*	*94*	*92*	*96*	*77*	*74*	*69*	*64*	*57*	*54*	*46*	*40*	*33*
13	111	109	105	100	97	95	89	87	81	79	75	72	78	61	59	55	51	47	44	37	35	29	25
	131	124	120	117	112	108	102	98	94	90	86	78	91	71	67	62	57	53	50	43	40	33	28
	160	*154*	*150*	*143*	*138*	*135*	*127*	*121*	*118*	*112*	*106*	*102*	*104*	*92*	*85*	*78*	*73*	*66*	*59*	*54*	*48*	*41*	*34*
14	119	116	110	108	112	100	94	92	87	84	80	84	72	68	63	60	54	50	49	42	37	32	26
	136	132	127	124	126	114	108	104	99	96	92	98	78	78	72	68	63	58	56	48	42	38	31
	169	*164*	*157*	*152*	*154*	*142*	*135*	*130*	*124*	*120*	*111*	*112*	*102*	*94*	*89*	*84*	*77*	*72*	*70*	*60*	*51*	*46*	*37*
15	130	123	117	111	111	110	100	99	91	87	87	80	75	72	66	65	60	52	48	45	45	34	30
	145	141	134	130	126	125	113	111	105	101	105	92	86	84	76	75	69	60	56	51	50	38	33
	180	*174*	*165*	*160*	*156*	*150*	*142*	*138*	*130*	*120*	*135*	*111*	*106*	*102*	*95*	*90*	*84*	*75*	*70*	*63*	*60*	*46*	*42*
16	130	136	122	118	114	112	104	100	94	112	87	84	79	76	69	66	61	64	51	48	41	40	30
	148	152	140	136	130	128	120	116	109	112	101	96	90	88	80	76	68	72	58	54	46	44	34
	185	*184*	*174*	*168*	*162*	*156*	*149*	*142*	*139*	*144*	*120*	*120*	*112*	*108*	*100*	*94*	*86*	*88*	*71*	*66*	*56*	*52*	*43*

Table A11. *(cont.)*

n	3	4	5	6	7	8	9	10	11	12	13	14	15	16	17	18	19	20	21	22	23	24	25
17	32	37	44	49	54	57	65	69	72	77	81	87	91	94	119	102	108	113	118	122	128	132	137
	35	44	49	56	61	65	74	77	83	89	94	99	105	109	136	118	125	130	135	141	146	150	156
	43	*53*	*61*	*68*	*75*	*81*	*92*	*97*	*104*	*111*	*118*	*124*	*130*	*139*	*153*	*150*	*157*	*162*	*168*	*175*	*181*	*187*	*192*
18	36	40	46	54	56	62	72	72	76	84	87	92	99	100	102	126	116	120	126	128	133	144	142
	39	46	51	66	64	72	81	82	87	96	98	104	111	116	118	144	127	136	144	148	151	162	161
	48	*56*	*63*	*78*	*81*	*88*	*99*	*104*	*108*	*120*	*121*	*130*	*138*	*142*	*150*	*180*	*160*	*170*	*177*	*184*	*189*	*198*	*201*
19	36	41	47	54	59	64	70	74	79	85	89	94	100	104	108	116	133	125	128	132	137	142	148
	40	49	56	61	68	73	80	85	92	98	102	108	113	120	125	127	152	144	147	151	159	162	168
	49	*57*	*67*	*77*	*85*	*91*	*99*	*104*	*114*	*121*	*127*	*135*	*142*	*149*	*157*	*160*	*190*	*171*	*183*	*189*	*197*	*204*	*211*
20	37	48	55	56	61	72	73	90	84	92	95	100	110	112	113	120	125	140	134	138	143	152	155
	41	52	60	66	72	80	83	100	95	104	108	114	125	128	130	136	144	160	154	160	163	172	180
	52	*64*	*75*	*80*	*87*	*100*	*103*	*120*	*117*	*128*	*135*	*142*	*150*	*156*	*162*	*170*	*171*	*200*	*193*	*196*	*203*	*212*	*220*
21	42	45	51	60	70	71	78	80	85	93	97	112	111	114	118	126	128	134	147	142	147	156	158
	45	52	60	69	77	81	90	91	101	108	112	126	126	130	135	144	147	154	168	163	170	177	182
	54	*64*	*75*	*84*	*98*	*100*	*111*	*118*	*124*	*132*	*138*	*154*	*156*	*162*	*168*	*177*	*183*	*193*	*210*	*205*	*212*	*222*	*225*
22	40	48	54	62	68	74	79	86	99	98	100	108	111	118	122	128	132	138	142	176	151	158	163
	46	56	62	70	77	84	91	98	110	110	117	124	130	136	141	148	151	160	163	198	173	182	188
	55	*66*	*76*	*88*	*97*	*106*	*111*	*120*	*143*	*138*	*143*	*152*	*160*	*168*	*175*	*184*	*189*	*196*	*205*	*242*	*217*	*228*	*234*

m

23	43	49	56	63	70	76	82	88	95	100	105	110	117	122	128	133	137	143	147	151	184	160	169
	47	57	65	73	79	89	94	101	108	113	120	127	134	140	146	151	159	163	170	173	207	183	194
	58	*69*	*81*	*91*	*99*	*107*	*117*	*125*	*132*	*138*	*150*	*157*	*165*	*174*	*181*	*189*	*197*	*203*	*212*	*217*	*253*	*228*	*242*
24	45	56	58	72	72	88	87	92	98	108	109	116	123	136	132	144	142	152	156	158	160	192	178
	51	60	67	78	83	96	99	106	111	132	124	132	141	152	150	162	172	172	177	182	183	216	204
	63	*76*	*82*	*96*	*103*	*120*	*123*	*130*	*138*	*156*	*154*	*164*	*174*	*184*	*187*	*198*	*204*	*212*	*222*	*228*	*228*	*264*	*254*
25	46	53	65	67	74	81	88	100	100	106	111	119	130	137	142	148	155	158	163	169	178	200	
	52	61	75	78	85	95	101	110	116	120	131	136	145	148	156	161	168	180	182	188	194	204	225
	64	*73*	*90*	*96*	*106*	*118*	*124*	*140*	*143*	*153*	*160*	*169*	*180*	*185*	*192*	*201*	*211*	*220*	*225*	*234*	*242*	*254*	*275*

Source: Reprinted from Siegel and Castellan (1988); reproduced with permission of McGraw-Hill, Inc. and *Journal of the American Statistical Association.* Copyright 1976 the American Statistical Association. All rights reserved.

Table A12. Critical values for the Kolmogorov-Smirnov two-sample, two-tailed test for small samples. The upper, middle and lower values are for alpha=0.10, 0.05 and 0.01, respectively

n	1	2	3	4	5	6	7	8	9	10	11	12	13	14	15	16	17	18	19	20	21	22	23	24	25
1																			19	20	21	22	23	24	25
2					10	12	14	16	18	18	20	22	24	24	26	28	30	32	32	34	36	38	38	40	42
								16	18	20	22	24	26	26	28	30	32	34	36	38	38	40	42	44	46
																			38	40	42	44	46	48	50
3			9	12	15	15	18	21	21	24	27	27	30	33	33	36	36	39	42	42	45	48	48	51	54
					15	18	21	21	24	27	30	30	33	36	36	39	42	45	45	48	51	51	54	57	60
									27	30	33	36	39	42	42	45	48	51	54	57	57	60	63	66	69
4				16	16	18	21	24	27	28	29	32	35	38	40	44	44	46	49	52	52	56	57	60	63
				16	20	20	24	28	28	30	33	36	39	42	44	48	48	50	53	56	59	62	64	68	68
						24	28	32	36	36	40	44	48	48	52	56	60	60	64	68	72	72	76	80	84
5					20	24	25	27	30	35	35	36	40	42	50	48	50	52	56	60	60	63	65	67	75
					25	25	28	30	35	40	39	43	45	46	55	54	55	60	61	65	69	70	72	76	80
					25	30	35	35	40	45	45	50	52	56	60	64	68	70	71	80	80	83	87	90	95
6						24	28	30	33	36	38	42	46	48	51	54	56	60	64	66	69	70	73	78	78
						24	30	34	39	40	43	48	52	54	57	60	62	72	70	72	75	78	80	90	88
						30	36	40	45	48	54	60	60	64	69	72	73	84	83	88	90	92	97	102	107
7							35	34	36	40	44	46	50	56	56	59	61	65	69	72	77	77	80	84	86
							42	40	42	46	48	53	56	63	62	64	68	72	76	79	91	84	89	92	97
							49	48	49	53	59	60	65	77	75	77	84	87	91	93	105	103	108	112	115

8	16	21	24	27	30	34	40	40	44	48	52	54	58	60	72	68	72	74	80	81	84	89	96	95
	16	21	28	30	34	40	46	48	48	53	60	62	64	67	80	77	80	82	88	89	94	98	104	104
		32	*35*	*40*	*48*	*55*	*56*	*60*	*64*	*68*	*72*	*76*	*81*	*88*	*88*	*94*	*98*	*104*	*107*	*112*	*115*	*128*	*125*
9	18	21	27	30	33	36	40	40	50	52	57	59	63	69	69	74	81	80	84	90	91	94	99	101
	18	24	28	35	39	42	46	54	53	59	63	65	70	75	78	82	90	89	93	99	101	106	111	114
		27	*36*	*40*	*45*	*49*	*55*	*63*	*63*	*70*	*75*	*78*	*84*	*90*	*94*	*99*	*108*	*107*	*111*	*117*	*122*	*126*	*132*	*135*
10	18	24	28	35	36	40	44	50	60	57	60	64	68	75	76	79	82	85	100	95	98	101	106	110
	20	27	30	40	40	46	48	53	70	60	66	70	74	80	84	89	92	94	110	105	108	114	118	125
		30	*36*	*45*	*48*	*53*	*60*	*63*	*80*	*77*	*80*	*84*	*90*	*100*	*100*	*106*	*108*	*113*	*130*	*126*	*130*	*137*	*140*	*150*
11	20	27	29	35	38	44	52	57	57	66	64	67	73	76	80	85	88	92	96	101	110	108	111	117
	22	30	33	39	43	48	53	59	60	77	72	75	82	84	89	93	97	102	107	112	121	119	124	129
		33	*40*	*45*	*54*	*59*	*64*	*70*	*77*	*88*	*86*	*91*	*96*	*102*	*106*	*110*	*118*	*122*	*127*	*134*	*143*	*142*	*150*	*154*
12	22	27	36	36	48	46	52	57	60	64	72	71	78	84	88	90	96	99	104	108	110	113	132	120
	24	30	36	43	53	53	60	63	66	72	84	81	86	93	96	100	108	108	116	120	124	125	144	138
		36	*44*	*50*	*60*	*60*	*68*	*75*	*80*	*86*	*96*	*95*	*104*	*108*	*116*	*119*	*126*	*130*	*140*	*141*	*148*	*149*	*168*	*165*
13	24	30	35	40	46	50	54	59	64	67	71	91	78	87	91	96	99	104	108	113	117	120	125	131
	26	33	39	45	52	56	62	65	70	75	81	91	89	96	101	105	110	114	120	126	130	135	140	145
		39	*48*	*52*	*60*	*60*	*65*	*72*	*78*	*84*	*91*	*95*	*117*	*104*	*115*	*121*	*127*	*131*	*138*	*143*	*150*	*156*	*161*	*166*	*172*

Table A12. (cont.)

																							m		
n	1	2	3	4	5	6	7	8	9	10	11	12	13	14	15	16	17	18	19	20	21	22	23	24	25
14		24	33	38	42	48	56	58	63	68	73	78	78	98	92	96	100	104	110	114	126	124	127	132	136
		26	36	42	46	54	63	64	70	74	82	86	89	112	98	106	111	116	121	126	140	138	142	146	150
			42	*48*	*56*	*64*	*77*	*76*	*84*	*90*	*96*	*104*	*104*	*126*	*123*	*126*	*134*	*140*	*148*	*152*	*161*	*164*	*170*	*176*	*182*
15		26	33	40	50	51	56	60	69	75	76	84	87	92	105	101	105	111	114	125	126	130	134	141	145
		28	36	44	55	57	62	67	75	80	84	93	96	98	120	114	116	123	127	135	138	144	149	156	160
			42	*52*	*60*	*69*	*75*	*81*	*90*	*100*	*102*	*108*	*115*	*123*	*135*	*133*	*142*	*147*	*152*	*160*	*168*	*173*	*179*	*186*	*195*
16		28	36	44	48	54	59	72	69	76	80	88	91	96	101	112	109	116	120	128	130	136	141	152	149
		30	39	48	54	60	64	80	78	84	89	96	101	106	114	128	124	128	133	140	145	150	157	168	167
			45	*56*	*64*	*72*	*77*	*88*	*94*	*100*	*106*	*116*	*121*	*126*	*133*	*160*	*143*	*154*	*160*	*168*	*173*	*180*	*187*	*200*	*199*
17		30	36	44	50	56	61	68	74	79	85	90	96	100	105	116	109	118	126	132	136	142	146	151	156
		32	42	48	55	62	68	77	82	89	93	100	105	111	116	124	136	133	141	146	151	157	163	168	173
			48	*60*	*68*	*73*	*84*	*88*	*99*	*106*	*110*	*119*	*127*	*134*	*142*	*143*	*170*	*164*	*166*	*175*	*180*	*187*	*196*	*203*	*207*
18		32	39	46	52	66	65	72	81	82	88	96	99	104	111	116	118	144	133	136	144	148	152	162	162
		34	45	50	60	72	72	80	90	92	97	108	110	116	123	128	133	162	142	152	159	164	170	180	180
			51	*60*	*70*	*84*	*87*	*94*	*108*	*108*	*118*	*126*	*131*	*140*	*147*	*154*	*164*	*180*	*176*	*182*	*189*	*196*	*204*	*216*	*216*
19	19	32	42	49	56	64	74	80	85	92	99	104	110	114	120	126	133	144	152	144	147	152	159	164	168
		36	45	53	61	70	82	89	94	102	108	114	121	127	133	141	152	160	171	163	169	177	183	187	
			38	*54*	*64*	*71*	*83*	*91*	*98*	*107*	*113*	*122*	*130*	*138*	*148*	*152*	*160*	*166*	*176*	*190*	*199*	*204*	*209*	*218*	*224*

20	20	34	42	52	60	66	72	80	84	100	96	104	108	114	125	128	132	136	144	160	154	160	164	172	180
		38	48	60	65	72	79	88	93	110	107	116	120	126	135	140	146	152	160	180	173	176	184	192	200
		40	*57*	*68*	*80*	*88*	*93*	*104*	*111*	*130*	*127*	*140*	*143*	*152*	*160*	*168*	*175*	*182*	*187*	*220*	*199*	*212*	*219*	*228*	*235*
21	21	36	45	52	60	69	77	81	90	95	101	108	113	126	126	130	136	144	147	154	168	163	171	177	182
		38	51	59	69	75	91	89	99	105	112	120	126	140	138	145	151	159	163	173	189	183	189	198	202
		42	*57*	*72*	*80*	*90*	*105*	*107*	*117*	*126*	*134*	*141*	*150*	*161*	*168*	*173*	*180*	*189*	*199*	*199*	*231*	*223*	*227*	*237*	*244*
22	22	38	48	56	63	70	77	84	91	98	110	110	117	124	130	136	142	148	152	160	163	168	173	182	189
		40	51	62	70	78	84	94	101	108	121	124	130	138	144	150	157	164	169	176	183	187	194	204	209
		44	*60*	*72*	*83*	*92*	*103*	*112*	*122*	*130*	*143*	*148*	*156*	*164*	*173*	*180*	*187*	*196*	*204*	*212*	*223*	*223*	*237*	*242*	*250*
23	23	38	48	57	65	73	80	89	94	101	108	113	120	127	134	141	146	152	159	164	171	173	207	183	195
		42	54	64	72	80	89	98	106	114	119	125	135	142	149	157	163	170	177	184	189	194	230	205	216
		46	*63*	*76*	*87*	*97*	*108*	*115*	*126*	*137*	*142*	*149*	*161*	*170*	*179*	*187*	*196*	*204*	*209*	*219*	*227*	*237*	*253*	*249*	*262*
24	24	40	51	60	67	78	84	96	99	106	111	132	125	132	141	152	151	162	164	172	177	182	183	216	204
		44	57	68	76	90	92	104	111	118	124	144	140	146	156	168	168	180	183	192	198	204	205	240	225
		48	*66*	*80*	*90*	*102*	*112*	*128*	*132*	*140*	*150*	*168*	*166*	*176*	*186*	*200*	*203*	*216*	*218*	*228*	*237*	*249*	*249*	*288*	*262*
25	25	42	54	63	75	78	86	95	101	110	117	120	131	136	145	149	156	162	168	180	182	189	195	204	225
		46	60	68	80	88	97	104	114	125	129	138	145	150	160	167	173	180	187	200	202	209	216	225	250
		50	*69*	*84*	*95*	*107*	*115*	*125*	*135*	*150*	*154*	*165*	*172*	*182*	*195*	*199*	*207*	*216*	*224*	*235*	*244*	*262*	*262*	*250*	*300*

Source: Reprinted from Siegel and Castellan 1988; reproduced with permission of McGraw-Hill, Inc.

Table A13. *Critical values for the Kolmogorov–Smirnov two-sample, two-tailed test for large samples*

Level of significance	Value of $D_{m,n}$ so large as to call for rejection of H_0 at the indicated level of significance, where $D_{m,n} = \text{maximum} \mid S_m(X) - S_n(X) \mid$
0.10	$1.22 \sqrt{\dfrac{m+n}{mn}}$
0.05	$1.36 \sqrt{\dfrac{m+n}{mn}}$
0.025	$1.48 \sqrt{\dfrac{m+n}{mn}}$
0.01	$1.63 \sqrt{\dfrac{m+n}{mn}}$
0.005	$1.73 \sqrt{\dfrac{m+n}{mn}}$
0.001	$1.95 \sqrt{\dfrac{m+n}{mn}}$

Source: From Siegel and Castellan (1988), reproduced with permission of McGraw-Hill, Inc., adapted from Smirnov (1948). Reproduced courtesy of the Institute of Mathematical Statistics.

Table A14. *Critical values for the Kruskal–Wallis one-way analysis of variance*

Sample size			Alpha				
n_1	n_2	n_3	0.10	0.05	0.01	0.005	0.001
2	2	2	4.25				
3	2	1	4.29				
3	2	2	4.71	4.71			
3	3	1	4.57	5.14			
3	3	2	4.56	5.36			
3	3	3	4.62	5.60	7.20	7.20	
4	2	1	4.50				
4	2	2	4.46	5.33			
4	3	1	4.06	5.21			
4	3	2	4.51	5.44	6.44	7.00	
4	3	3	4.71	5.73	6.75	7.32	8.02
4	4	1	4.17	4.97	6.67		
4	4	2	4.55	5.45	7.04	7.28	
4	4	3	4.55	5.60	7.14	7.59	8.32
4	4	4	4.65	5.69	7.66	8.00	8.65
5	2	1	4.20	5.00			
5	2	2	4.36	5.16	6.53		
5	3	1	4.02	4.96			
5	3	2	4.65	5.25	6.82	7.18	

Table A14. (*cont.*)

Sample size			Alpha				
n_1	n_2	n_3	0.10	0.05	0.01	0.005	0.001
5	3	3	4.53	5.65	7.08	7.51	8.24
5	4	1	3.99	4.99	6.95	7.36	
5	4	2	4.54	5.27	7.12	7.57	8.11
5	4	3	4.55	5.63	7.44	7.91	8.50
5	4	4	4.62	5.62	7.76	8.14	9.00
5	5	1	4.11	5.13	7.31	7.75	
5	5	2	4.62	5.34	7.27	8.13	8.68
5	5	3	4.54	5.71	7.54	8.24	9.06
5	5	4	4.53	5.64	7.77	8.37	9.32
5	5	5	4.56	5.78	7.98	8.72	9.68
Large samples			4.61	5.99	9.21	10.60	13.82

Note:

The absence of an entry in the extreme tails indicates that the distribution may not take on the necessary extremes values.

Source: From Siegel and Castellan (1988); reproduced with permission of McGraw-Hill, Inc.

Table A15. *Critical values of Q for Dunn's multiple comparison test.* k = *total number of samples*

					Alpha					
k	0.50	0.20	0.10	0.05	0.02	0.01	0.005	0.002	0.001
2	0.674	1.282	1.645	1.960	2.327	2.576	2.807	3.091	3.291
3	1.383	1.834	2.128	2.394	2.713	2.936	3.144	3.403	3.588
4	1.732	2.128	2.394	2.639	2.936	3.144	3.342	3.588	3.765
5	1.960	2.327	2.576	2.807	3.091	3.291	3.481	3.719	3.891
6	2.128	2.475	2.713	2.936	3.209	3.403	3.588	3.820	3.986
7	2.261	2.593	2.823	3.038	3.304	3.494	3.675	3.902	4.067
8	2.369	2.690	2.914	3.124	3.384	3.570	3.748	3.972	4.134
9	2.461	2.773	2.992	3.197	3.453	3.635	3.810	4.031	4.191
10	2.540	2.845	3.059	3.261	3.512	3.692	3.865	4.083	4.241
11	2.609	2.908	3.119	3.317	3.565	3.743	3.914	4.129	4.286
12	2.671	2.965	3.172	3.368	3.613	3.789	3.957	4.171	4.326
13	2.726	3.016	3.220	3.414	3.656	3.830	3.997	4.209	4.363
14	2.777	3.062	3.264	3.456	3.695	3.868	4.034	4.244	4.397
15	2.823	3.105	3.304	3.494	3.731	3.902	4.067	4.276	4.428
16	2.866	3.144	3.342	3.529	3.765	3.935	4.098	4.305	4.456
17	2.905	3.181	3.376	3.562	3.796	3.965	4.127	4.333	4.483
18	2.942	3.215	3.409	3.593	3.825	3.993	4.154	4.359	4.508
19	2.976	3.246	3.439	3.622	3.852	4.019	4.179	4.383	4.532
20	3.008	3.276	3.467	3.649	3.878	4.044	4.203	4.406	4.554
21	3.038	3.304	3.494	3.675	3.902	4.067	4.226	4.428	4.575
22	3.067	3.331	3.519	3.699	3.925	4.089	4.247	4.448	4.595
23	3.094	3.356	3.543	3.722	3.947	4.110	4.268	4.468	4.614
24	3.120	3.380	3.566	3.744	3.968	4.130	4.287	4.486	4.632
25	3.144	3.403	3.588	3.765	3.988	4.149	4.305	4.504	4.649

Source: From Zar (1984).

Table A16. *Critical values of S for Kendall's Coefficient of Concordance for small samples*

			N			Additional values for $N=3$	
k	3†	4	5	6	7	k	s
Values at the 0.05 level of significance							
3			64.4	103.9	157.3	9	54.0
4		49.5	88.4	143.3	217.0	12	71.9
5		62.6	112.3	182.4	276.2	14	83.8
6		75.7	136.1	221.4	335.2	16	95.8
8	48.1	101.7	183.7	299.0	453.1	18	107.7
10	60.0	127.8	231.2	376.7	571.0		
15	89.8	192.9	349.8	570.5	864.9		
20	119.7	258.0	468.5	764.4	1158.7		
Values at the 0.01 level of significance							
3			75.6	122.8	185.6	9	75.9
4		61.4	109.3	176.2	265.0	12	103.5
5		80.5	142.8	229.4	343.8	14	121.9
6		99.5	176.1	282.4	422.6	16	140.2
8	66.8	137.4	242.7	388.3	579.9	18	158.6
10	85.1	175.3	309.1	494.0	737.0		
15	131.0	269.8	475.2	758.2	1129.5		
20	177.0	364.2	641.2	1022.2	1521.9		

Note:
† Notice that additional critical values of s for $N=3$ are given in the right-hand column of this table.
Source: From Siegel (1956); reproduced with permission of McGraw-Hill, Inc. and courtesy of the Institute of Mathematical Statistics.

Table A17. *Probabilities for the two-tailed sign test. L = frequency of the less frequent sign. T = total frequency of both pluses and minuses. Tabular value is the probability of obtaining L or fewer of the less frequent sign, out of a total of T pluses and minuses*

T	0	1	2	3	4	5	6	7	8	9	10	11	12
5	062	376	1										
6	032	218	688	1									
7	016	124	454	1									
8	008	070	290	726	1								
9	004	040	180	508	1								
10	002	022	110	344	754	1							
11	000	012	066	226	548	1							
12	000	006	038	146	388	774	1						
13	000	004	022	092	266	582	1						
14	000	002	012	058	180	424	790	1					
15	000	000	008	036	118	302	608	1					
16		000	004	022	076	210	454	804	1				
17		000	002	012	050	144	332	630	1				
18		000	002	008	030	096	238	480	814	1			
19			000	004	020	064	168	360	648	1			
20			000	002	012	042	116	264	504	824	1		
21			000	002	008	026	078	189	384	664	1		
22				000	004	016	052	134	286	524	832	1	
23				000	002	010	034	094	210	404	678	1	
24				000	002	006	022	064	152	308	542	838	1
25				000	000	004	014	044	108	230	424	690	1

Source: From Robson (1973).

Table A18. *Critical values of T for the Wilcoxon matched-pairs signed-ranks test*

	Level of significance		
	0.025	0.01	0.005 (one-tailed)
N	0.05	0.02	0.01 (two-tailed)
6	0	—	—
7	2	0	—
8	4	2	0
9	6	3	2
10	8	5	3
11	11	7	5
12	14	10	7
13	17	13	10
14	21	16	13
15	25	20	16
16	30	24	20
17	35	28	23
18	40	33	28
19	46	38	32
20	52	43	38
21	59	49	43
22	66	56	49
23	73	62	55
24	81	69	61
25	89	77	68

Source: From Siegel 1956; reproduced with permission of McGraw-Hill, Inc. and American Cyanamid Company.

Table A19. *Critical values of Spearman's Rho correlation coefficient*

	Alpha								
	0.25	0.10	0.05	0.025	0.01	0.005	0.0025	0.001	0.0005 (one tailed)
N	0.50	0.20	0.10	0.05	0.02	0.01	0.005	0.002	0.001 (two-tailed)
4	0.600	1.000	1.000						
5	0.500	0.800	0.900	1.000	1.000				
6	0.371	0.675	0.829	0.886	0.943	1.000	1.000		
7	0.321	0.571	0.714	0.786	0.893	0.929	0.964	1.000	1.000
8	0.310	0.524	0.643	0.738	0.833	0.881	0.905	0.952	0.976
9	0.267	0.483	0.600	0.700	0.783	0.833	0.867	0.917	0.993
10	0.248	0.455	0.564	0.648	0.745	0.794	0.830	0.879	0.903
11	0.236	0.427	0.536	0.618	0.709	0.755	0.800	0.845	0.873
12	0.224	0.406	0.503	0.587	0.671	0.727	0.776	0.825	0.860
13	0.209	0.385	0.484	0.560	0.648	0.703	0.747	0.802	0.835
14	0.200	0.367	0.464	0.538	0.622	0.675	0.723	0.776	0.811
15	0.189	0.354	0.443	0.521	0.604	0.654	0.700	0.754	0.786
16	0.182	0.341	0.429	0.503	0.582	0.635	0.679	0.732	0.765
17	0.176	0.328	0.414	0.485	0.566	0.615	0.662	0.713	0.748
18	0.170	0.317	0.401	0.472	0.550	0.600	0.643	0.695	0.728
19	0.165	0.309	0.391	0.460	0.535	0.584	0.628	0.677	0.712
20	0.161	0.299	0.380	0.447	0.520	0.570	0.612	0.662	0.696
21	0.156	0.292	0.370	0.435	0.508	0.556	0.599	0.648	0.681
22	0.152	0.284	0.361	0.425	0.496	0.544	0.586	0.634	0.667
23	0.148	0.278	0.353	0.415	0.486	0.532	0.573	0.622	0.654
24	0.144	0.271	0.344	0.406	0.476	0.521	0.562	0.610	0.642
25	0.142	0.265	0.337	0.398	0.466	0.511	0.551	0.598	0.630

Table A19. (cont.)

					Alpha				
	0.25	0.10	0.05	0.025	0.01	0.005	0.0025	0.001	0.0005 (one tailed)
N	0.50	0.20	0.10	0.05	0.02	0.01	0.005	0.002	0.001 (two-tailed)
26	0.138	0.259	0.331	0.390	0.457	0.501	0.541	0.587	0.619
27	0.136	0.255	0.324	0.382	0.448	0.491	0.531	0.577	0.608
28	0.133	0.250	0.317	0.375	0.440	0.483	0.522	0.567	0.598
29	0.130	0.245	0.312	0.368	0.433	0.475	0.513	0.558	0.589
30	0.128	0.240	0.306	0.362	0.425	0.467	0.504	0.549	0.580
31	0.126	0.236	0.301	0.356	0.418	0.459	0.496	0.541	0.571
32	0.124	0.232	0.296	0.350	0.412	0.452	0.489	0.533	0.563
33	0.121	0.229	0.291	0.345	0.405	0.446	0.482	0.525	0.554
34	0.120	0.225	0.287	0.340	0.399	0.439	0.475	0.517	0.547
35	0.118	0.222	0.283	0.335	0.394	0.433	0.468	0.510	0.539
36	0.116	0.219	0.279	0.330	0.388	0.427	0.462	0.504	0.533
37	0.114	0.216	0.275	0.325	0.383	0.421	0.456	0.497	0.526
38	0.113	0.212	0.271	0.321	0.378	0.415	0.450	0.491	0.519
39	0.111	0.210	0.267	0.317	0.373	0.410	0.444	0.485	0.513
40	0.110	0.207	0.264	0.313	0.368	0.405	0.439	0.479	0.507
41	0.108	0.204	0.261	0.309	0.364	0.400	0.433	0.473	0.501
42	0.107	0.202	0.257	0.305	0.359	0.395	0.428	0.468	0.495
43	0.105	0.199	0.254	0.301	0.355	0.391	0.423	0.463	0.490
44	0.104	0.197	0.251	0.298	0.351	0.386	0.419	0.458	0.484
45	0.103	0.194	0.248	0.294	0.347	0.382	0.414	0.453	0.479
46	0.102	0.192	0.246	0.291	0.343	0.378	0.410	0.448	0.474
47	0.101	0.190	0.243	0.288	0.340	0.374	0.405	0.443	0.469
48	0.100	0.188	0.240	0.285	0.336	0.370	0.401	0.439	0.465
49	0.098	0.186	0.238	0.282	0.333	0.366	0.397	0.434	0.460
50	0.097	0.184	0.235	0.279	0.329	0.363	0.393	0.430	0.456

Source: From Siegel and Castellan, 1988; reproduced with permission of McGraw-Hill, Inc.; adapted from Zar, J.H. (1972). Reprinted with permission from *Journal of the American Statistical Association.* Copyright 1972 by the American Statistical Association. All

Table A20. *Critical values of Kendall's tau correlation coefficient.* N=*number of pairs of measurements. The τ values are the smallest values of τ significant at the 0.05 alpha level for different values of* N

N	τ	N	τ
5	0.80	18	0.33
6	0.69	19	0.33
7	0.63	20	0.31
8	0.57	21	0.31
9	0.53	22	0.29
10	0.49	23	0.29
11	0.45	24	0.28
12	0.43	25	0.27
13	0.41	26	0.27
14	0.39	27	0.26
15	0.37	28	0.25
16	0.35	29	0.25
17	0.35	30	0.25

Source: Adapted from Robson (1973).

Table A21. *Critical values of q for Dunnett's test*

df	\multicolumn{14}{c}{P = number of samples spanned in the comparison}

df	2	3	4	5	6	7	8	9	10	11	12	13	16	21
α_r=0.05														
5	2.57	3.03	3.29	3.48	3.62	3.73	3.82	3.90	3.97	4.03	4.09	4.14	4.26	4.42
6	2.45	2.86	3.10	3.26	3.39	3.49	3.57	3.64	3.71	3.76	3.81	3.86	3.97	4.11
7	2.36	2.75	2.97	3.12	3.24	3.33	3.41	3.47	3.53	3.58	3.63	3.67	3.78	3.91
8	2.31	2.67	2.88	3.02	3.13	3.22	3.29	3.35	3.41	3.46	3.50	3.54	3.64	3.76
9	2.26	2.61	2.81	2.95	3.05	3.14	3.20	3.26	3.32	3.36	3.40	3.44	3.53	3.65
10	2.23	2.57	2.76	2.89	2.99	3.07	3.14	3.19	3.24	3.29	3.33	3.36	3.45	3.57
11	2.20	2.53	2.72	2.84	2.94	3.02	3.08	3.14	3.19	3.23	3.27	3.30	3.39	3.50
12	2.18	2.50	2.68	2.81	2.90	2.98	3.04	3.09	3.14	3.18	3.22	3.25	3.34	3.45
13	2.16	2.48	2.65	2.78	2.87	2.94	3.00	3.06	3.10	3.14	3.18	3.21	3.29	3.40
14	2.14	2.46	2.63	2.75	2.84	2.91	2.97	3.02	3.07	3.11	3.14	3.18	3.26	3.36
15	2.13	2.44	2.61	2.73	2.82	2.89	2.95	3.00	3.04	3.08	3.12	3.15	3.23	3.33
16	2.12	2.42	2.59	2.71	2.80	2.87	2.92	2.97	3.02	3.06	3.09	3.12	3.20	3.30
17	2.11	2.41	2.58	2.69	2.78	2.85	2.90	2.95	3.00	3.03	3.07	3.10	3.18	3.27
18	2.10	2.40	2.56	2.68	2.76	2.83	2.89	2.94	2.98	3.01	3.05	3.08	3.16	3.25
19	2.09	2.39	2.55	2.66	2.75	2.81	2.87	2.92	2.96	3.00	3.03	3.06	3.14	3.23
20	2.09	2.38	2.54	2.65	2.73	2.80	2.86	2.90	2.95	2.98	3.02	3.05	3.12	3.22
24	2.06	2.35	2.51	2.61	2.70	2.76	2.81	2.86	2.90	2.94	2.97	3.00	3.07	3.16
30	2.04	2.32	2.47	2.58	2.66	2.72	2.77	2.82	2.86	2.89	2.92	2.95	3.02	3.11
40	2.02	2.29	2.44	2.54	2.62	2.68	2.73	2.77	2.81	2.85	2.87	2.90	2.97	3.06
60	2.00	2.27	2.41	2.51	2.58	2.64	2.69	2.73	2.77	2.80	2.83	2.86	2.92	3.00
120	1.98	2.24	2.38	2.47	2.55	2.60	2.65	2.69	2.73	2.76	2.79	2.81	2.87	2.95
∞	1.96	2.21	2.35	2.44	2.51	2.57	2.61	2.65	2.69	2.72	2.74	2.77	2.83	2.91

Source: From Glantz (1992); reproduced with permission of McGraw-Hill, Inc.; adapted from Dunnett (1964).

$\alpha_r = 0.01$

5	4.03	4.63	4.98	5.22	5.41	5.56	5.69	5.80	5.89	5.98	6.05	6.12	6.30	6.52
6	3.71	4.21	4.51	4.71	4.87	5.00	5.10	5.20	5.28	5.35	5.41	5.47	5.62	5.81
7	3.50	3.95	4.21	4.39	4.53	4.64	4.74	4.82	4.89	4.95	5.01	5.06	5.19	5.36
8	3.36	3.77	4.00	4.17	4.29	4.40	4.48	4.56	4.62	4.68	4.73	4.78	4.90	5.05
9	3.25	3.63	3.85	4.01	4.12	4.22	4.30	4.37	4.43	4.48	4.53	4.57	4.68	4.82
10	3.17	3.53	3.74	3.88	3.99	4.08	4.16	4.22	4.28	4.33	4.37	4.42	4.52	4.65
11	3.11	3.45	3.65	3.79	3.89	3.98	4.05	4.11	4.16	4.21	4.25	4.29	4.30	4.52
12	3.05	3.39	3.58	3.71	3.81	3.89	3.96	4.02	4.07	4.12	4.16	4.19	4.29	4.41
13	3.01	3.33	3.52	3.65	3.74	3.82	3.89	3.94	3.99	4.04	4.08	4.11	4.20	4.32
14	2.98	3.29	3.47	3.59	3.69	3.76	3.83	3.88	3.93	3.97	4.01	4.05	4.13	4.24
15	2.95	3.25	3.43	3.55	3.64	3.71	3.78	3.83	3.88	3.92	3.95	3.99	4.07	4.18
16	2.92	3.22	3.39	3.51	3.60	3.67	3.73	3.78	3.83	3.87	3.91	3.94	4.02	4.13
17	2.90	3.19	3.36	3.47	3.56	3.63	3.69	3.74	3.79	3.83	3.86	3.90	3.98	4.08
18	2.88	3.17	3.33	3.44	3.53	3.60	3.66	3.71	3.75	3.79	3.83	3.86	3.94	4.04
19	2.86	3.15	3.31	3.42	3.50	3.57	3.63	3.68	3.72	3.76	3.79	3.83	3.90	4.00
20	2.85	3.13	3.29	3.40	3.48	3.55	3.60	3.65	3.69	3.73	3.77	3.80	3.87	3.97
24	2.80	3.07	3.22	3.32	3.40	3.47	3.52	3.57	3.61	3.64	3.68	3.70	3.78	3.87
30	2.75	3.01	3.15	3.25	3.33	3.39	3.44	3.49	3.52	3.56	3.59	3.62	3.69	3.78
40	2.70	2.95	3.09	3.19	3.26	3.32	3.37	3.41	3.44	3.48	3.51	3.53	3.60	3.68
60	2.66	2.90	3.03	3.12	3.19	3.25	3.29	3.33	3.37	3.40	3.42	3.45	3.51	3.59
120	2.62	2.85	2.97	3.06	3.12	3.18	3.22	3.26	3.29	3.32	3.35	3.37	3.43	3.51
∞	2.58	2.79	2.92	3.00	3.06	3.11	3.15	3.19	3.22	3.25	3.27	3.29	3.35	2.42

Table A22. Values of $\log_2 n$ and $-p\log_2 p$. The entry is either n or p, depending on which column is read out

n or p	$\log_2 n$	$-p\log_2 p$	n or p	$\log_2 n$	$-p\log_2 p$
1	0.000	0.0664	51	5.672	0.4954
2	1.000	0.1129	52	5.700	0.4906
3	1.585	0.1518	53	5.728	0.4854
4	2.000	0.1858	54	5.755	0.4800
5	2.322	0.2161	55	5.781	0.4744
6	2.585	0.2435	56	5.807	0.4684
7	2.807	0.2686	57	5.833	0.4623
8	3.000	0.2915	58	5.858	0.4558
9	3.170	0.3127	59	5.883	0.4491
10	3.322	0.3322	60	5.907	0.4422
11	3.459	0.3503	61	5.931	0.4350
12	3.585	0.3671	62	5.954	0.4276
13	3.700	0.3826	63	5.977	0.4199
14	3.807	0.3971	64	6.000	0.4121
15	3.907	0.4105	65	6.022	0.4040
16	4.000	0.4230	66	6.044	0.3957
17	4.087	0.4346	67	6.066	0.3871
18	4.170	0.4453	68	6.087	0.3784
19	4.248	0.4552	69	6.109	0.3694
20	4.322	0.4644	70	6.129	0.3602
21	4.392	0.4728	71	6.150	0.3508
22	4.459	0.4806	72	6.170	0.3412
23	4.524	0.4877	73	6.190	0.3314
24	4.585	0.4941	74	6.209	0.3215
25	4.644	0.5000	75	6.229	0.3113

26	4.700	0.5053	76	6.248	0.3009
27	4.755	0.5100	77	6.267	0.2903
28	4.807	0.5142	78	6.285	0.2796
29	4.858	0.5179	79	6.304	0.2687
30	4.907	0.5211	80	6.322	0.2575
31	4.954	0.5238	81	6.340	0.2462
32	5.000	0.5260	82	6.358	0.2348
33	5.044	0.5278	83	6.375	0.2231
34	5.087	0.5292	84	6.392	0.2113
35	5.129	0.5301	85	6.409	0.1993
36	5.170	0.5306	86	6.426	0.1871
37	5.209	0.5307	87	6.443	0.1748
38	5.248	0.5304	88	6.459	0.1623
39	5.285	0.5298	89	6.476	0.1496
40	5.322	0.5288	90	6.492	0.1368
41	5.358	0.5274	91	6.508	0.1238
42	5.392	0.5256	92	6.524	0.1107
43	5.426	0.5236	93	6.539	0.0974
44	5.459	0.5211	94	6.555	0.0839
45	5.492	0.5184	95	6.570	0.0703
46	5.524	0.5153	96	6.585	0.0565
47	5.555	0.5120	97	6.600	0.0426
48	5.585	0.5083	98	6.615	0.0286
49	5.615	0.5043	99	6.629	0.0140
50	5.644	0.5000	100	6.644	0.0000

Note:
These values were originally prepared at the Operational Applications Laboratory. Complete tables are available in Air Force Cambridge Research Center, *Techmniocal Report 54–50.*
Source: From Meyer (1976).

Table A23. *Values of sine, cosine and tangent for direction (angular degrees) and time (0000 to 2400 hours)*

degrees	time	sine	cosine	tangent
0	0000	0	+1	0
1	0004	+0.0175	+0.9998	+0.0175
2	0008	+0.0349	+0.9994	+0.0349
3	0012	+0.0523	+0.9986	+0.0524
4	0016	+0.0698	+0.9976	+0.0699
5	0020	+0.0872	+0.9962	+0.0875
6	0024	+0.1045	+0.9945	+0.1051
7	0028	+0.1219	+0.9925	+0.1228
8	0032	+0.1392	+0.9903	+0.1405
9	0036	+0.1564	+0.9877	+0.1584
10	0040	+0.1736	+0.9848	+0.1763
11	0044	+0.1908	+0.9816	+0.1944
12	0048	+0.2079	+0.9781	+0.2126
13	0052	+0.2250	+0.9744	+0.2309
14	0056	+0.2419	+0.9703	+0.2493
15	0100	+0.2588	+0.9659	+0.2679
16	0104	+0.2756	+0.9613	+0.2867
17	0108	+0.2924	+0.9563	+0.3057
18	0112	+0.3090	+0.9511	+0.3249
19	0116	+0.3256	+0.9455	+0.3443
20	0120	+0.3420	+0.9397	+0.3640
21	0124	+0.3584	+0.9336	+0.3839
22	0128	+0.3746	+0.9272	+0.4040
23	0132	+0.3907	+0.9205	+0.4245
24	0136	+0.4067	+0.9135	+0.4452
25	0140	+0.4226	+0.9063	+0.4663
26	0144	+0.4384	+0.8988	+0.4877
27	0148	+0.4540	+0.8910	+0.5095
28	0152	+0.4695	+0.8829	+0.5317
29	0156	+0.4848	+0.8746	+0.5543
30	0200	+0.5000	+0.8660	+0.5774
31	0204	+0.5150	+0.8572	+0.6009
32	0208	+0.5299	+0.8480	+0.6249
33	0212	+0.5446	+0.8387	+0.6494
34	0216	+0.5592	+0.8290	+0.6745

Table A23. (*cont,*)

degrees	time	sine	cosine	tangent
35	0220	+0.5736	+0.8192	+0.7002
36	0224	+0.5878	+0.8090	+0.7265
37	0228	+0.6018	+0.7986	+0.7536
38	0232	+0.6157	+0.7880	+0.7813
39	0236	+0.6293	+0.7771	+0.8098
40	0240	+0.6428	+0.7660	+0.8391
41	0244	+0.6561	+0.7547	+0.8693
42	0248	+0.6691	+0.7431	+0.9004
43	0252	+0.6820	+0.7314	+0.9325
44	0256	+0.6947	+0.7193	+0.9657
45	0300	+0.7071	+0.7071	+1.000
46	0304	+0.7193	+0.6947	+1.036
47	0308	+0.7314	+0.6820	+1.072
48	0312	+0.7431	+0.6691	+1.111
49	0316	+0.7547	+0.6561	+1.150
50	0320	+0.7660	+0.6428	+1.192
51	0324	+0.7771	+0.6293	+1.235
52	0328	+0.7880	+0.6157	+1.280
53	0332	+0.7986	+0.6018	+1.327
54	0336	+0.8090	+0.5878	+1.376
55	0340	+0.8192	+0.5736	+1.428
56	0344	+0.8290	+0.5592	+1.483
57	0348	+0.8387	+0.5446	+1.540
58	0352	+0.8480	+0.5299	+1.600
59	0356	+0.8572	+0.5150	+1.664
60	0400	+0.8660	+0.5000	+1.732
61	0404	+0.8746	+0.4848	+1.804
62	0408	+0.8829	+0.4695	+1.881
63	0412	+0.8910	+0.4540	+1.963
64	0416	+0.8988	+0.4384	+2.050
65	0420	+0.9063	+0.4226	+2.145
66	0424	+0.9135	+0.4067	+2.246
67	0428	+0.9205	+0.3907	+2.356
68	0432	+0.9272	+0.3746	+2.475
69	0436	+0.9336	+0.3584	+2.605
70	0440	+0.9397	+0.3420	+2.747

Table A23. (*cont,*)

degrees	time	sine	cosine	tangent
71	0444	+0.9455	+0.3256	+2.904
72	0448	+0.9511	+0.3090	+3.078
73	0452	+0.9563	+0.2924	+3.271
74	0456	+0.9613	+0.2756	+3.487
75	0500	+0.9659	+0.2588	+3.732
76	0504	+0.9703	+0.2419	+4.011
77	0508	+0.9744	+0.2250	+4.331
78	0512	+0.9781	+0.2079	+4.705
79	0516	+0.9816	+0.1908	+5.145
80	0520	+0.9848	+0.1736	+5.671
81	0524	+0.9877	+0.1564	+6.314
82	0528	+0.9903	+0.1392	+7.115
83	0532	+0.9925	+0.1219	+8.144
84	0536	+0.9945	+0.1045	+9.514
85	0540	+0.9962	+0.0872	+11.43
86	0544	+0.9976	+0.0698	+14.30
87	0548	+0.9986	+0.0523	+19.08
88	0552	+0.9994	+0.0349	+28.64
89	0556	+0.9998	+0.0175	+57.29
90	0600	+1.000	0	∞
91	0604	+0.9998	−0.0175	−57.29
92	0608	+0.9994	−0.0349	−28.64
93	0612	+0.9986	−0.0523	−19.08
94	0616	+0.9976	−0.0698	−14.30
95	0620	+0.9962	−0.0872	−11.43
96	0624	+0.9945	−0.1045	−9.514
97	0628	+0.9925	−0.1219	−8.144
98	0632	+0.9903	−0.1392	−7.115
99	0636	+0.9877	−0.1564	−6.314
100	0640	+0.9848	−0.1736	−5.671
101	0644	+0.9816	−0.1908	−5.145
102	0648	+0.9781	−0.2079	−4.705
103	0652	+0.9744	−0.2250	−4.331
104	0656	+0.9703	−0.2419	−4.011
105	0700	+0.9659	−0.2588	−3.732
106	0704	+0.9613	−0.2756	−3.487

degrees	time	sine	cosine	tangent
107	0708	+0.9563	−0.2924	−3.271
108	0712	+0.9511	−0.3090	−3.078
109	0716	+0.9455	−0.3256	−2.904
110	0720	+0.9397	−0.3420	−2.747
111	0724	+0.9336	−0.3584	−2.605
112	0728	+0.9272	−0.3746	−2.475
113	0732	+0.9205	−0.3907	−2.356
114	0736	+0.9135	−0.4067	−2.246
115	0740	+0.9063	−0.4226	−2.145
116	0744	+0.8988	−0.4384	−2.050
117	0748	+0.8910	−0.4540	−1.963
118	0752	+0.8829	−0.4695	−1.881
119	0756	+0.8746	−0.4848	−1.804
120	0800	+0.8660	−0.5000	−1.732
121	0804	+0.8572	−0.5150	−1.664
122	0808	+0.8480	−0.5299	−1.600
123	0812	+0.8387	−0.5446	−1.540
124	0816	+0.8290	−0.5592	−1.483
125	0820	+0.8192	−0.5736	−1.428
126	0824	+0.8090	−0.5878	−1.376
127	0828	+0.7986	−0.6018	−1.327
128	0832	+0.7880	−0.6157	−1.280
129	0836	+0.7771	−0.6293	−1.235
130	0840	+0.7660	−0.6428	−1.192
131	0844	+0.7547	−0.6561	−1.150
132	0848	+0.7431	−0.6691	−1.111
133	0852	+0.7314	−0.6820	−1.072
134	0856	+0.7193	−0.6947	−1.036
135	0900	+0.7071	−0.7071	−1.000
136	0904	+0.6947	−0.7193	−0.9657
137	0908	+0.6820	−0.7314	−0.9325
138	0912	+0.6691	−0.7431	−0.9004
139	0916	+0.6561	−0.7547	−0.8693
140	0920	+0.6428	−0.7660	−0.8391
141	0924	+0.6293	−0.7771	−0.8098
142	0928	+0.6157	−0.7880	−0.7813

Table A23. (*cont,*)

degrees	time	sine	cosine	tangent
143	0932	+0.6018	−0.7986	−0.7536
144	0936	+0.5878	−0.8090	−0.7265
145	0940	+0.5736	−0.8192	−0.7002
146	0944	+0.5592	−0.8290	−0.6745
147	0948	+0.5446	−0.8387	−0.6494
148	0952	+0.5299	−0.8480	−0.6249
149	0956	+0.5150	−0.8572	−0.6009
150	1000	+0.5000	−0.8660	−0.5774
151	1004	+0.4848	−0.8746	−0.5543
152	1008	+0.4695	−0.8829	−0.5317
153	1012	+0.4540	−0.8910	−0.5095
154	1016	+0.4384	−0.8988	−0.4877
155	1020	+0.4226	−0.9063	−0.4663
156	1024	+0.4067	−0.9135	−0.4452
157	1028	+0.3907	−0.9205	−0.4245
158	1032	+0.3746	−0.9272	−0.4040
159	1036	+0.3584	−0.9336	−0.3839
160	1040	+0.3420	−0.9397	−0.3640
161	1044	+0.3256	−0.9455	−0.3443
162	1048	+0.3090	−0.9511	−0.3249
163	1052	+0.2924	−0.9563	−0.3057
164	1056	+0.2756	−0.9613	−0.2867
165	1100	+0.2588	−0.9659	−0.2679
166	1104	+0.2419	−0.9703	−0.2493
167	1108	+0.2250	−0.9744	−0.2309
168	1112	+0.2079	−0.9781	−0.2126
169	1116	+0.1908	−0.9816	−0.1944
170	1120	+0.1736	−0.9848	−0.1763
171	1124	+0.1564	−0.9877	−0.1584
172	1128	+0.1392	−0.9903	−0.1405
173	1132	+0.1219	−0.9925	−0.1228
174	1136	+0.1045	−0.9945	−0.1051
175	1140	+0.0872	−0.9962	−0.0875
176	1144	+0.0698	−0.9976	−0.0699
177	1148	+0.0523	−0.9986	−0.0524
178	1152	+0.0349	−0.9994	−0.0349

Table A23. (*cont,*)

degrees	time	sine	cosine	tangent
179	1156	+0.0175	−0.9998	−0.0175
180	1200	0	−1	0
181	1204	−0.0175	−0.9998	+0.0175
182	1208	−0.0349	−0.9994	+0.0349
183	1212	−0.0523	−0.9986	+0.0524
184	1216	−0.0698	−0.9976	+0.0699
185	1220	−0.0872	−0.9962	+0.0875
186	1224	−0.1045	−0.9945	+0.1051
187	1228	−0.1219	−0.9925	+0.1228
188	1232	−0.1392	−0.9903	+0.1405
189	1236	−0.1564	−0.9877	+0.1584
190	1240	−0.1736	−0.9848	+0.1763
191	1244	−0.1908	−0.9816	+0.1944
192	1248	−0.2079	−0.9781	+0.2126
193	1252	−0.2250	−0.9744	+0.2309
194	1256	−0.2419	−0.9703	+0.2493
195	1300	−0.2588	−0.9659	+0.2679
196	1304	−0.2756	−0.9613	+0.2867
197	1308	−0.2924	−0.9563	+0.3057
198	1312	−0.3090	−0.9511	+0.3249
199	1316	−0.3256	−0.9455	+0.3443
200	1320	−0.3420	−0.9397	−0.3640
201	1324	−0.3584	−0.9336	+0.3839
202	1328	−0.3746	−0.9272	+0.4040
203	1332	−0.3907	−0.9205	+0.4245
204	1336	−0.4067	−0.9135	+0.4452
205	1340	−0.4226	−0.9063	+0.4663
206	1344	−0.4384	−0.8988	+0.4877
207	1348	−0.4540	−0.8910	+0.5095
208	1352	−0.4695	−0.8829	+0.5317
209	1356	−0.4848	−0.8746	+0.5543
210	1400	−0.5000	−0.8660	+0.5774
211	1404	−0.5150	−0.8572	+0.6009
212	1408	−0.5299	−0.8480	+0.6249
213	1412	−0.5446	−0.8387	+0.6494
214	1416	−0.5592	−0.8290	+0.6745

Table A23. (*cont,*)

degrees	time	sine	cosine	tangent
215	1420	−0.5736	−0.8192	+0.7002
216	1424	−0.5878	−0.8090	+0.7265
217	1428	−0.6018	−0.7986	+0.7536
218	1432	−0.6157	−0.7880	+0.7813
219	1436	−0.6293	−0.7771	+0.8098
220	1440	−0.6428	−0.7660	+0.8394
221	1444	−0.6561	−0.7547	+0.8694
222	1448	−0.6691	−0.7431	+0.9004
223	1452	−0.6820	−0.7314	+0.9325
224	1456	−0.6947	−0.7193	+0.9657
225	1500	−0.7071	−0.7071	+1.000
226	1504	−0.7193	−0.6947	+1.036
227	1508	−0.7314	−0.6820	+1.072
228	1512	−0.7431	−0.6691	+1.111
229	1516	−0.7547	−0.6561	+1.150
230	1520	−0.7660	−0.6428	+1.192
231	1524	−0.7771	−0.6293	+1.235
232	1528	−0.7880	−0.6157	+1.280
233	1532	−0.7986	−0.6018	+1.327
234	1536	−0.8090	−0.5878	+1.376
235	1540	−0.8192	−0.5736	+1.428
236	1544	−0.8290	−0.5592	+1.483
237	1548	−0.8387	−0.5446	+1.540
238	1552	−0.8480	−0.5299	+1.600
239	1556	−0.8572	−0.5150	+1.664
240	1600	−0.8660	−0.5000	+1.732
241	1604	−0.8746	−0.4848	+1.804
242	1608	−0.8829	−0.4695	+1.881
243	1612	−0.8910	−0.4540	+1.963
244	1616	−0.8988	−0.4384	+2.050
245	1620	−0.9063	−0.4226	+2.145
246	1624	−0.9135	−0.4067	+2.246
247	1628	−0.9205	−0.3907	+2.356
248	1632	−0.9272	−0.3746	+2.475
249	1636	−0.9336	−0.3584	+2.605
250	1640	−0.9397	−0.3420	+2.747

degrees	time	sine	cosine	tangent
251	1644	−0.9455	−0.3256	+2.904
252	1648	−0.9511	−0.3090	+3.078
253	1652	−0.9563	−0.2924	+3.271
254	1656	−0.9613	−0.2756	+3.487
255	1700	−0.9659	−0.2588	+3.732
256	1704	−0.9703	−0.2419	+4.011
257	1708	−0.9744	−0.2250	+4.331
258	1712	−0.9781	−0.2079	+4.705
259	1716	−0.9816	−0.1908	+5.145
260	1720	−0.9848	−0.1736	+5.671
261	1724	−0.9877	−0.1564	+6.314
262	1728	−0.9903	−0.1392	+7.115
263	1732	−0.9925	−0.1219	+8.144
264	1736	−0.9945	−0.1045	+9.514
265	1740	−0.9962	−0.0872	+11.43
266	1744	−0.9976	−0.0698	+14.30
267	1748	−0.9986	−0.0523	+19.08
268	1752	−0.9994	−0.0349	+28.64
269	1756	−0.9998	−0.0175	+57.29
270	1800	−1	0	∞
271	1804	−0.9998	+0.0175	−57.29
272	1808	−0.9994	+0.0349	−28.64
273	1812	−0.9986	+0.0523	−19.08
274	1816	−0.9976	+0.0698	−14.30
275	1820	−0.9962	+0.0872	−11.43
276	1824	−0.9945	+0.1045	−9.514
277	1828	−0.9925	+0.1219	−8.144
278	1832	−0.9903	+0.1392	−7.115
279	1836	−0.9877	+0.1564	−6.314
280	1840	−0.9848	+0.1736	−5.671
281	1844	−0.9816	+0.1908	−5.145
282	1848	−0.9781	+0.2079	−4.705
283	1852	−0.9744	+0.2250	−4.331
284	1856	−0.9703	+0.2419	−4.011
285	1900	−0.9659	+0.2588	−3.732
286	1904	−0.9613	+0.2756	−3.487

Table A23. (*cont,*)

degrees	time	sine	cosine	tangent
287	1908	−0.9563	+0.2924	−3.271
288	1912	−0.9511	+0.3090	−3.078
289	1916	−0.9455	+0.3256	−2.904
290	1920	−0.9397	+0.3420	−2.747
291	1924	−0.9336	+0.3584	−2.605
292	1928	−0.9272	+0.3746	−2.475
293	1932	−0.9205	+0.3907	−2.356
294	1936	−0.9135	+0.4067	−2.246
295	1940	−0.9063	+0.4226	−2.145
296	1944	−0.8988	+0.4384	−2.050
297	1948	−0.8910	+0.4540	−1.963
298	1952	−0.8829	+0.4695	−1.881
299	1956	−0.8746	+0.4848	−1.804
300	2000	−0.8660	+0.5000	−1.732
301	2004	−0.8572	+0.5150	−1.664
302	2008	−0.8480	+0.5299	−1.600
303	2012	−0.8387	+0.5446	−1.540
304	2016	−0.8290	+0.5592	−1.483
305	2020	−0.8192	+0.5736	−1.428
306	2024	−0.8090	+0.5878	−1.376
307	2028	−0.7986	+0.6018	−1.327
308	2032	−0.7880	+0.6157	−1.280
309	2036	−0.7771	+0.6293	−1.235
310	2040	−0.7660	+0.6428	−1.192
311	2044	−0.7547	+0.6561	−1.150
312	2048	−0.7431	+0.6691	−1.111
313	2052	−0.7314	+0.6820	−1.072
314	2056	−0.7193	+0.6947	−1.036
315	2100	−0.7071	+0.7071	−1.000
316	2104	−0.6947	+0.7193	−0.9657
317	2108	−0.6820	+0.7314	−0.9325
318	2112	−0.6691	+0.7431	−0.9004
319	2116	−0.6561	+0.7547	−0.8693
320	2120	−0.6428	+0.7660	−0.8391
321	2124	−0.6293	+0.7771	−0.8098

Table A23. (*cont,*)

degrees	time	sine	cosine	tangent
322	2128	−0.6157	+0.7880	−0.7813
323	2132	−0.6018	+0.7986	−0.7536
324	2136	−0.5878	+0.8090	−0.7265
325	2140	−0.5736	+0.8192	−0.7002
326	2144	−0.5592	+0.8290	−0.6745
327	2148	−0.5446	+0.8387	−0.6494
328	2152	−0.5299	+0.8480	−0.6249
329	2156	−0.5150	+0.8572	−0.6009
330	2200	−0.5000	+0.8660	−0.5774
331	2204	−0.4848	+0.8746	−0.5543
332	2208	−0.4695	+0.8829	−0.5317
333	2212	−0.4540	+0.8910	−0.5095
334	2216	−0.4384	+0.8988	−0.4877
335	2220	−0.4226	+0.9063	−0.4663
336	2224	−0.4067	+0.9135	−0.4452
337	2228	−0.3907	+0.9205	−0.4245
338	2232	−0.3746	+0.9272	−0.4040
339	2236	−0.3584	+0.9336	−0.3839
340	2240	−0.3420	+0.9397	−0.3640
341	2244	−0.3256	+0.9455	−0.3443
342	2248	−0.3090	+0.9511	−0.3249
343	2252	−0.2924	+0.9563	−0.3057
344	2256	−0.2756	+0.9613	−0.2867
345	2300	−0.2588	+0.9659	−0.2679
346	2304	−0.2419	+0.9703	−0.2493
347	2308	−0.2250	+0.9744	−0.2309
348	2312	−0.2079	+0.9781	−0.2126
349	2316	−0.1908	+0.9816	−0.1944
350	2320	−0.1736	+0.9848	−0.1763
351	2324	−0.1564	+0.9877	−0.1584
352	2328	−0.1392	+0.9903	−0.1405
353	2332	−0.1219	+0.9925	−0.1228
354	2336	−0.1045	+0.9945	−0.1051
355	2340	−0.0872	+0.9962	−0.0875
356	2344	−0.0698	+0.9976	−0.0699

Table A23. (*cont,*)

degrees	time	sine	cosine	tangent
357	2348	−0.0523	+0.9986	−0.0524
358	2352	−0.0349	+0.9994	−0.0349
359	2356	−0.0175	+0.9998	−0.0175
360	2400	0	+1	0

Source: Abridged from Batschelet (1965). Copyright by American Institute of Biological Sciences.

Table A24. *Critical values of u for the V test*

n	P					
	0.10	0.05	0.01	0.005	0.001	0.0001
5	1.3051	1.6524	2.2505	2.4459	2.7938	3.0825
6	1.3009	1.6509	2.2640	2.4695	2.8502	3.2114
7	1.2980	1.6499	2.2734	2.4858	2.8886	3.2970
8	1.2958	1.6492	2.2803	2.4978	2.9164	3.3578
9	1.2942	1.6486	2.2856	2.5070	2.9375	3.4034
10	1.2929	1.6482	2.2899	2.5143	2.9540	3.4387
11	1.2918	1.6479	2.2933	2.5201	2.9672	3.4669
12	1.2909	1.6476	2.2961	2.5250	2.9782	3.4899
13	1.2902	1.6474	2.2985	2.5290	2.9873	3.5091
14	1.2895	1.6472	2.3006	2.5325	2.9950	3.5253
15	1.2890	1.6470	2.3023	2.5355	3.0017	3.5392
16	1.2885	1.6469	2.3039	2.5381	3.0075	3.5512
17	1.2881	1.6467	2.3052	2.5404	3.0126	3.5617
18	1.2877	1.6466	2.3064	2.5424	3.0171	3.5710
19	1.2874	1.6465	2.3075	2.5442	3.0211	3.5792
20	1.2871	1.6464	2.3085	2.5458	3.0247	3.5866
21	1.2868	1.6464	2.3093	2.5473	3.0279	3.5932
22	1.2866	1.6463	2.3101	2.5486	3.0308	3.5992
23	1.2864	1.6462	2.3108	2.5498	3.0335	3.6047

Table A24. (*cont.*)

n	P					
	0.10	0.05	0.01	0.005	0.001	0.0001
24	1.2862	1.6462	2.3115	2.5509	3.0359	3.6096
25	1.2860	1.6461	2.3121	2.5519	3.0382	3.6142
26	1.2858	1.6461	2.3127	2.5529	3.0402	3.6184
27	1.2856	1.6460	2.3132	2.5538	3.0421	3.6223
28	1.2855	1.6460	2.3136	2.5546	3.0439	3.6258
29	1.2853	1.6459	2.3141	2.5553	3.0455	3.6292
30	1.2852	1.6459	2.3145	2.5560	3.0471	3.6323
40	1.2843	1.6456	2.3175	2.5610	3.0580	3.6545
50	1.2837	1.6455	2.3193	2.5640	3.0646	3.6677
60	1.2834	1.6454	2.3205	2.5660	3.0689	3.6764
70	1.2831	1.6453	2.3213	2.5674	3.0720	3.6826
100	1.2826	1.6452	2.3228	2.5699	3.0775	3.6936
500	1.2818	1.6449	2.3256	2.5747	3.0877	3.7140
1000	1.2817	1.6449	2.3260	2.5752	3.0890	3.7165

Source: From Batschelet 1972.

Table A25. *Alpha levels of* h *for circular statistics two-sample runs test.* n,m=*sample sizes* (m≤n). P=*alpha level of significance*

		m											
h	n	3	4	5	6	7	8	9	10	11	12	13	14
2	5	P=0.143	0.071	0.038									
2	6	0.100	0.045	0.024	0.013								
4		0.600	0.409	0.262	0.175								
2	7	0.083	0.033	0.015	0.008	0.004							
4		0.583	0.333	0.197	0.121	0.077							
2	8	0.067	0.023	0.010	0.005	0.002	0.001						
4		0.533	0.279	0.152	0.088	0.051	0.032						
2	9	0.053	0.018	0.007	0.003	0.001	<0.001						
4		0.474	0.236	0.119	0.063	0.035	0.020	0.012					
6			0.745	0.510	0.343	0.231	0.157	0.106					
2	10	0.045	0.014	0.005	0.002	<0.001							
4		0.455	0.205	0.095	0.048	0.024	0.014	0.008	0.005				
6			0.699	0.453	0.286	0.179	0.117	0.077	0.051				
2	11	0.038	0.011	0.004	0.001	<0.001							
4		0.423	0.176	0.077	0.036	0.018	0.009	0.005	0.003	0.002			
6			0.670	0.407	0.242	0.145	0.088	0.055	0.035	0.023			
8				0.846	0.654	0.484	0.352	0.255	0.185	0.135			

Table A25. (*cont.*)

h	n						m						
		3	4	5	6	7	8	9	10	11	12	13	14
2	12	0.032	0.009	0.003	0.001	<0.001							
4		0.387	0.155	0.063	0.028	0.014	0.006	0.003	0.002	<0.001			
6			0.629	0.365	0.205	0.117	0.067	0.039	0.024	0.015	0.009		
8				0.819	0.603	0.428	0.293	0.202	0.140	0.098	0.068		
2	13	0.029	0.007	0.002	<0.001								
4		0.371	0.136	0.053	0.022	0.010	0.004	0.002	0.001	<0.001			
6			0.607	0.330	0.176	0.095	0.052	0.027	0.017	0.010	0.006	0.004	
8				0.792	0.561	0.378	0.251	0.164	0.110	0.074	0.050	0.034	
10					0.908	0.762	0.608	0.471	0.361	0.273	0.207	0.157	
2	14	0.024	0.006	0.002	<0.001								
4		0.350	0.122	0.044	0.017	0.007	0.003	<0.001					
6			0.576	0.299	0.151	0.078	0.038	0.022	0.012	0.007	0.004	0.002	0.001
8				0.766	0.521	0.336	0.197	0.135	0.086	0.055	0.036	0.024	0.016
10					0.888	0.723	0.515	0.417	0.306	0.223	0.162	0.119	0.087
2	15	0.022	0.005	0.001	<0.001								
4		0.326	0.108	0.037	0.014	0.006	0.002	0.001	<0.001				
6			0.554	0.272	0.131	0.064	0.033	0.017	0.009	0.005	0.003	0.002	<0.001
8				0.741	0.483	0.299	0.185	0.110	0.068	0.042	0.026	0.017	0.011
10					0.870	0.686	0.518	0.367	0.262	0.183	0.129	0.091	0.064

Source: From Batschelet (1981); adapted from Asano (1965)

Table A26. *Point estimation of k for the Watson–Williams two-sample test. n=sample size. r=length of the mean vector of the sample*

r

n	0.10	0.15	0.20	0.25	0.30	0.35	0.40	0.45	0.50	0.55	0.60	0.65	0.70	0.75	0.80	0.85	0.90	0.95
5	0	0	0	0	0	0	0	0.15	0.67	0.94	1.18	1.41	1.68	2.00	2.44	3.10	4.39	8.33
6	0	0	0	0	0	0	0	0.56	0.83	1.04	1.25	1.48	1.74	2.07	2.51	3.20	4.54	8.66
7		0	0	0	0	0	0.38	0.69	0.90	1.10	1.30	1.52	1.78	2.11	2.56	3.27	4.65	8.89
8	0	0	0	0	0	0	0.53	0.76	0.95	1.13	1.33	1.55	1.81	2.15	2.60	3.32	4.73	9.06
9	0	0	0	0	0	0.31	0.61	0.80	0.98	1.16	1.35	1.57	1.84	2.17	2.63	3.36	4.79	9.19
10	0	0	0	0	0	0.42	0.65	0.83	1.00	1.18	1.37	1.59	1.86	2.19	2.66	3.39	4.84	9.30
11	0	0	0	0	0	0.48	0.69	0.85	1.02	1.19	1.38	1.61	1.87	2.21	2.68	3.42	4.89	9.39
12	0	0	0	0	0.23	0.53	0.71	0.87	1.03	1.20	1.40	1.62	1.88	2.22	2.69	3.44	4.92	9.46
13	0	0	0	0	0.32	0.56	0.73	0.88	1.04	1.21	1.41	1.63	1.89	2.23	2.71	3.46	4.95	9.53
14	0	0	0	0	0.37	0.58	0.74	0.89	1.05	1.22	1.41	1.63	1.90	2.24	2.72	3.47	4.98	9.58
15	0	0	0	0	0.41	0.60	0.75	0.90	1.06	1.23	1.42	1.64	1.91	2.25	2.73	3.49	5.00	9.63
20	0	0	0	0.30	0.50	0.65	0.79	0.93	1.09	1.26	1.45	1.67	1.94	2.28	2.76	3.53	5.07	9.79
25	0	0	0	0.38	0.54	0.67	0.81	0.95	1.10	1.27	1.46	1.68	1.95	2.30	2.79	3.56	5.12	9.88
30	0	0	0.22	0.42	0.56	0.69	0.82	0.96	1.11	1.28	1.47	1.69	1.96	2.31	2.80	3.58	5.15	9.95
35	0	0	0.27	0.44	0.57	0.70	0.83	0.97	1.12	1.29	1.48	1.70	1.97	2.32	2.81	3.60	5.17	9.99
40	0	0	0.31	0.45	0.58	0.70	0.83	0.97	1.12	1.29	1.48	1.70	1.98	2.33	2.82	3.61	5.19	10.03
45	0	0.04	0.33	0.46	0.58	0.71	0.84	0.98	1.13	1.30	1.49	1.71	1.98	2.33	2.82	3.62	5.20	10.06
50	0	0.14	0.34	0.47	0.59	0.71	0.84	0.98	1.13	1.30	1.49	1.71	1.98	2.34	2.83	3.62	5.21	10.08
100	0	0.26	0.38	0.49	0.61	0.73	0.86	1.00	1.15	1.31	1.50	1.73	2.00	2.35	2.85	3.65	5.26	10.18
150	0.18	0.28	0.39	0.50	0.62	0.74	0.86	1.00	1.15	1.32	1.51	1.73	2.00	2.36	2.86	3.66	5.27	10.21
200	0.19	0.29	0.40	0.51	0.62	0.74	0.87	1.00	1.15	1.32	1.51	1.73	2.01	2.36	2.86	3.67	5.28	10.22
∞	0.20	0.30	0.41	0.52	0.63	0.75	0.87	1.01	1.16	1.33	1.52	1.74	2.01	2.37	2.87	3.68	5.31	10.27

Source: From Batschelet (1981)

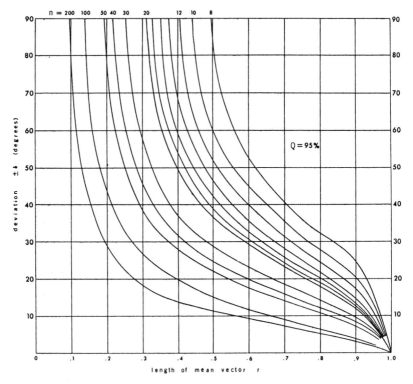

Fig. A1   Chart for determining a confidence interval for the mean angle with a 95%
confidence coefficient (from Batschelet, 1981).

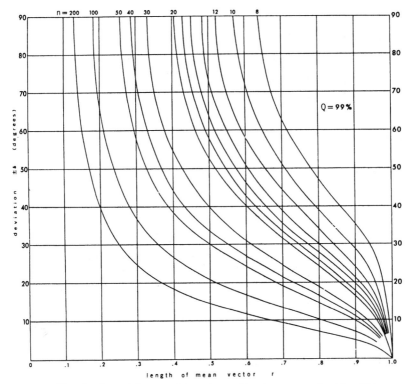

Fig. A2 Chart for determining a confidence interval for the mean angle with a 99%
confidence coefficient (from Batschelet, 1981).

Microcomputers in ethology: basic and advanced uses

James C. Ha

Microcomputers have become involved in almost every aspect of science. While psychologists have been in the forefront of applying microcomputers to the study of animal behavior, ethologists have lagged well behind. In this appendix, I will: 1. provide a brief introduction to microcomputers; 2. introduce some basic principles of microcomputer software; 3. briefly describe programming languages; 4. list four ways in which microcomputers can be applied to ethological questions; and 5. discuss sources of additional information on hardware and software.

WHAT IS A MICROCOMPUTER?

A microcomputer consists of several parts: a central processing unit ('CPU': an integrated-circuit microchip or 'microprocessor') acts as the 'brain' of the computer, while built-in, electronic random access memory ('RAM': measured in 'bytes' or characters of information) is used to hold program software and information while the computer is in use. The video screen and keyboard are used to exchange information between user and computer under the control of instructions stored in built-in read-only memory ('ROM'). Alternative input devices (mouse and trackball) have become popular, particularly for graphics-based applications and operating systems. Magnetic disks on either floppy plastic or hard metal platters are used for permanent storage of information. Floppy disks are produced in either a soft-case, 5.25' diameter form, or a hard-case, 3.5' diameter form, each suitable for different models of computers.

In its simplest form, the CPU executes a series of instructions which are permanently stored on a disk and temporarily stored for its use in RAM memory. These instructions are provided in 'software' called programs or applications. These instruc-

Table B.1. *A comparison of important microcomputer features, including typical values for older and more recent models. These are the extremes for these features; a typical (and quite satisfactory) machine will fall somewhere in the middle of these ranges*

Category	Older 1979		Newer 1994	Typical 1994
RAM capacity (bytes)	64K	↔	16M+	4M
Floppy disk capacity (bytes)	128K	↔	1.4M+	1.2–1.4M
Hard disk capacity (bytes)	20M	↔	850M	320M
Clock speed (MHz)	1	↔	66	25–33
Cost ($ in 1994)	200	↔	2500+	1000+

tions tell the CPU how and when to: 1. request information (from a disk file or from the keyboard); 2. perform various mathematical or logical operations on the information; and 3. display the results of these operations on the video screen (or to a 'hard-copy' paper printer). Various categories of software programs will be described later.

MICROCOMPUTER SPECIFICATIONS

Several specifications are used in describing microcomputers and these need to be considered when purchasing a computer (Table B.1). A microcomputer is generally defined as a computer dedicated to a single user, in contrast to larger institutional computers ('mainframes') which are shared simultaneously by many users. Microprocessors fall into two 'families': Intel's 80×86 (used in DOS-based machines) and Motorola's 680×0 line (used in Apple Co.'s Macintosh computers). The earliest models were the 8086 and 68000 and at this printing the Pentium ('80586') and 68040 are the state-of-the-art models. In general, microcomputers contain RAM capable of holding from 64 K(ilo) (65 536) to 32 M(ega) (32 768 000) bytes of information. A byte of information represents one character or digit and for technical reasons, a 'K' of information is actually 1024 characters. Floppy disk storage ranges from 128 Kbytes to 1.4 Mbytes, and hard disk storage ranges from 20 Mbytes to 850 Mbytes, or higher. 'Clock speed' determines the speed at which the CPU executes instructions. The earliest microcomputers ran at 1 MHz (1 000 000 cycles, or instructions, per second). More recent microcomputers run at speeds from 8–10 MHz to 66 MHz. When choosing among these features, a trade-off exists between better performance (more RAM, larger disks, faster CPU) and cost. Small systems with 64 K RAM, 180 K floppy disk storage, and 1 MHz CPU speed currently cost approximately $200. High performance 'workstations', at the upper end of the ranges described above, cost $10 000 and more.

A final factor to consider in purchasing a microcomputer is the 'operating system' (OS) or proprietary program which oversees the functioning of program software. Two operating systems are currently most popular: 1. IBM, Inc. and Microsoft Corp.'s DOS system; and 2. Apple, Inc.'s Macintosh system ('Finder'). In addition, DOS-based machines can run Microsoft's Windows program which emulates the Apple graphical interface and mouse-input support. While discussions of the pros and cons of these systems can come nearly to blows among aficionodos, the upshot is that DOS is faster, more flexible and powerful, and easier to program, while the Finder system is easier to learn and is more intuitive to use. The DOS/Windows combination falls in between. Much of the decision on operating system choices should be based on the availability of required program software, because software written for DOS or Windows will not run directly under Finder, and vice-versa. DOS-based computers are much more popular, and therefore a much wider selection of lower priced software is available.

TYPES OF MICROCOMPUTERS

Current models of microcomputers fall into five categories (Table B.2). First, IBM microcomputers and their numerous clones all use CPUs manufactured by the Intel Corporation, and use the DOS operating system; therefore, they can share software. Second is a group of microcomputers which utilize CPUs manufactured by the Motorola Corporation. These machines each use a proprietary operating system and cannot directly share program software. These include the Apple Macintosh (Finder OS), Commodore Amiga (Intuition OS), and Atari ST (Atari OS) computers. Intuition OS and Atari OS are similar in nature to Finder.

The third category includes discontinued microcomputers from the early 1980s with less memory and slower speeds, such as the Commodore 64 and 128, and the Apple IIE. These computers still play a valuable role in several types of research, as dedicated data collection devices and experiment controllers. The advantage to these machines is their inexpensive price. Several can be kept in the lab for easy replacement in case of breakdown, for less than the cost of one of the systems described above. With proper software, data collected on one of these computers can be easily exchanged with a more powerful 'data analysis' microcomputer, or even a large mainframe computer.

The fourth category of portable, laptop, and notebook computers comprise a special group of microcomputers which overlaps with those described above. Because of their small size and, in many cases, ability to run from built-in batteries, these computers are of special interest to field ethologists. Numerous DOS-based

Table B.2. *Types of microcomputer. See the text for a discussion of operating system pros and cons. Earlier microcomputers had very simple (and generally unnamed) operating systems, from which the PC/MS-DOS operating systems were developed*

Category	Operating System	Examples
IBM-compatible (Intel-based)	MS- or PC-DOS	IBM, Compaq, AST, many others
Motorola-based	Finder Intuition Atari	Apple Macintosh Commodore Amiga Atari ST
Older, discontinued		Commodore 64, 128 Apple II/e
Portables, laptops	MS- or PC-DOS Finder	IBM-compatible Apple Macintosh

machines exist in this category, including those manufactured by Sharp, Toshiba, NEC, Compaq, and Tandy. A portable form of the Macintosh computer has been produced by Apple, Inc. and several clones have appeared. Finally, a proven field computer, the TRS102, has been sold by Tandy/Radio Shack for several years. Originally manufactured as the TRS100, this computer has limited RAM, a slow speed, and a proprietary operating system which is incompatible with IBM or Apple computers. However, it is lightweight and compact, operates for a long time on a single battery charge, and is relatively inexpensive. For these reasons, the TRS102 is popular among field biologists. Data transfer to IBM- or Macintosh-compatible computers for sophisticated analysis is simple.

PROGRAM SOFTWARE FOR GENERAL USE

A computer without software is useless. Specialized software is available for scientific applications (e.g. data collection and statistical analysis) and specifically for ethological research (Chapter 9). These programs will be discussed below. For general use, there are five major categories of software used.

Wordprocessing

Wordprocessors allow the creation, editing and printing of text. A major advantage of wordprocessors over typewriters is the ability to store a previously created document for later recovery and editing. Wordprocessing constitutes the single largest use of microcomputers. Popular wordprocessing programs include Wordstar International's *Wordstar* (IBM), Wordperfect Corp.'s *Wordperfect* (IBM/Macintosh), and Microsoft's *Word* (IBM/Macintosh).

Database management

Database management software is designed to store information in 'fields'. In a study of animal movements, a single animal's identification, location, group number, weight and transmitter frequency are examples of fields. A collection of fields for a single block of information is a 'record' (e.g. all fields for a single animal is a record). A collection of records (e.g. all of the animals in the study) is a database. Information stored in a database can be searched for matches to user-entered text or numbers, sorted and printed in a predetermined format. Often, the information in a field can be manipulated or combined with other fields through mathematical or logical operations. Popular database programs include Ashton–Tate's *Dbase IV* (IBM), Microrim's *Rbase 3.0* (IBM), Borland International's *Paradox* (IBM), Microsoft's *Access* (IBM), Microsoft's *File* (Macintosh), and Acius' *4TH Dimension* (Macintosh).

Spreadsheets

Spreadsheet software was originally developed for accounting and business forcasting uses. It consists of a matrix of rows and columns. A 'cell' is located at the intersection of any specific row and column. The user may enter either numeric data, text or formulas in each cell. Formulas describe mathematical or statistical operations to be performed on a set of other cells. Spreadsheet software provides a quick and intuitive method of entering, manipulating, and displaying large amounts of numeric data. It functions well as a data entry program for larger and more complete statistical packages. Limited database management functions (e.g. sorting, simple searches) are usually available. However, spreadsheet software is not appropriate for large amounts of text information. Popular spreadsheet programs include Lotus Corp.'s *Lotus 1–2–3* (IBM; the industry standard), Borland International's *Quattro* (IBM) and Microsoft's *Excel* (IBM/Mackintosh).

Graphics

Graphics software provides a convenient method of producing scientific graphs and charts. Keyboard data entry is generally through a spreadsheet-type option, or data may be entered directly from a database management or spreadsheet file. Numerous options exist for controlling all aspects of the final product, including symbol type and size, axis ranges and marking, legends, labels and additional text. Finally, the graph may be printed to a printer or graphics plotter for final output. More sophisticated systems produce files which may be directly read by a film recorder, producing high-quality photographic slides. Popular graphics programs include Lotus Corp.'s *Freelance* (IBM), Jandel Inc.'s *Sigmaplot* (IBM), Software Publishing Corporation's *Harvard Graphics*, Microsoft's *Powerpoint* (Macintosh) and Computer Associate's *CricketGraph* (Macintosh).

Telecommunication

Telecommunication software comprises a more specialized but rapidly growing use for microcomputers. A modem, a hardware device providing a link using phone lines, enables users to access or share information with other computers, large or small. First, mainframe computers may be accessed through the microcomputer, when a microcomputer lacks the capabilities or software needed. By dialing a phone number and providing identification information, access to academic, government, or business mainframe computers can be obtained.

Second, through mainframe computers, an international network of communication has been established between researchers and, recently, the general public. Called the Internet, this system comprises the heart of the 'information superhighway' and allows messages and data to be sent between academic, military and commercial sites. Use of this communication link has been developed to a greater degree by some disciplines (notably computer science, molecular biology, and psychology). 'Electronic mail' has great potential for fast communication and dissemination of research results, already leading to electronic journals with peer-reviewed articles and interactive exchange of ideas (e.g. American Psychological Association's PSY-COLOQUY electronic journal). Two networks of ethologists and those with related interests, ABSnet and B-E-DIGEST, have been established to facilitate electronic information exchange. Contact the author at JCH@U>WASHINGTON.EDU for current information.

A third use for telecommunications involves large information databases (e.g. of scientific literature) which can be searched for a fee. Much as with academic mainframes, telecommunications software is used to establish contact with a remote mainframe computer holding the database information. Software held on the mainframe permits searching, extracting and displaying selected information. The most common use of these large database systems is for bibliographic information retrieval. Such printed databases as Bioabstracts and Zoological Records are now available for computerized search and retrieval, saving a considerable amount of library time at a cost that depends on the database searched and the mainframe system used. Systems such as BRS AfterDark (Maxwell Online, Inc.) are designed for academic use and permit searching after regular business hours at a significantly reduced cost. Generally, these systems are designed to be relatively easy to use, and training sessions are often free. These systems are often available through academic libraries, as well. (For additional discussion see section 4.3 on information resources.)

PROGRAMMING LANGUAGES

In theory, microcomputers for the masses will be simple to use and will never require the expertise to write the programs required to make a computer more than an expensive paperweight. In reality, the science of computers has not yet reached this goal. In many cases (when the use of microcomputers moves beyond simple word-processing), it becomes necessary to write software because the proper programs do not exist. At the very least, it may become necessary to understand the limitations of computers, and to be able to communicate one's requirements to a computer programmer. Therefore, most microcomputer users who move beyond the basics to apply the full power of computers to their work are advised to become reasonably proficient in a programming language. Ability to write even simple programs is a valuable skill, and the ability to communicate to an experienced programmer can result in a better product faster. In the following section, three categories of programming languages will be described: 1. simple languages for nonprofessionals; 2. older languages still in active use; and 3. advanced or professional languages.

Simple languages

Simple languages were designed to be used in teaching programming skills. In their newer, more advanced forms, they are quite adequate for all nonprofessional programming jobs. These languages include BASIC (Beginners All-purpose Symbolic Instruction Code) and Pascal (named after the mathematician Blaise Pascal). BASIC was, in its first forms, a very simple language but has become very powerful in newer versions, such as *TrueBASIC* (True BASIC, Inc.) and *QuickBASIC* (Microsoft, Inc.). It is quite simple to learn, using an English-like set of commands. It has become the most widely used programming language because it was shipped with all early microcomputers.

Pascal was written to teach structured programming, in which the organization of program sections is emphasized for clearer writing. As the evolution of these languages has occurred, they have become more similar. For example, the new forms of BASIC include structured programming concepts.

Older languages

Older languages, which have largely been surpassed in speed, compactness, and ease of use, include FORTRAN (FORmula TRANslation) and COBAL (COmmon Business Oriented Language). FORTRAN was written as a mainframe programming language for scientific formula manipulation. COBOL was a widely used mainframe language adapted for writing business software. Both languages are receiving less use today, but are still widely available on academic mainframe computers. Microcomputer versions of these languages are available.

Professional and specialized languages

More advanced, powerful languages have been developed recently. Many are designed for specific types of task, such as machinery control, mathematics, or microcomputer operating systems. These include C (a fast, compact, flexible, but difficult language used on microcomputers, mostly for writing operating systems), Forth (a 'fourth' generation language commonly used to control equipment), Modula-2 (a second generation form of Pascal), APL (a symbolic mathematics language) and ADA (a 'super' language developed for the US Defense Department).

APPLICATIONS IN ETHOLOGY

The application of microcomputers in ethological research, discussed in earlier chapters in this book, occurs in four areas: 1. data collection (see chapter 9); 2. data storage and manipulation (see Chapter 9); 3. data analysis (see Chapter 16); and 4. data presentation (graphics; see Chapter 18). For an overview of programming principles for laboratory behavior research, see Deni (1986).

SOURCES OF ADDITIONAL INFORMATION

Advances in microcomputing are occurring so quickly that it is difficult to stay current with the changes. The best advice is to locate a 'computer guru'. Once referred to as 'hackers' (before the term became tainted by illegal activities), these experts are often hobbyists, as well as professional users. They stay up-to-date and are often eager to impart information to neophytes (sometimes too eager to impart too much information . . . a 'guru' who can teach well is a real find!). They are a valuable resource.

Otherwise, magazines and journals provide the best ways of staying up to date. All microcomputer users are encouraged to read at least one magazine for their machine on a regular basis. There are several sources of information which can help the beginner as well as the advanced user.

For the beginner, there are three sources of introductory information. The first is the leading journal of microcomputing, *BYTE* (BYTE Publishing Inc.). While many of the articles are intended for the advanced user, *BYTE* provides a broad coverage of all microcomputer models, including excellent reviews, clear descriptions of new models, and valuable tutorial articles on a wide range of topics.

Machine-specific publications provide a second category of basic information. Publications such as *PC-World*, *PC-Resource*, *PC* (all IBM), and *MacWorld* (Macintosh) provide reviews and basic tutorials for beginners, as well as more advanced information for specific machine types. Publications also exist for Commodore, Atari, and Tandy 100/102 computers.

User groups are a third source of beginner information. Most medium to large cities contain numerous user groups, usually machine-specific. Many universities also sponsor user groups which may provide information more appropriate for the scientific user. Generally, help with mainframe computer usage is available through campus computer centers, which may also provide microcomputer support.

For more advanced users, *BYTE* magazine is a valuable source of information. Magazines and journals exist for specific languages, and for hardware design and maintenance. The journal *Behavioral Research Methods, Instruments, and Computers* is an excellent reference for specific applications in psychology (with many applications for ethologists). In particular, this journal is an excellent source of hardware designs and software for laboratory applications of microcomputers.

Several microcomputers 'weeklies' (*MacWeek*, *InfoWorld*, *PC Week*) are aimed primarily at microcomputer professionals. However, these publications contain news of software upgrades, special purchase arrangements, and an excellent set of software and hardware evaluations (particularly *InfoWorld*). These publications are sent free to professionals, and 'scientists' often qualify; check with your library or with the microcomputer specialists at an academic computer center.

Finally, a large number of books on microcomputers have appeared in recent years. Unfortunately, most books with specific information become outdated quickly. Two references which continue to be useful are *Microcomputers in Biology: A Practical Approach* by Ireland and Long (1984), and *Scientific and Engineering Applications with Personal Computers* by Annino and Driver (1986). Whilst some of the specific information is out of date, the discussion of basic concepts is useful and provides an excellent background for beginners. BASIC programs for scientific analyses are provided and are useful to more advanced users.

Ethics in ethology

Ethics in ethology involves the application of moral principles and standards of conduct to animal behavior research. It encompasses ethical procedures in conducting (e.g. humane care of animal subjects; see Appendix D), as well as reporting on research. Oral and written reports on the research must be an honest representation of the methods, results and the relative contribution of each researcher (e.g. Beveridge, 1950).

Ethologists must honestly represent not only what they have done (e.g. research reports and experience), but also the knowledge and skills they possess (e.g. expert testimony and consulting). In addition, they must apply generally accepted moral principles in their personal and professional interactions with students, colleagues and the general public; these include use of scientific integrity and avoidance of discrimination and harassment.

These and additional standards of conduct were incorporated into the Code of Ethics adopted by the Animal Behavior Society (ABS) in 1990 (a portion of this is reprinted below). Ethologists can also consult the series of articles on professional ethics compiled by Stamp, Dawkins and Gosling (1992), those written by Stuart Altmann, Chair of the ABS's Ethics Committee (*ABS Newsletter*, 1993, 38(4)) through 1995, 40(1) and the numerous volumes on bioethics, including Bulger *et al.* (1993).

CODE OF ETHICS
(Portion of the Code of Ethics adopted by the ABS in 1990)

All members of ABS:

1. Will conduct their professional affairs in an ethical manner as prescribed in this Code, will endeavor to protect the profession of animal behavior from misunderstanding and misrepresentation, and will cooperate with one another to assure the rapid and accurate interchange and dissemination of knowledge about animal behavior.
2. Will use their knowledge, skills, and training to assist in achieving the harmonious interaction of the human species with other organisms.

3 Will not represent themselves as spokespersons for the Society or imply Society endorsement except with prior approval of the Executive Committee.

4 Will avoid and discourage plagiarism and the dissemination of false, biased, exaggerated, or otherwise unwarranted statements and data concerning animal behavior.

5 Will offer professional advice only on those subjects in which they are qualified by virtue of professional training and experience.

6 Will neither harass nor request or accept inappropriate favor from students, employees, or colleagues.

7 Shall inform a prospective or current employer, colleague, student, or client of any professional or personal conflicts of interest that may impair their objectivity.

8 Shall neither seek employment, grants, or gain nor attempt to injure the reputation or opportunities for education or employment of students or colleagues by false, biased, or undocumented claims or accusations, or by any other malicious action.

9 Will not discriminate against students, employees, or colleagues on the basis of sex, creed, religion, race, color, national origin, life style, sexual orientation, economic status, organizational affiliation, or other irrelevant characteristics.

10 Shall carry out their work in accord with the Society's Policy, including the 'Guidelines for the Use of Animals in Research' and the 'Program for Certification of Applied Animal Behaviorists'.

11 Shall be obligated, when they have substantial evidence of a breach of this Code, to bring such conduct to the attention of the Chair of the Ethics Committee or any member of the Executive Committee.

Guidelines for the use of animals in research

One of the major ethical considerations for ethologists is the care and well-being of the animals under study. This is true not only for captive animals in enclosures and laboratories (e.g. Fraser, 1987; National Research Council, 1985, 1992; Poole, 1987; Public Health Service, 1986) but also for animals being studied in the field (Cuthill, 1991; Laurenson and Caro, 1994). The potential effects of handling and marking animals in field studies was discussed in Chapter 8.

Concerns for animal welfare in ethological research have been discussed relative to specific types of behavior, such as predation and aggression (Huntingford, 1984), infanticide and maternal aggression (Elwood, 1991), and the number of subjects used (Still, 1982; McConway, 1992). These and additional papers of importance to ethologists were reprinted by Dawkins and Gosling (1992).

Bateson (1986) proposed a 'decision cube' for assisting the researcher in deciding when to proceed with a research project. He recommended that the researcher evaluate: 1. the quality of the research; 2. the certainty of the benefit of the research; and 3. the extent of animal suffering. The researcher would obviously proceed when the quality and certainty of benefit are high and the suffering is low, and obviously not proceed when the quality and certainty of benefit are low and the suffering is high. Intermediate conditions require judgment by researchers and their colleagues.

In addition to our concern for humane treatment of animals, it is important to remember that descriptive and experimental research on the natural behavior of animals is valid only when the animals are in good physical and psychological health. Manipulation experiments on animals are valid only when the manipulation does not cause additional effects on the animal's ability to behave normally. That is, unwarranted animal suffering is not only ethically wrong, it can also contribute to invalid results.

In recognition of the importance of animal welfare in valid and humane ethological research, the Association for the Study of Animal Behaviour (ASAB) and the Animal Behavior Society (ABS) both have standing Animal Care committees which continuously assess standards of care and the conduct of researchers who have submitted papers for publication. Also, ethologists serve on boards to review scientific research proposals and committees to formulate national animal care guidelines, as well as producing their own animal care guidelines (Drickamer and Vessey, 1992). The joint ASAB and ABS *Guidelines for the Treatment of Animals in Behavioural Research and Teaching* are reproduced at the end of this appendix. Besides reading and following those guidelines, ethologists can consult additional general discussions of animal welfare (e.g. Dawkins, 1980), general concerns for ethologists (e.g. Martin and Bateson, 1993), and general guidelines for use of animals in research (e.g. Sigma Xi, Anonymous, 1992). Listed below are guidelines for the use of specific animal groups in research which have been adopted by professional societies in North America. Copies of the guidelines can be obtained from the societies listed.

- *Guidelines for the Use of Fishes in Field Research*
  Adopted jointly by:
  American Society of Ichthyologists and Herpetologists
  American Fisheries Society
  American Institute of Fisheries Research Biologists

- *Guidelines for the Use of Live Amphibians and Reptiles in Field Research*
  Adopted jointly by:
  American Society of Ichthyologists and Herpetologists
  The Herpetologists League
  Society for the Study of Amphibians and Reptiles

- *Guidelines for Use of Wild Birds in Research*
  Adopted jointly by:
  The American Ornithologists' Union
  Cooper Ornithological Society
  Wilson Ornithological Society
  (published as a supplement to *The Auk*, Vol. 105, No. 1, 1988)

- *Acceptable Field Methods in Mammalogy: Preliminary Guidelines Approved by the American Society of Mammalogists*
  (published as a supplement to the *Journal of Mammalogy*, Vol. 68, No. 4, 1987)

GUIDELINES FOR THE TREATMENT OF ANIMALS IN BEHAVIOURAL RESEARCH AND TEACHING
(published in *Animal Behaviour*, 1995, 49:277–282)

Behavioural studies are of great importance in increasing our understanding and appreciation of animals. In addition to providing knowledge about the diversity and complexity of behaviour in nature, such studies also provide information crucial to improvements in the welfare of animals maintained in laboratories, zoos and agricultural settings. The use of animals in behavioural research and teaching does, however, raise important ethical issues. While many behavioural studies are non-invasive and involve only observations of animals in their natural habitat, some research questions cannot be answered adequately without manipulation of animals. Studies of captive animals necessarily involve keeping animals in confinement, and manipulative procedures and surgery may be necessary to achieve the aims of the research. Studies of free-living animals in their natural habitats can cause disruption, particularly if feeding, capture, marking or experimental procedures are involved.

While the furthering of scientific knowledge is a proper aim and may itself advance an awareness of human responsibility towards animal life, the investigator must always weigh the potential gain in knowledge against any adverse consequences for the animals and populations under study. This is equally true for the evaluation of animal use in animal behaviour teaching activities. In fact, animal behaviour courses provide an excellent opportunity to introduce students to the ethical obligations a researcher accepts when animals are studied.

In order to help their members make what are sometimes difficult ethical judgements about the procedures involved in the study of animals, the Association for the Study of Animal Behaviour and The Animal Behaviour Society have formed Ethical and Animal Care committees, respectively. These committees jointly produced the following guidelines for the use of all those who are engaged in behavioural research and teaching activities involving vertebrate and invertebrate animals. These guidelines are general in scope, since the diversity of species and study techniques used in behavioural research precludes the inclusion of specific details about appropriate animal care and treatment. The guidelines will be used by the Editors of *Animal Behaviour* in assessing the acceptability of submitted manuscripts. Submitted manuscripts may be rejected by an Editor, after consultation with the Ethical or Animal Care Committees, if the content violates either the letter or the spirit of the guidelines. These guidelines supplement the legal requirements in the country and/or state or province in which the work is carried out. They should not be considered an imposition upon the scientific freedom of individual reseachers, but rather as helping to provide an ethical framework which each investigator may use in making decisions related to animal welfare.

1 LEGISLATION

Investigators are accountable for the care and wellbeing of animals used in their research and teaching activities, and must therefore abide by the spirit as well as the letter of relevant legislation. For those who reside in Great Britain, a summary of the laws designed to ensure the welfare of animals is given by Crofts (1989); detailed guidance on the operation of the Animals (Scientific Procedures) Act, 1986 is provided by the Home Office (1990). In the U.S.A., federal, state and local legislation and guidelines may apply. In particular, the care and use of many vertebrate laboratory animals are regulated under the Animal Welfare Act and its amendments and regulations (Code of Federal Regulations, Title 9) and/or the policies of the Public Health Service (NRC 1985; PHS 1986). Guidelines for farm animals used in research and teaching may also be applicable (Guide Development Committee 1988). In Canada, guidance can be obtained from the Guide to the Care and Use of Experimental Animals (Canadian Council on Animal Care 1992).

In Britain, lists of threatened species and laws aiming to protect them can be obtained from the International Union for the Conservation of Nature, Species Conservation Monitoring Unit (219C Huntingdon Road, Cambridge CB3 0DL, U.K.). In the U.S.A., information pertaining to the Endangered Species Act of 1973 may be found in the Code of Federal Regulations (Title 50, 1973). Lists of endangered species can be obtained from the Office for Endangered Species, U.S. Department of Interior, Fish and Wildlife Service (Washington, D.C. 20240), or from the Committee on the Status of Endangered Wildlife in Canada, Canadian Wildlife Service, (Environment Canada, Ottawa, Ontario K1A 0E7).

Investigators working in other countries must familiarize themselves with legislation both on animal welfare and on threatened and endangered species and conform with the spirit and letter of the laws. When submitting manuscripts to *Animal Behaviour*, all authors must confirm in their cover letter that they have adhered to the legal requirements of the country in which the study was conducted.

2 CHOICE OF SPECIES AND NON-ANIMAL ALTERNATIVES

Investigators should choose a species for study that is well suited for investigation of the questions posed. Choosing an appropriate subject usually requires knowledge of a species' natural history and phylogenetic level. Knowledge of an individual animal's previous experience, such as whether or not it has spent a lifetime in captivity, is also important. When research or teaching involves procedures or housing conditions that may cause pain, discomfort or stress to the animal, and when alternative species can be used, the researcher should employ the species which, in the opinion of the researcher and other qualified colleagues, is least likely to suffer (OTA 1986). Live animal subjects are generally essential in behavioural research, but non-animal alternatives such as video records from previous work or computer simulations can sometimes be used (Smyth 1978). Material of this kind also exists for teaching purposes and can be used instead of live animals to expand the range of behavioural subjects available to students.

3 NUMBER OF INDIVIDUALS

The researcher should use the smallest number of animals necessary and sufficient to accomplish the research goals, especially in studies which involve manipulations that are potentially detrimental to the animal or the population. The number of animals used in an experiment can often be dramatically reduced by pilot studies, good experimental design and the use of statistical tests that enable several factors to be examined simultaneously. Hunt (1980), Still (1982) and McConway (1992) discuss ways of reducing the number of animals used in experiments through alternative designs. Useful reference works are Cox (1958) and Cochran & Cox (1966).

4 PROCEDURES

Investigators are encouraged to discuss with colleagues both the scientific value of their research proposals as well as possible ethical considerations. There are several models for evaluating animal research which can be of use when making ethical decisions (Bateson 1986; Orlans 1987; Shapiro & Field 1988; Donnelly & Nolan 1990; Porter 1992). If procedures used in research or teaching involve exposure to painful, stressful or noxious stimuli, the investigator must consider whether the knowledge that may be gained is justified. Bateson (1991) discusses the assessment of pain and suffering. Additional information can be obtained from the U.S. National Academy of Sciences Publication, 'Recognition and Alleviation of Pain and Distress in Laboratory Animals' (NRC 1992), and from the American Veterinary Medical Association panel report on animal pain and distress (AVMA 1987). Researchers are urged to consider the use of alternative procedures before employing techniques that are likely to cause physical or psychological discomfort to the animal. Pain or suffering should be minimized both in duration and magnitude to the greatest extent possible under the requirements of the experimental design. Attention should be given to proper pre- and post-operative care in order to minimize preparatory stress and residual effects. Unless specifically contraindicated by the experimental design, procedures that are likely to cause pain or discomfort should be performed only on animals that have been adequately anaesthetized. Analgesics should be used after such procedures to minimize pain and distress whenever possible (Flecknell 1985; Benson et al. 1990).

The following more specific points may be of use.

a Fieldwork

Investigators studying free-living animals must take precautions to minimize interference with individuals as well as the populations and eco-systems of which they are a part. Capture, marking, radio-tagging, collection of physiological data such as blood or tissue samples or field experiments may not only have immediate effects on the animal, but may also have consequences such as a reduced probability of survival and reproduction. Investigators should consider the effects of such interference, and use less disruptive techniques such as individual recognition by the use of natural features rather than marking (Scott 1978) where possible. Cuthill (1991) discusses the ethical issues associated with field experiments, and recommends pilot investigations to assess potential environmental disruption and follow-up studies to detect and minimize persistent effects. Investigators should weigh the potential gain in knowledge from field studies against the adverse consequences of disruption for the animals used as subjects and also for other animals and plants in the ecosystem. When an experimental protocol requires that animals be removed from the population either temporarily or on a long-term basis, investigators should ensure that suffering or discomfort are minimized not only for the removed animals but for others dependent on them (e.g. dependent offspring). Removed individuals and their dependants must be housed and cared for appropriately. Sources of further information on field techniques are the books edited by Stonehouse (1978) and Amlaner & Macdonald (1980).

b Aggression, predation and intraspecific killing

The fact that the agent causing harm may be another non-human animal does not free the experimenter from the normal obligations to experimental animals. Huntingford (1984) and Elwood (1991) discuss the ethical issues involved and suggest ways to minimize suffering. Wherever possible, field studies of natural encounters should be used in preference to staged encounters. Where staged encounters are necessary, the use of models should be considered, the number of subjects should be kept to the minimum needed to accomplish the experimental goals, and the experiments made as short as possible. Suffering can also be reduced by continuous observation with intervention to stop aggression at predefined levels, and by providing protective barriers and escape routes for the subjects.

c Aversive stimulation and deprivation as motivational procedures

Aversive stimulation or deprivation can cause pain or distress to animals. To minimize suffering, the investigator should ascertain that there is no alternative way of motivating the animal, and that the levels of deprivation or aversive stimulation used are no greater than necessary to achieve the goals of the experiment. Alternatives to deprivation include the use of highly preferred foods and other rewards which may motivate even satiated animals. Use of minimal levels requires a knowledge of the technical literature in the relevant area: quantitative studies of aversive stimulation are reviewed by Church (1977) and Rushen (1986) and the behaviour of satiated animals in considered by Morgan (1974). Further comments on reducing distress due to motivational procedures are to be found in Lea (1979) and Moran (1975).

d Social deprivation, isolation and crowding

Experimental designs that require keeping animals in over-crowded conditions, or that involve social deprivation or isolation, may be extremely stressful to the animals involved. Because the degree of stress experienced by the animal can vary with species, age, sex, reproductive condition, developmental history and social status, the natural social behaviour of the animals concerned and their previous social experience must be considered in order to minimize such stress.

e Deleterious conditions

Studies aimed at inducing deleterious conditions in animals are sometimes performed in order to gain scientific knowledge of value to human or animal problems. Such conditions include inducing diseases, increasing parasite loads, and exposing animals to pesticides or homeostatic stressors. Where feasible, studies inducing a deleterious condition in animals should address the possible treatment or alleviation of the condition induced. Animals exposed to deleterious conditions that might result in suffering or death should be monitored frequently and, whenever possible, considering the aims of the research, humanely killed as soon as they show signs of distress. If the goals of the research allow it, the investigator should also consider experimental designs in which the deleterious condition is removed (e.g. removing rather than adding parasites as the experimental treatment) or in which naturally occurring instances of deleterious conditions are observed.

5 ENDANGERED SPECIES

All research on endangered or locally rare species must comply with relevant legislation and be coordinated with official agencies responsible for the conservation effort for the particular species under study. Legislation and sources of help in identifying endangered species have been outlined in Section 1. Members of threatened species should not be placed at risk except as part of a serious attempt at conservation. Observation alone can result in serious disturbance, including higher predation rates on nests or young, or their abandonment, and should only be undertaken after careful consideration of techniques and of alternative species. Investigators should also consider further adverse consequences of their work, such as opening up remote areas for subsequent access or teaching techniques of anaesthetization and capture which might be misused (e.g. by poachers).

6 PROCUREMENT OF ANIMALS

When it is necessary to procure animals either by purchase or donation from outside sources, only reputable suppliers should be used. For workers in the U.K., advice about purchasing animals may be obtained from the Laboratory Animal Breeder's Association, Charles River (U.K.) Ltd, Manson Research Centre, Manson Road, Margate, Kent CT9 4LP. In the U.S.A., information on licensed animal dealers can be obtained from the local office of the U.S. Department of Agriculture (U.S.D.A.). Other sources of information on laboratory animal suppliers in North America are the American Association for Laboratory Animal Science (70 Timber Creek Drive, Suite 5, Cordova, TN 38018) and the Canadian Association for Laboratory Animal Science (M524 Biological Sciences Building, University of Alberta, Edmonton, Alberta T6G 3E9). If animals are procured by capture in the wild, this must be done in as painless and humane a manner as possible and must comply with any relevant legislation. Individuals of endangered species or populations should not be taken from the wild unless they are part of an active conservation programme. So far as is possible, the investigator should ensure that those responsible for handling purchased, donated or wild-caught animals en route to the research facilities provide adequate food, water, ventilation and space, and do not impose undue stress.

7 HOUSING AND ANIMAL CARE

The researcher's responsibilities extend also to the conditions under which the animals are kept when not being studied. Caging conditions and husbandry practices must meet, at the very least, minimal recommended requirements of the country in which the research is carried out. Guidance can be obtained from the Universities Federation for Animal Welfare (U.F.A.W.) handbook (Poole 1987), the National Research Council guide (NRC 1985), the U.S.D.A. Animal Welfare Act Regulations (Code of Federal Regulations), the Guide for the Care and Use of Agricultural Animals in Agricultural Research and Teaching (Guide Development Committee 1988), and the Canadian Council on Animal Care's Guide to the Care and Use of Experimental Animals (1992).

Although these publications provide general guidance, the housing and care regimes established for the commonly used laboratory animals are not necessarily suitable for wild animals or for individuals of wild species born in captivity. Special attention may be required to enhance the comfort and safety of these animals. Normal maintenance of captive animals should incorporate, as much as possible, aspects of the natural living conditions deemed important to welfare and survival. Consideration should be given to providing features such as natural materials, refuges, perches and dust and water baths. Companions should be provided for social animals where possible, providing that this does not lead to suffering or injury. Frequency of cage cleaning should represent a compromise between the level of cleanliness necessary to prevent diseases and the amount of stress imposed by frequent handling and exposure to unfamiliar surroundings, odours, and bedding. Researchers in the United States should also ensure that the requirements outlined under the 1985 Amendment to the Animal Welfare Act to provide exercise for laboratory-housed dogs and to ensure the psychological well-being of captive non-human primates are met.

The nature of human–animal interactions during routine care and experimentation should be considered by investigators. Depending upon species, rearing history and the nature of the interaction, animals may perceive humans as conspecifics, predators or symbionts (Estep & Hetts 1992). Special training of animal care personnel can help in implementing procedures that foster habituation of animals to caretakers and researchers and minimize stress. Stress can also be reduced by training animals to cooperate with handlers and experimenters during routine husbandry and experimental procedures (Biological Council 1992).

8 FINAL DISPOSITION OF ANIMALS

When research projects or teaching exercises using captive animals are completed, it may sometimes be appropriate to distribute animals to colleagues for further study or breeding, if permitted by local legislation. However, if animals are distributed care must be taken to ensure that the same animals are not used repeatedly in stressful or painful experiments, and that they continue to receive a high standard of care. Animals should never be subjected to more than one major surgery unless it is an unavoidable element of a single experiment. Except as prohibited by national, federal, state, provincial, or local laws, researchers may release field-trapped animals if this is practical and feasible, especially if it is critical to conservation efforts. However, the researcher should assess whether releases into the wild might be injurious or detrimental both to the released animal and to existing populations in the area. Animals should be released at the site where they were trapped (unless conservation efforts dictate otherwise), and only when their ability to survive in nature has not been impaired and when they do not constitute a health or ecological hazard to existing populations. If animals must be killed subsequent to a study this must be done as humanely and painlessly as possible; death of the animals should be confirmed before their bodies are discarded. A veterinarian should be consulted for advice or methods of euthanasia that are appropriate for the particular species being used. Additional information on euthanasia methods can be found in the report of the AVMA panel on euthanasia (AVMA 1993).

9   OBTAINING FURTHER INFORMATION

There are a number of organizations that provide publications and detailed information about the care and use of animals. These include The Canadian Council on Animal Care (1105–151 Slater Street, Ottawa, Ontario, K1P 5H3 Canada), the Scientists Center for Animal Welfare (4805 St Elmo Avenue, Bethesda, MD 20814, U.S.A.), and the Universities Federation for Animal Welfare (8 Hamilton Close, South Mimms, Potters Bar, Hertfordshire EN6 3QD, U.K.). The Animal Welfare Information Center at the National Agricultural Library (Room 205, Beltsville, MD 20705, U.S.A.) publishes a series of bibliographies on special topics, and can also provide individualized database searches for investigators on potential alternatives, including techniques for replacement with non-animal models or alternative species, methods for reducing the total number of animals necessary to address the research question, and experimental refinements which can reduce pain and stress.

References

Ackerman, S. (1988). American Scientist interviews. *Am. Sci.* **76**:494–499.

Adams. R.M. and Markley R.P. (1978). Assessment of the accuracy of point and one–zero sampling techniques by computer simulation (unpublished manuscript). Paper presented at 1978 Animal Behavior Society Meeting, June 19–23, Seattle, Washington.

Adams, R.M. and MacDonald, J.D. (1987). Interobserver reliability checks: frequency of use in animal behavior and psychology. Paper presented at Animal Behavior Society annual meeting, June 1987, Williamstown, MA.

Adhikerana, A.S. and Slater, P.J.B. (1993). Singing interactions in coal tits, *Parus ater*: an experimental approach. *Anim. Behav.* **46**(6):1205–1211.

Alatalo, R.V., Lundbergh, A. and Sundberg, J. (1990). Can female preference explain sexual dichromatism in the pied flycatcher, *Ficedula hypoleuca*? *Anim. Behav.* **39**(2):244–252.

Alcock, J. (1970). Punishment levels and the response of black-capped chickadees (*Parus atricapillus*) to three kinds of artificial seeds. *Anim. Behav.* **18**(3):592–599.

Alcock, J. (1973). Cues used in searching for food by red-winged blackbirds (*Agelaius phoeniceus*). *Behaviour* **46**:174–188.

Alcock, J. (1989). *Animal Behavior: An Evolutionary Approach.* 4th edition. Sinauer Associates, Sunderland, MA. 596 pp.

Alcock, J. (1993). *Animal Behavior: An Evolutionary Approach.* 5th edition. Sinauer Associates, Sunderland, MA. 625 pp.

Alexander, R.D. (1975). The search for a general theory of behavior. *Behav. Sci.* **20**(2):77–100.

Alkon, P.U., Cohen, Y. and Jordan, P.A. (1989). Towards an acoustic biotelemetry system for animal behavior studies. *J. Wildl. Manage.* **53**(3):658–662.

Allan, S.E. and Simmons, A.M. (1994). Temporal features mediating call recognition in the green treefrog, *Hyla cinerea*: amplitude modulation. *Anim. Behav.* **47**(5):1073–1086.

Allee, W.C. (1938). *The Social Life of Animals.* Beacon Press, Boston. 233 pp.

Allen, A.A. (1934. Sex rhythm in the ruffed grouse (*Bonasa umbellus* L.) and other birds. *Auk* **51**:180–199.

Allport, S. (1986). *Explorers of the Black Box.* W.W. Norton Co., New York. 271 pp.

Altman, J. and Sudarshan, K. (1975). Positioned development of locomotion in the laboratory rat. *Anim. Behav.* **23**(4):896–920.

Altmann, J. (1974). Observational study of behavior: sampling methods. *Behaviour* **49**(3,4):227–265.

Altmann, J. (1980). *Baboon Mothers and Infants.* Harvard University Press, Cambridge, MA.

Altmann, J. (1984). Observational sampling methods for insect behavioral ecology. *Florida Entomologist* **67**(1):50–56.

Altmann, S.A. (1965). Sociobiology of rhesus monkeys II: stochastics of social communication. *J. Theor. Biol.* **8**:490–522.

Altmann, S.A. (1968a). Sociobiology of rhesus monkeys III: The basic communication network. *Behaviour* **32**:17–32.

Altmann, S.A. (1968b). Sociobiology of rhesus monkeys IV: testing Mason's hypothesis of sex differences in affective behaviour. *Behaviour* **32**(1–2):49–69.

Altmann, S.A. and Altmann, J. (1977). On the analysis of rates of behaviour. *Anim. Behav.* **25**(2):364–372.

Altmann, S.A. and Wagner, S.S. (1970). Estimating rates of behaviour from Hansen frequencies. *Primates* **2**:181–183.

Amlaner, C.J. (ed.). (1989). *Biotelemetry X: Proceedings of the Tenth International Symposium on Biotelemetry.* The University of Arkansas Press, Fayetteville.

Amlaner, C.J. and MacDonald, D.W. (eds.). (1980). *A Handbook on Biotelemetry and Radio Tracking.* Pergamon Press, New York.

Anasu, C. (1965). Runs test for a circular distribution and a table of probabilities. *Ann Inst. Stat. Math.* **17**: 331–346.

Anderka, F.W. and Angehrn, P. (1992). Transmitter attachment methods. In *Remote Monitoring and Tracking of Animals*, ed. I.G. Priede and S.M. Swift, pp. 135–146. Horwood (Prentice Hall), Englewood Cliffs, NJ. 708 pp.

Anderson, D.J. (1982). The home range: a new nonparametric estimation technique. *Ecology* **63**:103–112.

Anderson, N.H. (1961). Scales and statistics: parametric and nonparametric. *Psych. Bull.* **58**(4):305–316.

Anderson, S.A. and Harwood, J. (1985). Time budgets and topography: how energy reserves and terrain determine the breeding behaviour of grey seals. *Anim. Behav.* **33**(4):1343–1348.

Andrew, R.J. (1972). The information potentially available in mammalian displays. In *Nonverbal Communication*, ed. R.A. Hinde, pp. 179–204. Cambridge University Press, Cambridge.

Andrews, F.C. (1954). Asymptotic behavior of some rank tests for analysis of variance. *Ann. Math. Stat.* **25**:724–735.

Angerbjorn, A. and Becker, D. (1992). An automatic location system for wildlife telemetry. In *Remote Monitoring and Tracking of Animals*, ed. I.G. Priede and S.M. Swift, pp. 68–75. Horwood (Prentice Hall), Englewood Cliffs, NJ. 708 pp.

Annino, R. and Driver, R.D. (1986). *Scientific and Engineering Applications with Personal Computers.* John Wiley and Sons, New York. 577 pp.

Anonymous. (1970). *Basic Scientific Photography. Eastman Kodak Scientific Data Book* N-9, Rochester, New York. 40 pp.

Anonymous. (1977). *North American Bird Banding Manual.* Vol.II. US Fish and Wildlife Service and Canadian Wildlife Service, Washington, DC, USA and Ottawa, Ontario, Canada..

Anonymous. (1988). Seventy-five reasons to become a scientist. *Amer. Sci.* **76**:450–463.

Anonymous. (1989). *The Starkey Project: Automated animal tracking system.* (leaflet 2 pp.). Forestry and Range Sciences Laboratory, LaGrande, Oregon.

Anonymous. (1991). *North American Bird Banding Manual.* Vol.I (revised). US Fish and Wildlife Service and Canadian Wildlife Service. Washington, DC, USA and Ottawa, Ontario, USA.

Anonymous. (1992). Sigma Xi statement on the use of animals in research. *Am. Sci.* **80**:73–76.

Anonymous. (1995). Outdoor electronics: cat eyes. *Outdoor Life* **24**(1):12.

Ansell, A.D. (1967). Leaping and other movements in some cardiid bivalves. *Anim. Behav.* **15**(4):421–426.

Arber, A. (1985). *The Mind and the Eye.* Cambridge University Press, Cambridge 168 pp.

Archer, J. (1976). The organization of aggression and fear in vertebrates. In ed. P.P.G. Bateson and Klopfer, P.H. *Perspectives in Ethology,* vol. 2. pp. 231–298. Plenum Press, New York.

Asa, C.S. (ed.) (1991). *Biotelemetry Applications for Captive Animal Care and Research.* American Association of Zoological Parks and Aquariums, Wheeling, West Virginia. 59 pp.

Ashby, W.R. (1963). *An Introduction to Cybernetics.* John Wiley, New York. 295 pp.

Aspey, W.P. (1977a). RECDIS: A BASIC program for computing interindividual distances within a rectangular area. *Behav. Res. Method. Instrument.* **9**(1):26–27.

Aspey, W.P. (1977b). CIRCIS: A BASIC program for computing interindividual distances within a circular area. *Behav. Res. Method. Instrument.* **9**(1):50–51.

Aspey, W.P. (1977c). Wolf spider sociobiology: I. Agonistic display and dominance–subordinance relations in adult male *Schizocosa crassipes*. *Behaviour* **62**(1–2):103–141.

Aspey, W.P. and Blankenship, J.E. (1976). *Aplysia* behavioral biology: I. A multivariate analysis of burrowing in *A. brasiliana*. *Behav. Biol.* **17**:279–299.

Aspey, W.P. and Blankenship, J.E. (1977). Spiders and snails and statistical tales: application of multivariate analyses to diverse ethological data. In *Quantitative Methods in the Study of Behavior.* ed. B.A. Hazlett, pp. 75–120. Academic Press, New York. 222 pp.

Aspey, W.P. and Blankenship, J.E. (1978). Comparative ethometrics: congruence of different multivariate analyses applied to the same ethological data. *Behav. Proc.* **3**:173–195.

Avise, J.C. (1994). *Molecular Markers, Natural History and Evolution.* Chapman & Hall, London. 511 pp.

Axelrod, R. (1984). *The Evolution of Cooperation.* Basic Books, New York. 241 pp.

Baerends, G.P. (1976). The functional organization of behaviour. *Anim. Behav.* **24**(4):726–738.

Baerends, G.P. (1985). Two pillars of wisdom. In *Studying Animal Behavior: Autobiographies of the Founders,* ed. D.A. Dewsbury, pp. 13–40. University of Chicago Press, Chicago. 512 pp.

Baerends, G.P. and Kruijt, J.P. (1973). Stimulus selection. In *Constraints on learning,* ed. R.A. Hinde and J. Hinde, pp. 23–29. Academic Press, London.

Baerends, G.P. and Van der Cingel, N.A. (1962). On the phylogenetic origin of the snap display in the common heron (*Ardea cinerea L.*). *Symp. Zool. Soc. London* **8**:7–24.

Baerends, G.P., Brouwer, R. and Waterbolk, H.T. (1955). Ethological studies on *Lebistesreticulatus* (Peters): I. An analysis of the male courtship pattern. *Behaviour* **8**:249–334.

Baerends, G.P., Drent, R.H., Glas, P. and Groenewold, H. (1970). An ethological analysis of incubation behaviour in the herring gull. *Behaviour* suppl. **17**:135–235.

Bakeman, R. (1978). Untangling streams of behavior. In *Observing Behavior, vol. 2, Data Collection and Analysis Methods,* ed. G.P. Sackett, pp. 63–78. University Park Press, Baltimore.

Bakeman, R. (1983). Computing lag sequential statistics: the ELAG program. *Behav. Res. Method and Instrument.* **15**(5):530–535.

Bakeman, R. and Gottman, J.M. (1986). *Observing Interaction: An Introduction to Sequential Analysis.* Cambridge University Press, Cambridge. 221 pp.

Baker, R.R.(1980). Goal orientation by blindfolded humans after long-distance displacement: possible involvement of a magnetic sense. *Science* **210**:555–557.

Baker, R.R. (1987). Human navigation and magnetoreception: the Manchester experiments do replicate. *Anim. Behav.* **35**(3):691–704.

Balaban, E., Teillet, M. and Douarin, N .L. (1988). Application of the quail-chick chimera system to the study of brain development and behavior. *Science* **241**:1339–1342.

Baldock, N.M., Sibly, R.M. and Penning, P.D. (1988). Behaviour and seasonal variation in heart rate in domestic sheep, *Ovis gries*. *Anim. Behav.* **36**(1):35–43.

Balgooyen, T.G. (1976). *Behavior and ecology of the American kestrel (Falco sparverius L.) in the Sierra Nevada of California*. University of California, Publications in Zoology, vol. 103. Berkeley. 83 pp.

Balph, D.F. (1968). Behavioral responses of unconfined uinta ground squirrels to trapping. *J. Wildl. Manage.* **32**(4):778–794.

Barash, D.P. (1974). Mother–infant relations in captive woodchucks (*Marmota monax*). *Anim. Behav.* **22**(2)446–448.

Barki, A., Karplus, I. and Goren, M. (1992). Effects of size and morphotype on dominance hierarchies and resource competition in the freshwater prawn *Macrobrachium rosenbergii*. *Anim. Behav.* **44**(3):547–555.

Barlow, G.W. (1977). Modal action patterns. In *How Animals Communicate*, ed. T.A. Sebock, pp. 94–125. University of Indiana Pres, Bloomington. 1128 pp.

Barnard, C., Gilbert, F. and McGregor, P. (1993). *Asking Questions in Biology: Design, Analysis and Presentation in Practical Work*. Longman Scientific and Technical Publications, Harlow, England. 157 pp.

Batchelor, T.A. and McMillan, J.R. (1980). A visual marking system for nocturnal animals. *J. Wildl. Manage.* **44**(2):497–499.

Bateson, P.P.G. (1977). Testing an observer's ability to identify individual animals. *Anim. Behav.* **25**(1):247–248.

Bateson, P.P.G. (1983). Genes, environment and the development of behaviour. In *Genes, Development and Learning*, ed. T.R. Halliday and P.J.B. Slater, pp. 52–81. W.H. Freeman & Co., San Francisco. 246 pp.

Bateson, P.P.G. (1986). When to experiment on animals. *New Scient.* **20**:30–32.

Bateson, P.P.G. (1987). Epilogue: An ethological overview. In *Aims and Methods in Neuroethology*, ed. D.M. Guthrie, pp. 301–305. Manchester University Press, Manchester, England. 310 pp.

Batschelet, E. (1965). *Statistical Methods for the Analysis of Problems in Animal Orientation and Certain Biological Rhythms*. AIBS Monograph I. 57 pp.

Batschelet, E. (1972). Recent statistical methods for orientation data. In *Animal Orientation and Navigation*, ed. S.R. Galler, K. Schmidt-Koenig, G.J. Jacobs and R.E. Belleville, pp. 61–91. NASA SP262. US. Government Printing Office. Washington, DC. 606 pp.

Batschelet, E. (1981). *Circular Statistics in Biology*. Academic Press.,New York. 371 pp.

Bauer, H.H. (1992). *Scientific Literacy and the Myth of the Scientific Method*. University of Illinois Press, Urbana. 180 pp.

Baufle, J.M. and Varin, J.P. (1972). *Photographing Wildlife*. Oxford University Press, Oxford. 157 pp.

Beach, F.A. (1950). The snark was a boojum. *Am. Psychol.* **5**:115–124.

Beck, B.B. (1977). Kohler's chimpanzees – how did they really perform? *Zool. Garten* **47**:352–360.

Beer, C.G. (1977). What is a display? *Am. Zool.* **17**:155–165.

Beilharz, R.G. and Cox, D.F. (1967). Social dominance in swine. *Anim. Behav.* **15**(1):117–122.

Bekoff, A. (1978). A neuroethological approach to the study of the ontogeny of coordinated behavior. In *The Development of Behavior: Comparative and Evolutionary Aspects*, ed. G.M. Burghardt and M. Bekoff, pp. 19–41. Garland STPM Press, New York. 429 pp.

Bekoff, M. (1976). Animal play: problems and perspectives. In *Perspectives in Ethology*, vol. 2, ed. P.P.G. Bateson and P. Klopfer, pp. 165–188. Plenum, New York.

Bekoff, M. (1977a). Social communication in canids: evidence for the evolution of a stereotyped mammalian display. *Science* **197**:1097–1099.

Bekoff, M. (1977b). Quantitative studies of three areas of classical ethology: social dominance, behavioral taxonomy, and behavioral variability. In *Quantitative Methods in the Study of Animal Behavior*, ed. B. Hazlett, pp. 1–46. Academic Press, New York. 222 pp.

Bekoff, M. (1978a). A field study of the development of behavior in Adelie penguins: univariate and numerical taxonomic approaches. In *The Development of Behavior*, ed. G.M. Burghardt and M. Bekoff, pp. 177–202. Garland STPM Press, New York.

Bekoff, M. (1978b). Behavioral development in coyotes and eastern coyotes. In *Coyotes: Biology, Behavior and Management*, ed. M. Bekoff, pp. 97–126. Academic Press, New York. 384 pp.

Bekoff, M. (1979a). Behavioral acts: description, classification, ethogram analysis, and measurement. In *The Analysis of Social Interactions: Methods, Issues and Illustrations*, ed. R.B. Cairns, pp. 67–80. Lawrence Erlbaum Assoc. Hillsdale, NJ. 243 pp.

Bekoff, M. (1979b). Scent marking by free-ranging domestic dogs: olfactory and visual components. *Biol. Behav.* **4**:123–139.

Bekoff, M. and Corcoran, J. (1975). A method for the analysis of activity and spatial relations in animal groups. *Behav. Res. Method. Instrument.* **7**(6):569.

Bekoff, M. and Mech, L.D. (1984). Simulation analyses of space use: home range estimates, variability and sample size. *Behav. Res. Method. Instrument. Computers.* **16**(1):32–37.

Bekoff, M., Hill, H.L. and Mitton, J.B. (1975). Behavioral taxonomy of canids by discriminant function analyses. *Science* **190**:1223–1225.

Bell, C.R. (1979). Psychological aspects of probability and uncertainty. In *Uncertain Outcomes*, SP Media and Scientific Books, ed. C.R. Bell, pp. 5–21. New York. 204 pp.

Benedix Jr., J.H. (1994). A predictable pattern of daily activity by the pocket gopher *Geomys bursarius*. *Anim. Behav.* **48**(3):501–509.

Bentley, D.R. and Hoy, R.R. (1972). Genetic control of the neuronal network generating cricket (*Teleogryllus gryllus*) song patterns. *Anim. Behav.* **20**(3):478–492.

Bercovitch, F.B. (1988). Coalitions, cooperation and reproductive tactics among adult male baboons. *Anim. Behav.* **36**(4):1198–1209.

Berger, J. (1977). Organizational systems and dominance in feral horses in the Grand Canyon. *Behav. Ecol. Sociobiol.* **2**:131–146.

Berger, J. (1985). Interspecific interactions and dominance among wild great basin ungulates. *J. Mammal.* **66**(3):571–573.

Bergman, C.A. (1981). The glass of fashion. *Audubon* **83**(6):74–80.

Bergman, C.A. (1986). Audubon's guide to spotting scopes. *Audubon* **88**(4):95,96,98–109.

Bernstein, D.M. and Livingston, C. (1982). An interactive program for analysis of human behavior in a long-term continuous laboratory. *Behav. Res. Method. Instrument.* **14**(2):231–235.

Bernstein, I.S. (1970). Primate status hierarchies. In *Primate Behavior*, vol. 1. ed. L.A. Rosenblum, pp. 71–109. Academic Press, New York.

Beveridge, W.I.B. (1950). *The Art of Scientific Investigation.* Vintage Books, New York. 239 pp.

Blair, R.C. and Higgins, J.J. (1985). Comparison of the power of the paired samples *t* test to that of Wilcoxon's signed-ranks test under various population shapes. *Psych. Bull.* **97**(1):119–128.

Blair, W.F. (1953). Population dynamics of rodents and other small mammals. *Adv. Genet.* **5**:1–41.

Blaker, A.A. (1976). *Field Photography.* W.H. Freeman and Co., San Francisco. 451 pp.

Blalock, H.M. (1961). *Causal inferences in nonexperimental research.* University of North Carolina Press, Chapel Hill. 200 pp.

Bleiweiss, R. (1994). Behavioural and evolutionary implications of ultraviolet reflectance by gorgets of sunangel hummingbirds. *Anim. Behav.* **48**(4):978–981.

Boag, D.A. (1972). Effect of radio packages on behavior of captive red grouse. *J. Wildl. Manage.* **36**:511–518.

Bolles, R.C. (1988). The bathwater and everything. *Behav. Brain Sci.* **11**:449–450.

Borgia, G. (1986). Sexual selection in bowerbirds. *Sci. Am.* **254**(6):92–100.

Bouissou, M.F. (1972). Influence of body weight and presence of horns on social rank in domestic cattle. *Anim. Behav.* **20**(3):474–477.

Boyd, R. and Silk, J.B. (1983). A method for assigning cardinal dominance ranks. *Anim. Behav.* **31**(1):45–58.

Bradbury, J.W. and Nottebohm, F. (1969). The use of vision by the little brown bat, *Myotis lucifugus*, under controlled conditions. *Anim. Behav.* **17**(3):480–485.

Bradley, R. (1977). Making animal sound recordings. *Am. Birds* **31**(3):279–285.

Brander, R.B. and Cochran, W.M. (1969). Radio-location telemetry. In *Wildlife Management Techniques*, 3rd edn. ed. R.H. Giles, pp. 95–103. The Wildlife Society, Washington, DC.

Breitwisch, R. (1988). Sex differences in defence of eggs and nestlings by northern mocking-birds, *Mimus polyglottos. Anim. Behav.* **36**(1):62–72.

Brenowitz, E.A. and G.J. Rose. (1994). Behavioural plasticity mediates aggression in choruses of the Pacific treefrog. *Anim. Behav.* **47**(3):633–641.

Brockway, B.F. (1964). Ethological studies of the budgerigar (*Melopsittacus unclulatus*): non-reproductive behaviour. *Behaviour* **22**:193–222.

Brogdon, B. (1993). Space age trail-blazers. *Outdoor Life* **8**:53–55,88.

Bronowski, J. (1973). *The Ascent of Man.* Little, Brown, Boston. 448 pp.

Brooks, R.J. and Falls, J.B. (1975). Individual recognition by song in white-throated sparrows. III. Song features used in individual recognition. *Can. J. Zool.* **53**:1749–1761.

Broom, D.M. (1979). Methods of detecting and analyzing activity rhythms. *Biol. Behav.* **4**(1):201–211.

Brower, L.P. and Brower, J.V.Z. (1962). Investigations into mimicry. *Nat. Hist.* **71**:8–19.

Brower, L.P., Brower, J.V. and Cranston, F.P. (1965). Courtship behavior of the queen butterfly, *Danus gilippus berenice* (Cramer). Zoologica **50**:1-39.

Brown, C.H. and Waser, P.M. (1984). Hearing and communication in blue monkeys. *Anim. Behav.* **32**(1):66–75.

Brown, G.S. and Gass, C.L. (1993). Spatial association learning by hummingbirds. *Anim. Behav.* **46**(3):487–497.

Brown, J.L. (1963). Aggressiveness, dominance and social organization in the Steller's jay. *Condor* **65**:460–484.

Brown, J.L. (1975). *The Evolution of Behavior.* W.W. Norton, New York. 761 pp.

Bub, H., Hamerstrom, F. and Wuertz-Schaefer, K. (1991). *Bird Trapping and Bird Banding.* Cornell Univeristy Press, Ithaca, New York. 448 pp.

Buchler, E.R. (1976). The use of echolocation by the wandering shrew (*Sorex vagrans*). *Anim. Behav.* **24**(4):858–873.

Buckley, P.A. and Hancock, J.T., Jr. (1968). Equations for estimating and a simple computer program for generating unique color- and aluminum-band sequences. *Bird-Banding* **39**:123–129.

Bulger, R.E., Heitman, E. and Reiser, S.J. (1993). *The Ethical Dimensions of the Biological Sciences.* Cambridge University Press, Cambridge. 294 pp.

Bullock, R.E. (1974). Functional analysis of locomotion in pronghorn antelope. In *The Behaviour of Ungulates and its Relation to Management*, ed. V. Geist and F. Walther, p. 274–305. International Union Conservation Natural Resources. Publication, New Series, 24.

Burgess, J.W. (1979). Measurement of spatial behavior: methodology applied to rhesus monkeys, neon tetras, communal and solitary spiders, cockroaches, and gnats in open fields. *Behav. Neural., Biol.* **26**:132–160.

Burghardt, G.M. (1973). Instinct and innate behaviour: toward an ethological psychology. In *The Study of Behavior: Learning, Motivation, Emotion and Instinct*, ed. J.A. Nevin and G.S. Reynolds. Scott, Foresman and Co., Glenview, IL.

Burington, R.S. (1948). *Handbook of Mathematical Tables and Formulas.* 3rd edn. McGraw-Hill. New York. 296 pp.

Burke, T., Davies, N.B., Bruford, M.W. and Hatchwell, B.J. (1989). Parental care and mating behavior of polyandrous dunnocks *Prunella modularis* related to paternity by DNA fingerprinting. *Nature* **338**:249–251.

Burley, N.T. (1981). Sex ratio manipulation and selection for attractiveness. *Science* **211**:721–722.

Burley, N.T. (1988). Wild zebra finches have band-color preferences. *Anim. Behav.* **36**(4):1235–1237.

Burley, N.T., Enstrom, D.A. and Chitwood, L. (1994). Extra-pair relations in zebra finches: differential male success results from female tactics. *Anim. Behav.* **48**(5):1031–1041.

Butlin, R.K., Hewitt, G.M. and Webb, S.F. (1985). Sexual selection for intermediate optimum in *Chorthippus brunneus* (Orthoptera: Acrididae). *Anim. Behav.* **33**(4):1281–1292.

Cade, W. (1975). Acoustically orienting parasitoids: fly phonotaxis to cricket song. *Science* **190**:1312–1313.

Cairns, S.J. and Schwager, S.J. (1987). A comparison of association indices. *Anim. Behav.* **35**(5):1454–1469.

Camhi, J.M. (1984). *Neuroethology: Nerve Cells and the Natural Behavior of Animals.* Sinauer Associates, Sunderland, MA. 416 pp.

Campbell, D.J. (1990). Resolution of spatial complexity in a field sample of singing crickets *Teleogryllus commodus* (Walker) (Gryllidae): a nearest-neighbor analysis. *Anim. Behav.* **39**(6):1051–1057.

Campbell, D.T. (1956). Perception as substitute trial and error. *Psych. Rev.* **63**:330–342.

Campbell, D.T. and Blake, R. (1977). Animal awareness? *Am. Sci.* **65**:146–147.

Campbell, R.C. (1974). *Statistics for Biologists*, 2nd edn. Cambridge University Press, Cambridge. 385 pp.

Capen, D.E. (1978). Time-lapse photography and analysis of behavior of nesting white-faced ibises. *Wading Birds, Res. Rep.* **7**:41–43.

Carey, M. and Nolan, V. Jr. (1975). Polygyny in indigo buntings: a hypothesis tested. *Science* **190**:1296–1297.

Carlier, C. and Noirot E. (1965). Effects of previous experience on maternal retrieving by rats. *Anim. Behav.* **13**(4):423–426.

Carpenter, C.C. and Grubitz, G. (1961). Time–motion study of the lizard. *Ecology* **42**:199–200.

Carthy, J.D. (1966). *The Study of Behaviour.* Arnold, London. 57 pp.

Castellan, N.J. (1979). The analysis of behavior sequences. In *The Analysis of Social Interactions*, ed. R.B. Cairns, pp. 81–116. Lawrence Erlbaum Assoc., Hillsdale, NJ. 243 pp.

Catania, A.C. and Harnad, S. (1984). Canonical papers of B.S. Skinner: selection by consequences. *Behav. Brain Sci.* **7**:477–510.

Cates, R.G. and Orians, G.H. (1975). Successional status and the palatability of plants of generalized herbivores. *Ecology* **56**:410–418.

Celhoffer, L., Boukyois, C. and Minde, K. (1977). The DCR-II event recorder: a portable high-speed digital cassette system with direct computer access. *Behav. Res. Method. Instrument.* **9**(5):442–446.

Chaiken, M., Bohner, J. and Marler, P. (1993). Song acquisition in European starlings, *Sturnus vulgaris*: a comparison of the songs of live-tutored, tape-tutored, untutored, and wild-caught males. *Anim. Behav.* **46**(6):1079–1090.

Chalmers, N.R. (1968). The social behaviour of free living Mangabeys in Uganda. *Folia Primatol.* **8**:263–281.

Chapman, F.M. (1935). The courtship of Gould's manakins (*Manacus vitellinus vitellinus*) on Barro Colorado Island, Canal Zone. *Bull. Am. Mus. Nat. Hist.* **68**:471–525.

Chase, I.D. (1974). Models of hierarchy formation in animal societies. *Behav. Sci.* **19**:374–382.

Chase, I.D. (1982). Dynamics of hierarchy formation: the sequential development of dominance relationships. *Behaviour* **80**:218–240.

Chatfield, C. (1973). Statistical inference regarding Markov chain models. *Appl. Stat.* **22**:7–20.

Cheesman, C.L. and Mitson, R.B. (eds.). (1982). *Telemetric Studies of Vertebrates.* Zoological Society of London Series No. 49. Academic Press, New York. 388 pp.

Cheney, D.L. and Seyfarth, R.M. (1990). *How Monkeys See the World. Inside the Mind of Another Species.* University of Chicago Press, Chicago. 377 pp.

Cherfas, J.J. (1980). Signals for food: reinforcers or informants? *Science* **209**:1552–1553.

Chomsky, N. (1957). *Syntactic Structures.* Mouton. The Hague. 116 pp.

Choudhury, S. and Black, J.M. (1993). Mate-selection behaviour and sampling strategies in geese. *Anim. Behav.* **46**(4):747–757.

Churchland, P.S. and Sejnowski, T.J. (1988). Perspectives on cognitive neuroscience. *Science* **242**:741–745.

Cigliano, J.A. (1993). Dominance and den use in *Octopus bimaculoides*. *Anim. Behav.* **46**(4):677–684.

Clapperton, K. (1989). Scent-marking behaviour of the ferret, *Mustela furo L. Anim. Behav.* **38**(3):436–446.

Clark, P.J. and Evans, F.C. (1954). Distance to nearest neighbor as a measure of spatial relationships in populations. *Ecology* **35**:445–453.

Clayton, D. (1976). The effects of pre-test conditions on social facilitation of drinking in ducks. *Anim. Behav.* **24**:125–134.

Cleveland, W.S. (1985). *The Elements of Graphing Data*. Wadsworth Advanced Books and Software, Monterey, CA. 323 pp.

Cleveland, W.S. (1993). *Visualizing Data*. Hobart Press, Summit, NJ.

Clutton-Brock, T.H. (1989). Mammalian mating systems. *Proc. R. Soc. Lond. Ser. B.* **236**:339–372.

Clutton-Brock, T.H., Guinness, F.E. and Albon, S.D. (1982). *Red Deer: Behavior and Ecology of Two Sexes*. University of Chicago Press, Chicago. 378 pp.

Clutton-Brock, T.H., Albon, S.D. and Guinness, F.E. (1986). Great expectations: dominance, breeding success and offspring sex ratios in red deer. *Anim. Behav.* **34**(2):460–471.

Cocatre-Zilgien, J.H. and Delcomyn, F. (1993). A new method for depicting animal step patterns. *Anim. Behav.* **45**(4):820–824.

Cochran, W.W. (1980). Wildlife telemetry. In *Wildlife Management Techniques Manual*, 4th edn, ed. S.D. Schemnitz, pp. 507–520. The Wildlife Society, Washington, DC. 722 pp.

Cochran, W.W., Warner, D.W., Tester, J.R. and Kuechle, V.B. (1965). Automatic radio-tracking system for monitoring animal movements. *BioScience* **15**(2):98–100.

Cohen, J.A. (1960). A coefficient of agreement for nominal scales. *Educ. Psychol. Measure.* **20**:37–46.

Cohen, J.A. (1988). *Statistical Power Analysis for the Behavioral Sciences*. 2nd edn. Lawrence Erlbaum Assoc. Hillsdale, NJ. 567pp.

Cole, L.C. (1949). The measurement of interspecific association. *Ecology* **30**(4):411–424.

Colgan, P. (1978). Modeling. In *Quantitative Ethology*, ed. P. Colgan, pp. 313–326. John Wiley, New York. 364 pp.

Collias, N.E. (1962). Social development in birds and mammals. *Roots of Behavior*, ed. E.L. Bliss, pp. 264–273.. Hafner Publishing Co., New York. 339 pp.

Collins, D.A. (1984). Spatial patterns in a troop of yellow baboons (*Papio cynocephalus*) in Tanzania. *Anim. Behav.* **32**(2):536–553.

Comrey, A.L. (1973). *A First Course in Factor Analysis*. Academic Press, New York. 316 pp.

Conger, R.D. and McLeod, D. (1977). Describing behavior in small groups with the Datamyte event recorder. *Behav. Res. Method. Instrument.* **9**(5):418–424.

Conner, W.E. and Masters, W.M. (1978). Infrared video viewing. *Science* **199**:1004.

Connor, R.C. and Smolker, R.S. (1985). Habituated dolphins (*Tursiops sp.*) in Western Australia. *J. Mammal.* **66**(2):398–400.

Cooley, W.W. and Lohnes, P.R. (1971). *Multivariate Data Analysis*. John Wiley, New York. 364 pp.

Cooper, H.M. and Charles-Dominique, P. (1985). A microcomputer data acquisition system: a study of activity in the bat. *J. Wildl. Manage.* **49**(4):850–854.

Coughlin, D.J., Strickler, J.R. and Sanderson, B. (1992). Swimming and search behaviour in clownfish, *Amphiprion perideeraion*, larvae. *Anim. Behav.* **44**(3):427–440.

Coulson, J.C. and Wooller, R.D. (1984). Incubation under natural conditions in the Kittiwake gull, *Rissa tridactyla*. *Anim. Behav.* **32**(4):1204–1215.

Crawford, J.D. (1984). Orientation in a vertical plane: the use of light cues by an orb-weaving spider, *Araneus diadematus* Clerk. *Anim. Behav.* **32**(1):162–171.

Cresswell, W. (1993). Escape responses by redshanks, *Tringa totanus*, on attack by avian predators. *Anim. Behav.* **46**(3):609–611.

Crews, D. (1977). The annotated anole: studies on the control of lizard reproduction. *Am. Sci.* **65**:428–434.

Crocker, D.R. (1981). Anthropomorphism: bad practice, honest prejudice? *New Sci.* **91**(1262):159–162.

Crockett, C.M. (1996). Data collection in the zoo setting, emphasizing behavior. In *Wild Mammals in Captivity*, ed. D.G. Kleiman, M.E. Allen, K.V. Thompson, S. Lumpkin and H. Harris, pp. 545–565. University of Chicago Press, Chicago, IL.

Crook, J.H. and Butterfield, P.A. (1970). Gender role in the social system of Quelea. In *Social Behaviour in Birds and Mammals*, ed. J.H. Crook, pp. 211–248. Academic Press, New York. 492pp.

Crook, J.H., Ellis, J.E. and Goss-Custard, J.D. (1976). Mammalian social systems: structure and function. *Anim. Behav.* **24**(2):261–274.

Croze, H. (1970). Searching image in carrion crows. *Z. Tierpsychol. Beih.* 5. 85 pp.

Crump, M.L. (1988). Aggression in harlequin frogs: male-male competition and a possible conflict of interest between the sexes. *Anim. Behav.* **36**(4):1064–1077.

Curio, E. (1975). The functional organization of anti-predator behaviour in the pied flycatcher: a study of avian visual perception. *Anim. Behav.* **23**(1):1–115.

Cuthill, I. (1991). Field experiments in animal behaviour: methods and ethics. *Anim. Behav.* **42**(6):1007–1014.

Dale, P.S. (1976). *Language Development: Structure and Function*, 2nd edn. Holt, Rinehart and Winston, New York. 358 pp.

Dane, B. and Van der Kloot, W.G. (1964). An analysis of the display of the goldeneye duck (*Bucephala clangula L.*). *Behaviour* **22**:282–328.

Darling, F. (1937). *A Herd of Red Deer*. Oxford University Press, Oxford. 215 pp.

Davey, C.C., Fullagar, P.J. and Kogon, C. (1980). Marking rabbits for individual identification and a use for betalights. *J. Wildl. Manage.* **44**(2):494–497.

Davey, G. (1981). *Animal Learning and Conditioning*. University Park Press, Baltimore. 488 pp.

Davey, G. (1989). *Ecological Learning Theory*. Routledge. London. 392 pp.

Davies, N.B. (1991). Mating systems. In *Behavioural Ecology*, 3rd edn, ed. J.R. Krebs and N.B. Davies, pp.263–294. Blackwell Scientific Publications. London. 482 pp.

Davies, N.B. and Brooke, M. de L. (1988). Cuckoos versus reed warblers: adaptations and counteradaptations. *Anim. Behav.* **36**(1):262–284.

Davis, D.E. (1964). The hormonal control of aggressive behavior. In *Proc. 13th Int. Ornithol. Congr. The Am. Ornithologists' Union*, ed. C.G. Sibley, pp. 994–1003, American Ornithologists Union, Washington DC.

Davis, E.A. and Hopkins, C.D. (1988). Behavioural analysis of electric signal localization in the electric fish, *Gymnotus carapo* (Gymnotiformes). *Anim. Behav.* **36**(6):1658–1671.

Davis, G.J. and Lussenhop, J.F. (1970). Roosting of starlings (*Sturnus vulgaris*): a function of light and time. *Anim. Behav.* **18**(2):362–365.

Davis, H. and Balfour, D. (Eds.). (1992). *The Inevitable Bond*. Cambridge University Press, Cambridge. 589 pp.

Davis, J.M. (1975). Socially induced flight reaction in pigeons. *Anim. Behav.* **23**(3):597–601.

Davis, R.O. (1986). The Personal Acoustics Lab (PAL): a microcomputer-based system for digital signal acquisition, analysis and synthesis. *Computer Method. and Program. Biomed.* **23**(1986):199–210.

Davis, W.J. (1984). Letter to the editors. *Am. Sci.* **73**:125.

Dawkins, M.S. (1971). Perceptual changes in chicks: another look at the 'search image' concept. *Anim. Behav.* **19**(3):566–574.

Dawkins, M.S. (1980). *Animal Suffering. The Science of Animal Welfare.* Chapman and Hall, London. 149 pp.

Dawkins, M.S. (1986). *Unravelling Animal Behavior.* Longman, Essex, England. 159 pp.

Dawkins, M.S. and Gosling, L.M. (eds.). (1992). *Ethics in Research on Animal Behaviour.* Published for the Association for the Study of Animal Behaviour and The Animal Behavior Society by Academic Press, London. 64 pp.

Dawkins, M.S., Halliday, T.R. and Dawkins, R. (eds.). (1992). *The Tinbergen Legacy.* Chapman and Hall, New York. 146 pp.

Dawkins, R. (1976a). Hierarchical organization: a candidate principle for ethology. In *Growing Points in Ethology*, ed. P.P.G. Bateson and R.A. Hinde, pp. 7–54. Cambridge University Press, London.

Dawkins, R. (1976b). *The Selfish Gene.* Oxford University Press, Oxford. 224 pp.

Dawkins, R. (1984). Replicators, consequences, and displacement activities. *Behav. Brain Sci.* **7**:486–487.

Dawkins, R. and Dawkins, M. (1973). Decisions and the uncertainty of behaviour. *Behaviour* **45**(1–2):83–103.

Dawkins, R. and Dawkins, M. (1976). Hierarchical organization and postural facilitation: rules for grooming in flies. *Anim. Behav.* **24**(4):739–755.

Day, G.I., Schemnitz, S.D. and Taber, R.D. (1980). Capturing and marking wild animals. In *Wildlife Management Techniques*, ed. S.D. Schemnitz, p.61–88. The Wildlife Society, Washington, DC.

DeGhett, V.J. (1978). Hierarchical cluster analysis. In *Quantitative Ethology*, ed. P. Colgan, pp. 115–144, John Wiley, New York. 364 pp.

DeGhett, V.J. (1980). Maternal behavior of the Mongolian gerbil. In *Maternal Influences and Early Behavior*, ed. R.W. Bell and W.P. Smotherman, pp. 57–85. SP Medical & Scientific Books, New York.

Delgado, R.R. and Delgado, J.M.R. (1962). An objective approach to measurement of behavior. *Philos. Sci.* **29**:253–268.

Dellmeier, G.R., Friend, T.H. and Gbur, E.E. (1985). Comparisons of four methods of calf confinement. II. Behavior. *J. Anim. Sci.* **60**(5):1102–1109.

Denenberg, V.H. (1976). *Statistics and Experimental Design for Behavioral and Biological Researchers.* Hemisphere Publ. Co., Washington, DC. 344 pp.

Denes, P.B. and Pinson, E.N. (1973). *The Speech Chain: The Physics and Biology of Spoken Language.* Anchor Press, Garden City, New York. 217 pp.

Deni, R. (1986). *Programming Microcomputers for Psychology Experiments.* Wadsworth Publ. Co., Belmont, CA. 262 pp.

Dethier, V.G. (1962). *To Know a Fly.* Holden-Day, San Francisco. 119 pp.

Dethier, V.G. (1966). Insects and the concept of motivation. *Nebraska Symposium on Motivation* **1966**:105–136.

Dethier, V.G. and Bodenstein, D. (1958). Hunger in the blowfly. *Z. Tierpsychol. Beih.* **15**:129–140.

DeVore, I. and Hall, K.R.L. (1965). Baboon ecology. In *Primate Behavior*, ed. I. De Vore, pp. 20–52. Holt, Rinehart and Winston, New York. 654 pp.

Dewsbury, D.A. (1975). Filming animal behavior. In *Animal Behavior in Laboratory and Field*, 2nd edn, ed. E.D. Price and A.W. Stokes, pp. 13–15. W.H. Freeman Co., San Francisco. 130 pp.

Dewsbury, D.A. (ed.) (1985). *Studying Animal Behavior. Autobiographies of the Founders.*

University of Chicago Press, 512 pp.

Diakow, C. (1975). Motion picture analysis of rat mating behavior. *J. Comp. Physiol. Psychol.* **88**(2):704–712.

Dice, L.R. (1945). Measures of the amount of ecologic association between species. *Ecology* **26**:297–302.

Dilger, W.C. (1962). The behavior of lovebirds. *Sci. Am.* **206**(1):88–98.

Dingle, H. (1969). A statistical and information analysis of aggressive communication in the mantis shrimp *Gonodactylus bredini* Manning. *Anim. Behav.* **17**(3):561–575.

Dingle, H. (1972). Aggressive behavior in somatopods and the use of information theory in the analysis of animal communication. In *Behavior of Marine Animals, vol. 1. Invertebrates*, ed. H.E. Winn and B.L. Olla, pp. 126–156. Plenum Press, New York.

Dixon, K.R. and Chapman, J.A. (1980). Harmonic mean measure of animal activity areas. *Ecology* **61**:1040–1044.

Dixon, W.J. (1954). Power under normality of several nonparametric tests. *Ann. Math. Stat.* **25**:610–614.

Donnelly, K.P. (1978). Simulations to determine the variance and edge effect of total nearest-neighbour distance. In *Simulation Studies in Archaeology*, ed. I. Hodder, pp. 91–95. Cambridge University Press, Cambridge.

Drew, C.J. and Hardman, M.L. (1985). *Designing and Conducting Behavioral Research.* Pergamon Press, New York. 305 pp.

Drickamer, L.C. and Vessey, S.H. (1992). *Animal Behavior.* 3rd edition. Wm. C. Brown Publishers, Dubuque, Iowa. 479 pp.

Drori, D. and Folman, Y. (1967). The sexual behaviour of male rats unmated to 16 months of age. *Anim. Behav.* **15**(1):20–24.

Drummond, H. (1981). The nature and description of behavior patterns. In *Perspectives in Ethology, Vol. IV. Advantages of Diversity*, ed. P.P.G. Bateson and P.H. Klopfer, pp. 1–33. Plenum Press, New York.

Dudzinski, M.L. and Norris, J.M. (1970). Principal component analysis as an aid for studying animal behaviour. *Forma et Functio* **2**:101–109.

Dunbar, R.I.M. (1976). Some aspects of research design and their implications in the observational study of behaviour. *Behaviour* **58**(1–2):78–98.

Duncan, I.J.H. and Wood-Gush, D.G.M. (1972). An analysis of displacement preening in the domestic fowl. *Anim. Behav.* **20**:68–71.

Dunn, J.E. and Gipson, P.S. (1977). Analysis of radio telemetry data in studies of home range. *Biometrics* **33**:85–101.

Dunnett, C.W. (1964). New tables for multiple comparisons with a control. *Biometrics* **20**:482–491.

Dwyer, T.J. (1975). Time budget of breeding gadwalls. *Wilson Bull.* **87**(3):335–343.

Dyson, M.L., Henzi, S.P. and Passmore, N. I. (1994). The effect of changes in the relative timing of signals during female phonotaxis in the reed frog, *Hyperolius marmoratus*. *Anim. Behav.* **48**(3):679–685.

Eales, L.A. (1985). Song learning in zebra finches; some effects of song model availability on what is learnt and when. *Anim. Behav.* **33**(4):1293–1300.

Eden, S.F. (1987). Dispersal and competitive ability in the magpie: an experimental study. *Anim. Behav.* **35**(3):764–772.

Edwards, A.L. (1985). *Experimental Design in Psychological Research*, 5th edn. Harper and

Row, New York. 584 pp.

Ehrenfeld, D.W. and Carr, A. (1967). The role of vision in the sea-finding orientation of the green turtle (*Chelonia mydas*). *Anim. Behav.* **15**(1):25–36.

Eibl-Eibesfeldt, I. (1963). Angeborenes und Erworbenes im Verhalten einiger Sauger. *Z. Tierpsychol.* **20**:705–754.

Eibl-Eibesfeldt, I. (1972). Similarities and differences between cultures in expressive movements. In *Non-verbal Communication*, R.E. Hinde, pp. 297–314. Cambridge University Press, London. 441 pp.

Eigen, M. and Winkler, R. (1981). *Laws of the Game: How Principles of Nature Govern Chance.* Alfred A. Knopf, New York. 347 pp.

Eiriksson, T. (1994). Song duration and female response behaviour in the grasshopper *Omocestus viridulus*. *Anim. Behav.* **47**(3):707–712.

Eiseley, L. (1964). *The Unexpected Universe.* Harcourt, Brace, Jovanovich, New York. 239 pp.

Eisenberg, J.F. (1963). The behavior of heteromyid rodents. *University of California Publ. Zool.* **69**:1–100.

Eisenberg, J.F. (1966). The social organization of mammals. *Handb. Zool.* **10**(7):1–92.

Eisenberg, J.F. (1967). A comparative study in rodent ethology with an emphasis on evolution of social behavior. *Proc. US Natl. Mus.* **122**:1–51.

Eisner, T., Aneshansley, D.J. and Eisner, M. (1988). Ultraviolet viewing with a color television camera. *BioScience* **38**:496–498.

Eisner, T. and Wilson, E.O. (1975). *Animal Behavior: Readings from Scientific American.* W.H. Freeman and Co., San Francisco. 339 pp.

Ekman, J. and Sklepkovych, B. (1994). Conflict of interest between sexes in Siberian jay winter flocks. *Anim. Behav.* **48**(2):485–487.

Elowson, A.M. and Snowdon, C.T. (1994). Pygmy marmosets, *Cebuella pygmaea*, modify vocal structure in response to changed social environment. *Anim. Behav.* **47**(6):1267–1277.

Elwood, R.W. (1991). Ethical implications of studies on infanticide and maternal aggression in rodents. *Anim. Behav.* **42**(5):841–849.

Ely, C.R. (1987). An inexpensive device for recording animal behavior. *Wildl. Soc. Bull.* **15**:264–265.

Emery, L. and Wydoski, R. (1987). *Marking and Tagging Aquatic Animals: An Indexed Bibliography.* US Dept. Int., US Fish and Wildlife Service, Resource Publication 165, Washington, DC. 57 pp.

Emlen, J.T., Jr. (1958). The art of making field notes. *Jack-Pine Warbler* **36**(4):178–181.

Emlen, S.T. (1967). Migratory orientation in the indigo bunting, *Passerina cyanea*. I. Evidence for use of celestial cues. *Auk* **84**:309–342.

Emlen, S.T. (1970). Celestial rotation: its importance in the development of migratory orientation. *Science* **170**:1198–1201.

Emlen, S.T. (1972). An experimental analysis of the parameters of bird song eliciting species recognition. *Behaviour* **41**(1–2):130–171.

Emlen, S.T. and Oring, L.W. (1977). Ecology, sexual selection and the evolution of mating systems. *Science* **197**:215–223.

Emlen, S.T., Wiltschko, W., Demong N., Wiltschko, R. and Bergman, S. (1976). Magnetic direction finding: evidence for its use in migratory indigo buntings. *Science* **193**:505–508.

Enquist, M., Plane, E. and Roed, J. (1985). Aggressive communication in fulmars (*Fulmarus* *glacialis*) competing for food. *Anim. Behav.* **33**(3):1007–1020.

Esch, H. (1967). The evolution of bee language. *Sci. Am.* **216**(4):96–104.

Estep, D.Q. and Bruce, K.E.M. (1981). The concept of rape in non-humans: A critique. *Anim. Behav.* **29**(4):1272–1273.

Estes, R.D. (1967). Predators and scavengers. *Nat. Hist.* **76**(2,3):20–29,38–47.

Ettlinger, D.M.T. (Ed.). (1974). *Natural History Photography.* Academic Press, New York. 389 pp.

Evans, H.E. (1957). *Studies on the Comparative Ethology of Digger Wasps of the Genus Bembix.* Comstock Publ. Assoc., Ithaca, NY. 248 pp.

Everitt, B.S. (1977). Cluster analysis and miscellaneous techniques. In *Analysis in Behavioural Research*, ed. A.E. Maxwell, pp. 136–152. Chapman and Hall, London. 164 pp.

Ewert, J.P. (1980). *Neuroethology: An Introduction to the Neurophysiological Fundamentals of Behavior.* Springer-Verlag. New York. 342 pp.

Ewert, J.P. (1987). Neuroethology of releasing mechanisms: prey catching in toads. *Behav. Brain Sci.* **10**:337–405.

Fagen, R.M. (1977). Quantitative ethology: new results in catalog and sequence analysis (unpubl.). Paper presented at Animal Behavior Society Meeting, Penn State University, University Park, Pennsylvania, June 5–10, 1977.

Fagen, R.M. and Goldman, R.N. (1977). Behavioural catalogue analysis methods. *Anim. Behav.* **25**(2):261–274.

Fagen, R.M. and Young, D.Y. (1978). Temporal patterns of behaviors: durations, intervals, latencies, and sequences. In *Quantitative Ethology*, ed. P.W. Colgan, pp. 79–114. John Wiley and Sons, New York. 364 pp.

Fager, E.W. (1957). Determination and analysis of recurrent groups. *Ecology* **38**:586–595.

Fagerstone, K.A. and Johns, B.E. 1987. Transponders as permanent identification markers for domestic ferrets, black-footed ferrets, and other wildlife. *J. Wildl. Manage.* **51**(2):294–297.

Falls, J.B. and Brooks, R.J. (1975). Individual recognition by song in white-throated sparrows. II. Effects of location. *Can. J. Zool.* **53**:1412–1420.

Fantino, E. and Logan, C.A. (1979). *The Experimental Analysis of Behavior.* W.H. Freeman and Co., San Francisco. 559 pp.

Farr, J. and Herrnkind, W.F. (1974). A quantitative analysis of social interaction of the guppy, *Poecilia reticulata* (Pisces: Poecilidiae) as a function of population density. *Anim. Behav.* **22**(3):582–591.

Federer, W.T. (1955). *Experimental Design: Theory and Application.* MacMillan Co., New York. 544 pp.

Feldman, S.E. and Kluger, E. (1963). Short cut calculation of the Fisher–Yates 'exact test'. *Psychometrika* **28**:289–291.

Fenner, P. and Armstrong, M.C. (1981). *Research: A Practical Guide to Finding Information.* William Kaufmann, Los Altos, CA 174 pp.

Fenton, M.B. (1970). A technique for monitoring bat activity with results obtained from different environments in southern Ontario. *Can. J. Zool.* **48**:847–851.

Fenton, M.B., Jacobson, S.L. and Stone, R.N. (1973). An automatic ultrasonic sensing system for monitoring activity of some bats. *Can.J. Zool.* **51**:291–299.

Fentress, J.C. (1990). The categorization of behavior. In *Interpretation and Explanation in the Study of Animal Behavior. Vol. I. Interpretation, Intentionality, and Communcation*, ed. M. Bekoff and D. Jamieson, pp. 7–34. Westview Press, Boulder, CO. 505 pp.

Fentress, J.C. and Stillwell, F.P. (1973). Grammar of movement sequence in inbred mice. *Nature* **244**:52–53.

Ferster, C.B. and Skinner, B.F. (1957). *Schedules of Reinforcement.* Appleton–Century–Crofts, New York. 739 pp.

Festa-Bianchet, M. (1991). The social system of bighorn sheep: grouping patterns, kinship and female dominance rank. *Anim. Behav.* **42**(1):71–82.

Field, R. (1976). Application of a digitizer for measuring sound spectrograms. *Behav. Biol.* **17**:579–583.

Fienberg, S.E. (1972). On the use of Hansen frequencies for estimating rates of behavior. *Primates* **13**:323–326.

Fincke, O.M. (1985). Alternative mate-finding tactics in a non-territorial damselfly (Odonata: Coenagrionidae). *Anim. Behav.* **33**(4):1124–1137.

Finney, D.J. (1960). *An Introduction to the Theory of Experimental Design.* University of Chicago Press, Chicago. 222 pp.

Fisher, J.A. (1990). The myth of anthropomorphism. In *Interpretation and Explanation in the Study of Animal Behavior. Vol. I. Interpretation, Intentionality, and Communication,* ed. M. Bekoff and D. Jamieson, pp. 96–116. Westview Press, Boulder, CO. 505 pp.

Flegg, J.J.M. (1972). *Binoculars, Telescopes & Cameras for the Birdwatcher.* 4th ed. The British Trust for Ornithology, Hertfordshire, England. 40 pp.

Fleishmann, L.J. (1988). Sensory influences on physical design of a visual display. *Anim. Behav.* **36**(5):1420–1424.

Fleiss, J.L. (1981). *Statistical Methods for Rates and Proportions.* John Wiley & Sons, New York. 223 pp.

Fleiss, J.L., Cohen, J. and Everitt, B.S. (1969). Large sample standard errors of kappa and weighted kappa. *Psychol. Bull.* **72**:323–327.

Flowers, J.H. (1982). Some simple Apple II software for the collection and analysis of observational data. *Behav. Res. Method. Instrument.* **14**(2):241–249.

Flowers, J.H. and Leger, D.W. (1982). Personal computers and behavioral observation: an introduction. *Behav. Res. Method. Instrument.* **14**(2):227–230.

Franck, F. (1973). *The Zen of Seeing.* Vintage Books, New York. 130 pp.

Franck, F. (1979). *The Awakened Eye.* Vintage Books, New York. 147 pp.

Fraser, A.F. (1987). *From Animal Behaviour to Animal Bioethics.* Memorial University Press, Newfoundland, Canada. 96 pp.

Fraser, J. and Nelson, M.C. (1984). Communication in the courtship of a Madagascan hissing cockroach. I. Normal courtship. *Anim. Behav.* **32**(1):194–203.

Free, J.B. (1967). Factors determining the collection of pollen by honeybee foragers. *Anim. Behav.* **15**(1):134–144.

Frey, D.F. and Pimentel, R.A. (1978). Principle component analysis and factor analysis. In *Quantitative Ethology,* ed. P. Colgan, pp. 219–245. John Wiley and Sons, New York. 364 pp.

Fruchter, B. (1954). *Introduction to Factor Analysis.* D. van Nostrand Co., New York. 280 pp.

Fryer, T.B., Miller, H.A. and Sandler, H. (Eds.) (1976). *Biotelemetry III. Third International Symposium, Pacific Grove, California.* Academic Press, New York. 381 pp.

Fuiman, L.A. and Webb, P.W. (1988). Ontogeny of routine swimming activity and performance in zebra danios (Teleostei: Cyprinidae). *Anim. Behav.* **36**(1):250–261.

Fujioka, M. (1985). Sibling competition and siblicide in asynchronously-hatching broods of the cattle egret *Bubulcus ibis. Anim. Behav.* **33**(4):1228–1242.

Fuller, J.L. (1967). Effects of the albino gene upon behaviour of mice. *Anim. Behav.* **15**(4):467–470.

Futuyma, D.J. (1986). *Evolutionary Biology.* 2nd edn. Sinauer Associates, Sunderland, MA. 600 pp.

Gadagkar, R. and Joshi, N.V. (1983). Quantitative ethology of social wasps: time-activity budgets and caste differentiation in *Ropalidia marginata* (Lep.) (Hymenoptera: Vespidae). *Anim. Behav.* **31**(1):26–31.

Gaioni, S.J. and Evans, C.S. (1984). The use of rate or period to describe temporally patterned stimuli. *Anim. Behav.* **32**(3):940–941.

Gaito, J. (1959). Non-parametric methods in psychological research. *Psych. Reports* **5**:115–125.

Gans, C. (1978). All animals are interesting! *Am. Zool.* **18**(1):3–9.

Garcia, J. and Koelling, R.A. (1966). Relation of cue to consequence in avoidance learning. *Psychonomic Sci.* **4**:123–124.

Garcia, J., Clarke, J.C. and Hankins, W.G. (1973). Natural responses to scheduled rewards. In *Perspectives in Ethology*, ed. P.P.G. Bateson, and P.H. Klopfer, pp. 1–41. Plenum Press, New York. 336 pp.

Gardner, R.A. and Gardner, B.T. (1988). Feedforward versus feedbackward: An ethological alternative to the law of effect. *Behav. Brain Sci.* **11**:429–493.

Gass, C.L. (1977). A digital encoder for field recording of behavioral, temporal, and spatial information in directly computer-accessible form. *Behav. Res. Method. Instrument.* **9**(1):5–11.

Gass, C.L. (1985). Behavioral foundations of adaptation. In *Perspectives in Ethology*, Vol. 6, ed. P.P.G. Bateson and P.H. Klopfer, pp. 63–107. Plenum Publishing Co., New York.

Gauch, H.G. (1993). Prediction, parsimony and noise. *Am. Sci.* **81**:468–478.

Gaultier, J.P. (1980). Biotelemetry of the vocalizations of a group of monkeys. In *A Handbook of Biotelemetry and Radiotracking*, ed. C.J. Amlaner Jr. and D.W. Macdonald, pp. 535–544. Pergamon Press, Oxford, UK.

Geffen, E. and Macdonald, D.W. (1992). Small size and monogamy: spatial organization of Blanford's foxes, *Vulpes cana. Anim. Behav.* **44**(6):1123–1130.

Geist, V. (1971). *Mountain Sheep: A Study in Behaviour and Evolution.* University of Chicago Press, Chicago. 383 pp.

Gerhardt, H.C., Dyson, M.L., Tanner, S.D. and Murphy, C.G. (1994). Female frogs do not avoid heterospecific calls as they approach conspecific calls: implications for mechanisms of mate choice. *Anim. Behav.* **47**(6):1323–1332.

Gerth, J.M., Lewis, C.M., Stine, W.W. and Maple, T.L. (1982). Evaluation of two computerized data collection devices for research in zoos. *Zoo. Biol.* **1**:65–70.

Getty, T. (1981). Terrestrial behavior of eastern chipmunks (*Tamias striatus*): encounter avoidance and spatial time-sharing. *Ecology* **62**:915–921.

Getz, L.L. (1972). Social structure and aggressive behavior in a population of *Microtus pensylvanicus. J. Mammal.* **53**:310–317.

Ghent, A.W. (1979). Some considerations governing the selection of appropriate statistical procedures. I. Questions of numerical scale and research interest. *The Biologist* **61**(2):59–73.

Ghiselin, M.T. (1973). Darwin and evolutionary psychology. *Science* **179**:964–968.

Gibbons, J.D. (1993). *Nonparametric Statistics: An Introduction.* Sage Publications, London. 87pp.

Gilbert, B.K. and Hailman, J.P. (1966). Uncertainty of leadership-rank in Fallow deer. *Nature* **209**:1041–1042.

Giles, N. and Huntingford, F.A. (1984). Predation risk and inter-population variation in antipredator behaviour in the three-spined stickleback, *Gasterosteus aculeatus* L. *Anim. Behav.* **32**(1):264–275.

Gillingham, M.P. and Bunnell, F.L. (1985). Reliability of motion-sensitive radio collars for estimating activity of black-tailed deer. *J. Wildl. Manage.* **49**(4):951–958.

Ginsberg, J.R. and Young, T.P. (1992). Measuring association between individuals or groups in behavioural studies. *Anim. Behav.* **44**(2):377–379.

Giraldeau, L., Kramer, D.L., Deslandes, I. and Lair, H. (1994). The effect of competitors and distance on central place foraging eastern chipmunks, *Tamias striatus.* *Anim. Behav.* **47**(3):621–632.

Given, M.F. (1993). Male response to female vocalizations in the carpenter frog, *Rana virgatipes. Anim. Behav.* **46**(6):1139–1149.

Glantz, S.A. (1992). *Primer of Biostatistics.* 3rd edn. McGraw-Hill, New York. 440 pp.

Godwin, J. (1994). Behavioural aspects of protandrous sex change in the anemonefish, *Amphiprion melanopus*, and endocrine correlates. *Anim. Behav.* **48**(3):551–567.

Golani, I. (1973). Non-metric analysis of behavioral interaction sequences in captive jackals (*Canis aureus L.*). *Behaviour* **44**(1–2):89–112.

Golani, I. (1976). Homeostatic motor processes in mammalian interactions: a choreography of display. In *Perspectives in Ethology*, Vol. 2, ed. P.P.G. Bateson and P.H. Klopfer, pp. 69–134. Plenum Press, New York. 340 pp.

Goodenough, J., McGuire, B. and Wallace, R.A. (1993). *Perspectives on Animal Behavior.* John Wiley & Sons, New York. 764 pp.

Goodman, L.A. (1968). The analysis of crossclassified data: independence, quasi-independence and interactions in contingency tables with or without missing entries. *J. Am. Stat. Assoc.* **63**:1091–1131.

Gordon, W.A., Foree, D. and Eckerman, D.A. (1983). Using an Apple II computer for real-time control in a behavioral laboratory. *Behav. Res. Method. Instrument.* **15**:158–166.

Goss-Custard, J.D. (1977). Optimal foraging and the size selection of worms by redshank, *Tringa totanus*, in the field. *Anim. Behav.* **25**(1):10–29.

Goss-Custard, J.D. and Sutherland, W.J. (1984). Feeding specializations in oystercatchers. *Anim. Behav.* **32**(1):299–301.

Gotmark, F. and Ahlund, M. (1984). Do field observers attract nest predators and influence nesting success of common eiders? *J. Wildl. Manage.* **48**(2):381–387.

Gottman, J.M. (1978). Nonsequential data analysis techniques in observational research. In *Observing Behavior. Vol. 2. Data Collection and Analysis Methods*, ed. G.P. Sackett, pp. 45–61. University Park Press, Baltimore, MD. 110 pp.

Gottman, J.M. and Bakeman, R. (1979). The sequential analysis of observational data. In *Social Interaction Analysis: Methodological Issues*, ed. M.E. Lamb, S.J. Suomi and G.R. Stephenson, pp. 185–206. University of Wisconsin Press, Madison.

Gottman, J.M. and Notarius, C. (1978). Sequential analysis of observational data using Markov chains. In *Single Subject Research*, ed. T.R. Kratochwill, pp. 237–285. Academic Press, New York. 316 pp.

Gottman, J.M. and Roy, A.K. (1990). *Sequential Analysis: A Guide for Behavioral Researchers.*

Cambridge University Press, Cambridge, England. 275 pp.

Gould, J.L. (1982). *Ethology: The Mechanisms and Evolution of Behavior.* W.W. Norton & Co., New York. 544 pp.

Gould, J.L. and Able, K.P. (1981). Human homing: an elusive phenomenon. *Science* **212**:1061–1063.

Gould, J.L. and. Marler, P. (1987). Learning by instinct. *Sci. Am.* **256**(1):74–85.

Govindarajulu, Z. (1976). A brief survey of non-parametric statistics. *Communications in Statistics – Theory and Methods*, **A5**, 429–453 (2.5).

Grandin, T. (1989). Effect of rearing environment and environmental enrichment on behavior and neural development in young pigs. Doctoral Thesis. University of Illinois, Urbana–Champaign. 201 pp.

Grant, E.C. and Mackintosh, J.H. (1963). A comparison of the social postures of some common laboratory rodents. *Behaviour* **21**:246–259.

Grant, T.R. (1973). Dominance and association among members of a captive and a free-ranging group of grey kangaroos (*Macropus giganteus*). *Anim. Behav.* **21**(3):449–456.

Grant, T.R. (1987). A behavioural study of a beagle bitch and her litter during the first three weeks of lactation. *J. Small Anim. Pract.* **28**(1):992–1003.

Green, J.S., Golightly, R.T. Jr., Lindsey, S. L. and LeaMaster, B.R. (1985). Use of radio transmitter implants in wild canids. *Great Basin Naturalist* **45**(3):567–570.

Greenwood, R.J. and Stewart, A.B. (1973). Influence of radio-packs on captive mallards and blue-winged teal. *J. Wildl. Manage.* **37**:3–9.

Grier, J.W. and Burk, T. (1992). *Biology of Animal Behavior*, 2nd edn. Mosby – Year Book. St. Louis, MO. 890 pp.

Griffin, B. and Adams, R. (1983). A parametric model for estimating prevalence, incidence and mean bout duration from point sampling. *Am. J. Primatol.* **4**:261–271.

Griffin, D.R. (1976). *The Question of Animal Awareness: Evolutionary Continuity of Mental Experience.* The Rockefeller University Press, New York. 135 pp.

Griffin, D.R. (1984a). *Animal Thinking.* Harvard University Press, Cambridge, MA. 237 pp.

Griffin, D.R. (1984b). Animal thinking. *Am. Sci.* **72**:456–464.

Grobecker, D.B. and Pietsch, T.W. (1979). High-speed cinematographic evidence for ultrafast feeding in Antennariid anglerfishes. *Science* **205**: 1161–1162.

Guhl, A.M. (1953). Social behavior of the domestic fowl. Kansas State College, *Agr. Exp. Sta. Tech. Bull 73*. Manhattan, KS.

Gust, D.A. and Gordon, T.P. (1993). Conflict resolution in sooty mangabeys. *Anim. Behav.* **46**(4):685–694.

Gust, D.A. and Gordon, T.P. (1994). The absence of a matrilineally based dominance system in sooty mangabeys, *Cercocebus torquatus atys*. *Anim. Behav.* **47**(3):589–594.

Guthrie, D.M. (1980). *Neuroethology: An Introduction.* J. Wiley. New York. 221 pp.

Guthrie, D.M. (ed.) (1987). *Aims and Methods in Neuroethology.* Manchester University Press, Manchester, England. 310 pp.

Guttman, L. (1966). The nonmetric breakthrough for the behavioral sciences. *Proceedings of the Second National Conference on Data Processing.* Rehovot, Israel.

Guttman, R., Lieblich, I. and Naftali, G. (1969). Variation in activity scores and sequences in two inbred mouse strains, their hybrids, and back crosses. *Anim. Behav.* **17**(2):335–374.

Guynn Jr., D.C., Davis, J.R. and von Recum, A.F. (1987). Pathological potential of intraperitoneal transmitter implants in beavers. *J. Wildl. Manage.* **51**(3):605–606.

Ha, J.C., Lehner, P.N. and Farley, S.D. (1990). Risk-prone foraging behaviour in captive gray jays (*Perisoreus canadensis*). *Anim. Behav.* **39**(1):91–96.

Haccou, P. and Meelis, E. (1992). *Statistical Analysis of Behavioural Data.* Oxford University Press, Oxford. 396pp.

Hailman, J.P. (1967). The ontogeny of an instinct. The pecking response in chicks of the laughing gull (*Larus africilla* L.) and related species. *Behaviour* (suppl. 15). 159 pp.

Hailman, J.P. (1969). How an instinct is learned. *Sci. Am.* **221**(6):98–106.

Hailman, J.P. (1971). The role of stimulus-orientation in eliciting the begging response from newly-hatched chicks of the laughing gull (*Larus atricilla*). *Anim. Behav.* **19**(2):328–335.

Hailman, J.P. (1973). Fieldism. *BioScience* **23**(3):149.

Hailman, J.P. (1975). The scientific method: *modus operandi* or supreme court? *Am. Biol. Teach.* **37**:309–310.

Hailman, J.P. (1977). *Optical Signals: Animal Communication and Light.* Indiana University Press, Bloomington. 362 pp.

Hailman, J.P. and Sustare, B.D. (1973). What a stuffed toy tells a stuffed shirt. *BioScience* **23**(11):644–651.

Hall-Craggs, J. (1979). Sound spectrographic analysis: suggestions for facilitating auditory imagery. *Condor* **81**:185–192.

Halliday, T.R. (1975). An observational and experimental study of sexual behavior in the smooth newt, *Triturus vulgaris* (Amphibia: Salamandridae). *Anim. Behav.* **23**(2):291–322.

Hamilton, W.J., III. (1966). Social aspects of bird orientation mechanisms. In *Animal Orientation and Navigation*, Proc. 27th Ann. Biol. Colloquium, May 6–7, 1966, ed. R.L. Storm, p.57–71. Oregon State University Press, Corvallis. 125 pp.

Hanenkrat, F.T. (1977). *Wildlife Watcher's Handbook.* Winchester Press, New York. 241 pp.

Hansen, E.W. (1966). The development of maternal and infant behavior in the rhesus monkey. *Behaviour* **27**:107–149.

Hargrove, D.S. and Martin, T.A. (1982). Development of a microcomputer system for verbal interaction analysis. *Behav. Res. Method. Instrument.* **14**(2):236–239.

Harris, R.B., Fancy, S.G., Douglas, D.C.,. Garner, G.W, Amstrup, S.C., McCabe, T.R. and Pank, L.F. (1990). *Tracking Wildlife by Satellite: Current Systems and Performance.* U.S. Dept. of the Interior, Fish and Wildlife Service Technical Report No.30. 52 pp.

Harthoorn, A.M. (1976). *The Chemical Capture of Animals: A Guide to the Chemical Restraint of Wild and Captured Animals.* Bailliere Tindall, London. 416 pp.

Hartmann, D.P. (1972). Notes on methodology: 1. On choosing an interobserver reliability measurement. (unpublished manuscript). University of Utah. Salt Lake City.

Harvey, P.H. and Pagel, M.D. (1991). *The Comparative Method in Evolutionary Biology.* Oxford University Press, Oxford. 239 pp.

Hausfater, G. (1975). Dominance and reproduction in baboons (*Papio cynocephalus*). *Contrib. Primatol.* **7**:1–150.

Hausfater, G. (1977). Tail carriage in baboons (*Papio cynocephalus*): relationship to dominance rank and age. *Folia Primatol.* **27**(1):41–59.

Havkin, Z. and Fentress, J.C. (1985). The form of combative strategy in interactions among wolf pups (*Canis lupus*). *Z. Tierpsychol.* **68**:177–200.

Hayne, D.W. (1949). Calculation of size of home range. *J. Mammal* **39**(2):190–206.

Hazlett, B.A. and Bach, C.E. (1977). Predicting behavioral relationships. In *Quantitative Methods in the Study of Animal Behavior*, ed. B.A. Hazlett, p.121–144. Academic Press, New York. 222 pp.

Hazlett, B.A. and Bossert, W.H. (1965). A statistical analysis of the aggressive communication systems of some hermit crabs. *Anim. Behav.* **13**(2,3):357–373.

Heath, R.G.M. (1987). A method for attaching transmitters to penguins. *J. Wildl. Manage.* **51**(2):399–401.

Heat-Moon, W.L. (1991). *Prairyerth.* Houghton Mifflin Co., Boston. 624 pp.

Heiligenberg, W. (1965). A quantitative analysis of digging movements and their relationship to aggressive behaviour in Cichlids. *Anim. Behav.* **13**:163–170.

Heimstra, N.W. and Davis, R.T. (1962). A simple recording system for the direct observation technique. *Anim. Behav.* **10**:202–210.

Heinrich, B. (1971). The effect of leaf geometry on the feeding behaviour of the caterpillar of *Manduca sexta* (Sphingidae). *Anim. Behav.* **19**(1):119–124.

Heinroth, O. (1911). Beitrage zur Biologie, namentlich Ethologie und Psychologie der Anatiden. In: *Proc. V. Int. Ornith. Congr. Berlin, 1910*, pp. 598–702. .

Henderson, R.W. (1988). EVENTOLOG: A tool for observational research. *Acad. Computing* **2**(6):36,47.

Hensler, G. L., Klugman, S.S. and Fuller, M.R. (1986). Portable microcomputers for field collection of animal behavior data. *Wildl. Soc. Bull.* **14**:189–192.

Heppner, F. (1965). Sensory mechanisms and environmental clues used by the American robin in locating earthworms. *Condor* **67**(3):247–256.

Herzog, P.W. (1979). Effects of radio-marking on behavior, movements, and survival of spruce grouse. *J. Wildl. Manage.* **43**(2):316–323.

Hess, E.H. (1962). Imprinting and the 'critical period' concept. In *Roots of Behavior*, ed. E.L. Bliss. pp. 254–263. Haffner, New York.

Hess, E.H. (1972). 'Imprinting' in a natural laboratory. *Sci. Am.* **227**(2):24–31.

Hews, D.K. (1988). Alarm response in larval western toads, *Bufo boreas*: release of larval chemicals by a natural predator and its effect on predator capture efficiency. *Anim. Behav.* **36**(1):125–133.

Hildebrand, M. (1965). Symmetrical gaits of horses. *Science* **150**:701–709.

Hildebrand, M. (1977). Analysis of asymmetrical gaits. *J. Mammal.* **58**(2):131–156.

Hile, M.G. (1991). Hand-held behavioral observations: The Observer. *Behav. Assessment* **13**:187–196.

Hill, S.B. and Clayton, D.H. (1985). *Wildlife After Dark: A Review of Nocturnal Observation Techniques.* Occasional Paper No. 17, James Ford Bell Museum of Natural History. University of Minnesota. Minneapolis. 23 pp.

Hinde, R.A. (1954). Changes in responsiveness to a constant stimulus. *Brit. J. Anim. Behav.* **2**(1):41–55.

Hinde, R.A. (1966). *Animal Behaviour: A Synthesis of Ethology and Comparative Psychology.* McGraw-Hill Book Co., New York. 534 pp.

Hinde, R.A. (1970). *Animal Behaviour: A Synthesis of Ethology and Comparative Psychology.* 2nd edn. McGraw-Hill Book Co., New York. 876 pp.

Hinde, R.A. (1973). On the design of checksheets. *Primates* **14**:393–406.

Hinde, R.A. (1975). The concept of function. In *Function and Evolution in Behaviour*, ed. G.P. Baerends, C. Beer and A. Manning, pp. 3–15. Clarendon Press, Oxford. 393pp.

Hinde, R.A. (1982). *Ethology: Its Nature and Relations with other Sciences.* Oxford University Press, Oxford. 320 pp.

Hinde, R.A. and Spencer-Booth, Y. (1967). Behaviour of socially living rehsus monkeys in their first two and a half years. *Anim. Behav.* **15**(1):169–196.

Hinde, R.A. and Stevenson-Hinde, J. (eds.). (1973). *Constraints on Learning: Limitations and Predispositions.* Academic Press, New York. 488 pp.

Hinde, R.A. and Stevenson-Hinde, J. (1976). Towards understanding relationships: dynamic stability. In *Growing Points in Ethology*, ed. P.P.G. Bateson and R.A. Hinde, pp.451–479. Cambridge University Press, Cambridge. 548 pp.

Hineline, P.N. (1988). Feeding, forward and backward: Mostly red herrings. *Behav. Brain Sci.* **11**:456–457.

Hoffman, A.A. (1987). A laboratory study of male territoriality in the sibling species *Drosophila melanogaster* and *D. simulans. Anim. Behav.* **35**(3):807–818.

Hoffman, H.S. and Ratner, A.M. (1973). A reinforcement model of imprinting: implications for socialization in monkeys and men. *Psych. Rev.* **80**(6):527–544.

Hofstadter, D.R. (1979). *Godel, Escher, Bach: An Eternal Golden Braid.* Basic Books, New York. 777 pp.

Hogan, J.A. and Boxel, F.V. (1993). Causal factors controlling dustbathing in Burmese red junglefowl: some results and a model. *Anim. Behav.* **46**(4):627–635.

Holekamp, K.E. and Smale, L. (1993). Ontogeny of dominance in free-living spotted hyaenas: juvenile rank relations with other immature individuals. *Anim. Behav.* **46**(3):451–466.

Holldobler, B. and Wilson, E.O. (1990). *The Ants.* Belknap (Harvard University Press). Cambridge, MA. 732 pp.

Hollenbeck, A.R. (1978). Problems of reliability in observational research. In *Observing Behavior, Vol. II. Data Collection and Analysis Methods*, ed. G.P. Sackett, pp. 79–98. University Park Press, Baltimore, MD. 110 pp.

Holm, R.A. (1978). Techniques of recording observational data. . In *Observing Behavior. Vol.II. Data Collection and Analysis Methods*, ed. G.P. Sackett, pp. 99–108. University Park Press, Baltimore, MD. 110 pp.

Hopkins, C.D., Rossetto, M. and Lutjen, A. (1974). A continuous sound spectrum analyzer for animal sounds. *Z. Tierpsychol. Beih.* **34**:313–320.

Horii, Y. (1974). Digital sound spectrograms with simultaneous plotting of intensity and fundamental frequency for speed study. *Behav. Res. Method. Instrument.* **6**(1):55.

Horn, S.W. and Lehner, P.N. (1975). Scotopic sensitivity in the coyote (*Canis latrans*). *J. Comp. Physiol. Psychol.* **89**(9):1070–1076.

Horrocks, J. and Hunte, W. (1983). Maternal rank and offspring rank in vervet monkeys: an appraisal of the mechanisms of rank acquisition. *Anim. Behav.* **31**(3):772–782.

Hotelling, H. (1958). The statistical method and the philosophy of science. *Am. Stat.* **12**(5):9–14.

Hotelling, H. and Pabst, M.R. (1936). Rank correlation and tests of significance involving no assumption of normality. *Ann. Math. Stat.* **7**:29–43.

Howard, R.D. (1988). Sexual selection on male body size and mating behaviour in American toads, *Bufo americanus. Anim. Behav.* **36**(6):1796–1808.

Howell, D.C. (1992). *Statistical Methods in Psychology*, 3rd edn. Duxbury Press, Boston, MA. 693 pp.

Hoyle, G. (1984). The scope of neuroethology. *Behav. Brain Sci.* **7**:367–412.

Huck, U.W. and Price, E.O. (1976). Effect of the post-weaning environment on the climbing behavior of wild and domestic Norway rats. *Anim. Behav.* **24**(2):364–371.

Hughes, B.O. and Duncan, I.J.H. (1988). The notion of ethological 'need', models of motivation and animal welfare. *Anim. Behav.* **36**:1696–1707.

Hunsaker, D. (1962). Ethological isolating mechanisms in the *Sceloperus torguatus* group of lizards. *Evolution* **16**:62–74.

Huntingford, F.A. (1976). The relationship between anti-predator behaviour and aggression among conspecifics in the three-spined stickleback, *Gasterosteus aculeatus. Anim. Behav.* **24**(2):245–260.

Huntingford, F.A. (1982). Do inter- and intraspecific aggression vary in relation to predation pressure in sticklebacks? *Anim. Behav.* **30**(3):909–916.

Huntingford, F.A. (1984). *The Study of Animal Behaviour.* Chapman and Hall, London. 411 pp.

Huntingford, F.A. and Turner, A.K. (1987). *Animal Conflict.* Chapman and Hall. London. 448 pp.

Hurlbert, S.H. (1984). Pseudoreplication and the design of ecological field experiments. *Ecol. Monogr.* **54**(2):187–211.

Hutt, S.J. and Hutt, C. (1970). *Direct Observation and Measurement of Behavior.* Charles C. Thomas. Springfield, IL. 224 pp.

Huxley, J. (1968). *Courtship Habits of the Great Crested Grebe.* Jonathan Cape. London. 98 pp. (Reprint of: The courtship habits of the Great Crested Grebe (*Podiceps cristatus*); with an addition to the theory of natural selection. *Proc. Zool. Soc. London.* 1914, **35**:491–562.)

Ireland, C.R. and Long, S.P. (1984). *Microcomputers in Biology: A Practical Approach.* IRL Press (Oxford University Press), New York 324 pp.

Iverson, I.H. and Lattal, K.A. (eds.). (1991). *Experimental Analysis of Behavior. Parts 1 and 2.* Elsevier, Amsterdam. 724 pp.

Jardine, N. and Sibson, R. (1968). The construction of hierarchic and nonhierarchic classifications. *Comput.* J. **11**:177–184.

Jardine, N. and Sibson R. (1971). *Mathematical Taxonomy.* John Wiley & Sons, New York. 286 pp.

Jarman, P.J. (1974). The social organization of antelope in relation to their ecology. *Behaviour* **48**:215–267.

Jarvi, T. and Bakken, M. (1984). The function of the variation in the breast stripe of the great tit (*Parus major*). *Anim. Behav.* **32**(2):590–596.

Jennrich, R.I. and Turner, F.B. (1969). Measurement of non-circular home range. *J. Theor. Biol.* **22**:227–237.

Jessop, N. (1970). *Biosphere: A Study of Life.* Prentice-Hall. Englewood Cliffs, NJ. 954 pp.

Johnson, C.H. and. Hasting, J.W. 1986. The elusive mechanisms of the circadian clock. *Am. Sci.* **74**:29–37.

Johnston, T.D. (1981). Selective costs and benefits in the evolution of learning. In *Advances in the Study of Behavior*, ed. J.S. Rosenblatt, R.A. Hinde, C. Beer and M.C. Busnel, pp. 65–106. Academic Press, New York.

Johnston, T.D. (1985). Introduction: conceptual issues in the ecological study of learning. In *Issues in the Ecological Study of Learning*, ed. T.D. Johnston and A.T. Pietrewicz, pp. 1–24. Lawrence Erlbaum Associates, Hillsdale, NJ. 451 pp.

Jones, E.N. and Sherman, I.J. (1983). A comparison of meadow vole home ranges derived from grid trapping and radiotelemetry. *J. Wildl. Manage.* **47**(2):558–561.

Kalinoski, R. (1975). Intra- and interspecific aggression in house finches and house sparrows. *Condor* **77**:375–384.

Kallman, H.J. (1986). A commodore 64–based experimental psychology laboratory. *Behav. Res. Method. Instrument. Computers* **18**:222–227.

Kandel, E.R. (1977). *Cellular Basis of Behavior.* W.H. Freeman & Co., San Francisco. 725 pp.

Kaufman, C. and Rosenblum, L.A. (1966). A behavioral taxonomy for *Macaca nemestrina* and *Macaca radiata*: based on longitudinal observation of family groups in the laboratory. *Primates* **8**(2):205–252.

Kaufmann, J.H. (1983). On the definitions and functions of dominance and territoriality. *Biol. Rev.* **58**:1–20.

Kazdin, A.E. (1982). *Single-case Research Designs.* Oxford University Press, Oxford. 368 pp.

Keeton, W.T. (1974). The mystery of pigeon homing. *Sci. Am.* **231**(6):96–98, 101–107.

Kelly, J.G. (1967). Naturalistic observations and theory confirmation: an example. *Hum. Dev.* **20**:212–222.

Kelly, J.G. (1969). Naturalistic observations in contrasting social environments. In *Naturalistic Viewpoints in Psychological Research*, ed. E.P. Willems and H.L. Rausch, pp. 183–199. Holt, Rinehart and Winston, New York. 294pp.

Kendall, M.G. (1948). *Rank Correlation Methods.* Charles Griffin and Co., London. 199 pp.

Kennedy, J.S. (1992). *The New Anthropomorphism.* Cambridge University Press, Cambridge. 208 pp.

Kenward, R.E. (1987). *Wildlife Radio Tagging: Equipment, Field Technique and Data Analysis.* Academic Press, San Diego. 222 pp.

Kerfoot, W.B. (1967). The lunar periodicity of *Sphecodogastra texana*, a nocturnal bee (Hymenoptera: Halictidae). *Anim. Behav.* **15**(4):479–486.

Kerlinger, F.N. (1964). *Foundations of Behavioral Research.* Holt, Rinehart and Winston, New York. 739 pp.

Kerlinger, F.N. (1973). *Foundations of Behavioral Research.* 2nd edn. Holt, Rinehart and Winston, New York. 741 pp.

Kessel, E.L. (1955). The mating activities of balloon flies. *Syst. Zool.* **4**:97–104.

Kieras, D.E. (1981). Effective ways to dispose of unwanted time and money with a laboratory computer. *Behav. Res. Method. Instrument.* **13**:145–148.

King, M.B. and Duvall, D. (1990). Praire rattlesnake seasonal migrations: episodes of movement, vernal foraging and sex differences. *Anim. Behav.* **39**(5):924–935.

King, M.G. (1965). The effect of social context on dominance capacity of domestic hens. *Anim. Behav.* **13**(1):132–133.

Kinsey, K.P. (1976). Social behaviour in confined populations of the Allegheny woodrat, *Neotoma floridana magister*. *Anim. Behav.* **24**(1):181–187.

Kirk, R.E. (1968). *Experimental Design: Procedures for the Behavioral Sciences.* Brooks/Cole Publ. Co., Belmont, CA. 577 pp.

Kleerekoper, H. (1969). *Olfaction in Fishes.* Indiana University Press, Bloomington. 222 pp.

Kleiman, D.G. (1974). Activity rhythms in the giant panda *Ailuropoda melanoleucca*: An example of the use of checksheets for recording behaviour data in zoos. *Int. Zoo Yearbk.* **14**:165–169.

Klein, M.S. (1955). Ionophone on haut-parleur ionique. In *Collogue sur L'acoustigue des Orthopteres*, ed. R.G. Busnel, pp. 46–49. Institut National de la Recherche Agronomique, Paris.

Klingel, H. (1965). Notes on the biology of the plains zebra *Equus guagga boehmi* Matschie. *East Afr. Wildl. J.* **3**:86–88.

Klisz, A.M. (1995). *The Video Source Book*, 2 vols. Gale Research Inc., 835 Penobscot Bldg., Detroit, MI.

Klopfer, P. (1963). Behavioral aspects of habitat selection: the role of early experience. *Wilson Bull.* **75**:15–22.

Knight, R.L., Temple, S.A., Olesen, K.G. and Stearns, C.R. (1985). An inexpensive event recorder for field studies. *Wildl. Soc. Bull.* **13**(3):332–335.

Kodric-Brown, A. (1988). Effects of sex-ratio manipulation on territoriality and spawning success of male pupfish, *Cyprinodon pecosensis. Anim. Behav.* **36**(4):1136–1144.

Koeppl, J.W., Slade, N.A. and Hoffman, R.S. (1975). A bivariate home range model with possible application to ethological data analysis. *J. Mammal.* **56**(1):81–90.

Koeppl, J.W., Slade, N.A., Harris, K.S. and Hoffman, R.S. (1977). A three-dimensional home range model. *J. Mammal.* **58**(2):213–220.

Kolata, G. (1985). Birds, brains and the biology of song. *Science* **85**(12):58–63.

Konishi, M. (1989). Birdsong for neurobiologists. *Neuron* **3**:541–549.

Konishi, M., Takahashi, T., Wagner, H., Sullivan, W. E. and Carr, C.E. (1988). Neurophysiological and anatomical substrates of sound localization in the owl. In *Auditory Function: Neurobiological Bases of Hearing*, ed. G.M. Edelman, W.E. Gall and W.M. Cowan, pp. 721–745. John Wiley & Sons, New York.

Konishi, M., Emlen, S.T., Ricklefs, R.E. and Wingfield, J.C. (1989). Contributions of bird studies to biology. *Science* **246**:465–472.

Korschgen, C.E., Maxon, S.J. and Kuechle, V.B. (1984). Evaluation of implanted radio transmitters in ducks. *J. Wildl. Manage.* **48**(3):982–987.

Kovacs, K.M. (1987). Maternal behaviour and early behavioural ontogeny of harp seals, *Phoca groenlandica. Anim. Behav.* **35**(3):844–855.

Kraemer, H.C. (1979). One–zero sampling in the study of primate behavior. *Primates* **20**(2):237–244.

Kraemer, H.C. and Thiemann, S. (1987). *How Many Subjects? Statistical Power Analysis in Research.* Sage Publications, Newbury Park, England. 120 pp.

Kraemer, H.C., Alexander, B., Clark, C., Busse, C. and Riss, D. (1977). Empirical choice of sampling procedures for optimal research design in the longitudinal study of primate behavior. *Primates* **18**:825–833.

Kramer, G. (1952). Experiments on bird orientation. *Ibis* **94**:265–285.

Kramer, M. and Schmidhammer, J. (1992). The chi-squared statistic in ethology: use and misuse. *Anim. Behav.* **44**(5):833–841.

Kratochwill, T.R. (1978). *Single Subject Research: Strategies for Evaluating Change.* Academic Press, New York. 316pp.

Kratochwill, T.R. and Wetzel, R.J. (1977). Observer agreement, credibility, and judgement: Some considerations in presenting observer agreement data. *J. Appl. Behav. Anal.* **10**(1):133–139.

Krauss, R.M., Morrel-Samuels, P. and Hochberg, J. (1988). VIDEOLOGGER: a computerized multichannel event recorder for analyzing videotapes. *Behav. Res. Meth. & Instrument.* **20**(1):37–40.

Krauth, J. (1988). *Distribution-free Statistics: An Application-oriented Approach.* Elsevier, Amsterdam. 381 pp.

Krebs, H.A. (1975). The August Krogh principle: 'For many problems there is an animal on which it can be most conveniently studied'. *J. Exp. Zool.* **194**:221–226.

Krebs, J.R. (1971). Territory and breeding density in the great tit, *Parus major* L. *Ecology* **52**(1):2–22.

Krebs, J.R. (1974). Colonial nesting and social feeding as strategies for exploiting food resources in the great blue heron (*Ardea herodias*). *Behaviour* **51**:99–134.

Krebs, J.R. (1977). Review of: *The Question of Animal Awareness. Nature* **266**:792.

Krebs, J.R., Kacelnik, A. and Taylor, P. 1978. Test of optimal sampling by foraging great tits. *Nature* **275**:27–31.

Kreithen, M.L. and Keeton, W.T. (1974a). Detection of changes in atmospheric pressure by the homing pigeon. *J. Comp. Physiol.* **89**:73–82.

Kreithen, M.L. and Keeton, W.T. (1974b). Detection of polarized light by the homing pigeon, *Columba livia. J. Comp. Physiol.* **89**:83–92.

Krogh, A. (1929). Progress in physiology. *Am. J. Physiol.* **90**:243–251.

Kroodsma, D.E. (1977). A re-evaluation of song development in the song sparrow. *Anim. Behav.* **25**(2):390–399.

Kruijt, J.P. (1964). *Ontogeny of Social Behaviour in Burmese Red Junglefowl* (Gallus gallus spadiceous). Brill, Leiden. 201 pp.

Kruskal, J.B. (1964). Nonmetric multidimensional scaling: a numerical method. *Psychometrika* **29**:115–129.

Kruuk, H. (1972). *The Spotted Hyena: A Study of Predation and Social Behavior.* University of Chicago Press, Chicago. 335 pp.

Kucera, T.E. (1978). Social behavior and breeding system of the desert mule deer. *J. Mammal.* **59**(3):463–476.

Kummer, H. (1968). *Social Organisation of Hamadryas Baboons. A Field Study.* University of Chicago Press, Chicago; Basil, S.Karger, *Bibl. Primatol.* No. 6. 189 pp.

Kummer, H. (1984). From laboratory to desert and back: A social system of Hamadryas baboons. *Anim. Behav.* **32**(4):965–971.

Kunz, T.H. and Brock, C.E. (1975). A comparison of mist nets and ultrasonic detectors for monitoring flight activity of bats. *J. Mammal.* **56**(4):907–911.

Lack, D.L. (1943). *The Life of the Robin.* H.F. and G. Witherby. London. 224 pp.

Lair, H. (1987). Estimating the location of the focal center in red squirrel home ranges. *Ecology* **68**:1092–1101.

L'Amour, L. (1982). *The Cherokee Trail.* Bantam Books, New York. 178 pp.

Landsman, R.E. (1993). Sex differences in external morphology and electric organ discharges in imported *Gnathonemus petersii* (Mormyriformes). *Anim. Behav.* **46**(3):417–429.

Larkin, R.P. (1977). Reactions of migrating birds to sounds broadcast from the ground (unpubl.). Paper presented Animal Behavior Society Meeting. Penn State University. University Park, PA, June 5–10, 1977.

Larkin, R.P. and Sutherland, P.J. (1977). Migrating birds respond to Project Seafarer's electromagnetic field. *Science* **195**:777–779.

Laundre, J.W., Reynolds, T.D., Knick, S.T. and Ball, I.J. (1987). Accuracy of daily point relocations in assessing real movement of radio-marked animals. *J. Wildl. Manage.* **51**(4):937–940.

Laurenson, M.K. and Caro, T.M. (1994). Monitoring the effects of non-trivial handling in free-living cheetahs. *Anim. Behav.* **47**(3):547–557.

Lawhon, D.K. and Hafner, M.S. (1981). Tactile discriminatory ability and foraging strategies in kangaroo rats and pocket mice (Rodentia: Heteromyidae). *Oecologia* **50**:303–309.

Lawrence, E.S. (1985). Evidence for search image in blackbirds (*Turdus merula* L.): short-term learning. *Anim. Behav.* **33**(3):929–937.

Layne, J.N. (1967). Evidence for the use of vision in diurnal orientation of the bat *Myotis austroriparius. Anim. Behav.* **15**(4):409–415.

Lazo, A. (1994). Social segregation and the maintenance of social stability in a feral cattle population. *Anim. Behav.* **48**(5):1133–1141.

Ledley, R.S. (1965). *Use of computers in Biology and Medicine.* McGraw-Hill, New York. 965 pp.

Lehner, P.N. (1976). Coyote howls and Wynne-Edward's hypothesis revisited (unpubl.). Paper presented at theAnimal Behavior Society Meeting. Boulder, Colorado. June 20–25, 1976.

Lehner, P.N. (1979). *Handbook of Ethological Methods.* Garland STPM Press, New York. 403 pp.

Lehner, P.N. (1982). Differential vocal responses of coyotes to 'Group Howl' and 'Group Yip-Howl' playbacks. *J. Mammal.* **63**(4):675–679.

Lehner, P.N. (1987). Design and execution of animal behavior research: an overview. *J. Anim. Sci.* **65**:1213–1219.

Lehner, P.N. (1992). Sampling methods in behavior research. *Poultry Sci.* **71**(4):643–649.

Lehner, P.N. and Dennis, D.S. (1971). Preliminary research on the ability of ducks to discriminate atmospheric pressure changes. In *Orientation: Sensory Basis*, ed. H.E. Adler, pp. 98–109 . Annals NY Acad. Sci., Vol. 188. 408 pp.

Lehrman, D.S. (1955). The perception of animal behavior. In *Group Processes: Transactions of the First Conference*, ed. B. Schaffner, pp. 259–267. Josiah Macy, Jr. Foundation, New York. 334 pp.

Lemen, C.A. and Freeman, P.W. (1985). Tracking animals with fluorescent pigments: a new technique. *J. Mammal.* **66**(1):134–136.

Lemon, R.E. and Chatfield, C. (1971). Organization of song in cardinals. *Anim. Behav.* **19**(1):1–17.

Leuthold, W. (1977). *African Ungulates: A Comparactive Review of their Ethology and Behavioral Ecology. Zoophysiology and Ecology. Vol. 8.* Springer-Verlag, New York. 307 pp.

Leuze, C.C.K. (1980). The application of radio-tracking and its effects on the behavioural ecology of the water vole, *Arvicola terrestris* (Lacepede). In *A Handbook of Biotelemetry and Radio-tracking*, ed. C.J. Amlaner and D.W. Macdonald, pp. 361–366. Pergamon Press, Oxford.

Lewontin, R.C. (1979). Sociobiology as an adaptionist program. *Behav. Sci.* **24**:5–14.

Lightbody, J.P. and Weatherhead, P.J. (1987). Polygyny in the yellow-headed blackbird: female choice versus male competition. *Anim. Behav.* **35**(6):1670–1684.

Lindauer, M. (1985). Personal recollections of Karl von Frisch. In *Experimental Behavioral Ecology and Sociobiology*, ed. B. Holldobler and M. Lindauer, pp. 5–7. Sinauer Associates, Sunderland, MA 488pp.

Lindzey, G., Thiessen, D.D. and Tucker, A. (1968). Development and hormonal control of territorial marking in the male mongolian gerbil (*Meriones unguiculatus*). *Dev. Psychobiol.* **1**(2):97–99.

Line, S., Markowitz, H. and Carlson, E. (1987). A portable computer/bar code system for recording behavioral observations. Paper presented at the Animal Behavior Society annual meeting. Williams College. Williamstown, MA.

Lingoes, J.C. (1966). An IBM-7090 program for Guttman–Lingoes multidimensional scalogram analysis–I. *Behav. Sci.* **11**:76–78.

Linn, I.J. (1978). Radioactive techniques for small mammal marking. In *Animal Marking: Recognition Marking of Animals in Research*, ed. B. Stonehouse, pp. 177–191.

University Park Press, Baltimore, MD. 224 pp.

Lissaman, P.B.S. and Shollenberger, C.A. (1970). Formation flight of birds. *Science* **168**:1003–1005.

Littlejohn, M.J. and Martin, A.A. (1969). Acoustic interaction between two species of Leptodactylid frogs. *Anim. Behav.* **17**(4):785–791.

Lockard, J.S. (1976). Small interval timer for observational studies. *Behav. Res. Method. Instrument.* **8**(5):478.

Lockie, J.D. (1966). Territory in small carnivores. In *Play, Exploration and Territory in Mammals. Symp. Zool. Soc. London no. 18*, ed. P.A. Jewell and C. Loizos, pp. 143–165. Academic Press, London. 280pp.

Logsdon, T. (1992). *The Navstar Global Positioning System.* Van Nostrand Reinhold. New York. 256 pp.

Long, F.M. (ed.). (1977). *Proceedings of the First International Conference on Wildlife Biotelemetry.* Laramie, Wyoming, July 27–29, 1977. 159 pp.

Lorenz, K. (1935). Companionship in bird life: fellow members of the species as releasers of social behavior. In *Instinctive Behavior*, ed. C.E. Schiller, pp. 83–128. International Universities Press, New York. 328 pp.

Lorenz, K. (1941). Verliechends Bewegungsstudien an Anatinen. *J. Ornithol.* **89** (suppl.):194–294.

Lorenz, K. (1950). The comparative method in studying innate behaviour patterns. *Symp. Soc. Exp. Biol.* **4**:221–268.

Lorenz, K. (1951–1953). Comparative studies on the behavior of Anatinae. *Avic. Mag.* **57**:157–182; **58**:8–17, 61–72, 86–94, 172–184; **59**:24–34, 80–91.

Lorenz, K. (1960a). Methods of approach to the problems of behavior. In *The Harvey Lectures 1958–59*, pp. 60–103. Academic Press, New York.

Lorenz, K. (1960b). Foreward. In *The Herring Gull's World*, revised edn., ed. N. Tinbergen, pp. xi-xii. Harper and Row, New York. 255 pp.

Lorenz, K. (1965). *Evolution and Modification of Behavior.* University of Chicago Press, Chicago. 121 pp.

Lorenz, K. (1974). Analogy as a source of knowledge. *Science* **185**:229–234.

Lorenz, K. (1981). *The Foundations of Ethology.* Springer-Verlag. New York. 380 pp.

Lorenz, K. (1991). *Here am I – Where are You: The Behaviour of the Greylag Goose.* Harcourt Brace Jovanovich. New York. 270 pp.

Losey, G.S., Jr. (1977). The validity of animal models: a test for cleaning symbiosis. *Biol. Behav.* **2**(3):223–238.

Losey, G.S., Jr. (1978). Information theory and communication. In *Quantitative Ethology*, ed. P. Colgan, pp. 43–78. John Wiley, New York. 364 pp.

Losito, M.P., Mirarchi, R.E. and Baldassarre, G.A. (1988). New techniques for time-activity studies of avian flocks in view-restricted habitats. Paper presented at the Animal Behavior Society annual meeting. Missoula, Montana.

Lott, D.F. (1975). Protestations of a field person. *BioScience* **25**(5):328.

Lott, D.F. and Minta, S.C. (1983). Random individual association and social group instability in American bison (*Bison bison*). *Z. Tierpsychol.* **61**:153–172.

Ludwig, J.A. and Reynolds, J.F. (1988). *Statistical Ecology: A Primer on Methods and Computing.* Wiley, New York. 337 pp.

Luria, S.E. (1984). *A Slot Machine, A Broken Test Tube.* Harper and Row, Publishers. New York. 228 pp.

Lyons, D.M., Price, E.O. and Moberg, G.P. (1988). Individual differences in temperament of domestic dairy goats: constancy and change. *Anim. Behav.* **36**:1323–1333.

McBride, G. (1976). The study of social organizations. *Behaviour* **59**(1–2):96–115.

McConway, K. (1992). The number of subjects in animal behavior experiments: is Still right? In *Ethics in Research on Animal Behaviour*, ed. M.S. Dawkins and L.M. Gosling, pp. 35–38. Published for the Association for the Study of Animal Behaviour and The Animal Behavior Society by Academic Press, London. 64 pp.

McFarland, D. (1971). *Feedback Mechanisms in Animal Behavior.* Academic Press, New York. 279 pp.

McFarland, D. (1976). How animal behavior became a science. *New Sci.* **72**(1027): 376,377,379.

McFarland, D. (1985). *Animal Behavior. Psychobiology, Ethology and Evolution.* The Benjamin/Cummings Publishing Co., Menlo Park, CA. 576 pp.

McFarland, D. and Houston, A. (1981). *Quantitative Ethology: The State Space Approach.* Pitman Books, London. 204 pp.

McGregor, P.K. (Ed.) (1992). *Playback and Studies of Animal Communication. NATO Advanced Science Institute Series A*, Vol. 228. Plenum Press, New York. 231 pp.

McGregor, P.K. and Krebs, J.R. (1984). Song learning and deceptive mimicry. *Anim. Behav.* **32**(1):280–287.

McKaye, K.R., Mughogho, D.E. and Stauffer, J.R., Jr. (1994). Sex-role differentiation in feeding and defence of young by a biparental catfish, *Bagrus meridionalis*. *Anim. Behav.* **48**(3):587–596.

McKinney, F. (1975). The evolution of duck displays. In *Function and Evolution in Behavior*, ed. G. Baerends, C. Beer and A. Manning, pp. 331–357. Clarendon Press, Oxford. 393 pp.

McNamara, J. and Houston, A. (1980). The application of statistical decision theory to animal behaviour. *J. Theor. Biol.* **85**:673–690.

McPartland, R.J., Foster, F.G. and Kupfer, D.J. (1976). A computer-compatible multichannel event counting and digital recording system. *Behav. Res. Method. Instrument.* **8**(3):299–301.

McPherson, J.M. (1988). Preferences of cedar waxwings in the laboratory for fruit species, colour and size: a comparison with field observations. *Anim. Behav.* **36**(4):961–969.

MacArthur, R.H. (1958). Population ecology of some warblers of northeastern coniferous forests. *Ecology* **39**:599–619.

MacCracken, J.G. , Steigers, W.D. Jr., Helm, D. and Mayer, P.V. (1984). Evaluation of an electronic data-collection device. *Wildl. Soc. Bull.* **12**(2):189–193.

Macdonald, D. and Amlaner, C. (1981). Listening to wildlife. *New Sci.* **89**(1241):466–469.

Machlis, L. (1977). An analysis of the temporal patterning of pecking in chicks. *Behaviour* **63**(1–2):1–70.

Machlis, L., Dood, P.W.D. and Fentress, J.C. (1985). The pooling fallacy: problems arising when individuals contribute more than one observation to the data set. *Z. Tierpsychol.* **68**:201–214.

Mackay, R.S. (1993). *Bio-medical Telemetry: Sensing and Transmitting Biological Information from Animals and Man.* IEEE Press, New York. 540 pp.

Mackintosh, N.J. (1983). General principles of learning. In *Genes, Development and Learning*, ed. T.R. Halliday and P.J.B. Slater, pp. 149–177. W.H. Freeman and Co., New York. 246 pp.

Malafant, K.W.J. and Tweedie, R.L. (1982). Computer production of kinetograms. *Appl. Anim. Ethol.* **8**:179–187.

Mangel, M. and Clark, C.W. (1988). *Dynamic Modeling in Behavioral Ecology.* Princeton University Press, Princeton, NJ. 308 pp.

Mankovich, N.J. and Banks, E.M. (1982). An analysis of social interaction and the use of space in a flock of domestic fowl. *Appl. Anim. Ethol.* **9**:177–193.

Manly, B.F.J. (1986). *Multivariate Statistical Methods: A Primer.* Chapman and Hall, London, 159 pp.

Mann, H.B. and Whitney, D.R. (1947). On a test of whether one of two random variables is stochastically larger than the other. *Ann. Math. Stat.* **18**:52–54.

Manning, A. and Dawkins, M.S. (1992). *An Introduction to Animal Behaviour.* 4th edn. Cambridge University Press, Cambridge. 196 pp.

Marchington, J. and Clay, A. (1974). *An Introduction to Bird and Wildlife Photography.* Faber & Faber, London. 149 pp.

Marion, W.R. and Shamis, J.D. (1977). An annotated bibliography of bird marking techniques. *Bird-Banding* **48**(1):42–61.

Marler, P. (1975). Observation and description of behavior. In *Animal Behavior in Laboratory and Field*, 2nd edn., ed. E.O. Price and A.W. Stokes, pp. 2–4. W.H. Freeman and Co., San Francisco. 130 pp.

Marler, P. and Hamilton, W.J. III. (1966). *Mechanisms of Animal Behavior.* John Wiley & Sons, Inc., New York. 771 pp.

Marler, P. and Isaac, D. (1960). Physical analysis of a simple bird song as exemplified by the chipping sparrow. *Condor* **62**:124–135.

Marler, P. and Peters, S. (1977). Selective vocal learning in a sparrow. *Science* **198**:519–521.

Marques, D.M. and Valenstein, E.S. (1977). Individual differences in aggressiveness of female hamsters: response to intact and castrated males and females. *Anim. Behav.* **25**(1):131–139.

Marsden, H.M. (1968). Agnostic behaviour of young rhesus monkeys after changes induced in social rank of their mothers. *Anim. Behav.* **16**(1):38–44.

Marshall, J.C. (1965). The syntax of reproductive behaviour in the male pigeon. Medical Research Council Psycholinguistics Unit Report, Oxford.

Marshall, J.T. (1977). Audiospectrograms with pitch scale: a universal 'language' for presenting bird songs graphically. *Auk* **94**:150–152.

Martin, P. and Bateson, P. (1986). *Measuring Behaviour: An Introductory Guide.* Cambridge University Press, Cambridge. 200 pp.

Martin, P. and Bateson, P. (1993). *Measuring Behaviour: An Introductory Guide*, 2nd edn. Cambridge University Press, Cambridge. 222 pp.

Marzluff, J.M. (1988). Vocal recognition of mates by breeding pinyon jays, *Gymnorhinus cyanocephalus. Anim. Behav.* **36**(1):296–298.

Mason, W.A. (1960). The effects of social restriction on the behaviour of rhesus monkeys. *J. Comp. Physiol. Psychol.* **53**:582–589.

Mason, W.A. (1968). Naturalistic and experimental investigations of the social behavior of monkeys and apes. In *Primates: Studies in Adaptation and Variability*, ed. P.C. Jay, pp. 398–419. Holt, Rinehart and Winston, New York. 529 pp.

Massey, A. (1988). Sexual interactions in red-spotted newt populations. *Anim. Behav.* **36**(1):205–210.

Matthews, G.V.T. (1951). The experimental investigation of navigation in homing pigeons. *J.*

Exp. Biol. **28**:508–536.

Matthews, G.V.T. and Cook, W.A. (1977). The role of landscape features in the 'nonsense' orientation of the mallard. *Anim. Behav.* **25**(2):508–517.

Matthews, L.R. and Ladewig, J. (1994). Environmental requirements of pigs measured by behavioral demand functions. *Anim. Behav.* **47**(3):713–719.

Matzkin, M.A. (1975). *Super 8mm movie making.* Amphoto, Garden City, New York. 96 pp.

Maurus, M. and Pruscha, H. (1973). Classification of social signals in squirrel monkeys by means of cluster analysis. *Behaviour* **47**:106–128.

Maxim, P.E. (1976). An interval scale for studying and quantifying social relations in pairs of rhesus monkeys. *J. Exp. Psychol.: General* **105**:123–147.

Maxwell, A.E. (ed.). (1977). *Multivariate Analysis in Behavioural Research.* Chapman and Hall, London. 164 pp.

May, B., Moody, D.B. and Stebbins, W.C. (1988). The significant features of Japanese macaque coo sounds: a psychophysical study. *Anim. Behav.* **36**(5):1432–1444.

Maynard Smith, J. (1982). *Evolution and the Theory of Games.* Cambridge University Press, Cambridge. 224 pp.

Maynard Smith, J. (1984). Science and myth. *Nat. Hist.* **93**(11):11–12,14,17–18,20,22,24.

Mayr, E. (1974). Behavior programs and evolutionary strategies. *Am. Sci.* **62**:650–659.

Mech, L.D. (1983). *Handbook of Animal Radio-tracking.* University of Minnesota Press, Minneapolis, MN. 107 pp.

Mech, L.D., Heezen, K.L. and Siniff, D.B. (1966). Onset and cessation of activity in cottontail rabbits and showshoe hares in relation to sunset and sunrise. *Anim. Behav.* **14**(4):410–413.

Mech, L.D., Chapman, R.C.. Cochran, W.W, Simmons, L. and Seal, U.S. (1984). Radio-triggered anesthetic-dart collar for recapturing large mammals. *Wildl. Soc. Bull.* **12**(1):69–73.

Medawar, P.B. (1960). *Uniqueness of the Individual.* Basic Books, New York. 192 pp.

Medawar, P.B. (1979). *Advice to a Young Scientist.* Harper and Row, New York. 109 pp.

Meddis, R. (1973). *Elementary Analysis of Variance for the Behavioural Sciences.* John Wiley & Sons, New York. 129 pp.

Mendel, F.C. (1985). Use of hands and feet of three-toed sloths (*Bradypus variegatus*) during climbing and terrestrial locomotion. *J. Mammal.* **66**(2):359–366.

Mendl, M. (1988). The effects of litter-size variation on the development of play behaviour in the domestic cat: litters of one and two. *Anim. Behav.* **36**(1):20–34.

Menzel, E.W. Jr. (1969). Naturalistic and experimental approaches to primate behavior. In *Naturalistic Viewpoints in Psychological Research*, ed. E.P. Willems and H.L. Rausch, pp. 78–121. Holt, Rinehart and Winston, New York. 294 pp.

Mertz, D.B. and McCauley, D.E. (1980). The domain of laboratory ecology. *Synthese* **43**:95–110.

Mesce, K.A. (1993). The shell selection behaviour of two closely related hermit crabs. *Anim. Behav.* **45**(4):659–671.

Meserve, P.L. (1977). Three-dimensional home ranges of cricetid rodents. *J. Mammal.* **58**(4):549–558.

Metzgar, L. (1967). An experimental comparison of screech owl predation on resident and transient white-footed mice (*Peromyscus leucopus*). *J. Mammal.* **48**(3):387–391.

Meyer, K. (1989). *How to Shit in the Woods.* Ten Speed Press, Berkeley, CA. 77 pp.

Meyer, M.E. (1964). Discriminative basis for astronavigation in birds. *J. Comp. Physiol. Psychol.* **58**(3):403–406.

Meyer, M.E. (1976). *A Statistical Analysis of Behavior*. Wadsworth Publishing Co., Belmont, CA. 408 pp.

Michener, G.R. (1980). The measurement and interpretation of interaction rates: an example with adult Richardson's ground squirrels. *Biol. of Behav.* **5**:371–384.

Milinski, M. (1984). Competitive resource sharing: an experimental test of a learning rule for ESSs. *Anim. Behav.* **32**(1):233–242.

Miller, D.B. (1985). Methodological issues in the ecological study of learning. In *Issues in the Ecological Study of Learning*, ed. T.D. Johnston and A.T. Pietrewicz, pp. 73–95. Lawrence Erlbaum Assoc., Hillsdale, NJ. 451 pp.

Miller, D.B. (1994). Social context affects the ontogeny of instinctive behaviour. *Anim. Behav.* **48**(3):627–634.

Miller, D.B. and Blaich, C.F. (1984). Caution in the use of period in bioacoustic analysis: reply to Gaioni and Evans. *Anim. Behav.* **32**(3):941–942.

Mitani, J.C., Grether, G.F., Rodman, P.S. and Priatna, D. (1991). Associations among wild oran-utans: sociality, passive aggregations or chance? *Anim. Behav.* **42**(1):33–46.

Mitchell, G. and Clark, D.L. (1968). Long term effects of social isolation in nonsocially adapted rhesus monkeys. *J. Gen. Psychol.* **113**:117–128.

Moffet, M.W. (1990). Dance of the electronic bee. *Nat. Geogr.* **177**(1):134–140.

Mohr, C.O. and Stumpff, W.A. (1966). Comparison of methods for calculating areas of animal activity. *J. Wildl. Manage.* **30**(2):293–304.

Moller, A.P. (1987). Variation in badge size in male house sparrows *Passer domesticus*: evidence for status signalling. *Anim. Behav.* **35**(6):1637–1644.

Montgomery, G.G. and Sunquist, M.E. (1974). Contact-distress calls of young sloths. *J. Mammal.* **55**:211–213.

Mood, A.M. (1954). On the aymptotic efficiency of certain non-parametric two-sample tests. *Ann. Math. Stat.* **25**:514–522.

Moore, F.R. (1977). Geomagnetic disturbance and the orientation of nocturnally migrating birds. *Science* **196**:682–684.

Morgan, B.J.T., Simpson, M.J.A., Haanby, J.P. and Hall-Craggs, J. (1976). Visualizing interaction and sequential data in animal behaviour: theory and application of cluster-analysis methods. *Behaviour* **56**(1–2):1–43.

Morgan, E. and Cordiner, S. (1994). Entrainment of a circa-tidal rhythm in the rock-pool blenny *Lipophrys pholis* by simulated wave action. *Anim. Behav.* **47**(3):663–669.

Moriarity, J.J. and McComb, W.C. (1982). A fiber optics system for tree cavity inspection. *Wildl. Soc. Bull.* **10**(2):173–174.

Morton, T.L. (1993). Gregarious behavior in large mammals: modeling, methodology, and application. Ph.D dissertation. Utah State University. Logan, Utah. 123 pp.

Moss, C. (1975). *Portraits in the Wild*. Houghton Mifflin Co., Boston. 363 pp.

Moss, R. and Watson, A. (1980). Inherent changes in the aggressive behaviour of a fluctuating red grouse *Lagopus scotius* population. *Ardea* **68**:113–119.

Mrosovsky, N. and Shettleworth, S.J. (1975). On the orientation circle of the leatherback turtle, *Dermochelys coriacea*. *Anim. Behav.* **23**(3):568–591.

Muller-Schwarze, D. (1968). Locomotion in animals. In *Animal Behavior in Laboratory and Field*, ed. A.W. Stokes, pp. 13–19. W.H. Freeman and Co., San Francisco. 198 pp.

Multiple authors. (1992). Software multiple review. The Observer: software for behavioral research, version 2.0. *Ethol. Ecol. Evol.* **4**:401–416.

Murchison, C. (1935). The experimental measurement of a social hierarchy in *Gallus domesti-*

cus: IV. Loss of body weight under conditions of mild starvation as a function of social dominance. *J. Gen. Psychol.* **12**:296–312.

Myers, J.P. (1983). Space, time and the pattern of individual associations in a group-living species: sanderlings have no friends. *Behav. Ecol. Sociobiol.* **12**:129–134.

Myrberg, A.A., Jr. and Gruber, S.J. (1974). The behavior of the bonnethead shark, *Sphyrna tiburo. Copeia* **1974**:358–374.

National Research Council. (1985). *Guide for the Care and Use of Laboratory Animals. A Report of the Institute of Laboratory Animal Resource Committee on the Care and Use of Laboratory Animals.* NIH publication no. 85–23, US Department of Health and Human Services,Washington, DC.

National Research Council. (1992). *Recognition and Alleviation of Pain and Distress in Laboratory Animals. A report of the Committee on Pain and Distress in Laboratory Animals.* Institute of Laboratory Animal Resources, Commission on Life Science, National Research Council, National Academy Press, Washington, DC.

Nice, M.M. (1937). Studies in the life history of the song sparrow. I. *Trans. Linn. Soc. New York.* **4**:1–247.

Nice, M.M. (1943). Studies in the life history of the song sparrow. II. *Trans. Linn. Soc. New York.* **6**:1–328.

Nicoletto, P.F. (1993). Female sexual response to condition-dependent ornaments in the guppy, *Poecilia reticulata. Anim. Behav.* **46**(3):441–450.

Nie, N.I., Hull, H., Jenkins, J., Steinbrenner, K. and Bent, D. (1975). *Statistical Package for the Social Sciences.* McGraw-Hill Book Co., New York. 675 pp.

Nielsen, E.T. (1958). The method of ethology. In *Proc. 10th Int. Congr. Entomol.*, Vol. 2, ed. E.C. Becker, pp. 563–565. International Congress of Entomology, Ottowa, Canada.

Nisbet, I.C.T. and Drury, W.H. (1968). Short-term effects of weather on bird migration: a field study using multivariate statistics. *Anim. Behav.* **16**(4):496–530.

Nisbett, A. (1977). *Konrad Lorenz: A Biography.* Harcourt Brace Jovanovich, New York. 240 pp.

Nolan, V. (1978). The ecology and behavior of the praire warbler, *Dendroica discolor. Ornith. Monograph.* 26. American Ornithologists' Union, Washington, DC. 595pp.

Noldus, L.P.J.J. (1991). The Observer: a software system for collection and analysis of observational data. *Behav. Res. Method., Instrument. Computers* **23**(3):415–429.

Noldus, L.P.J.J., van de Loo, E.L.H.M. and Timmers, P.H.A. (1989). Computers in behavioral research. *Nature* **341**:767–768.

Norman, G.R. and Streiner, D.L. (1986). *PDQ Statistics.* B.C. Decker, Inc., Toronto, Canada. 172 pp.

Nottebohm, F. (1989). From bird song to neurogenesis. *Sci. Am.* **260**(2):74–79.

Notterman, J.M. (1973). Discussion of 'On-line computers in the animal laboratory'. *Behav. Res. Method. Instrument.* **5**(2):129–131.

Nyby, J., Wysocki, C., Whitney, G. and Dizinno, G. (1977). Pheromonal regulation of male mouse ultrasonic courtship (*Mus musculus*). *Anim. Behav.* **25**:33–341.

Oberhauser, K.S. (1988). Male monarch butterfly spermatophore mass and mating strategies. *Anim. Behav.* **36**(5):1384–1388.

Obin, M.S. and Vander Meer, R.K. (1988). Sources of nestmate recognition cues in the imported fire ant *Solenopsis invicta* Buren (Hymenoptera:Formicidae). *Anim. Behav.* **36**(5):1361–1370.

O'Connell, S.M. and Cowlishaw, G. (1994). Infanticide avoidance, sperm competition and mate choice: the function of copulation calls in female baboons. *Anim. Behav.* **48**(3):687–694.

O'Dell, J.W. and Jackson, D.E. (1986). Notes – The Commodore 64 and an interface system for controlling operant chambers. *Behav. Res. Method., Instrument. Computers* **18**:339–341.

Oden, N. (1977). Partitioning dependence in nonstationary behavioral sequences. In *Quantitative Methods in the Study of Animal Behavior*, ed. B.A. Hazlett, pp. 203–220. Academic Press, New York. 222 pp.

Odum, E.P. and Kuenzler, E.J. (1955). Measurement of territory and home range size in birds. *Auk* **72**:128–137.

Orcutt, F.S. (1967). Oestrogen stimulation of nest material preparation in the peach-faced lovebird (*Agapornis roseicollis*). *Anim. Behav.* **15**(4):471–478.

Orford, H.J.L., Perrin, M.R. and Berry, H.H. (1989). Contraception, reproduction and demography of free-ranging Etosha lions (*Panthera leo*). *J. Zool. Lond.* **216**:717–734.

Overall, J.E. and Free, S.M. (1972). Multidimensional scaling based on a subset of objects or variables. *Psychometric Laboratory Report, No. 30*. University Texas Medical Branch of Galveston.

Overall, J.E. and Klett, C.J. (1972). *Applied Multivariate Analysis*. McGraw-Hill Book Co., New York. 500 pp.

Owen-Smith, R.N. (1974). The social system of the white rhinoceros. In *The Behaviour of Ungulates and its Relation to Management, Vol. 1*, ed. V. Geist and F. Walther, pp. 341–351. IUCN Publ. 24. Morges, Switzerland. 511 pp.

Packer, C. (1994). *Into Africa*. University of Chicago Press, Chicago. 277 pp.

Paige, K.N., Mink, L.A. and McDaniel, V.R. (1985). A broadband ultrasonic field detector for monitoring bat cries. *J. Wildl. Manage.* **49**(1):11–13.

Palmer, R.S. (1962). Handbook of North American Birds. 5 vols. Yale University Press, New Haven, CN.

Parker, J.W. (1972). A mirror and pole device for examining high nests. *Bird Banding* **43**(3):216–218.

Parker, P.G., Waite, T.A. and Decker, M.D. (1995). Kinship and association in communally roosting black vultures. *Anim. Behav.* **49**(2):395–401.

Parker, R.E. (1979). *Introductory Statistics for Biology*, 2nd edn. University Park Press, Baltimore, MD. 122 pp.

Partridge, L. (1983). Genetics and behaviour. In *Genes, Development and Behaviour*, ed. T.R. Halliday and P.J.B. Slater, pp. 11–51. W.H. Freeman & Co., San Francisco. 246 pp.

Patterson, I.J. (1977). Aggression and dominance in winter flocks of shelduck *Tadorna tadorna* (L.). *Anim. Behav.* **25**(2):447–459.

Pavlov, I.P. (1927). *Dvadtsatiletnii opyt obektivnogo izucheniia vysshei nervnoi deiatelnosti zhivotnykh*. Oxford University Press, London. (Translated and edited by Anrep, G.V. 1960. *Conditioned Reflexes. An Investigation of the Physiological Activity of the Cerebral Cortex*. Dover Publications, New York). 430 pp.

Payne, R.B. and Payne, L.L. (1993). Song copying and cultural transmission in indigo buntings. *Anim. Behav.* **46**(6):1045–1065.

Peeke, H.V.S. and Petrinovich, L. (1984). Approaches, constructs, and terminology for the study of response change in the intact organism. . In *Habituation, Sensitization and Behavior*, ed. H.V.S. Peeke and L. Petrinovich, pp. 1–14. Academic Press, New York. 488 pp.

Pengelley, E.T. and Asmundson, S.T. (1971). Annual biological clocks. *Sci. Am.* **224**(4):72–79.

Pennycuick, C.J. (1978). Identification using natural markings. In *Animal Marking: Recognition Marking of Animals in Research*, ed. B. Stonehouse, pp. 147–159. University Park Press, Baltimore, MD. 224 pp.

Pennycuick, C.J. and Rudnai, J.A. (1970). A method of identifying individual lions *Panthera leo* with an analysis of the reliability of identification. *J. Zool. Lond.* **160**:497–508.

Penzhorn, B.L. (1984). A long-term study of social organization and behaviour of Cape Mountain zebras *Equus zebra zebra*. *Z. Tierpsychol.* **64**:97–146.

Perdeck, A.C. (1958). Two types of orientation in migrating starlings, *Sturnus vulgaris* L. and chaffinches, *Fringilla coelebs* L., as revealed by displacement experiments. *Ardea* **46**:1–37.

Perry, M.C. (1981). Abnormal behavior of canvasbacks equipped with radio transmitters. *J. Wildl. Manage.* **45**(3):786–789.

Petko-Seus, P.A., Hastings, B.C., Hammitt, W.E. and Pelton, M.R. (1985). Public attitudes towards collars and ear markers on wildlife. *Wildl. Soc. Bull.* **13**:283–286.

Pettingill, O.S. (1970). *Ornithology in Laboratory and Field*. 4th edn. Burgess Publishing Co., Minneapolis, MN. 524 pp.

Pietrewicz, A.T. and Kamil, A.C. (1977). Visual detection of cryptic prey by blue jays (*Cyanocitta cristata*). *Science* **195**:580–582.

Pietrewicz, A.T. and Kamil, A.C. (1979). Search image formation in the blue jay (*Cyanocitta cristata*). *Science* **204**:1332–1333.

Pimentel, R.A. and Frey, D.F. (1978). Multivariate analysis of variance and discriminant analysis. In *Quantitative Ethology*, ed. P. Colgan pp. 247–274. John Wiley and Sons, New York. 364 pp.

Pitcher, T.J. (1973). The three-dimensional structure of schools in the minnow, *Phoxinus phoxinus* (L.). *Anim. Behav.* **21**(4):673–685.

Plotkin, H.C. (1988). Learning and evolution. In *The Role of Behavior in Evolution*, ed. H.C. Plotkin, pp. 133–164. MIT Press, Cambridge, MA. 198 pp.

Polak, M. (1994). Large-size advantage and assessment of resource holding potential in male *Polistes fuscatus* (F.) (Hymenoptera: Vespidae). *Anim. Behav.* **48**(5):1231–1234.

Poole, J.H. (1989). Announcing intent: the aggressive state of musth in African elephants. *Anim. Behav.* **37**(1):140–152.

Poole, T.B. (Ed.) (1987). *The UFAW Handbook on the Care and Management of Laboratory Animals*, 6th edn. Longman Scientific and Technical, Harlow.

Potash, L.M. (1972). A signal detection problem and possible solution in Japaneze quail (*Coturnix coturnix japonica*). *Anim. Behav.* **20**(1):192–195.

Powers, R. (1994). The medium is the message. *Audubon* **96**(5):78–79.

Poysa, H. (1991). Measuring time budgets with instantaneus sampling: a cautionary note. *Anim. Behav.* **42**(2):317–318.

Poysa, H. (1994). Group foraging, distance to cover and vigilance in the teal, *Anas crecca*. *Anim. Behav.* **48**(4):921–928.

Premack, D. (1965). Reinforcement theory. In *Nebraska Symposium on Motivation*, Vol. 13, ed. D. Levine, pp. 123–180. University of Nebraska Press, Lincoln. 344 pp.

Priede, I.G. and Swift, S.M. (eds.). (1992). *Remote Monitoring and Tracking of Animals*. Horwood (Prentice Hall), Englewood Cliffs, NJ. 708 pp.

Pringle, J.W.S. (1951). On the parallel between learning and evolution. *Behaviour* **3**:174–215.

Prinz, K. and Wiltschko, W. (1992). Migratory orientation of pied flycatchers: interaction of stellar and magnetic information during ontogeny. *Anim. Behav.* **44**(3):539–545.

Public Health Service. (1986). Public Health Service Policy on Humane Care and Use of Laboratory Animals. Washington, DC. US Department of Health and Human Services. Available from: Office for Protection from Research Risks, Building 31, Room, 4B09, NIH, Bethesda, MD 20892.

Pulliam, H.R. and Dunford, C. (1980). *Programmed to Learn: An Essay on the Evolution of Culture.* Columbia University Press, New York. 138 pp.

Rajecki, D.W. (1973). Imprinting in precocial birds: interpretation, evidence, and evaluation. *Psych. Bull.* **79**:48–58.

Ralph, M.R. and Menaker, M. (1988). A mutation of the circadian system in golden hamsters. *Science* **241**:1225–1227.

Ralph, M.R., Foster, R.G., Davis, F.C. and Menaker, M. (1990). Transplanted suprachiasmatic nucleus determines circadian period. *Science* **247**:975–978.

Ralston, S.L. (1977). The social organization of two herds of domestic horses. Unpubl. MS Thesis. Colorado State University. 100 pp.

Ramakka, J.M. (1972). Effects of radio-tagging on breeding behavior of male woodcock. *J. Wildl. Manage.* **36**:1309–1312.

Ramsay, M.A. and Stirling, I. (1986). Long-term effects of drugging and handling free-ranging polar bears. *J. Wildl. Manage.* **50**:619–626.

Randall, J. (1994). Discrimination of foot-drumming signatures by kangaroo rats, *Dipidomys spectabilis. Anim. Behav.* **47**(1):45–54.

Rayfield, F. (1982). Computer technology – experimental control and data acquisition with BASIC in the Apple computer. *Behav. Res. Method. Instrument.* **14**:409–411.

Regelmann, K. (1984). Competitive resource sharing: a simulation model. *Anim. Behav.* **32**(1):226–232.

Reichert, R.J. and Reichert, E. (1961). *Binoculars and Scopes: How to Choose, Use and Photograph through Them.* Chilton Co., Book Div., Philadelphia. 128 pp.

Reid, M.L. (1987). Costliness and reliability in the singing vigour of Ipswich sparrows. *Anim. Behav.* **35**(6):1735–1743.

Remsen, J.V. (1977). On taking field notes. *Am. Birds* **31**:946–953.

Renouf, D. (1989). Sensory function in the harbor seal. *Sci. Am.* **260**(4):90–95.

Reynierse, J.H. (1968). Effects of temperature and temperature change on earthworm locomotor behaviour. *Anim. Behav.* **16**(4):480–484.

Reynierse, J.H. and Toeus, J.N. (1973). An ideal signal generator for time-sampling observation procedures. *Behav. Res. Method. Instrument.* **5**(1):57–58.

Rhine, R.J. and Ender, P.B. (1983). Comparability of methods used in the sampling of primate behavior. *Am. J. Primatol.* **5**:1–15.

Rhine, R.J. and Flanigan, M. (1978). An empirical comparison of one-zero, focal-animal, and instantaneous methods of sampling spontaneous primate social behavior. *Primates* **19**(2):353–361.

Rice, J.O. and Thompson, W.L. (1968). Song development in the indigo bunting. *Anim. Behav.* **16**(4):462–469.

Richards, S.M. (1974). The concept of dominance and methods of assessment. *Anim. Behav.* **22**(4):914–930.

Richardson, J.M.L. (1994). Shoaling in White Cloud Mountain minnows, *Tanichthys albonubes*: effects of predation risk and prey hunger. *Anim. Behav.* **48**(3):727–730.

Ridgway, R. (1912). *Color Standards and Color Nomenclature.* Published by the author. Washington, DC. 43pp.

Riechert, S.E. (1984). Games spiders play. III. Cues underlying context-associated changes in agonistic behaviour. *Anim. Behav.* **32**(1):1–15.

Riechert, S.E. and Hedrick, A.V. (1993). A test for correlation among fitness-linked behavioural traits in the spider *Agelenopsis aperta* (Araneae, Agelenidae). *Anim. Behav.* **46**(4):669–675.

Riley, H.T., Bryant, D.M., Carter, R.E. and Parkin, D.T. (1995). Extra-pair fertilizations and paternity defence in house martins, *Delichon urbica. Anim. Behav.* **49**(2):495–509.

Rioch, D.M. (1967). Discussion of agonistic behavior. In *Social Communication Among Primates*, ed. S.A. Altmann, pp. 115–122. University Chicago Press, Chicago.

Ripley, B.D. (1979). Test of 'randomness' for spatial point patterns. *J. R. Stat. Soc. Ser. B* **41**:368–374.

Ristau, C.A. (1986). Do animals think? In *Animal Intelligence*, ed. R.J. Hoage and L. Goldman, pp. 165–185. Smithsonian Institution Press, Washington. 207 pp.

Ritchie, M.G. and Kyriacou, C.P. (1994). Genetic variability of courtship song in a population of *Drosophila melanogaster. Anim. Behav.* **48**(2):425–434.

Roberts, G. (1994). When to scan: an analysis of predictability in vigilance sequences using autoregression models. *Anim. Behav.* **48**(3):579–585.

Robinson, L.J. (1989). *Outdoor Optics*. Lyons and Burford Publishers, New York. 146 pp.

Robinson, S.R. and Smotherman, W.P. (1987). Environmental determinants of behaviour in the rat fetus. II. The emergence of synchronous movement. *Anim. Behav.* **35**(6):1652–1662.

Robson, C. (1973). *Experiment, Design and Statistics in Psychology*. Penguin Books, Baltimore. 174 pp.

Rodenhouse, N.L. and Best, L.B. (1983). A portable observation tower-blind. *Wildl. Soc. Bull.* **11**(3):293–297.

Rodgers, R.S. and Rosebrugh, R.D. (1979). Computing a grammar for sequences of behavioral acts. *Anim. Behav.* **27**(3):737–749.

Rohlf, F.J. (1968). Stereograms in numericial taxonomy. *Syst. Zool.* **17**:246–255.

Rohlf, F.J. and Sokal, R.R. (1981). *Statistical Tables*. 2nd edn. W.H. Freeman and Co., San Francisco. 219 pp.

Rohwer, S. (1977). Status signaling in Harris sparrows: Some experiments in deception. *Behaviour* **61**(1–2):107–129.

Roper, T.J. (1983). Learning as a biological phenomenon. In *Genes, Development and Learning*, ed. T.R. Halliday and P.J.B. Slater, pp. 178–212. W.H. Freeman and Co., San Francisco. 246 pp.

Rosenberg, A. (1984). Fitness, reinforcement, underlying mechanisms. *Behav. Brain Sci.* **7**:495–496.

Rosenblum, L.A. (1978). The creation of a behavioral taxonomy. In *Observing Behavior: Vol. 2. Data Collection and Analysis Methods*, ed. G.P. Sackett, pp. 15–24. University Park Press, Baltimore. 110 pp.

Rosenblum, L.A., Kaufman, I.C. and Stynes, A.J. (1964). Individual distance in two species of macaque. *Anim. Behav.* **12**(2–3):338–342.

Rosenthal, R. (1976). *Experimenter Effects in Behavioral Research*. Halsted Press, New York. 500 pp.

Rosenthal, R. and Rubin, D.B. (1978). Interpersonal expectancy effects: the first 345 studies. *Behav. Brain Sci.* **1**:377–415.

Rosenthal, R. and Rubin, D.B. (1985). Statistical analysis: summarizing evidence versus establishing facts. *Psych. Bull.* **3**:527–529.

Roth, C.E. (1982). *The Wildlife Observer's Guidebook.* Prentice-Hall, New York. 239 pp.

Rothstein, S.I., Yokel, D.A. and Fleischer, R.C. (1988). The agonistic and sexual functions of vocalizations of male brown-headed cowbirds, *Molothrus ater. Anim. Behav.* **36**(1):73–86.

Royall Jr., W.C., Guarino, J.L. and Bray, O.E. (1974). Effect of color on retention of leg streamers by red-winged blackbirds. *Western Bird Bander* **49**(4):64–65.

Royama, T. (1970). Factors governing the hunting behaviour and selection of seed by the great tit (*Parus major* L.). *J. Anim. Ecol.* **39**:619–668.

Rudnai, J. (1973). *The Social Life of the Lion.* Washington Square East Publishers, Wallingford, PA. 122 pp.

Ruff, R.L. (1969). Telemetered heart rates of free-living Uinta ground squirrels in response to social interactions. Unpublished PhD dissertation. Utah State University. Logan. 71 pp.

Rushen, J. (1984). Should cardinal dominance ranks be assigned? *Anim. Behav.* **32**(3):932–933.

Ryden, H. (1975). *God's Dog.* Coward, McCann and Geoghegan.New York. 288 pp.

Sackett, G.P. (1974). A nonparametric lag sequential analysis for studying dependency among responses in observational scoring systems. Unpublished manuscript. University of Washington.

Sackett, G.P. (1978). Measurement in observational research. In *Observing Behavior. Vol. 2. Data Collection and Analysis Methods,* ed. G.P. Sackett, pp. 25–45. University Park Press, Baltimore, MD. 110 pp.

Sackett, G.P. (1979). The lag sequential analysis of contingency and cyclicity in behavioral interaction research. In *Handbook of Infant Development,* ed. J.D. Osofsky, pp. 623–649. John Wiley & Sons, New York. 954 pp.

Sackett, G.P. (1980). Lag sequential analysis as a data reduction technique in social interaction research. In ed. D.B. Sawin, R.C. Hawkins, L.O. Walker and P.H. Penticuff, *Exceptional Infant. (Vol.4): Psychological Risks in Infant–Environment Transactions,* pp. 300–340. Brunner/Mazel. New York.

Sackett, G.P., Stephenson, E. and Ruppenthal, G.C. (1973). Digital data acquisition systems for observing behavior in laboratory and field settings. *Behav. Res. Method. Instrument.* **5**(4):344–348.

Sackett, G.P., Ruppenthal, G.C. and Clark. J. (1978). Introduction: An overview of methodological and statistical problems in observational research. In *Observing Behaviour. Vol. 2. Data Collection and Analysis Methods,* ed. G.P. Sackett, pp. 1–14. University Park Press, Baltimore, MD. 110 pp.

Sackett, G.P., Holm, R., Crowley, C. and Henkins, A. (1979). A FORTRAN program for lag sequential analysis of contingency and cyclicity in behavioral interaction data. *Behav. Res. Method. Instrument.* **11**(3):366–378.

Sade, D.S. (1966). Ontogeny of social relations in a group of free-ranging rhesus monkeys (*Macaca mulatta* Zimmerman). Unpublished PhD dissertation. University of California, Berkeley.

Sales, G. and Pye, D. (1974). *Ultrasonic Communication by Animals.* Halsted Press, New York. 281 pp.

Samuel, M.D. and Garton, E.O. (1985). Home range: a weighted normal estimate and tests of underlying assumptions. *J. Wild. Manage.* **49**:513–519.

Sanderson, G.C. (1966). The study of mammal movements – a review. *J. Wildl. Manage.*

30:215–235.

Sargent, A.B. (1972). Red fox spatial characteristics in relation to waterfowl predation. *J. Wildl. Manage.* **36**:225–236.

Sauer, E.G.F. (1957). Die Sternenorientierung nachtlich ziehender Grasmucken (*Sylvia africapilla, borin*, and *curruca* L.). *Z. Tierpsychol. Beih.* **14**:29–70.

Savidge, J.A. and Seibert, T.F. (1988). An infrared trigger and camera to identify predators of artificial nests. *J. Wildl. Manage.* **52**(2):291–294.

Schaffner, B. (ed.). (1955). *Group Processes: Transactions of the First Conference.* Josiah Macy, Jr., Foundation, New York. 334 pp.

Schaller, G.B. (1972). *The Serengeti Lion: A Study of Predator–Prey Relations.* University of Chicago Press, Chicago. 480 pp.

Schaller, G.B. (1973). *Golden Shadows, Flying Hooves.* Alfred A. Knopf. New York. 287 pp.

Scheller, R.H. and Axel, R. 1984. How genes control an innate behavior. *Sci. Am.* **250**(3):54–62.

Schemnitz, S.D. (ed.) (1980). *Wildlife Management Techniques Manual.* 4th edn. The Wildlife Society, Washington, DC. 686 pp.

Schemnitz, S.D. and Giles, R.H. Jr. (1980). Instrumentation. In *Wildlife Management Techniques Manual*, 4th. edn., ed. S.D. Schemnitz, pp. 499–504. The Wildlife Society, Washington, DC.

Schladweiler, J.L. and Ball, I.J., Jr. (1968). Telemetry bibliography emphasizing studies of wild animals under natural conditions. *Bell Museum of Natural History Technical Report No. 15.* 31 pp.

Schleidt, W.M. (1973). Tonic communication: continual effects of discrete signs in animal communication systems. *J. Theor. Biol.* **42**:359–386.

Schleidt, W.M. (1982). Stereotyped feature variables are essential constituents of behavior patterns. *Behaviour* **79**(2–4):230–238.

Schleidt, W.M., Yakalis, G., Donnelly, M. and McGarry, J. (1984). A proposal for a standard ethogram, exemplified by an ethogram of the bluebreasted quail (*Coturnix chinensis*). *Z. Tierpsychol.* **64**:193–220.

Schmidt-Koenig, K. (1961). Die sonne als Kompassim Heim-orientierungsytem der Brieftauben. *Z. Tierpsychol. Beih.* **18**:221–224.

Schmitt, J.C. (1977). Factor analysis in BASIC for minicomputers. *Behav. Res. Method. Instrument.* **9**(3):302–304.

Schneirla, T.C. (1950). The relationship between observation and experimentation in the field study of behavior. *Ann. NY Acad. Sci.* **51**(6):1022–1044.

Schulz, J.H. and Ludwig, J.R. (1985). A possible cause of premature loss for deer fawn transmitters. *J. Mammal.* **66**(4):811–812.

Scott, J.P. (Ed.). (1950). Methodology and techniques for the study of animal societies. *Ann. NY Acad. Sci.* **51**(96):1001–1122.

Scott, J.P. (1963). *Animal Behavior.* Anchor Books, Doubleday and Co., Garden City, New York. 331 pp.

Scott, J.P. and Fuller, J.L. (1965). *Genetics and the Social Behavior of the Dog.* University of Chicago Press, Chicago. 468 pp.

Scott, K.G. and Masi, W.S. (1977). Use of the Datamyte in analyzing duration of infant visual behaviors. *Behav. Res. Method. Instrument.* **9**(5):429–433.

Scoville, R. and Gottlieb, G. (1978). The calculation of repetition rate in avian vocalizations. *Anim. Behav.* **26**(3):962–963.

Seal, H.L. (1964). *Multivariate Statistical Analysis for Biologists.* Methuen, London. 209 pp.

Seligman, M.E.P. (1970). On the generality of the laws of learning. *Psychol. Rev.* **77**(5):406–418.

Sharman, M. and Dunbar, R.I.M. (1982). Observer bias in selection of study group in baboon field studies. *Primates* **23**(4):567–573.

Shepard, R.N. (1980). Multidimensional scaling, tree-fitting and clustering. *Science* **210**:390–397.

Shepard, R.N., Romney, A.K. and Nerlove, S. (1972). *Multidimensional Scaling*, Vol. 1. Seminar Press, London. 261 pp.

Sherry, D.F., Krebs, J.R. and Cowie, R.J. (1981). Memory for the location of stored food in marsh tits. *Anim. Behav.* **29**(4):1260–1266.

Shettleworth, S.J. (1972). Constraints on learning. *Adv. Stud. Behav.* **4**:1–68.

Shettleworth, S.J. (1978a). Reinforcement and the organization of behavior in golden hamsters: sunflower seed and nest paper reinforcers. *Anim. Learning and Behav.* **6**(3):352–362.

Shettleworth, S.J. (1978b). Reinforcement and the organization of behavior in golden hamsters: Punishment of three action patterns. *Learning Motiv.* **9**:99–123.

Shettleworth, S.J. (1983). Function and mechanism in learning. In *Biological Factors in Learning*, M.D. Zeiler and P. Harzem, pp. 1–39. John Wiley and Sons, New York. 410 pp.

Short, L.L., Jr. (1970). Bird listing and the field observer. *Calif. Birds* **1**:143–145.

Short, R. and Horn, J. (1984). Some notes on factor analysis of behavioral data. *Behaviour* **90**:203–214.

Sibly, R.M., Nott, H.M.R. and Fletcher, D.J. (1990). Splitting behaviour into bouts. *Anim. Behav.* **39**(1):63–69.

Siegel, S. (1956). *Nonparametric Statistics for the Behavioral Sciences.* McGraw-Hill, New York. 312 pp.

Siegel, S. and Castellan, N.J., Jr. (1988). *Nonparametric Statistics for the Behavioral Sciences*, 2nd edn. McGraw-Hill, New York. 399 pp.

Siegfried, W.R., Frost, P.G.H., Ball, I.J. and McKinney, D.F. (1977). Effects of radio packages on African black ducks. S. *Afr. J. Wildl. Res.* **7**:37–40.

Siever, R. (1968). Science: observation, experimental, historical. *Am. Sci.* **56**(1):70–77.

Sigmund, K. (1993). *Games of Life.* Oxford University Press, Oxford. 244 pp.

Silk, J.B. and Boyd, R. (1984). Response to J. Rushen. *Anim. Behav.* **32**(3):933–934.

Silver, H. and Silver, W.T. (1969). Growth and behavior of the coyote-like canid of Northern New England with observations on canid hybrids. *Wildl. Monogr. No. 17.* The Wildlife Society, Washington DC. 41 pp.

Simmons, J.A. (1971). The sonar receiver of the bat. *Ann. N Y Acad. Sci.* **188**:161–174.

Simon, H.A. (1966). Thinking by computers. In *Mind and Cosmos*, ed. R.G. Colodny. University of Pittsburgh Press, Pittsburgh, PA. 362 pp.

Simpson, M.J.A. and Simpson, A.E. (1977). One–zero and scan methods for sampling behaviour. *Anim. Behav.* **25**(3):726–731.

Simpson, S.J., Simmonds, M.S.J., Wheatley, A.R. and Benays, E.A. (1988). The control of meal termination in the locust. *Anim. Behav.* **36**(4):1216–1227.

Skinner, B.F. (1953). *Science and Human Behavior.* Macmillan, New York. 461 pp.

Skinner, B.F. (1981). Selection by consequences. *Science* **213**:501–504.

Skinner, B.F. (1988). Signs and countersigns. *Behav. Brain Sci.* **11**:466–467.

Slade, N.A. (1976). Analysis of social structure from multiple capture data. *J. Mammal.* **57**(4):790–795.

Slater, L.E. (ed.). (1965). Biotelemetry. *BioScience* **15**(2):79–121.

Slater, P.J.B. (1973). Describing sequences of behavior. In *Perspectives in Ethology*, Vol.1, ed. P.P.G. Bateson and P.H. Klopfer, pp. 131–154. Plenum Press, New York. 336 pp.

Slater, P.J.B. (1974). Bouts and gaps in behaviour of zebra finches, with special reference to preening. *Rev. Comp. Anim.* **8**:47–61.

Slater, P.J.B. (1978). Data collection. In *Quantitative Ethology*. ed. P. Colgan, pp. 7–24. John Wiley and Sons, New York. 364 pp.

Slater, P.J.B. (1981). Individual differences in animal behavior. In *Perspectives in Ethology*, Vol.4, ed. P.P.G. Bateson and P.H. Klopfer, pp. 35–49. Plenum Press, New York. 249 pp.

Slater, P.J.B. and Lester, N.P. (1982). Minimizing errors in splitting behaviour into bouts. *Behaviour* **79**:153–161.

Slater, P.J.B. and Ollason, J.C. (1972). The temporal pattern of behaviour in isolated male zebra finches: Transition analysis. Behaviour **42**:248–269.

Slotow, R., Alcock, J. and Rothstein, S.I. (1993). Social status signalling in white-crowned sparrows: an experimental test of the social control hypothesis. *Anim. Behav.* **46**(5):977–989.

Smirnov, M. (1948). Tables for estimating the goodness of fit of empirical distributions. *Ann. Math. Stat.* **19**:280–281.

Smith, A.T. and Dobson, F.S. (1994). A technique for evaluation of spatial data using asymmetrical weighted overlap values. *Anim. Behav.* **48**(6):1285–1292.

Smith, D.G. (1972). The role of the epaulets in the red-winged blackbird (*Agelaius phoeniceus*) social system. *Behaviour* **41**(3–4):251–268.

Smith, D.G. and Spencer, D.A. (1976). A simple pole and mirror device. *N. Am. Bird Bander* **1**(4):175.

Smith, E.O. and Begeman, M. L. (1980). BOXES: Behavior observation recording and editing system. *Behav. Res. Method. Instrument.* **12**(1):1–7.

Smith, K. (1953). Distribution-free statistical methods and the concpet of power efficiency. In *Research Methods in the Behavioural Sciences*, ed. L. Festinger and D. Katz, pp. 536–577. Dryden, New York. 660 pp.

Smith, N.G. (1967). Visual isolation in gulls. *Sci. Am.* **217**(4):94–102.

Smith, R.J.F. and Hoar, W.S. (1967). The effects of prolactin and testosterone on the parental behaviour of the male stickleback *Gasterosteus aculeatus*. *Anim. Behav.* **15**(2–3):342–352.

Smith, W. and Hale, E.B. (1969). Modification of social rank in the domestic fowl. In *Animal Social Psychology*, ed. R. B. Zajonc, pp. 297–300. John Wiley & Sons, New York. 325 pp.

Smith, W.J. (1968). Message-meaning analyses. In *Animal Communication. Techniques of Study and Results of Research*, ed. T.A. Sebeok, pp. 44–60. University of Indiana Press, Bloomington. 686 pp.

Smithe, F.B. (1972). *Naturalist's Color Guide. Parts I, II.* Amer. Museum of Natural History, New York. 229 pp.

Sneath, P.H. and Sokal, R. (1973). *Numerical Taxonomy*. W.H. Freeman and Co., San Francisco. 573 pp.

Snedecor, G.W. (1946). *Statistical Methods.* The Iowa State College Press, Ames, Iowa. 485 pp.

Sokal, R.R. and Rohlf, F.J. (1962). The comparison of dendrograms by objective methods. *Taxon* **11**:33–40.

Sokal, R.R. and Rohlf, F.J. (1969). *Biometry: The Principles and Practice of Statistics in Biological Research*. W.H. Freeman and Co., San Francisco. 776 pp.

Sokal, R.R. and Rohlf, F.J. (1981). *Biometry: The Principles and Practice of Statistics in Biological Research*. 2nd ed. W.H. Freeman and Co., San Francisco. 859 pp.

Sordahl, T.A. (1986). Evolutionary aspects of avian distraction display: Variation in American avocet and black-necked stilt antipredator behavior. In *Deception: Perspectives on Human and Nonhuman Deceit*, ed. R.W. Mitchell and N.S. Thompson, pp. 87–112. State University of New York Press, Albany, New York. 388 pp.

Sorenson, L.G. (1994). Forced extra-pair copulation and mate guarding in the white-cheeked pintail: timing and trade-offs in an asynchronously breeding duck. *Anim. Behav.* **48**(3):519–533.

Sorensen, R. (1991). Thought experiments. *Am. Sci.* **79**:250–263.

Southern, L.K. and Southern, W.E. (1983). Responses of ring-billed gulls to cannon-netting and wing-tagging. *J. Wildl. Manage.* **47**(1):234–237.

Southern, W.E. (1975). Orientation of gull chicks exposed to Project Sanguine's electromagnetic field. *Science* **189**:143–145.

Southwood, T.R.E. (1966). *Ecological Methods*. Methuen and Co., London. 391 pp.

Sparling, D.W. and Williams, J.D. 1978. Multivariate analyses of avian vocalizations. *J. Theor. Biol.* **74**(4):83–107.

Spence, I. (1978). Multidimensional scaling. In *Quantitative Ethology*, ed. P. Colgan, pp. 175–217. John Wiley and Sons, New York. 364 pp.

Staddon, J.E.R. (1972). A note on the analysis of behavioural sequences in *Columba livia*. *Anim. Behav.* **20**(2):284–292.

Staddon, J.E.R. (1980). *Limits to Action: The Allocation of Individual Behavior*. Academic Press, New York. 308 pp.

Staddon, J.E.R. (1983). *Adaptive Behavior and Learning*. Cambridge University Press, Cambridge. 555 pp.

Stamp Dawkins, M. and Gosling, L.M. (Eds.) (1992). *Ethics in Research on Animal Behaviour. Readings from Animal Behaviour*. Academic Press, London. 64 pp.

Stapanian, M.A., Higgins, J.J. and Smith, C.C. (1982). Statistical tests for visitation patterns on grids. *Ecology* **63**(6):1972–1974.

Steele, R.H. and Partridge, L. (1988). A courtship advantage for small males in *Drosophila subobscura*. *Anim. Behav.* **36**(4):1190–1197.

Stefanski, R.A. (1967). Utilization of the breeding territory in the black-capped chickadee. *Condor* **69**(3):259–267.

Steinberg, J.B. (1977). Information theory as an ethological tool. In *Quantitative Methods in the Study of Animal Behavior*, ed. B. Hazlett, pp. 47–74 . Academic Press, New York. 222 pp.

Stephenson, G.R. and Roberts, T.W. (1977). The SSR System 7: A general encoding system with computerized transcription. *Behav. Res. Method. Instrument.* **9**(3):434–441.

Stevens, S.S. (1946). On the theory of scales of measurement. *Science* **103**:677–680.

Stewart, R.E. and Aldrich, J.W. (1951). Removal and repopulation of breeding birds in a spruce-fir forest community. *Auk* **68**:471–482.

Stickel, L.F. (1954). A comparison of certain methods of measuring ranges of small mammals. *J. Mammal.* **35**(1):1–15.

Still, A.W. (1982). On the number of subjects used in animal behaviour experiments. *Anim.*

Behav. **30**(3):873–880.

Stokes, A.W. (1962). Agonistic behavior among blue tits at a winter feeding station. *Behaviour* **19**:118–138.

Stonehouse, B. (ed.). (1978). *Animal Marking: Recognition Marking of Animals in Research.* University Park Press, Baltimore, MD. 224 pp.

Stout, J.F. and Brass, M.E. (1969). Aggressive communication by *Larus glaucescens.* II. Visual communication. *Behaviour* **34**(1–2):42–52.

Stricklin, W.R., Graves, H.B. and Wilson, L.L. (1977). DISTANGLE: a Fortran program to analyze and simulate spacing behavior of animals. *Behav. Res. Method. Instrument.* **9**(4):367–370.

Studd, M.V. and Robertson, R.J. (1985). Evidence for reliable badges of status in territorial yellow warblers (*Dendroica petechia*). *Anim. Behav.* **33**(4):1102–1113.

Styles, B.C. (1980). The MICRO system. In *Museum Documentation Association Occasional Paper No.4*, ed. J.D. Stewart, pp. 43–46. Museum Documentation Association, Duxford, England.

Suga, N. (1990). Biosonar and neurocomputation in bats. *Sci. Am.* **262**:60–68.

Sullivan, E.E. and Morton, T.L. (1994). Accuracy of observers in determining individual animal locations in artificial-deer groups. Unpublished paper presented at the Animal Behavior Society annual meeting. University of Washington. Seattle, WA.

Sustare, B.D. (1978). Systems diagrams. In *Quantitative Ethology*, ed. P. Colgan, pp. 275–311. John Wiley and Sons, New York. 364 pp.

Sutton, M. (1985). Patterns of spacing in a coral reef fish in two habitats on the Great Barrier Reef. *Anim. Behav.* **33**(4):1332–1337.

Svendsen, G.E. and Armitage, K.B. (1973). Mirror image stimulation applied to field behavior studies. *Ecology* **54**:623–627.

Swaddle, J.P. and Cuthill, I.C. (1994). Preference for symmetric males by female zebra finches. *Nature* **367**:165–166.

Swed, F.S. and Eisenhart, C. (1943). Tables for testing randomness of grouping in a sequence. *Ann. Math. Stat.* **14**:83–86.

Swihart, R.K. (1992). Home range attributes and spatial structure of woodchuck populations. *J. Mammal.* **35**:1–15.

Swihart, R.K. and Slade, N.A. (1985). Influence of sampling interval on estimates of home-range size. *J. Wildl. Manage.* **49**(4):1019–1025.

Syme, G.J. (1974). Competitive orders as measures of social dominance. *Anim. Behav.* **22**(4):931–940.

Symonds, R.J. and Unwin, D.M. (1982). The use of a microcomputer to collect activity data. *Physiol. Entomol.* **7**:91–98.

Tacha, T.C. (1988). *Social Organization of Sandhill Cranes from Midcontinental North-America.* Wildlife Monograph No. 99. 37 pp.

Taillade, M. (1992). Animal tracking by satellite. In *Remote Monitoring and Tracking of Animals*, ed. I.G. Priede and S.M. Swift, pp. 149–160. Horwood (Prentice Hall), Englewood Cliffs, NJ. 708 pp.

Tapp, J. and Walden, T. (1993). PROCORDER: a professional tape control, coding, and analysis system for behavioral research using videotape. *Behav. Res. Method. Instrument. Computers.* **25**(1):53–56.

Tarpy, R. (1982). *Principles of Animal Learning and Behavior.* Scott, Foresman & Co., Glenview, IL 403 pp.

Thomas, G. (1977). The influence of eating and rejecting prey items upon feeding and food searching behaviour in *Gasterosteus aculleatus* L. *Anim. Behav.* **25**(1):52–66.

Thompson, H.R. (1956). Distribution of distance to *n*th neighbour in a population of randomly distributed individuals. *Ecology* **37**:391–394.

Thompson, W.L. (1979). Suggestions for preparing audiospectrograms for publication. *Condor* **81**:220–221.

Thomson, K.S. (1984). Marginalia: The literature of science. *Am. Sci.* **72**:185–187.

Thornhill, R. and Alcock, J. 1983. *The Evolution of Insect Mating Systems.* Harvard University Press, Cambridge, MA. 547 pp.

Thorpe, W.H. (1979). *The Origins and Rise of Ethology.* Praeger Publishers, New York. 174 pp.

Timberlake, W. (1983). The functional organization of appetitive behavior systems and learning. In *Advances in Analysis of Behavior. Vol. 3. Biological Factors in Learning*, ed. M.D. Zeiler and P. Harzem, pp. 177–221. John Wiley, Chichester, England.

Timberlake, W. and Lucas, G. A. (1989). Behavior systems and learning: from misbehavior to general principles. In *Contemporary Learning Theories: Instrumental Conditioning Theory and The Impact of Biological Constraints on Learning*, ed. S.B. Klein and R.R. Mowrer pp. 237–275. Lawrence Erlbaum Assoc., Hillsdale, NJ. 293 pp.

Tinbergen, L. (1960). The natural control of insects in pinewoods. I. Factors influencing the intensity of predation by songbirds. *Arch. Neerl. Zool.* **13**:265–343.

Tinbergen, N. (1950). The hierarchical organization of nervous mechanisms underlying instinctive behaviour. *Symp. Soc. Exp. Biol.* **4**:305–312.

Tinbergen, N. (1951). *The Study of Instinct.* Oxford University Press, New York. 228 pp.

Tinbergen, N. (1953). *Social Behaviour in Animals.* John Wiley, New York. 150 pp.

Tinbergen, N. (1958). *Curious Naturalists.* Basic Books, New York. 301 pp.

Tinbergen, N. (1959). Comparative studies of the behaviour of gulls (Laridae): A progress report. *Behaviour* **15**:1–70.

Tinbergen, N. (1960a). The evolution of behaviour in gulls. *Sci. Am.* **203**(6):118–126, 128, 130.

Tinbergen, N. (1960b). *The Herring Gull's World.* Harper and Row, New York. 255 pp.

Tinbergen, N. (1963). On aims and methods of ethology. *Z. Tierpsychol. Beih.* **20**:410–433.

Tinbergen, N. (1965). *Animal Behavior.* Time-Life. New York. 199 pp.

Tinbergen, N. (1972). *The Animal in its World*, 2 vols. Harvard University Press, Cambridge, MA. 343 pp., 231 pp.

Tinbergen, N. (1974). *Curious Naturalists.* revised edn. Penguin Education, Harmondsworth, Middlesex, England. 269 pp.

Tinbergen, N. and Kruyt, W. (1938). Uber die Orientierung des Bienenwolfes (*Philanthus triangulum* Fabr.) III. Die Bevorzugung bestimmter Wegmarken. *Z. V. Physiol.* **25**:292–334. In *The Animal in its World*, Vol.I , ed. N. Tinbergen, pp. 146–196. Harvard University Press, Cambridge, MA. 343 pp.

Tinbergen, N. and Kuenen, D. J. (1939). Uber die ausloesenden und die richtunggebenden Resizsituationen der Sperrbewegung von jungen Drosseln (*Turdus m. merula* L. and *T. e. ericetorium* Turton). *Z. Tierpsychol. Beih.* **3**:37–60.

Tinbergen, N. and Perdeck, A.C. (1950). On the stimulus situation releasing the begging response in the newly hatched herring gull chick (*Larus argentatus* Pont.). *Behaviour* **3**(1):1–39.

Toates, F. (1980). *Animal Behaviour: A Systems Approach.* John Wiley and Sons, New York. 299 pp.

Toates, F. (1988). Feedforward and (not versus) feedbackward. *Behav. Brain Sci.* **11**:474–475.

Tokarz, R.R. (1985). Body size as a factor determining dominance in staged agonistic encounters between male brown anoles (*Anolis sagrei*). *Anim. Behav.* **33**(3):746–753.

Tokarz, R.R. and Beck, J.W. Jr. (1987). Behaviour of the suspected lizard competitors *Anolis sagrei* and *Anolis carolinensis*: an experimental test for behavioural interference. *Anim. Behav.* **35**(3):722–734.

Tolman, E. (1958). *Behavior and Psychological Man.* University of California Press, Berkeley. 269 pp.

Tomaru, M. and Oguma, Y. (1994). Differences in courtship song in the species of the *Drosophila auraria* complex. *Anim. Behav.* **47**(1):133–140.

Topoff, H. (ed.). (1987). *The Natural History Reader in Animal Behavior.* Columbia University Press, New York. 245 pp.

Topoff, H. and Zimmerli, E. (1993). Colony takeover by a socially parasitic ant, *Polyergus breviceps*: the role of chemicals obtained during host-queen killing. *Anim. Behav.* **46**(3):479–486.

Torgerson, L. (1977). Datamyte 900. *Behav. Res. Method. Instrument.* **9**(5):405–406.

Tourtellot, M.K. (1992). Software Review: *The Observer. J. Insect. Behav.* **5**(3):415–416.

Trochim, W.M.K. (1976). The three-dimensional graphic method for quantifying body position. *Behav. Res. Method. Instrument.* **8**(1):1–4.

Trotter, J.R. (1959). An aid to field observation. *Anim. Behav.* **7**(1–2):107.

Tufte, E.R. (1983). *The Visual Display of Quantitative Information.* Graphics Press, Cheshire, CN. 197 pp.

Turner, E.R.A. (1964). Social feeding in birds. *Behaviour* **24**(1–2):1–46.

Tyler, S. (1979). Time-sampling: a matter of convention. *Anim. Behav.* **27**(3):801–810.

Ulinski, P.S. (1972). Tongue movements in the common boa (*Constrictor constrictor*). *Anim. Behav.* **20**(2):373–382.

Unwin, D.M. and Martin, P. (1987). Recording behaviour using a portable microcomputer. *Behaviour* **101**:87–100.

Upton, G.J.J. and Fingleton, B. (1985). *Spatial Data Analysis by Example. Vol. 1. Point Pattern and Quantitative Data. Vol. 2. Categorical and Directional Data.* Wiley, New York.

Van Abeelen, J.H.F. (1966). Effects of genotype on mouse behaviour. *Anim. Behav.* **14**(2–3):218–225.

Van Der Kloot, N. and Morse, M.J. (1975). A stochastic analysis of the display behavior of the red-breasted merganser (*Mergus serrator*). *Behaviour* **54**(3–4):181–216.

Van Der Vlugt, M.J., Kruk, M.R., Van Erp, A.M.M. and Geuze, R.H. (1992). CAMERA: a system for fast and reliable acquisition of multiple ethological records. *Behav. Res. Method. Instrument. Computers.* **24**(2):147–149.

Van Elteren, P. and Noether, G.E. (1959). The asymptotic efficiency of the χ^2 test for a balanced incomplete block design. *Biometrika* **46**:475–477.

Van Hoof, J.A.R.A.M. (1970). A component analysis of the structure of the social behaviour of a semi-captive chimpanzee group. *Experientia* **26**:549–550.

Van Hoof, J.A.R.A.M. (1982). Categories and sequences of behavior: methods of description and analysis. In *Handbook of Methods in Non-Verbal Behavior Research*, ed. K. R. Scherer and P. Ekman pp. 362–439. Cambridge University Press, Cambridge. 593 pp.

Van Tets, G.G. (1965). A comparative study of some social communication patterns in the Pelecaniformes. *Ornithol. Monogr.* **2**:1–88.

Vastrade, F.M. (1987). Spacing behaviour of free-ranging domestic rabbits, *Oryctolagus cuniculus* L. *Appl. Anim. Behav. Sci.* **18**:185–195.

Verberne, G. and Leyhausen, P. (1976). Marking behaviour of some Viverridae and Felidae:

time-interval analysis of the marking pattern. *Behaviour* **58**(3–4):192–253.

Vives, S.P. (1988). Parent choice by larval convict cichlids, *Cichlasoma nigrofasciatum* (Cichlidae, Pisces). *Anim. Behav.* **36**(1):11–19.

Von Frisch, K. (1953). *The Dancing Bees*. Harcourt, Brace and Co., New York. 182 pp.

Von Holst, E. and Von Saint Paul, U. (1963). On the functional organization of drives. *Anim. Behav.* **11**:1–20.

Walbott, H.G. (1982). Technical appendix: audiovisual recording: Procedures, equipment, and troubleshooting. In *Handbook of Methods in Non-verbal Behavior Research*, ed. K.R. Scherer and P. Ekman, pp. 542–579. Cambridge University Press, New York. 593 pp.

Walcott, C. and Green, R.P. (1974). Orientation of homing pigeons altered by a change in the direction of an applied magnetic field. *Science* **184**:180–182.

Walker, J.T. (1985). *Using Statistics for Psychological Research: An Introduction.* Holt, Rinehart and Winston, New York. 585 pp.

Walker, T.J. and Wineriter, S.A. (1981). Marking techniques for recognizing individual insects. *Florida Entomol.* **64**(1):18–29.

Wallace, R.A. (1973). *The Ecology and Evolution of Animal Behavior*. Goodyear Publishing Co., Pacific Palisades, CA. 342 pp.

Walther, F.R. (1978). Behavioral observations on oryx antelope (*Oryx beisa*) invading Serengeti National Park, Tanzania. *J. Mammal.* **59**(2):243–260.

Walther, F.R. (1984). *Communication and Expression in Hoofed Animals.* Indiana University Press, Bloomington. 423 pp.

Ward, P.I. (1988). Sexual dichromatism and parasitism in British and Irish freshwater fish. *Anim. Behav.* **36**(4):1210–1215.

Waser, P.M. (1975a). Diurnal and nocturnal strategies of the bushbuck *Tragelaphus scriptus* (Pallas). *East Afr. Wildl. J.* **13**:49–63.

Waser, P.M. (1975b). Experimental playbacks show vocal mediation of inter-group avoidance in a forest monkey. *Nature* **255**(5503):56–58.

Waser, P.M. (1982). Primate polyspecific associations: do they occur by chance? *Anim. Behav.* **30**(1):1–8.

Waser, P.M. and Wiley, R.H. (1979). Mechanisms and evolution of spacing in animals. In *Handbook of Behavioral Neurobiology. Vol.3. Social Behavior and Communication.* ed. P. Marler and J. Vandenbergh, pp. 159–223. Plenum Press, New York.

Watt, K.E.F. (1966). Ecology in the future. In *Systems Analysis in Ecology*, ed. K.E.F. Watt, pp. 253–267. Academic Press, New York. 276 pp.

Watt, P.J. and Young, S. (1994). Effect of predator chemical cues on *Daphnia* behaviour in both horizontal and vertical planes. *Anim. Behav.* **48**(4):861–869.

Weary, D. and Weisman, R. (1993). Software review: *SoundEdit v. 2.0.3. Anim. Behav.* **45**(2):417–418.

Weatherhead, P.J. (1986). How unusual are unusual events? *Am. Nat.* **128**(1):150–154.

Wecker, S.C. (1964). Habitat selection. *Sci. Am.* **221**(4):109–116.

Weeden, J.S. (1965). Territorial behavior of the tree sparrow. *Condor* **67**:193–209.

Wehner, R. and Rossel, S. (1985). The bee's celestial compass – A case study in behavioral neurobiology. In *Experimental Behavioral Ecology and Sociobiology*, ed. B. Holldobler and M. Lindauer, pp. 11–53. Sinauer Associates, Inc., Sunderland, MA. 488 pp.

Weidmann, U. and Darley, J. (1971). The role of the female in the social display of mallards.

Anim. Behav. **19**(2):287–298.

Weigensberg, I. and Fairbairn, D.J. (1994). Conflicts of interest between the sexes: a study of mating interactions in a semiaquatic bug. *Anim. Behav.* **48**(4):893–901.

Weihs, D. and Katzir, G. (1994). Bill sweeping in the spoonbill, *Platalea leucordia*: evidence for a hydrodynamic function. *Anim. Behav.* **47**(3):649–654.

Weisman, R.G. and Dodd, P.W.D. (1980). Classical conditioning and evolution. In *Comparative Psychology: An Evolutionary Analysis of Animal Behavior*, ed. M.R. Denny, pp. 64–83. John Wiley & Sons, New York. 496 pp.

Wells, M.C. (1977). The relative importance of the senses in coyote predatory behavior. Unpublished PhD dissertation. Colorado State University. Fort Collins. 118 pp.

Wells, M.C. and Lehner, P.N. (1978). The relative importance of the distance senses in coyote predatory behaviour. *Anim. Behav.* **26**(1):251–258.

Welkowitz, J., Ewen, R.B. and Cohen, J. (1976). *Introductory Statistics for the Behavioral Sciences*. Academic Press, New York. 316 pp.

Westman, R.S. (1977). Environmental languages and the functional bases of animal behavior. In *Quantitative Methods in the Study of Animal Behavior*, ed. B.A. Hazlett, pp. 145–201 Academic Press, New York. 222 pp.

Westneat, D.F. (1995). Paternity and paternal behaviour in the red-winged blackbird, *Agelaius phoeniceus*. *Anim. Behav.* **49**(1):21–35.

White, G.C. and Garrott, R.A. (1990). *Analysis of Wildlife Radio-tracking Data*. Academic Press, San Diego, CA. 383 pp.

Whiten, A. and Barton, R.A. (1988). Demise of the checksheet: using off-the-shelf minature hand-held computers for remote fieldwork applications. *Trends Ecol. Evol.* **3**:146–148.

Whitfield, D.P. (1986). Plumage variability and territoriality in breeding turnstones *Aremaria interpres*: status signalling or individual recognition? *Anim. Behav.* **34**(5):1471–1482.

Wiens, J.A., Martin, S.G., Holthaus, W.R. and Iwen, F.A. (1970). Metronome timing in behavioral ecology studies. *Ecology* **51**(2):350–352.

Wiepkema, P.R. (1961). An ethological analysis of the reproductive behaviour of the bitterling (*Rhodeus amarus* Bloch). *Arch. Neerl. Zool.* **14**:103–199.

Wildhaber, M.L., Green, R.F. and Crowder, L.B. (1994). Bluegills continuously update patch giving-up times based on foraging experience. *Anim. Behav.* **47**(3):501–513.

Wildi, E. (1973). *16mm Movie Making*. Peterson Publishing Co., Los Angeles, CA. 80 pp.

Wiley, R.H. (1973). The strut display of male sage grouse: a 'fixed' action pattern. *Behaviour* **47**:129–152.

Wiley, R.H. (1983). The evolution of communication: Information and manipulation. In *Animal Behaviour. II. Communication*, ed. T.R. Halliday and P.J.B. Slater, pp. 156–215. W.H. Freeman and Co., San Francisco. 225 pp.

Willems, E.P. and Raush, H.L. (1969). Introduction. In *Naturalistic Viewpoints in Psychological Research*, ed. E.P. Willems and H.L. Raush, pp. 1–20. Holt, Rinehart and Winston, New York. 294 pp.

Wilson, E.O. (1962). Chemical communication among workers of the fire ant *Solenopsis saevissima* (Fr. Smith): 1, the organization of mass-foraging; 2, an information analysis of the odour trail; 3, the experimental induction of social responses. *Anim. Behav.* **10**(1,2):134–164.

Wilson, E.O. (1975). *Sociobiology*. Harvard University Press, Cambridge, MA. 697 pp.

Wilson, E.O. (1994). *Naturalist.* Island Press, Washington, DC. 380pp.

Wiltschko, W. and Wiltschko, R. (1972). Magnetic compass of European robins. *Science* **176**:62–64.

Winkler, D.W. (1994). Anti-predator defence by neighbours as a responsive amplifier of parental defence in tree swallows. *Anim. Behav.* **47**(3):595–605.

Wittenberger, J.F. (1979). The evolution of mating systems in birds and mammals. In *Handbook of Behavioral Neurobiology. Vol.3. Social Behavior and Communication,* ed. P. Marler and J. Vandenbergh, pp. 271–349. Plenum Press, New York.

Wittenberger, J.F. (1981). *Animal Social Behavior.* Duxbury Press, Boston, MA. 722 pp.

Woakes, A.J. and Butler, P.J. (1975). An implantable transmitter for monitoring heart rate and respiratory frequency in diving ducks. *Biotelemetry* **2**:153–160.

Wolach, A.H., Roccaforte, P., Van Berschot, S.N. and McHale, M.A. (1975). Converting an electronic calculator into a combination stopwatch-calculator. *Behav. Res. Method. Instrument.* **7**(6):549–551.

Wolcott, T.G. (1977). Optical tracking and telemetry for nocturnal field studies. *J. Wildl. Manage.* **41**(2):309–312.

Wolff, D.D. and Parsons, M.L. (1983). *Pattern Recognition Approach to Data Interpretation.* Plenum, New York. 219 pp.

Woodin, M.C. (1983). A portable umbrella blind for observing wildlife. *Wildl. Soc. Bull.* **11**(1):72–73.

Wursig, B. and Wursig, M. (1977). The photographic determination of group size, composition, and stability of coastal porpoises (*Tursiops truncatus*). *Science* **198**:755–756.

Yasukawa, K. and Bick, E.I. (1983). Dominance hierarchies in dark-eyed juncos (*Junco hyemalis*): a test of a game-theory model. *Anim. Behav.* **31**(2):439–448.

Yoerg, S.I. (1994). Development of foraging behaviour in the Eurasian dipper, *Cinclus cinclus,* from fledging until dispersal. *Anim. Behav.* **47**(3):577–588.

Young, E. (Ed.). (1975). *The Capture and Care of Wild Animals.* Curtis Books, Hollywood, FL. 224 pp.

Young, S. and Getty, C. (1987). Visually guided feeding behaviour in the filter feeding cladoceran, *Daphnia magna. Anim. Behav.* **35**(2):541–548.

Yotsumato, N. (1976). The daily activity rhythm in a troop of wild Japanese monkeys. *Primates* **17**(2):183–204.

Zar, J.H. (1972). Significance testing of the Spearman rank correlation coefficient. *J. Am. Stat. Assoc.* **67**:578–580.

Zar, J.H. (1984). *Biostatistical Analysis.* 2nd edn. Prentice-Hall, Englewood Cliffs, New Jersey. 718 pp.

Zeki, S. (1993). *A Vision of the Brain.* Blackwell Scientific Publications, London. 366 pp.

Index

LIBRARY UNIVERSITY COLLEGE CHESTER